Flexible Forming for Fluid Architecture

Arno Pronk

Flexible Forming for Fluid Architecture

Springer

Arno Pronk
The Built Environment
Eindhoven University of Technology
Eindhoven, The Netherlands

ISBN 978-3-030-71553-3 ISBN 978-3-030-71551-9 (eBook)
https://doi.org/10.1007/978-3-030-71551-9

This Springer imprint is published by the registered company Springer Nature Switzerland AG
The registered company address is: Gewerbestrasse 11, 6330 Cham, Switzerland

Preface

For Bilbao, the Guggenheim Museum is the international cultural icon. The success of this sculptural building for the reputation of Bilbao made that every city would like to have at least one archetypical iconic building. These types of buildings have no functional necessity, are always challenging, complex and expensive. They tend to be irregular, dynamic and geometrically complex and therefore the biggest problem is how to make them.

Buildings like the Bilbao Guggenheim are often described as "digital", "free-form" or "blob" architecture. However, in "free-form" architecture technique and material have become the limiting factors. Because of these limitations, the word "free" is misleading. Therefore, I propose to use the word "fluid": fluid architecture consists of unique buildings with curved surfaces. In fluid architecture one has to perceive fluidity: there is no absolute demand for a physical fluidity as long as the audience will perceive the building as fluid. The highlights of fluid architecture from the past 20 years clearly demonstrate that there is an area of tension between idea and technique; there is not only beauty of the designs, but also the tremendous complexity with regard to construction. The aim of this book is to support people dealing with fluid architecture with a better understanding of the main aspects of fluid architecture and to help them find a personal, suitable combination of architectural concepts and technical possibilities.

The building industry is accustomed to working with large tolerances compared to other industries. The combination of curved surfaces, unique products and large tolerances makes fluid architecture an interesting application for flexible moulding. These moulds can be used to create complex shapes. The cast materials will make a phase change from liquid to rigid. Therefore, the materials water/ice, polymer composite, concrete and glass were researched in combination with flexible moulding. This research resulted in a matrix with 25,000 theoretical possibilities for each material. 200 of these could be feasible. 49 case studies were researched and described. They are intended as a helpful instrument for both researchers and designers. It might help them to find the ultimate combination of perceived fluid architecture, material and flexible moulding techniques.

The case studies in this research resulted in several patents for moulding techniques, world records for the highest tower in ice and largest span in ice, a prototype of a laminated glass dome and several systems of composites and concrete with inflatables and mechanically pre-stressed membranes. This book is an important step in the design and realization of fluid architecture.

Eindhoven, The Netherlands Arno Pronk

Contents

About the Author

Arno Pronk graduated as an Architect at the TU Delft Faculty of Architecture in 1994. During his studies he explored both sculptural art and technical innovation. After his studies, not only did he work as an Architect but also as a Designing Technician. His technical interest had been developed further during a two-year post-graduate course at the chair Product Development (TU Delft). In cooperation with Professor Lichtenberg (TU/e) he developed a patented flooring system, which was successfully introduced on the Dutch market and used in several high-profile buildings (among others Kraanspoor in Amsterdam and the European Patent office by Jean Novel). He had two years' experience as an Assistant Professor at TU Delft, 12 years of experience in the same job at the Eindhoven University of Technology (0.8 FTE) and in addition in the past seven years as a lecturer on material science at KU Leuven Faculty of Architecture (0.2 FTE). This Ph.D. research on flexible moulding for fluid architecture connects fluid architecture with technology. It is a summary of the intensive and long-lasting research in flexible forming with different techniques and materials. It has resulted in the realization of several projects and technologies (Please see the project list for more information). In addition to practical projects, he has organized several international workshops and conferences. The biggest ones were Adaptables (2006), Bridges Leeuwarden (2008) and IASS Amsterdam (2015). In 2017 he became the chair of IASS workgroup 21 about advanced manufacturing and materials (Fig. 1).

Fig. 1 Portrait of Arno Pronk 2013, painted by Donovan Spaanstra. Photo by Arno Pronk (cover picture)

Chapter 1
Research Approach

1.1 Introduction

"*The straight line belongs to man. The curved line belongs to God.*" This quotation from Gaudi (Gruber 2007) is an example of the historical appreciation of curvature by some designers and architects throughout history. For the realization of their ideas, architects have a constant demand for new ways of expression in forms, techniques and materials. The designer could look for a new innovative way to solve a functional problem, but he can also be inspired by the possibilities of a new material, form or technique and look for its application in architecture.

1.2 Problem Area

The tension between idea and technique took an interesting turn during the mid-nineteen-nineties. The new possibilities offered by modelling software (Kalay 2004) that were developed since then have led to the revival of curvature in architecture in all its wealth and beauty. The rendering software developed since the mid-nineties is able to show a realistic image of a very complex building, even if it would be impossible to build it. Looking at the architectural highlights over the past 15 years, it is clear that this area of tension between idea and technique has brought us the so-called "fluid architecture". Kas Oosterhuis, Zaha Hadid, Greg Lynn, and Frank O. Gehry are some of the pioneers operating in the field of fluid architecture. Their building designs can be described as irregular, dynamic and geometrically complex (Pottmann et al. 2007).

Beside questions concerning the technical realization, the question concerning the motivation for fluid architecture can be raised. The *why* question is interesting but not the most demanding. For this research, it is sufficient to state that these buildings are designed, that there are clients who want to explore them and contractors and engineers who want to calculate and build them. Therefore, this research will focus

A. Pronk, *Flexible Forming for Fluid Architecture*,
https://doi.org/10.1007/978-3-030-71551-9_1

on *what* fluid architecture is and *how* it can be realized. To understand what fluid architecture is, it is necessary to go back to the source. In Chaps. 3–5 this is researched and highlighted. For clarification, the why question for fluid architecture will get some attention, too.

1.3 The Problem

In the twentieth century, most utility buildings had an orthogonal structure and building envelope. The building envelope typically consisted of mass-produced prefabricated elements. The structure was in most cases made on site. The standardization of prefabricated orthogonal elements has led to an increase in quality and a reduction of costs. It has also led to mass production resulting in a loss of identity and in uniformity in the built environment.

An examination of recently built fluid architecture (Chap. 4) shows that most of it has a prefabricated doubly curved skin supported by a (semi-)orthogonal or at least a facetted structure. The technical basis is the same as for orthogonal buildings. However, to realize this, the prefabrication process to make fluid architecture developed tremendously from mass production to very sophisticated, just-in-time-shape-and-place production. To quote Majowiecki, an Italian structural engineer of many fluid buildings: "*in the old days, we built [orthogonal buildings] to get us from here to there, nowadays from point A to point B and, bit-by-bit, because we are becoming increasingly sophisticated, this is what we end up with [...] this is no longer architecture, but simply industrial design*" (Vday Majowiecki EN, beppegrillo.it).

In spite of what has been achieved, the construction of fluid architecture still causes many problems, is expensive and time consuming. Martijn Veltkamp researched several structures related to fluid architecture. He states that "*shell and membrane structures fall outside the scope of [his] research*" (Veltkamp 2007). It is regrettable that his excellent work did not include this field of structures, because the nature of these structures is much closer to fluid forms. There are many reasons for looking at shell and membrane structures in order to realize fluid architecture and this will be explained in Chaps. 6 (The Possibilities of Form-active and Flexible Moulds for Structures) and 7 (Induction).

In my opinion, further research on the integration of structure and surface of prefabricated elements, or on further automation of the production process will yield predictable and limited results. Research should instead focus on curved buildings throughout history before developing new methods and techniques. Shell and membrane structures have a doubly curved morphology and have been built and developed for many years. Together with fluid architecture, they are the only building typology with a doubly curved surface. Therefore, it will be interesting to make an inventory of the techniques developed for shell and membrane structures and research if and how they can be used in the creation of fluid architecture.

1.4 Field of Research

In the twentieth century, elegant irregular curved shell structures were produced, with forces and shape in harmony. With the rise in the price of labour, the high cost of scaffolding and formwork for these concrete shells became an important problem. Frei Otto and his team tried to solve this problem by introducing grid shells in 1962: shells of timber laths or steel members/mesh coverings (Hennicke 1994). Another solution was the use of inflatable formwork to construct thin curved shells. It was introduced as early as 1942 by W. Neff (1942) and was developed further by several others. These inflatable moulds were applied to construct regularly shaped domes. Despite their apparent regularity, domes have doubly curved surfaces and, in many cases, the outer surface is smoothened as is common in fluid architecture. The knowledge developed for shells made with membrane moulds in the 1950s, 1960s and 1970s is hardly used in the construction of fluid architecture. The similarity in form in the so-called flexible and form-active structures (Engel 1967) and fluid architecture is obvious. In the past, the scaffolding problem was solved with an inflatable formwork. The present research will focus on the question of how flexible and form-active structures can be used as a moulding material in the construction of fluid architecture. In the materials Chaps. 8–11, old, existing and new techniques are compared, researched, developed and combined in case studies.

1.5 Research Questions

W. Neff, H. Isler and others proved that form-active and flexible structures can be used as a mould to make domes or dome-like shell structures. F. Otto and others proved that form-active and flexible structures can be manipulated in many ways. They are adequate for making irregular and flexible structures. Can these two findings be combined to make fluid architecture?

How can form-active, flexible structures be used as a mould for the construction of irregular curved buildings or building elements?

Antoni Gaudi can be seen as the first architect/engineer researching form-active moulds. He investigated hanging structures as the reverse of a load-bearing frame. His ideas were based on the theories postulated by Moseley (1835: 293–313), Mery (1840: 133–134) and the graphical method of Rankine (1858) and Maxwell (Flood et al. 2014). Gaudi used the concept of hanging structures such as catenaries in a completely original way: he integrated the structural design in the process of architectural design. These hanging structures gave him a mirror image of the "truss line" of a load-bearing frame. Form-active structures were used as a model and helped him to understand the relationship between the direction of forces and the form of the structure. The design of the structure was a direct result of the method used (Molema 1987, 2008; Huerta 2006).

Others, like Wallace Neff (1942), Heinz Isler (Chilton 2000) Mark West (West and Araya 2009) and Dante Bini (1969), used form-active structures in a far more direct way; they used form-active moulds for the production of building structures, elements or components. This research follows the line of the last group.

Within a design process, the search for optimal form is called 'form finding'. The form finding of the authors above is both technical and structural. In fluid architecture the form finding is technical and architectural and the appearance of the building is leading. The technical optimization of a building needs to respect that image. This implies a technical and structural challenge in the realization of fluid architecture. There is a similarity in form between the highly optimized shells, the membrane structures and fluid architecture. In this research, the techniques for the construction of shell and membrane structures are transferred to fluid architecture. Structural engineers like to research the structural optimization of fluid architecture. In contrast with the classical engineering approach, this research will focus on the construction of fluid architecture and the building surface as an architectural object with a structural behaviour that has to be optimized.

The research topic is the use of flexible moulds to realize fluid architecture. It is not the intention to give one concept to construct fluid architecture, because there are too many different options. The research aim is to explore and present the use of flexible moulding methods to realize fluid architecture. The moulding methods will be methodology categorized. The architect can make his choice and add new combinations. In this research, new methods and techniques will be developed and added to existing combinations.

The goal is:

- to improve the realization of the architectural intention of fluid architecture;
- to improve integrated solutions between structure and envelope;
- to improve the quality of fluid buildings in general; and
- to have a more efficient construction method with less environmental impact.

Bibliography

Beppegrillo.it. Vday Majowiecki EN. www.beppegrillo.it/eng/immagini/vday_majowiecki_EN.pdf. Accessed 9 Apr 2011

Bini D (1969) Method for erecting structures. US Patent 3462521

Chilton JC (2000) The engineer's contribution to contemporary architecture: Heinz Isler. In: The engineer's contribution to contemporary architecture (ECCA). Thomas Telford Ltd and Riba Publications, London

Engel H (1967) Tragsysteme (structure systems). Hatje Cantz, Stuttgart

Flood R, McCartney M, Whitaker A (eds) (2014) James Clerk Maxwell: perspectives on his life and work. In: Contemporary physics 55(4). Oxford University Press, Oxford, pp 356–356

Groot AD de (1961) Methodologie: grondslagen van onderzoek en denken in de gedragswetenschappen. Mouton & Co., 's Gravenhage

Gruber F (2007) Einsatz klassisch geometrischer Methoden in Architektur und Computational Aesthetics. Department of Geometry, University of Applied Arts Vienna, Vienna

Hennicke JW (1994) Light and spacious: aspects of the design and construction of spacious light structures. Temes de Disseny 10:214–224
Huerta S (2006) Structural design in the work of Gaudí. Architectural Sci Rev 49(4):324–339
Kalay YE (2004) Architecture's new media: principles, theories, and methods of computer-aided design. The MIT Press, Cambridge, MA
Lynn G (ed) (1993) Folding in architecture. In: Architectural design profile 102. Wiley, Chichester
Lynn G (1999) Animate form. Princeton Architectural Press, New York
Mery E (1840) Mémoire sur l'équilibre des voûtes en berceau. Annales de Pont et chaussées 50(70):133–134
Molema J (1987) Antoni Gaudí, een weg tot oorspronkelijkheid. TU Delft, Delft
Molema J (2008) Gaudí: the construction of dreams. Episode Publishers, Rotterdam
Moseley H (1835) On the equilibrium of the arch. Cambridge Philos Trans 5:293–313
Neff W (1942) Method and apparatus for constructing shell-form structures. US patent 2388701 A
Pottmann H, Asperl A, Hofer M, Kilian A (2007) Architectural geometry. Bentley Institute Press, Exton PA
Rankine WJM (1858) A manual of applied mechanics. Charles Griffin, London
Veltkamp M (2007) Freeform structural design. Delft University Press, Delft
West M, Araya R (2009) Fabric formwork for concrete structures and architecture. In: International conference on textile composites and inflatable structures, 5–7 October. Barcelona

Chapter 2
Form

2.1 Introduction

The focus of this book is on the application of forms, with a special interest placed in complex forms as used in fluid architecture. In Chaps. 3–5 the morphology of fluid architecture is investigated. One of the conclusions will be that fluid architecture is irregular, doubly curved and smooth. Chap. 6 will review the relationship between membrane forms and structure. But first, form itself and form in relation to membranes needs to be researched.

There has been a long and impressive tradition in research on geometry. It may be necessary to explain some terms and principles used for fluid architecture and mesh generation in relation to doubly curved surfaces. This chapter will focus on "free" forms and forms resulting from the manipulation of membranes.

2.2 "Free-Form" Surfaces

Free or amorphous forms fall outside the mathematical description offered by geometry. A way to deal with this problem is to cut the free form into pieces that can be handled. Frei Otto's theory: 'objects as surfaces' (Otto and Schaur 1979: 21) distinguishes single-, dual-, and multi-faced bodies. Below is a matrix with the classification of Frei Otto on single-, dual- or multi-faced bodies at the horizontal axis and the classification of Gauss on curvature at the vertical axis. The bodies in the matrix are limited to one type of surface (Fig. 2.2).

A. Pronk, *Flexible Forming for Fluid Architecture*,
https://doi.org/10.1007/978-3-030-71551-9_2

2.3 Single-Faced Bodies

A single-faced body is an object whose surface is not distorted by points or lines. An egg is a typical example of a single-faced body. A Möbius strip has a continuous surface, therefore it is a single-faced object. In this case it is not a body because it has no volume. Single-faced bodies in combination with one type of curvature have a very specific form. If the surface is synclastic it will be a sphere or ellipsoid. If the surface is anticlastic it must be a Möbius strip. An oloid is a single-faced body with an monoclastic surface, defined by H. Dirnbock and H. Stachel as the unit "of two circles kA, kB in perpendicular planes such that each circle contains the centre of the other" (Dirnbock and Stachel 1997: 105–118). The convex hull of these circles is called an oloid. Zeroclastic surfaces always have more than one face. In the column Single-faced bodies of Fig. 2.2 three different types of shapes are presented: Anticlastic form (Möbius strip), Synclastic body (ellipsoid) and Monoclastic body (oloid).

2.4 Dual-Faced Bodies

A dual-faced body separates the object into two continues surfaces. A lens shape is a typical example of a dual-faced body. The surface can be either singly curved or doubly curved. In the column dual-faced bodies of Fig. 2.2 the different types of shapes are presented: a flat surface, The translation of two irregularly curved lines along a straight line, a lens with two non-symmetrical ellipsoids and two hyperboloid surfaces.

2.5 Multi-Faced Body

A multi-faced body is divided in several surfaces. The different surfaces can have any geometry: zero, mono-, anti- and synclastic. A cube is a typical multi-faced body, made of six zeroclastic surfaces. There are five completely regular volumes with plane (zeroclastic) surfaces. They are the Platonic solids named tetrahedron, hexahedron, octahedron, dodecahedron and icosahedron. The names refer to the Greek and mean: four faces, six faces, eight faces, twelve faces and twenty faces (Fig. 2.1).

In the column multi-faced bodies of Fig. 2.2 the different types of shapes are presented: Zeroclastic multi-faced body, Monoclastic multi-faced body, Conical monoclastic multi-faced surface, Synclastic multi-faced object and anticlastic multi-faced objects.

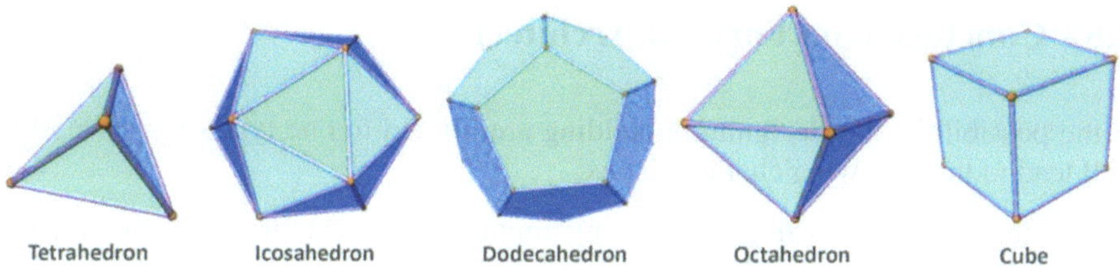

Fig. 2.1 The platonic solids

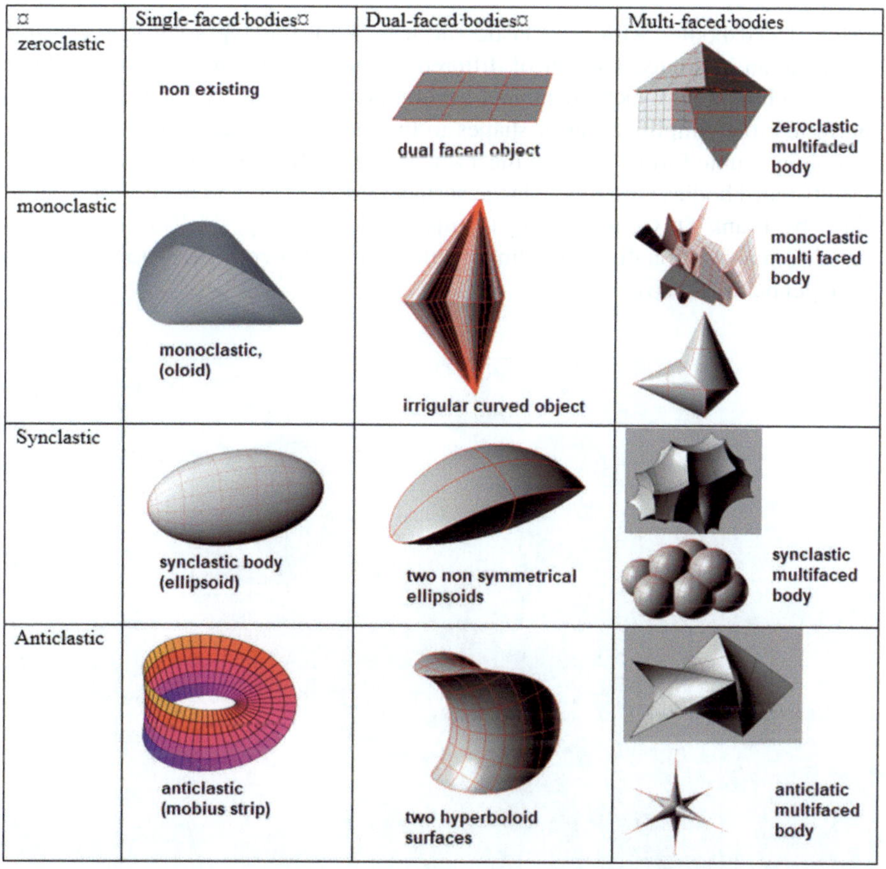

Fig. 2.2 The matrix with the classification of Frei Otto on single-, dual-, or multi-faced bodies at the horizontal axis and the classification of Gauss on curvature at the vertical axis

2.6 Conclusion on Form in Architecture

If the possibilities for designing a building are limited to one type of surface, this will lead to the following characteristics:

- Single-faced bodies will give too many limitations. Except for some experiments, this typology is not used for buildings (Fig. 2.3 Spherical tree house).
- Dual-faced bodies are possible but are limited to some exceptions. Like inflatables, dome-like structures and lens-like structures (Fig. 2.4 Evoluon in Eindhoven, the Netherlands).
- Multi-faced bodies with zeroclastic surfaces. With this typology it is possible to make an unlimited number of different bodies. The mainstream of buildings can be classified as such. Technical developments in materials and construction have led to a simplification of shapes to the most elementary forms within this typology: cubic forms (Fig. 2.5 the Twin Towers in New York).
- Multi-faced bodies with monoclastic surfaces. Most of the so-called 'fluid' buildings by Frank O. Gehry belong to this typology (Chap. 2). This typology is used to make rationalized buildings with seemingly free-form shapes (Fig. 2.6 Guggenheim Bilbao).

Fig. 2.3 (Free Spirit) Tree house, Vancouver Island, Canada

Fig. 2.4 Evoluon Eindhoven

Fig. 2.5 Twin towers New York

Fig. 2.6 Guggenheim museum Bilbao

- Multi-faced bodies with synclastic surfaces. Most of the time, this type is used for inflatable structures in combination with a cable structure or a space frame (Fig. 2.7 Water Cube Beijing).
- Multi-faced bodies with anticlastic surfaces. Most of the time this type is used for prestressed tensile structures or cable structures (Fig. 2.8 Munich Olympic Stadium).

According to the matrix above, the possibilities are limited as long as the forms have to stay within the boundaries of the 12 fields of the matrix above.

Chaps. 3–5 will define 'free-form' architecture as being irregular, doubly curved and smooth. Single-faced bodies and Dual-faced bodies pose too many restrictions. With the exception of the anticlastic Möbius strip, they are too regular for 'free-form' architecture. Multi-faced bodies with zeroclastic surfaces have unlimited possibilities as exemplified by the so-called 'deconstructive architecture' (Chap. 3). 'Free-form' architecture can be seen as the smooth version of deconstructive architecture. It is difficult to bend rigid materials into doubly curved panels but it is relatively easy to bend rigid materials into monoclastic surfaces. Therefore, many buildings with the suggestion of a free form are made out of multi-faced bodies with monoclastic surfaces.

Multi-faced bodies with anticlastic and synclastic surfaces can be made with flexible or elastic materials. In the words of H. Engel, "Form-active structure systems are systems of flexible, non-rigid matter, in which the redirection of forces is effected by a self-found Form design and characteristic Form stabilization" (Engel 1967). The so-called flexible and form-active structures (Chap. 6) are inflatables, pre-tensioned

Fig. 2.7 Water Cube Beijing

Fig. 2.8 Munich Olympic Stadium

Fig. 2.9 Church in Maassluis

membranes and cable structures. Flexible and form-active structures can be made irregular, smooth and doubly curved. Flexible and form-active structures have been used mostly for temporary buildings and buildings with limited functionality like canopies. There is a tendency to create long-term flexible and form-active buildings with the functionality of a permanent building, including insulation, heating and other installations. The first permanent building with a mechanically (pre-)stressed surface was Dulles airport by E. Saarinen. More recent examples include the Catholic church in Maassluis by Royal Haskoning (Fig. 2.9) and the airport building in Bangkok by Murphy/Jahn Architects (Fig. 2.10).

The possibilities with flexible form-active buildings are limited. Most clients doubt the '*firmitas*' of such buildings. Another problem with flexible form-active structures concerns the limited possibilities to shape a building in combination with the limited understanding of many architects of the structural dos and don'ts of this structural typology.

If the doubly curved typologies in the matrix above are combined, it is possible to connect synclastic and anticlastic forms. In that way the freedom to shape buildings becomes endless. This freedom by combining different form typologies yields an endless horizon of possibilities and freedom in design. This liberation of forms leads to the truly 'free' form. Every form realized in a building or sculpture is limited by the possibilities of the materials and techniques that will be used. Chaps. 3–5 set out more reasons for combining doubly curved forms and the problems in realizing them. Because of these limitations, the word 'free' is not used here, but is replaced by 'fluid' (Chap. 3). Below, there will be more about forms and how to deal with

Fig. 2.10 Bangkok Airport

them in a way that makes them appear like free forms, whereas, in fact, they are highly rationalized.

In fluid architecture it is a common technique to use a multi-faced zeroclastic body to approach the doubly curved surface as in Figs. 2.11 and 2.12.

Depending on the curvature, the panels have to be made smaller. The bigger the curvature the smaller the panels have to be. To avoid the waste of material it is preferred to use quadruple panels. The surface of the panels can be curved and varied in thickness to make a continuous surface on top of the multi-faced zeroclastic body. It is possible to make almost every facetted surface out of zeroclastic triangular elements because three points will always be in one plane. Four points, however, will not automatically form a quadruple zeroclastic element. Zeroclastic quadruple

Fig. 2.11 Free-form surface divided into zeroclastic quadruple elements

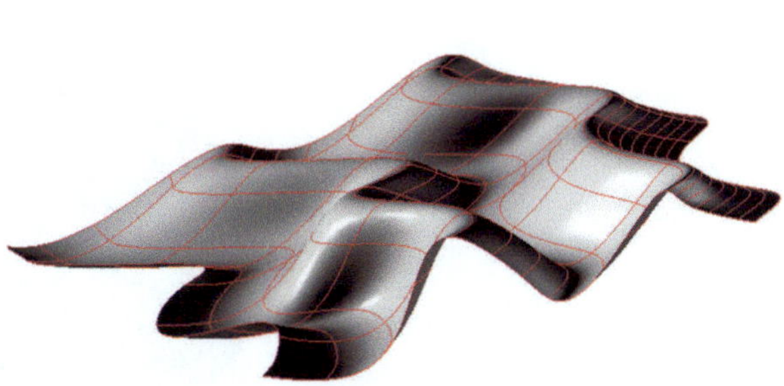

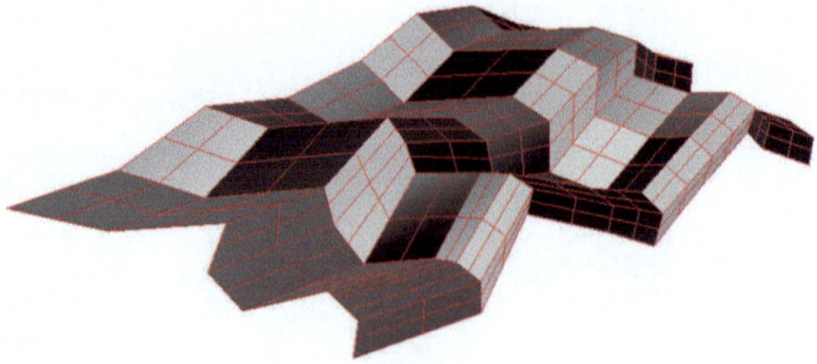

Fig. 2.12 Free-form surface divided into doubly curved elements

elements are preferred in production and engineering for their efficiency in the use of material.

Synclastic and anticlastic forms can be combined as a doubly curved (irregular) translational surface. This surface can be formed with the translation of a variating curve along a negatively or positively curved line. The advantage of a translation is that the four 'corner' points of a translation are in the same surface. Even when the line is scaled during the translation, the four corner points of an element stay within the same plane. This is of great benefit in the production and engineering of elements for 'curved' buildings (Glymph et al. 2004: 187–202). Irregular ruled surfaces have the appearance of a 'free' form. Gehry frequently uses this way of optimization in his designs.

Figure 2.13 is a translation of a curve along a curved line.

Figure 2.14 is made with the curvatures of Fig. 2.13. the curve is scaled during the translation. In both figures, the four corner points of every element stay within a plane surface.

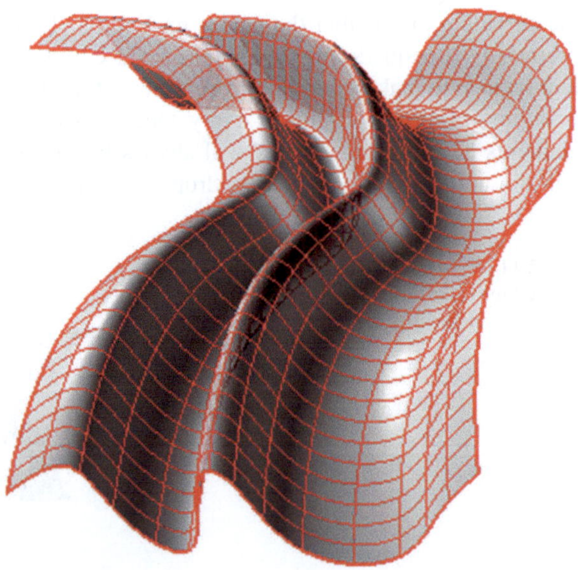

Fig. 2.13 Translation of a curve along a curved line

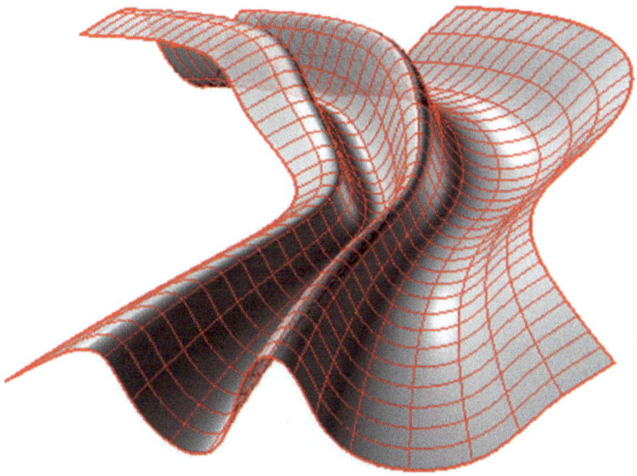

Fig. 2.14 Scaled surface based on Fig. 2.13

2.7 Membranes

The focus of this research is on the forms that can be made and material that can be used for flexible moulding. In many cases the flexibility is achieved by a membrane or foil. Chaps. 5 and 6 will investigate if membrane moulding is suitable to construct fluid architecture. This paragraph will research which forms can be made with membranes and how to do it. There has been a long and impressive tradition in research on geometry and form finding of membranes.

Although the title uses the term "membrane", I prefer to use the term "form-active". Form-active is used in the way defined by H. Engel: "*Form-active structure systems are systems of flexible, non-rigid matter, in which the redirection of forces is effected by a self-found form design and characteristic form stabilisation*" (Engel 1967). In practice, form-active and flexible structures are prestressed membranes, inflatable membranes, chains and cable structures. It is possible to use this type of structures for moulding. They have a doubly curved surface and the ability to make non-repetitive surfaces. This way of moulding is pre-eminently suitable for fluid architecture and all kinds of doubly curved building elements.

In this chapter, I will consistently connect colours to a certain form. The most important colours are:

Yellow for zeroclastic;
Green for monoclastic;
Blue for synclastic; and
Red for anticlastic (Figs. 2.15, 2.16, 2.17, and 2.18).

Fig. 2.15 Zeroclastic

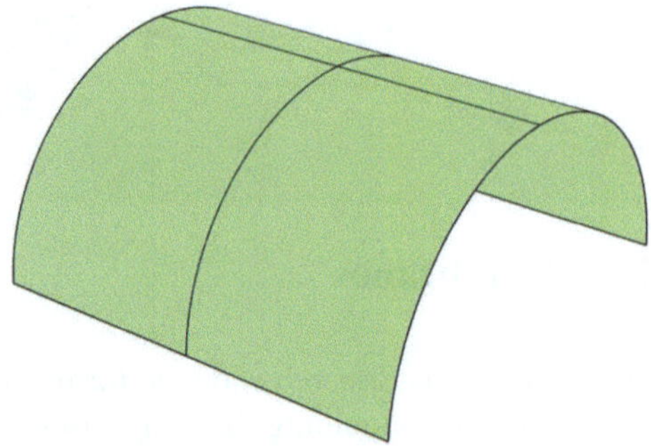

Fig. 2.16 Monoclastic

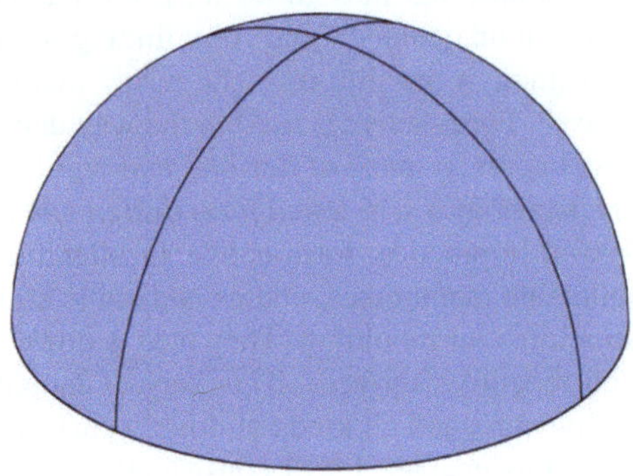

Fig. 2.17 Synclastic

2.7.1 Mechanically Stressed Surface Structures

Mechanically (pre-)stressed surface structures are always anticlastic except for structures having all the corner points in the same plane and having all the forces on and in the structure within the surface of that plane, in that case the form is zeroclastic. When elastic materials are used, the surface of the model is formed within a number of high points, low points and borders. The surfaces are not absolutely minimal. The geometry of the plane is determined by the extent to which the material is tensioned

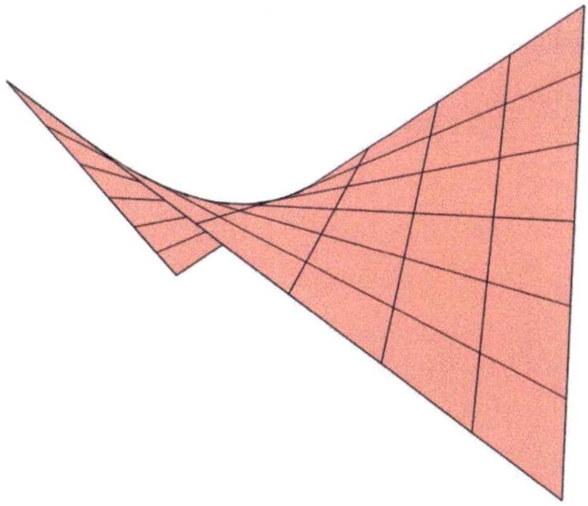

Fig. 2.18 Anticlastic surfaces

Fig. 2.19 Synclastic balloon

in one or more directions. The degree of tension in the different directions can be changed and will influence the surface of the membrane. If the ratio of tension in the different directions is more than 1:2, then depending on the material properties and border conditions a membrane might wrinkle.

2.7.2 *Inflatable Membranes*

Inflatable membrane structures contain a circular cross-section in at least one direction. In most cases, they are synclastic (Fig. 2.19). Under certain conditions, it is also possible to contain monoclastic (Fig. 2.20) and even anticlastic (Fig. 2.21) surfaces. The different surfaces that can be made with inflatables depend on the circular cross-section in relation to other circular cross-sections.

Fig. 2.20 Monoclastic balloon

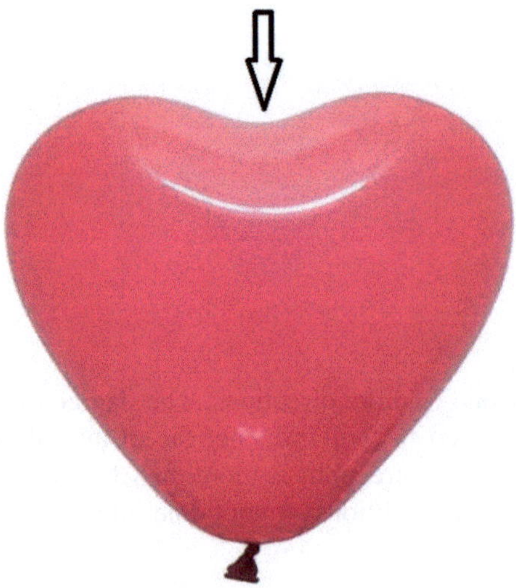

Fig. 2.21 Balloon with an anticlastic part

The two parameters that influence the form are:

1. The angle of the cross-section to the other cross-section; and
2. The radius of the circle.

Both parameters 1 and 2 can be divided into three subgroups:

1a One centre point;
1b Parallel circular sections perpendicular to a straight centreline; and
1c Non-parallel circular sections perpendicular to a central curvature.

2a Fixed radius;
2b Linear changing radius; and
2c Non-linear changing radius.

Parameters 1 and 2 are combined in the table in Fig. 2.22 and will lead to 7 ways to curve the surface of an inflatable. The table gives examples of every category.

Below and in the Figs. 2.44, 2.55, 2.58, and 2.60, there are drawings of inflatable objects with particular curved surfaces (Fig. 2.28).

Object 2.23 with one centre is always a sphere. Spheres have a synclastic surface. Objects with one centre point in combination with a changing radius do not exist. Objects 2.24 and 2.26 with a parallel circular section on a straight line have always a monoclastic surface. Objects 2.25 and 2.27 have a synclastic and an anticlastic part following the curved centreline. At the 'inner side' of the centre curve the surface is anticlastic; at the 'outer side', the surface is synclastic. Object 2.29 has a non-linear changing radius. If the radius changes in a progressive way, the surface is anticlastic. If the radius changes in a regressive way, the surface is synclastic. Object 2.30 and 2.31 is a combination of a curved centreline and a non-linear changing

The form of inflatables	Synclastic	Monoclastic	Synclastic	Monoclastic	Synclastic	Synclastic	Synclastic
			Anticlastic		Anticlastic	Anticlastic	Anticlastic
One centerpoint							
Parallel circular section on a centerline							
Non-Parallel circular section on a central curve							
Fixed radius							
Linear changing radius							
Non-Linear changing radius							
Name	Sphere	Cylinder	Torus	Cone			
Figure	2.23	2.24	2.25	2.26	2.27	2.28	2.29 2.30

Fig. 2.22 Overview with shapes

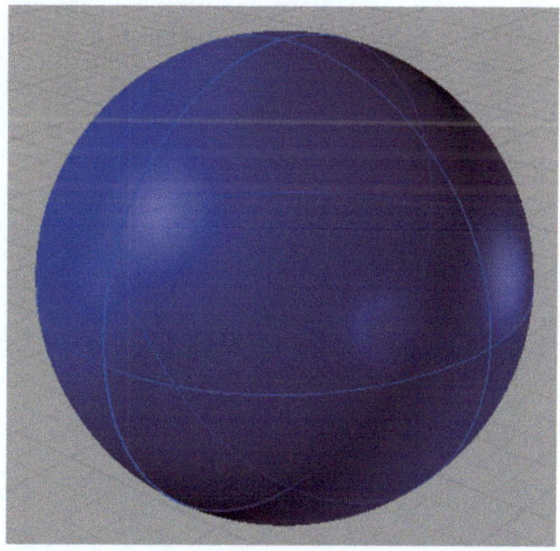

Fig. 2.23 Sphere

Fig. 2.24 Cylinder

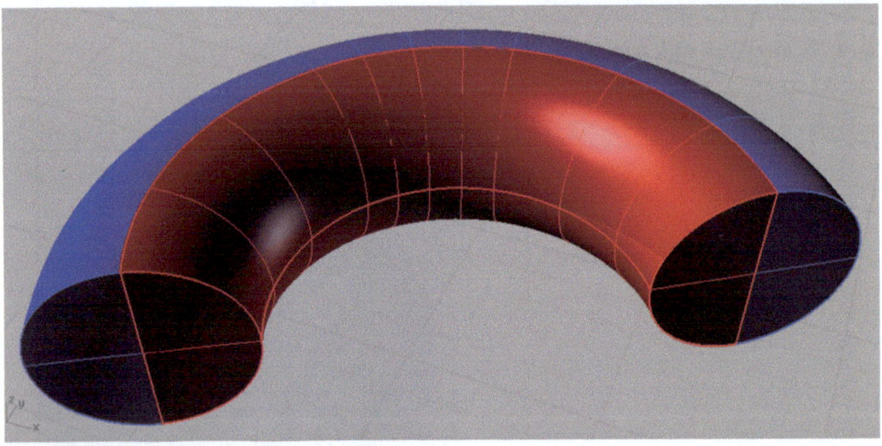

Fig. 2.25 Torus

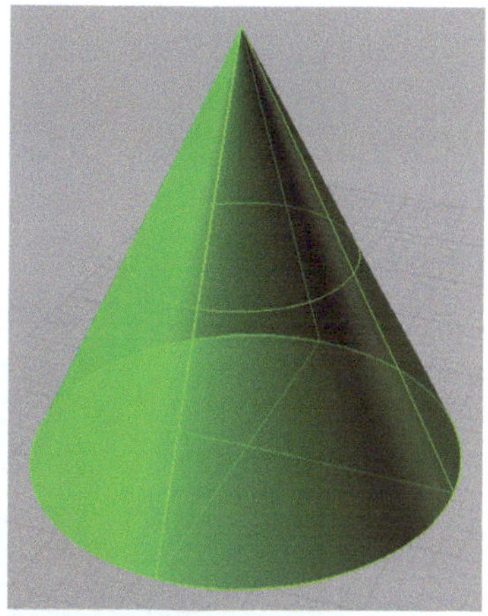

Fig. 2.26 Cone

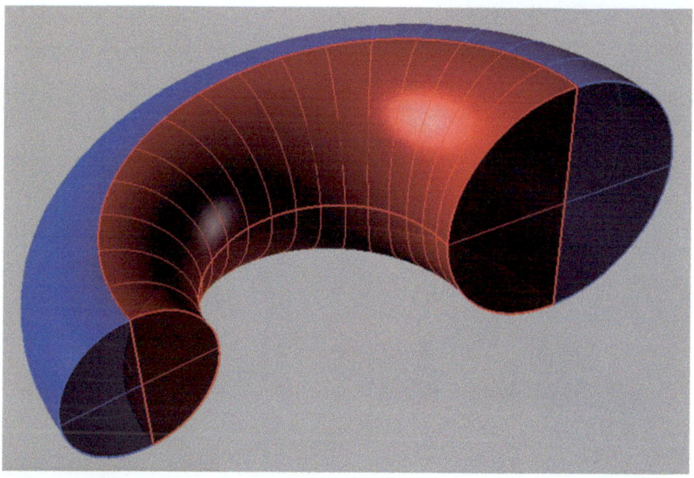

Fig. 2.27 Curved cone

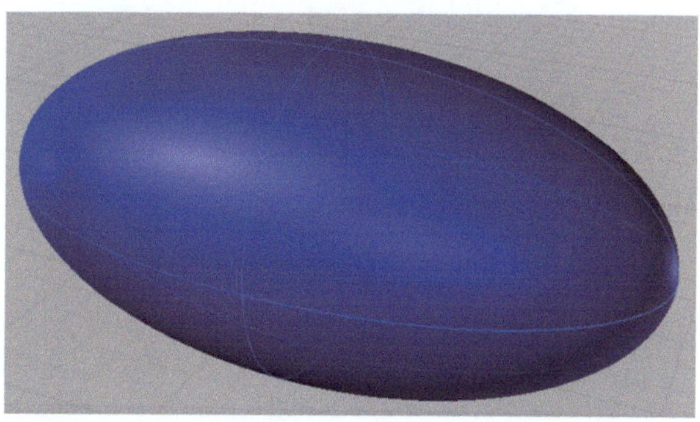

Fig. 2.28 Ellipsoid

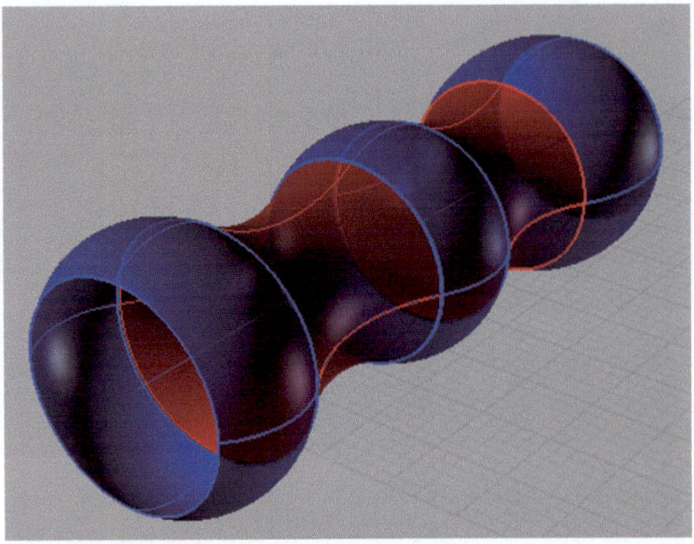

Fig. 2.29 Linear ruled surface

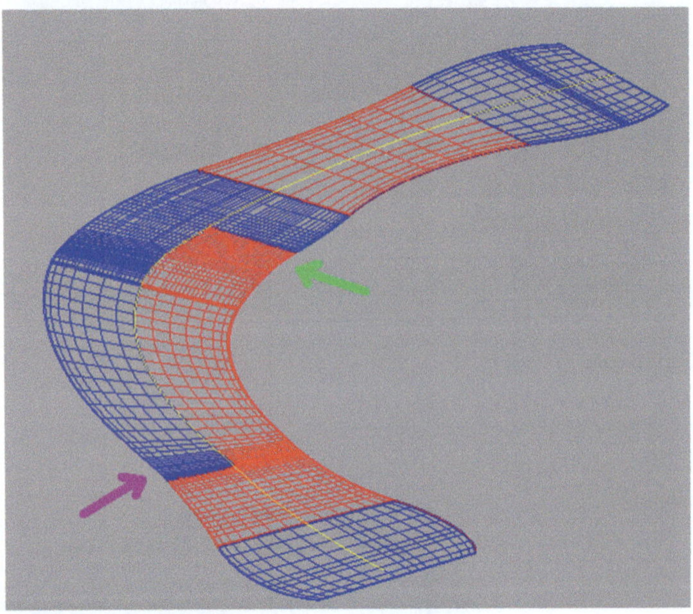

Fig. 2.30 2D Curved ruled surface

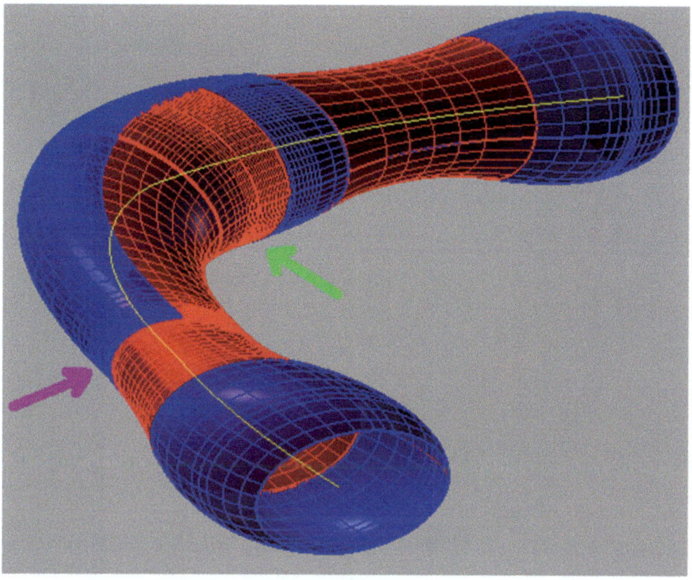

Fig. 2.31 3D Curved ruled surface

Curvature of inflatable with a non-parallel circular section on a central curve with a non-linear changing radius					
"Innersite" of the center curve	Anticlastic				
"Outersite" of the center curve	Synclastic				
Progressive changing radius	Anticlastic				
Regressive changing radius	Synclastic				
Result of the combination		Anticlastic	Anti/synclastic	Synclastic	Anti/synclastic

Fig. 2.32 Table of inflatable curvatures

radius. Both features will give a double curvature. They can stimulate or inhibit each other. They stimulate each other in the combination of outer side of the centre curve and regressive curvature to form a synclastic surface. In the case of the inner side of the centre curve and progressively changing radius they stimulate each other to form an anticlastic surface. In the other two combinations (inner side of the curve—regressively changing radius and outer side of the curve and progressively changing radius) the feature with the strongest surface curvature will determine the result. Below, there is a table with the features and curvatures. The colours in the table correspond with the colour of the inflatable object below. The green arrow in object 2.30 and 2.31 shows the place where a synclastic surface changes into an anticlastic surface as a result of the strong curvature of the centreline, although the change of the radius is regressive. The purple arrow in object G shows the place where an anticlastic surface changes into a synclastic surface as a result of the strong curvature of the centreline, although the change of the radius is progressive (Figs. 2.32 and 2.33).

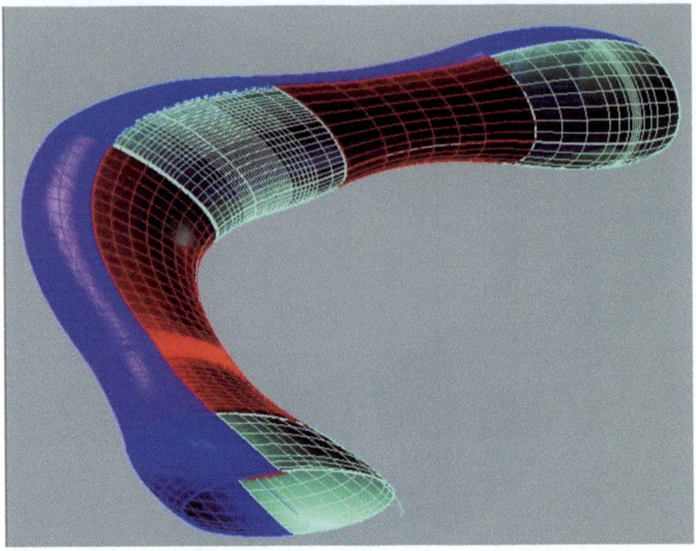

Fig. 2.33 Model of inflatable object with several curvatures

	Soap film	Elastic membrane/film
Boundary	flexible or rigid closed boundaries	flexible of rigid points and flexible or rigid boundaries
Minimal surface	absolute minimal surface	differ from the minimal surface
Form	anticlastic/synclastic	anticlastic/synclastic
Surface manipulation	a) depends on the boundary conditions b) pressure on the surface c) material behaviour	a) depends on the boundary conditions b) pressure on the surface c) material behaviour d) depends on the direction and force of the pre-tension in the membrane

Fig. 2.34 Table: Comparison between physical methods, soap film versus elastic membrane/film

2.8 Form Finding

Form finding can be split into physical form finding and analytical form finding, as discussed below.

2.8.1 Physical Form Finding

Like every design process, the form finding of tensile structures is an iterative process, i.e., the conditions can be changed over and over again until an optimum solution is reached. Finding a final solution is an interplay between the decisions taken by the designer and the form-finding process. The form-finding process that is used, may well affect the results. The table in Fig. 2.34 compares two experimental form-finding methods.

2.8.2 *Analytical Form Finding of Membrane Structures*

Until the 1970s, physical form finding was used for engineering tensile structures. The implementation of the model has to be as precise as possible. It is a time-consuming process with a risk of many inaccuracies. After the knowledge gained during the construction of the Olympic Stadium in Munich (1972), computer programs were introduced for the engineering and form finding of membrane structures.

In 1974, H.J. Schek published the paper "The force density method for form-finding and computations of general networks", the fundamental theory for the analytical approach of form finding, the force density method. The computer program Easy and, more recently, other programs use this method for form-finding membrane structures. In the analytical form-finding process, according to Klaus Linkwitz, two phases can be identified.

Phase 1

A number of design studies, non-materialized equilibrium models, have to be performed. The process is almost analogous to the soap film method. Within the boundaries a minimal surface will be generated. The membrane can be seen as the discretization of a cable net. There are a number of parameters that influence the geometry of the equilibrium surface. By varying these parameters, the geometry (curvature) of the surface can be influenced. The parameters are:

a. the position of boundaries, the high points and low points;
b. the curvature of the surface;
c. proportional force density in the different parts of the surface;
d. orientation of the network; and
e. external forces/load cases.

In the first phase the equilibrium geometry of the surface is found. In this phase it is possible to do preliminary studies of load cases, deformation, and stress contribution.

Phase 2

The second phase consists of the materialization of the equilibrium surface. By introduction of the material properties, it is possible to have a complete analysis of the load cases, deformation and stress contribution. The geometry of the surface will be influenced by the material properties (Fig. 2.35).

	Phase 1	Phase 2
conditions	closed boundaries and/or points	equilibrium (Phase 1)
shape	equilibrium	materialized equilibrium
shape influence	boundaries, high points, low points the curvature of the surface proportional force density orientation of the network external forces/load cases	material properties

Fig. 2.35 Process for the analytical form finding of membrane structures

2.8.3 *Dynamic Relaxation*

In the force density method, there is a ratio between the force and the length of a line. This is specified by the designer. Beside the force density method, it is possible to calculate the deformation of membranes with the dynamic relaxation method. The dynamic relaxation method is based on "discretizing the continuum under consideration by lumping the mass at nodes and defining the relationship between nodes in terms of stiffness. The system oscillates about the equilibrium position under the influence of loads. An iterative process is followed by simulating a pseudo-dynamic process in time, with each iteration based on an update of the geometry" (Pronk ADC 2012).

2.9 The Manipulation of Membranes

The surface of a membrane can be manipulated in different ways. This paragraph will give an overview of the different ways a membrane can be influenced. The use and interest in membranes for moulding has been growing over the last 10 years. If designers ask for the dos and don'ts concerning the geometry of membranes, it is hard to give a clear and simple answer. Therefore, this paragraph is written to give a complete overview of the possibilities for the manipulation of membranes. It is presented in a matrix with 84 icons representing the 84 ways to manipulate a membrane. The overview aims to be a helpful instrument for designers and researchers working with membranes.

2.9.1 *84 Ways to Manipulate a Membrane*

If single-layer form-active surfaces in force equilibrium are combined with their Gaussian curvature, the following five basic form-active morphologies can be distinguished:

- prestressed membrane or cable structure with an anticlastic surface (in figures referred to as "AS");
- prestressed membrane or cable structure with a zeroclastic surface (in figures referred to as "ZS");
- inflatable with a synclastic surface (in figures referred to as "SI");
- inflatable with a monoclastic surface (in figures referred to as "MI"); and
- inflatable with an anticlastic surface (in figures referred to as "AI").

The form-active surfaces in force equilibrium can be manipulated in four different ways:

- by changing the (pre-)stress in a certain area, indicated by numbers 1–5;
- by an external load, indicated by numbers 6–10;

- by pushing other form-active surfaces into a membrane, indicated by numbers 11–15;
- by pushing or pulling a rigid object into a membrane, indicated by numbers 16–19.

Those four ways will be subdivided as well. Although it is not possible to make every shape imaginable, it is possible to come close to any desired shape. Instead of describing the formal geometry of surfaces, we give an overview of the different parameters for manipulating membranes and their major geometric effects. The overview is presented in four matrices (Figs. 2.44, 2.55, 2.58, and 2.60). The first matrix shows the results of changing the prestress, the second the load cases, the third pushing other form-active surfaces into a membrane and the fourth pushing or pulling a rigid element into a membrane. In the columns, the five basic form-active morphologies are given. In the rows you find the different ways to manipulate as mentioned above. The legend shows the meaning of the colours and symbols. The symbols in the legend represent the increase or decrease of the absolute value of the Gaussian curvature, defined as the product of the two principal curvatures ($|K| = k1 \times k2$). There are some exceptional cases where the 3D effect is not represented by a change in the Gaussian curvature. In that case $|k1 \times k2| = |k1a \times k2a|$, the variation in the principal curves, is described. Not every combination leads to a manipulation with a 3D effect on the surface and some combinations will give equal results. For example, pushing or pulling into a zeroclastic membrane will give the same deformation, but mirrored. There are 5 × 19 combinations, represented in the four matrices. If you take out the combinations without a result of deformation or having the same result as other combinations, there are 84 (20+21+24+19) quintessentially different ways to manipulate a membrane.

2.9.2 The Experiments by Linkwitz

The form of a membrane in equilibrium is the result, in principle, of the tensile forces in the membrane and (boundary) cables. The force-density method was introduced to find the geometry and to model the behaviour of the membrane, cables and forces (Schek 1974: 115–134; Linkwitz and Schek 1971: 145–158). Linkwitz and a group of students did some experimental research with regard to the ratio of the force densities in relation to geometry (Linkwitz 1999: 709–718, 2002: 159–167). For a better understanding of geometry, we researched the experiments by Linkwitz with the help of a force density program (Tess3D). The net is octagonal to ease comparison of the pictures and to stimulate the deflection in the net structure. I am aware that for structural reasons it is better to have a diagonal net structure.

Figure 2.36 shows a perpendicular net structure with a force density in the boundary of 1. In Fig. 2.37, the force density in the boundary and the network is 5. The equilibrium shape of the surface is the same. The conclusion is that if the ratio in force density of the boundary and the network is the same, the shape of the structure will also be the same.

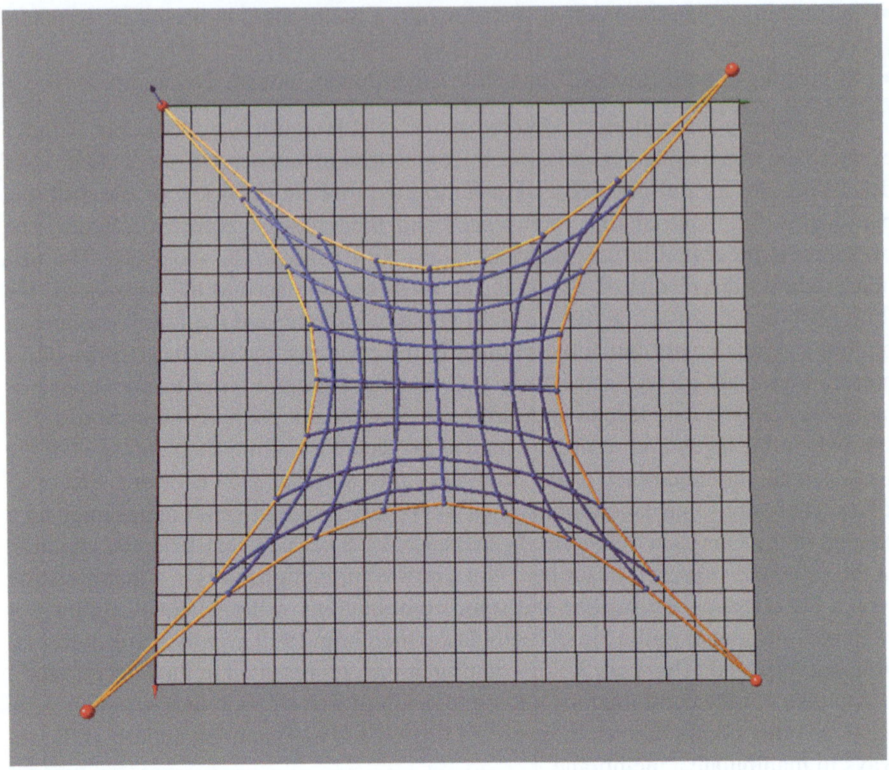

Fig. 2.36 The influence of the size of the force density on the equilibrium shape

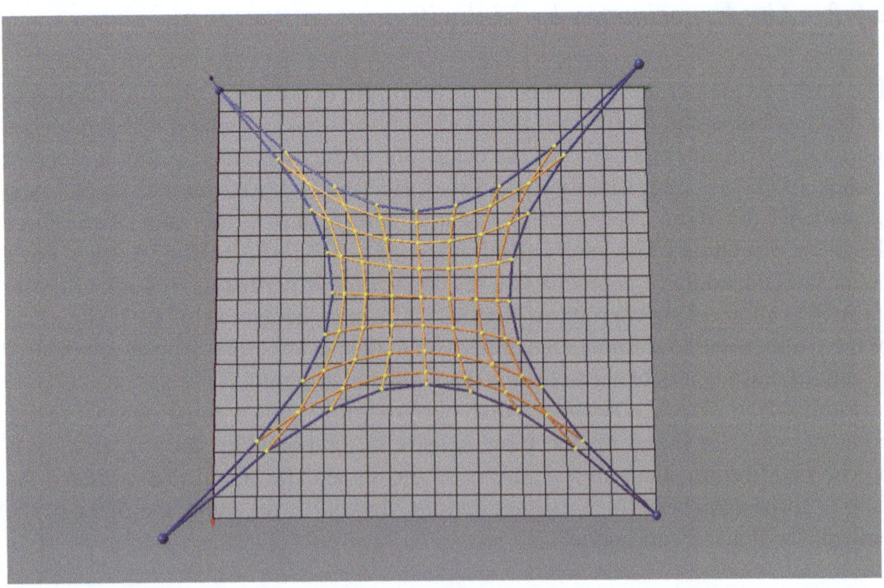

Fig. 2.37 The influence of the size of the force density on the equilibrium shape

The forces in the boundary are proportionately bigger than in the network. Therefore, the curvature of the boundary is smaller and the surface of the network increases. Conclusion: if the force density in the boundary compared to the network is bigger, the network will become bigger and the curvature in the network and boundary will become smaller.

The force density in the network is relatively big compared to the boundary cables. The curvature of the boundary cables is lower and the surface of the network is smaller. The above figures show that the shape of the equilibrium depends on the proportional relationship between the force densities in boundaries and network.

Below, there's a summary of the conclusions by Linkwitz.

- If the ratio in force density of the boundary cables and the network is equal, the shape of the structure will be the same.
- If the force density in the boundary compared to the network is bigger, the network will become bigger and the curvature in the network and boundary will decrease.
- The shape of the equilibrium depends on the proportional relationship between the force densities in boundaries and network (Figs. 2.38, 2.39, 2.40, and 2.41).

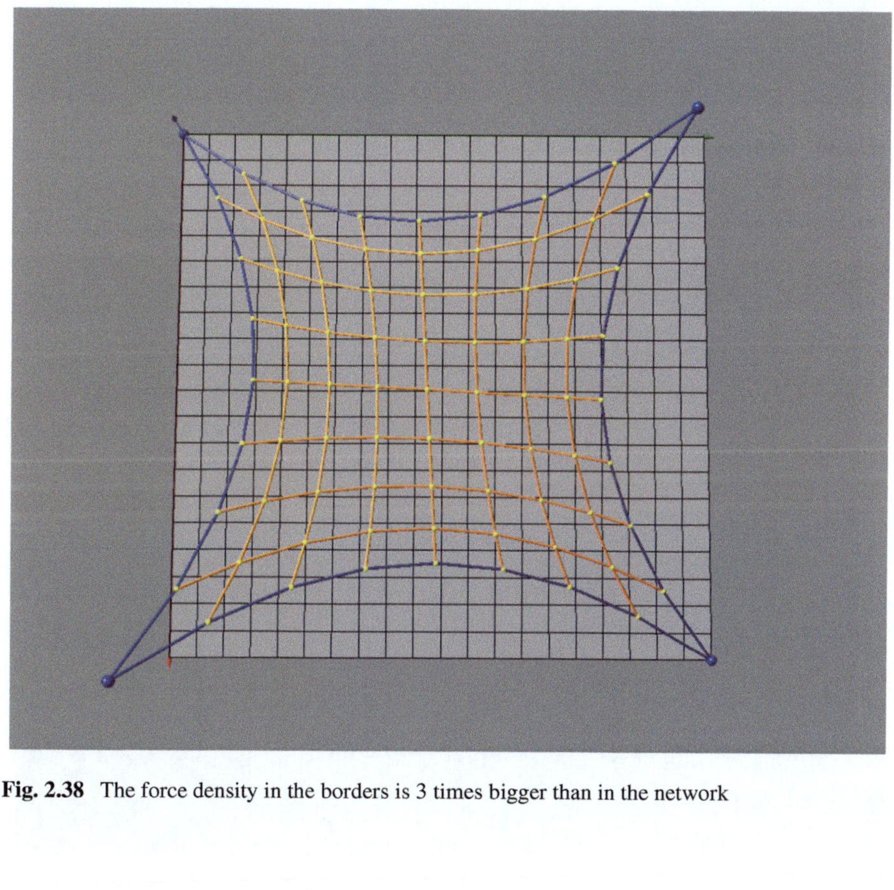

Fig. 2.38 The force density in the borders is 3 times bigger than in the network

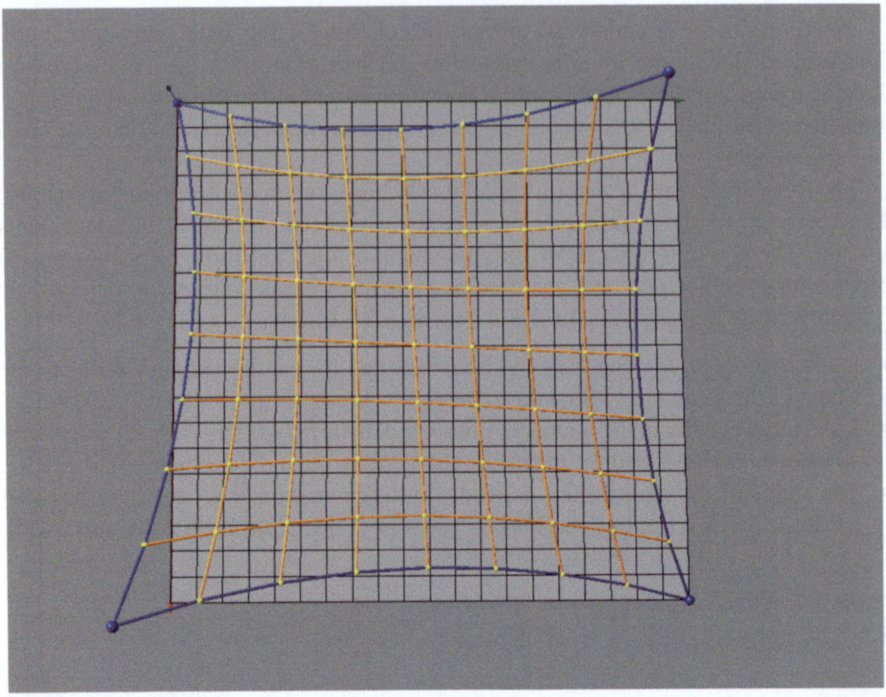

Fig. 2.39 The force density in the borders is 10 times bigger than in the network

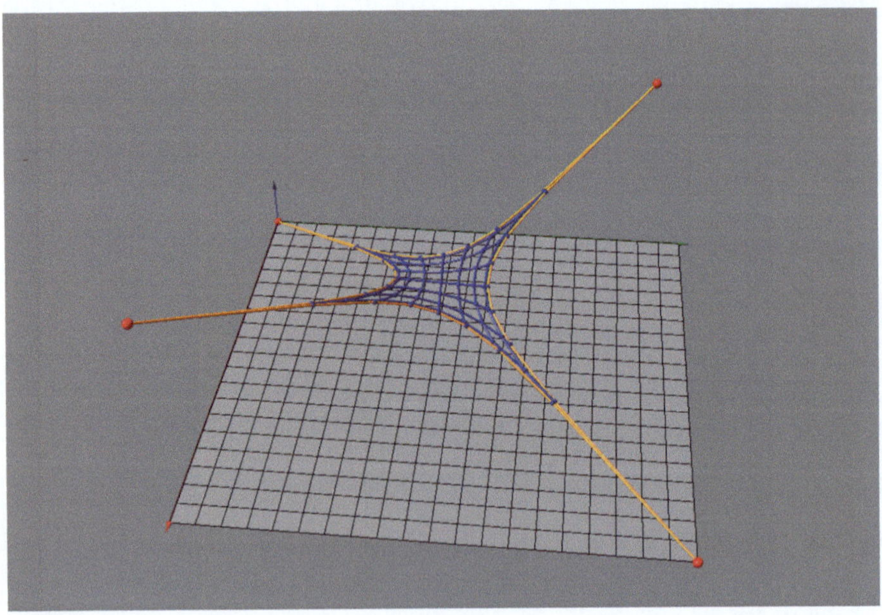

Fig. 2.40 The force density in the boundary is 3 times smaller than in the network

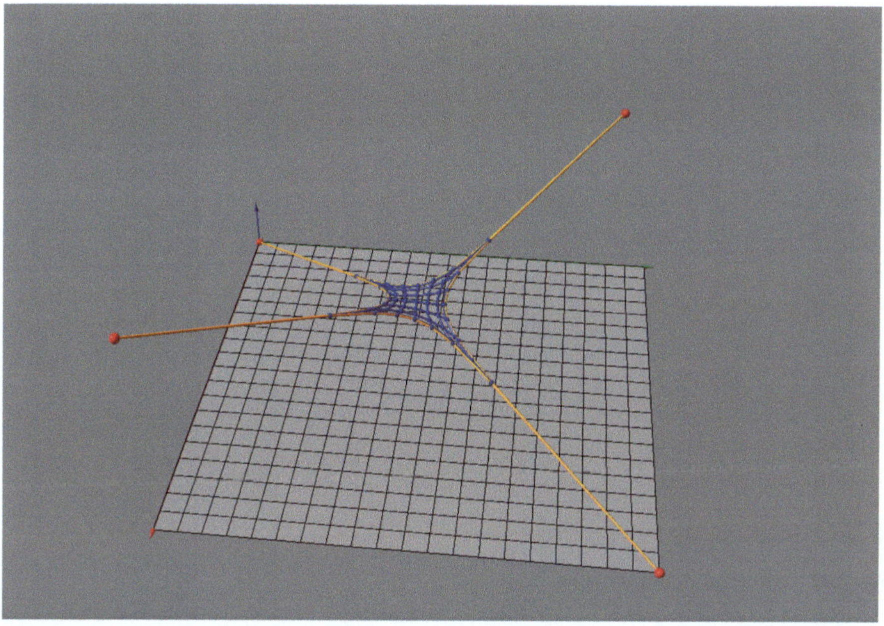

Fig. 2.41 The force density in the boundary is 6 times smaller than in the network

2.9.3 Changing the Prestress in a Certain Area of the Membrane

2.9.3.1 Local Prestressed Areas in Anticlastic Surfaces

Figure 2.42 shows equal force density proportions in cable and net. The force density in the cable over the surface of the net in this figure is 0. In Fig. 2.43, the force density is 5.

In Fig. 2.44, AS1, the examples AS2, AS3, AS4 and AS5 show anticlastic membrancs with a change in local prestressed areas.

In case of a change in a local prestressed distributed force within a surface, there will be a new state of force equilibrium. AS1 shows a cable with a higher prestress. This cable can be seen as a new boundary. Figure 2.43 shows that if the ratio between boundary and membrane becomes bigger, the curvature decreases and the area increases. AS1 is identical. The higher the stress in a cable, the less curvature there will be in membrane and cable. A prestressed cable on or in a mechanically prestressed membrane will decrease the anticlastic curvature of the membrane and will give a curved folding line at the position of the cable.

AS3 shows an area with a higher prestress. In that case the curvature perpendicular to the stressed area will increase. Outside this area and in the direction of the stress, the curvature will decrease.

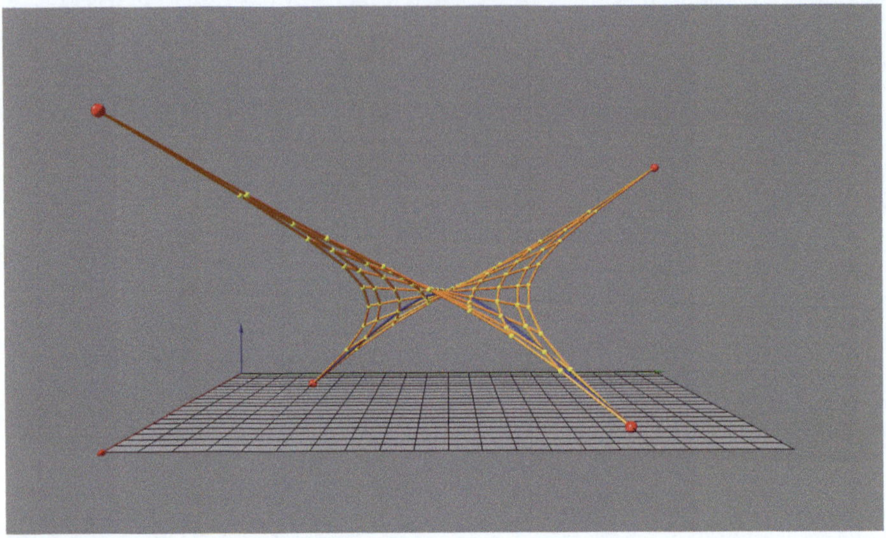

Fig. 2.42 Two equilibrium surfaces

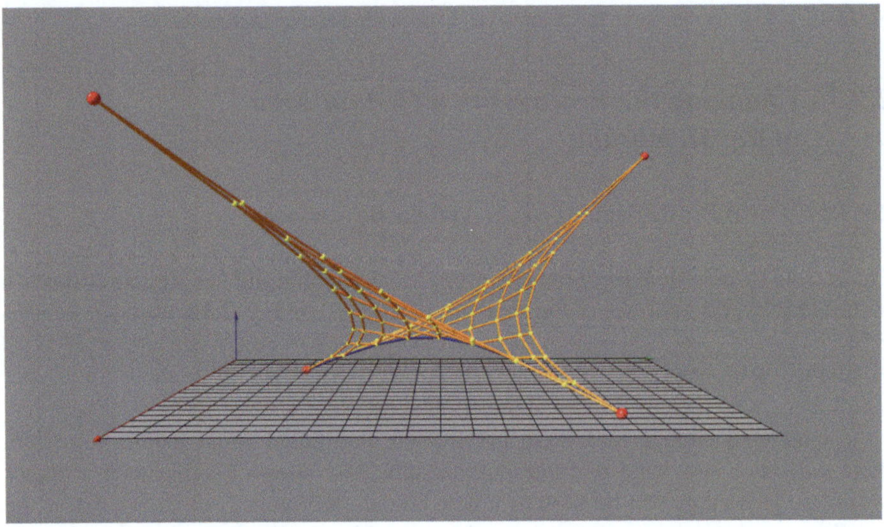

Fig. 2.43 Two equilibrium surfaces

AS2 shows a local area with prestress in both directions. In the part with the higher prestress the curvature will increase, in the other parts the curvature will decrease. AS4 and AS5 show a membrane with a local area with a relief of the prestress. In those parts the opposite occurs of what takes place in the membranes with a local higher prestress.

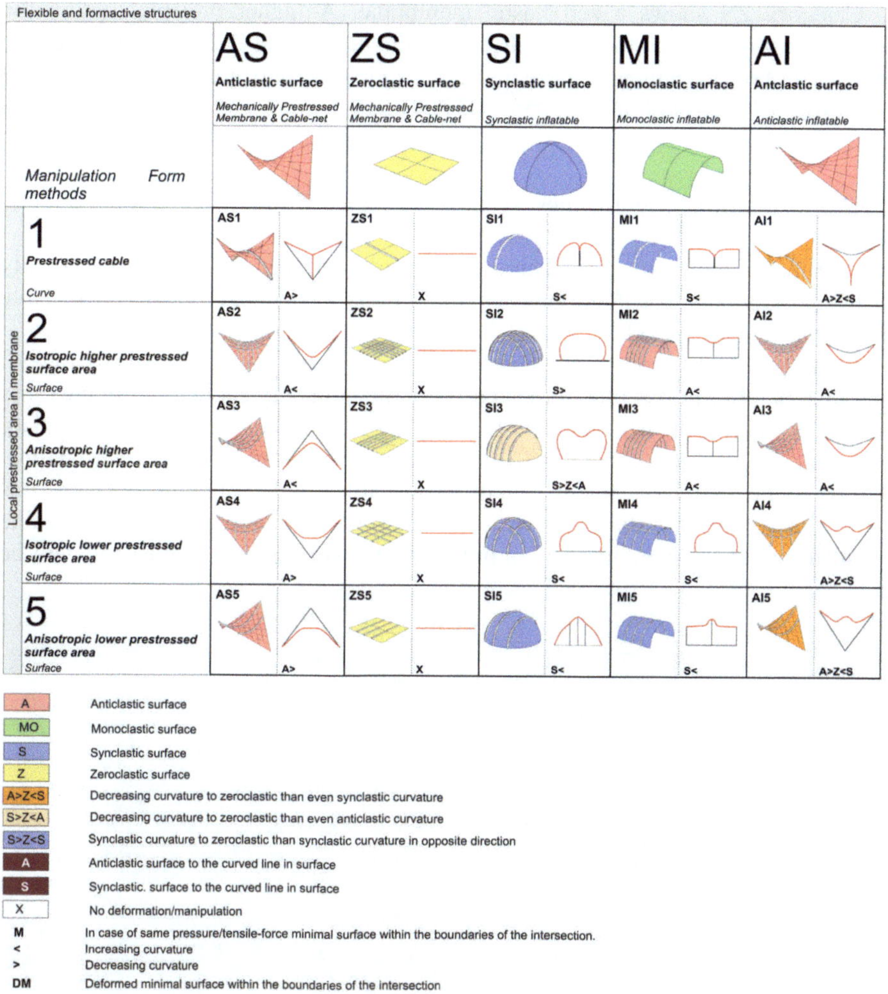

Fig. 2.44 A table of changing prestresses in a certain area

In case the stress in a prestressed surface is locally lower in one direction, the curvature in that direction will be higher. Perpendicular to this, the curvature in that area will be lower. The curvature adjacent to this area will increase.

2.9.3.2 Local Prestressed Areas in Zeroclastic Membranes

Within Fig. 2.44, the column of category ZS represents: Zeroclastic membranes while the examples in SI1, MI1 and AI1 show inflatable membranes with a prestressed cable.

Category ZS shows a zeroclastic membrane. A higher prestress within the membrane will not lead to a curvature; wrinkling may occur if the ratio between the tensions in both directions is more than a multiple of 2, depending on the boundary conditions (Seffen et al. 2000: 2055–2073).

2.9.3.3 Inflatable Membranes with a Prestressed Cable

Figures 2.45, 2.46, and 2.47: Three equilibrium surfaces with the same equally distributed upload. The force density proportions in boundary cable, cross cable and net have been varied.

Figures 2.45, 2.46, and 2.47 show a prestressed cable on or in the membrane of an inflatable. In Fig. 2.45 the force density in the boundary cable is 12, in the cross cable it is 1 and in the net cable it is 1. In Fig. 2.46 the force density in the boundary cable is 12, in the cross cable it is 12 and in the net cable it is 1. In Fig. 2.47 the force density in the boundary cable is 1, in the cross cable it is 12 and in the net cable it is 1. In all cases the result is a synclastic curvature of the membrane. When the cable is pulled out of the membrane of an inflatable, the surface next to the cable will be anticlastic.

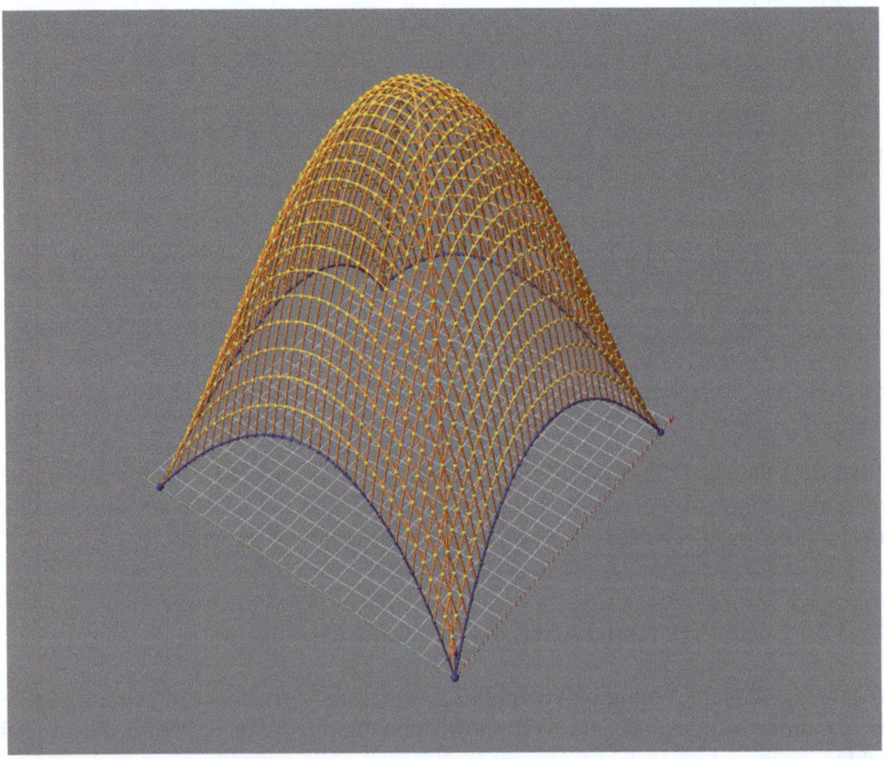

Fig. 2.45 Surface with upload

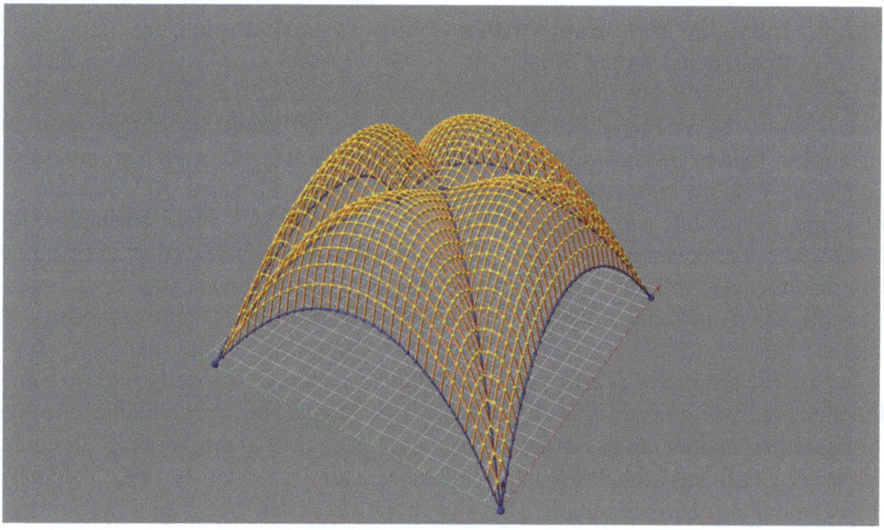

Fig. 2.46 Surface with upload and cable

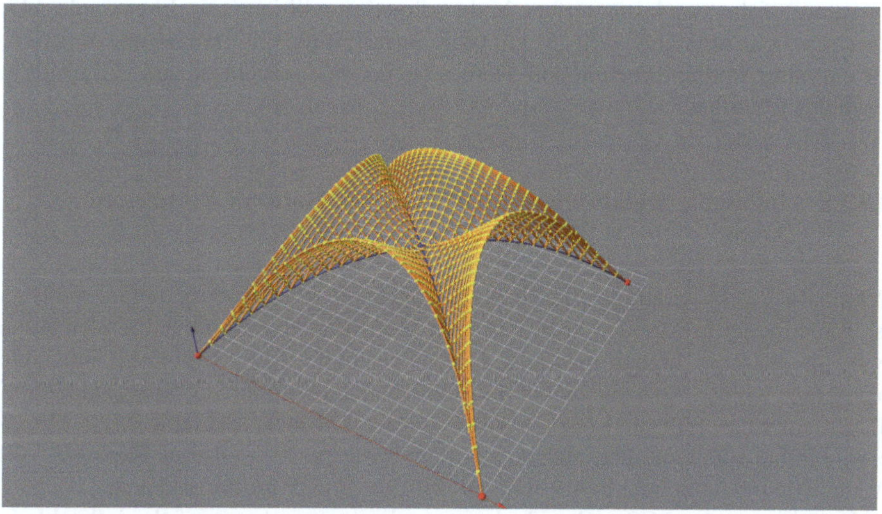

Fig. 2.47 Surface without boundary cables and upload

2.9.3.4 Inflatable Membranes with a Change of Anisotropic Prestressed Areas

Within Fig. 2.44, example SI3 shows a Synclastic inflatable membrane with an anisotropic higher prestressed area. Example MI3 shows a Monoclastic inflatable membrane with an anisotropic higher prestressed area. Example AI3: Anticlastic inflatable membrane with an anisotropic higher prestressed area. Example SI5: Synclastic inflatable membrane with an anisotropic lower prestressed area. Example MI5: Monoclastic inflatable membrane with an anisotropic lower prestressed area. Example AI5: Anticlastic inflatable membrane with an anisotropic lower prestressed area.

In case the stress in a prestressed or inflatable membrane is locally higher in one direction, the curvature in that direction will be lower, but perpendicularly to the curvature in that area the curvature will increase. Adjacent to this area the curvature will decrease. In case the stress in an inflatable with a synclastic surface is locally higher in one direction (example SI3), the curvature in that direction will decrease. It can become zeroclastic and then even anticlastic. The curvature of an anticlastic surface as in example AI3 will increase. A monoclastic surface will become anticlastic in the middle and adjacent to this area synclastic (example MI3). In examples SI5 and MI5, the stress in an inflatable surface is locally lower in one direction; the curvature of the synclastic surface in that direction will increase. In example AI5, the stress in an inflatable with an anticlastic surface is locally lower in one direction; the curvature in that direction will decrease to become zeroclastic and will result in a synclastic surface.

2.9.3.5 Inflatable Membranes with a Change in Isotropic Prestressed Areas

Within Fig. 2.44, example SI2 shows a Synclastic inflatable membrane with an isotropic higher prestressed area. Example MI2: Monoclastic inflatable membrane with an isotropic higher prestressed area. Example AI2: Anticlastic inflatable membrane with an isotropic higher prestressed area. Example SI4: Synclastic inflatable membrane with an isotropic lower prestressed area. Example MI4: Monoclastic inflatable membrane with an isotropic lower prestressed area. Example AI4: Anticlastic inflatable membrane with an isotropic higher prestressed area.

In the examples MI2 and AI2, the stress in the membrane of the inflatable is locally higher in both directions; for AI2, the curvature in that area will increase and for MI2 it will change from monoclastic to increasingly anticlastic. In example SI2, the synclastic curvature will decrease. In examples SI4, MI4 and AI4 the opposite occurs: the stress in the membrane of the inflatable is lower; the curvature in that area will increase, or change into a synclastic surface. The surface area adjacent to the lower stressed area will decrease or even become anticlastic.

2.9.4 *External Loads on a Structure in Equilibrium*

2.9.4.1 Curvature and Deformation

Below, there are some more experiments by Linkwitz and a group of students. In this research, the experiments by Linkwitz were redone in the program Tess3D. The influence of the force density on the deformation of the equilibrium surface as a result of a vertical point load is shown in Figs. 2.48 and 2.49.

The main curvature in a network will be influenced by the relative position of its boundary. There is a relation between the curvature of a network and the behaviour of that network under load: the lower the curvature, the higher the deformation and vice versa Figs. 2.50, 2.51, and 2.52.

The experiments by Linkwitz show the possibilities to manipulate the form of pre-tensioned membranes and the behaviour under load. Below, there are some more experiments with the following combinations:

- an equally spread upload in combination with one-point load; and
- an equally spread upload in combination with some cables.

The force density proportions in border cable (12) and net (1) are equal. In Fig. 2.53, the surface is pulled down in one point with a force of 25. In Fig. 2.54, the surface is pulled upwards in one point with a force of 19.

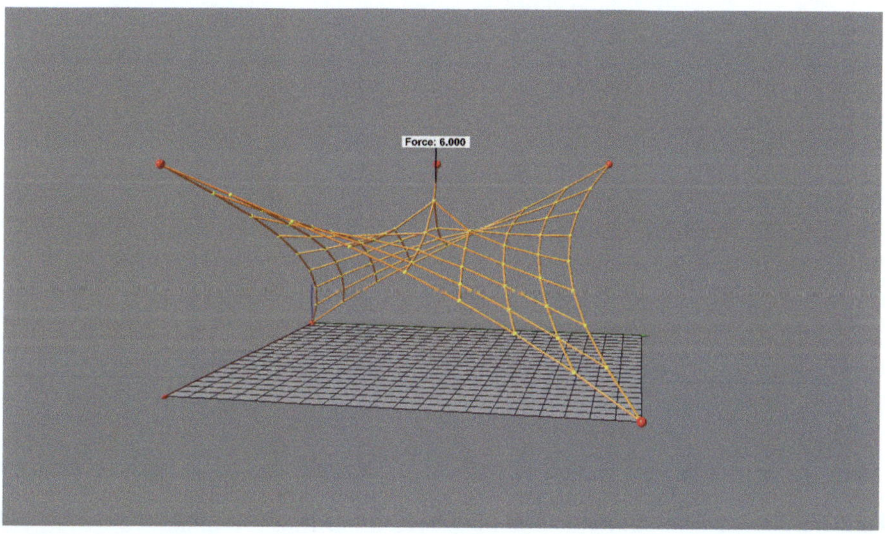

Fig. 2.48 The force in the boundary = 5, in the network = 1; on the network there is a vertical point load with a value of 6

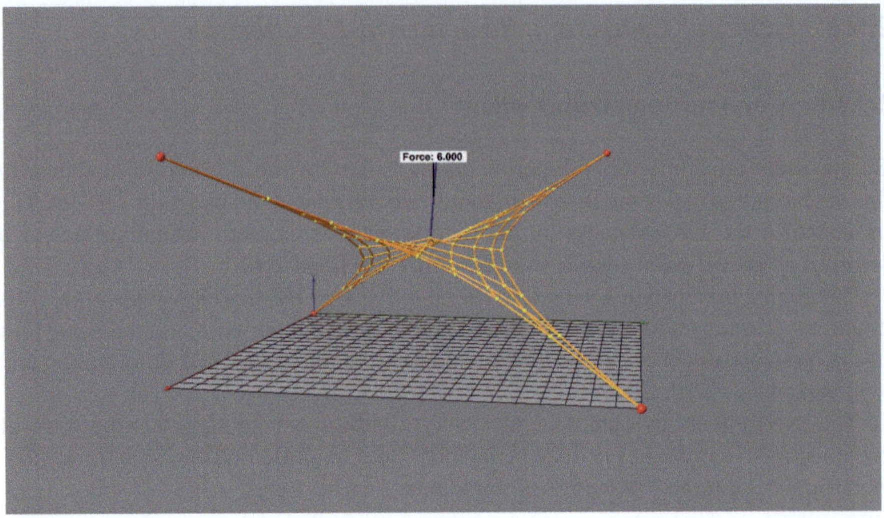

Fig. 2.49 The force in the boundary = 5, in the network = 5; the vertical point load has a force equivalent of 6. The figures show that the higher the force density in the network, the smaller the deformation will be

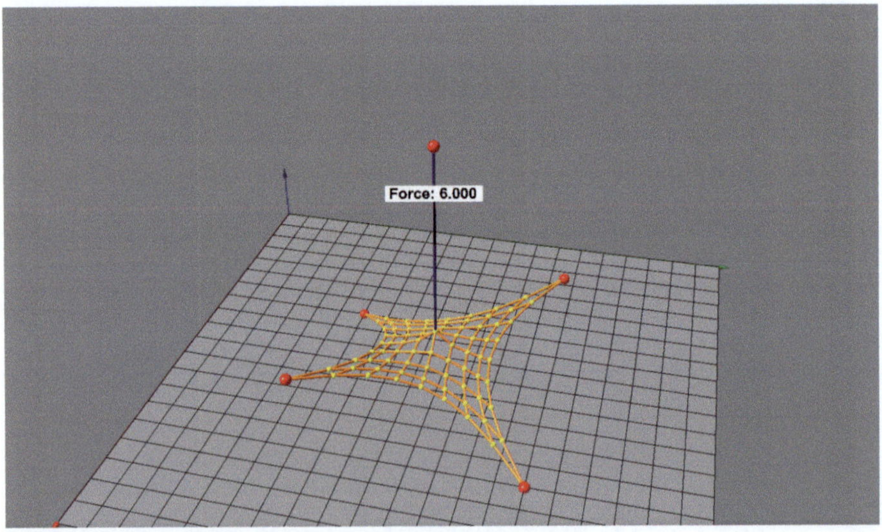

Fig. 2.50 Two equilibrium surfaces with equal force density proportions and vertical point load but a different curvature

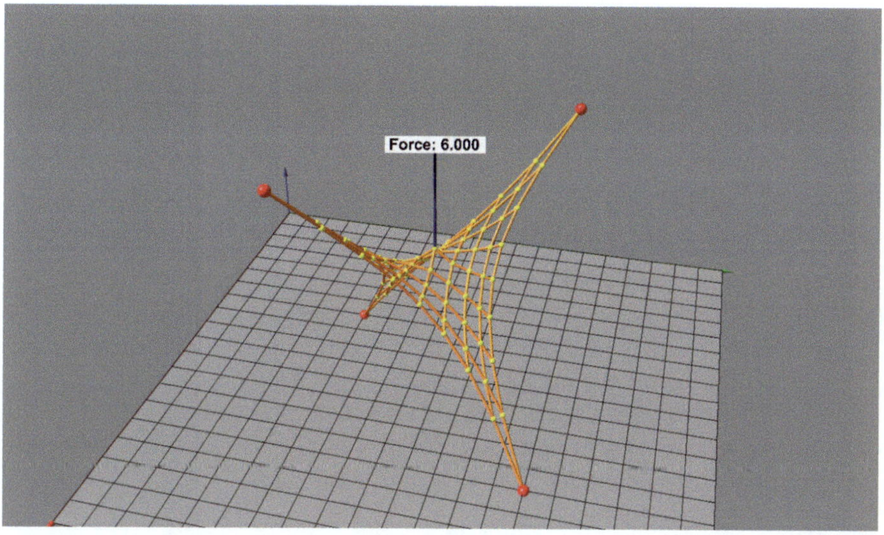

Fig. 2.51 Two equilibrium surfaces with equal force density proportions and vertical point load but a different curvature

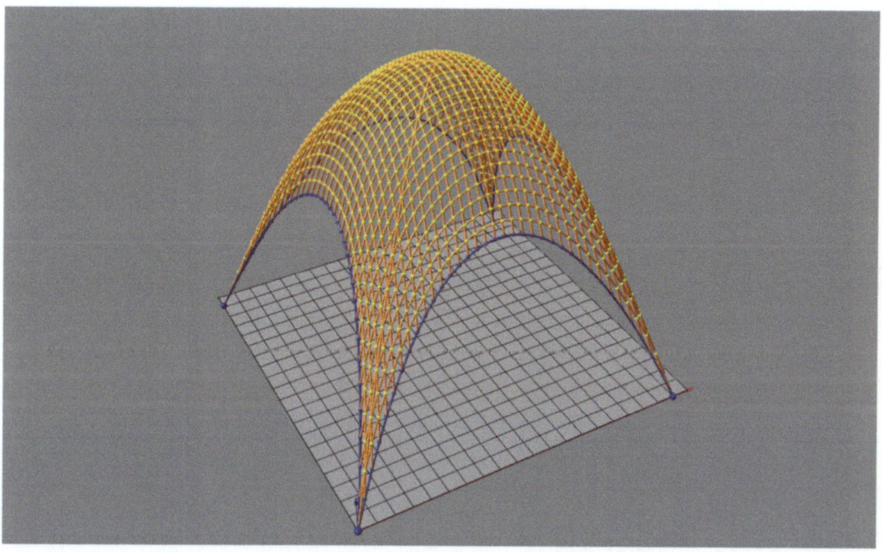

Fig. 2.52 Three equilibrium surfaces with an equally spread upload

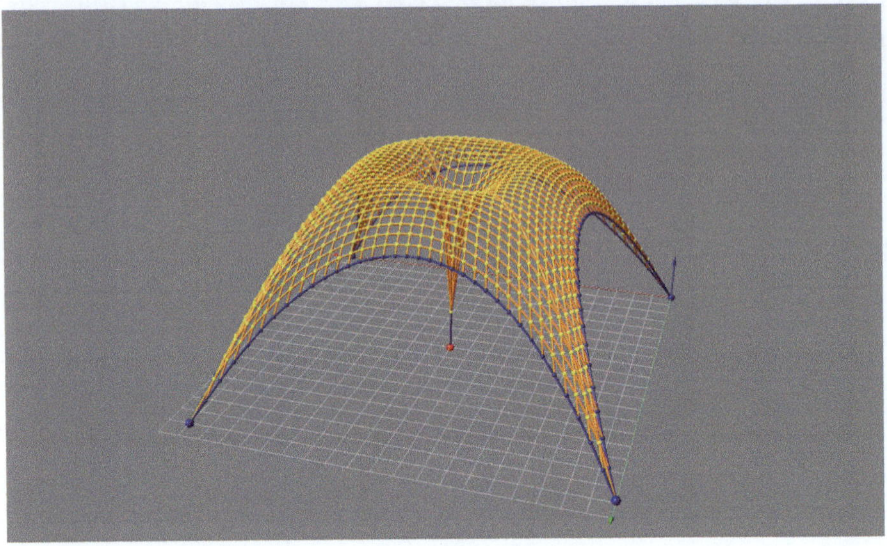

Fig. 2.53 Three equilibrium surfaces with an equally spread upload

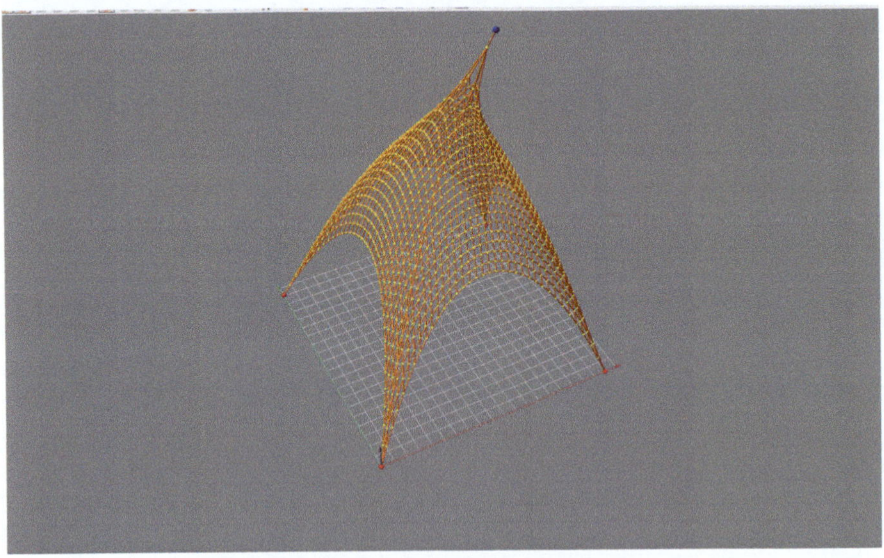

Fig. 2.54 Three equilibrium surfaces with an equally spread upload

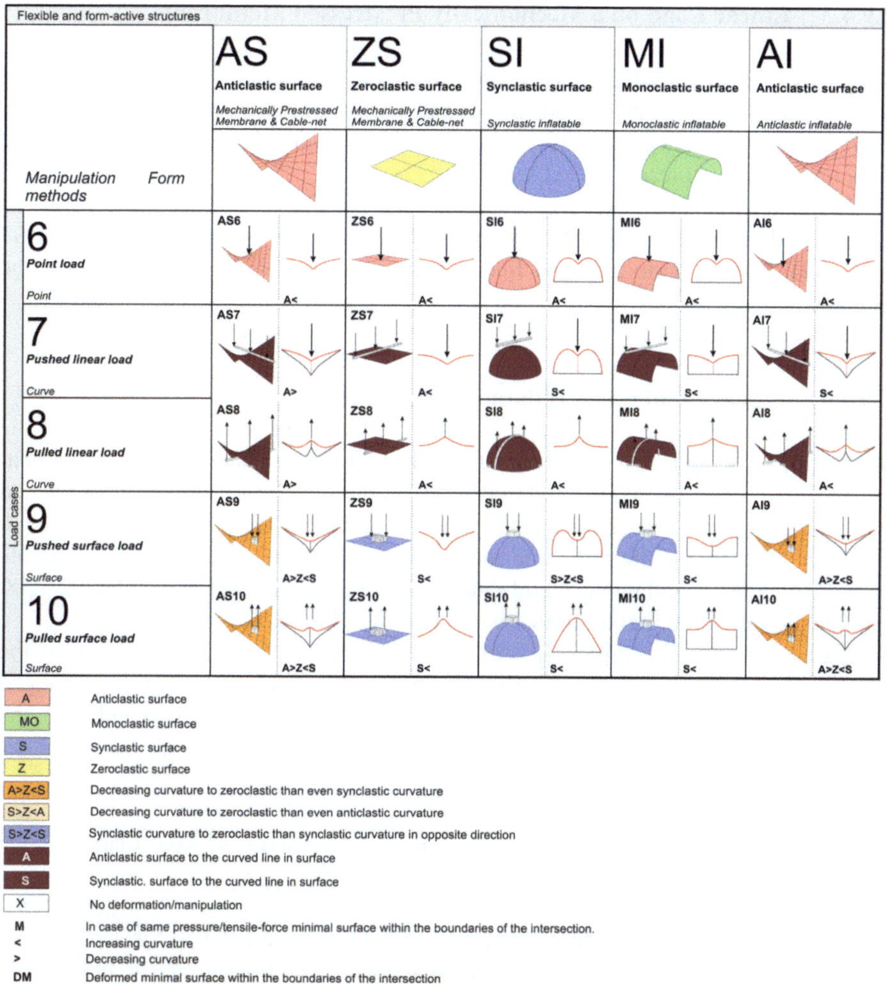

Fig. 2.55 Table: External load

2.9.4.2 Point Load

Within Fig. 2.55, example AS6 represents an Anticlastic membrane with a point load. Example ZS6: Zeroclastic membrane with a point load. Figure SI6: Synclastic inflatable membrane with a point load. Example MI6: Monoclastic inflatable membrane with a point load. Example AI6: Anticlastic inflatable membrane with a point load. In case of an external load, there will be a new state of equilibrium. A prestressed surface in equilibrium will always result in an anticlastic surface in the place of contact with another point.

2.9.4.3 Linear Load on a Mechanically Prestressed Membrane

Within Fig. 2.55, example AS7 shows an Anticlastic membrane with a linearly pushed load. Example AS8: Anticlastic membrane with a linearly pulled load. Example ZS7: Zeroclastic membrane with a linearly pushed load. Example ZS8: Zeroclastic membrane with a linearly pulled load.

For anticlastic prestressed membranes and zeroclastic surfaces, the result of pulling and pushing is the same (examples AS7 and AS8 as well as examples ZS7 and ZS8): it will create a curved line in the surface. The changed surface will be anticlastic.

2.9.4.4 Linear Load on an Inflatable Membrane

Within Fig. 2.55, example SI7 shows a Synclastic inflatable membrane with a linear pushed load. Example MI7: Monoclastic inflatable membrane with a linear pushed load. Example AI7: Anticlastic inflatable membrane with a linear pushed load. Example SI8: Synclastic inflatable membrane with a linear pulled load. Example MI8: Monoclastic inflatable membrane with a linear pulled load. Example AI8: Anticlastic inflatable membrane with a linear pulled load.

A linear load on an inflatable will result in a curved line in the surface of the membrane. Pushed from the outside (examples SI7, MI7 and AI7), the changed surface will be synclastic. Pulled from the outside (examples SI8, MI8 and AI8), the surface will be anticlastic.

2.9.4.5 Local Surface Load on a Mechanically Prestressed Membrane

A surface load on an anticlastic surface will always decrease the anticlastic curvature. If the load on the surface is high enough, it will lead to a synclastic surface. A zeroclastic membrane will always form a synclastic surface. For anticlastic prestressed membranes and zeroclastic surfaces the result for pulling and pushing is the same (examples AS9 and AS10 as well as examples ZS9 and ZS10).

2.9.4.6 Local Surface Load on an Inflatable Membrane

The surface load on an inflatable will always lead to a local synclastic surface if the load is high enough. In examples AI9 and AI10 the anticlastic curvature will decrease to zeroclastic before it becomes synclastic. In example SI9 the synclastic curvature decreases to zeroclastic and becomes synclastic in the opposite direction. At the borders of a synclastic surface to its inverse synclastic curvature there will be

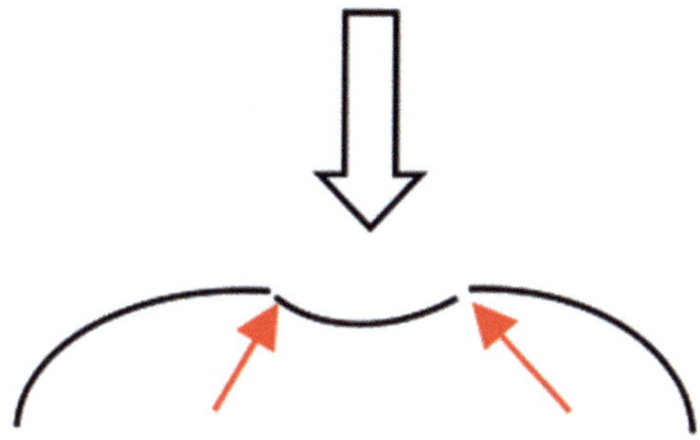

Fig. 2.56 Sections of a synclastic membrane with a locally pushed surface load

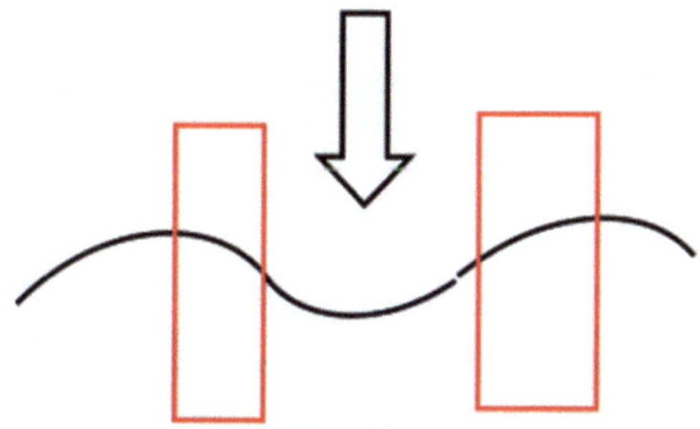

Fig. 2.57 Sections of a synclastic membrane with a locally pushed surface load

a folding line (Fig. 2.56) or anticlastic area (curved line within the rectangular part of Fig. 2.57).

Example SI9: Synclastic membrane with a locally pushed surface load. Example MI9: Monoclastic membrane with a locally pushed surface load. Example AI9: Anticlastic membrane with a locally pushed surface load. Example SI10: Synclastic membrane with a locally pulled surface load. Example MI10: Monoclastic membrane with a locally pulled surface load. Example AI10: Anticlastic membrane with a locally pulled surface load.

2.9.5 Pushing Form-Active Surfaces in Equilibrium into Another Form-Active Surface in Equilibrium

Pushing form-active surfaces in equilibrium into another form-active surface in equilibrium will transform the two surfaces into a new state of equilibrium based on the forces and form of the two individual surfaces (Fig. 2.58).

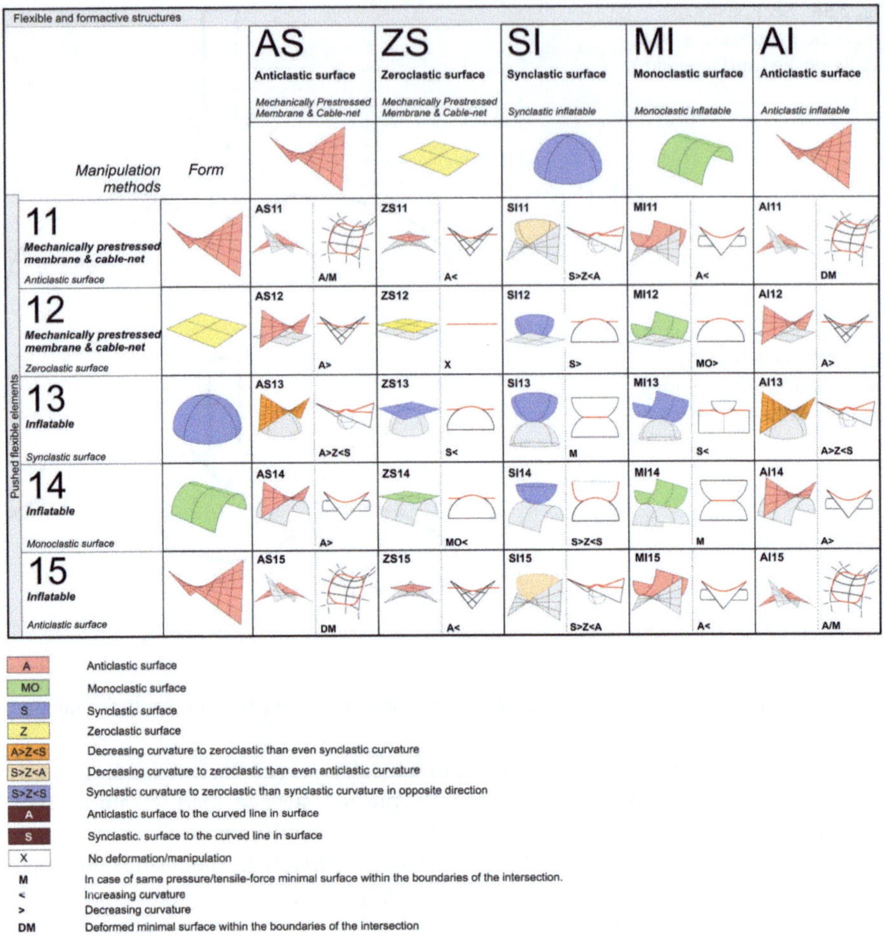

Fig. 2.58 Table: Pushing other form-active surfaces into the surface

The general rules are:

- At the place of contact between both surfaces the angle between both surfaces is zero (Fig. 2.59).
- If membrane structures are combined, a continuous fluid surface develops at the place where both surfaces touch.
- At the place of contact of surfaces, the membrane with the highest tension determines the form of the surface.
- If the tension in both surfaces is equal, the surface will form a minimal surface within the boundaries of the connected area. That is an anticlastic or zeroclastic surface.
- In all the cases where a membrane is pushed into a zeroclastic membrane (examples AS12, SI12, MI12 and AI12) the curvature of the membrane will decrease.

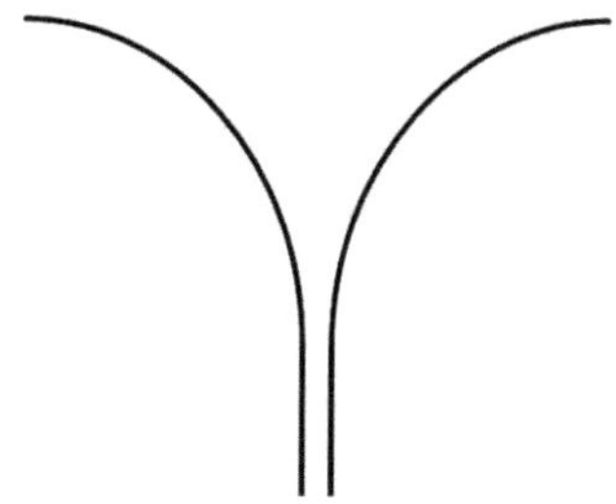

Fig. 2.59 Two pushed inflatables

2.9.5.1 Pushing an Anticlastic Mechanically Prestressed Membrane into Another Membrane

Within Fig. 2.58, example AS11 shows an Anticlastic mechanically prestressed membrane pushed into another anticlastic mechanically prestressed membrane. Example AS12: Anticlastic mechanically prestressed membrane pushed into a zeroclastic mechanically prestressed membrane. Example AS13: Anticlastic mechanically prestressed membrane pushed into a synclastic inflatable membrane. Example AS14: Anticlastic mechanically prestressed membrane pushed into a monoclastic inflatable membrane. Example AS15: Anticlastic mechanically prestressed membrane pushed into an anticlastic inflatable membrane.

In examples AS11 and AS12 the contact area will be anticlastic. If the prestress in both membranes is equal, the surface will form a minimal surface within the boundaries of the contact area. In examples AS14 and AS15 the contact area will be anticlastic. If the prestress in both membranes is the same, the surface will form a minimal surface within the boundaries of the contact area. If not, it will differ from a minimal surface due to the overpressure at one side. In example AS13 the contact area is anticlastic and decreases to zeroclastic to become synclastic if the tension in the synclastic inflatable membrane and the pushing force are strong enough.

2.9.5.2 Pushing a Zeroclastic Prestressed Membrane Against Another Membrane

Within Fig. 2.58, example ZS11 shows a Zeroclastic mechanically prestressed membrane pushed against an anticlastic mechanically prestressed membrane. Example ZS12: Zeroclastic mechanically prestressed membrane pushed against another zeroclastic mechanically prestressed membrane. Example ZS13: Zeroclastic mechanically prestressed membrane pushed against a synclastic inflatable membrane. Example ZS14: Zeroclastic mechanically prestressed membrane pushed against a monoclastic inflatable membrane. Example ZS15: Zeroclastic mechanically prestressed membrane pushed against an anticlastic inflatable membrane.

In all the examples the zeroclastic membrane will adapt to the surface of the membrane against which it is pushed. It will also decrease the curvature of the curved membrane below. In example ZS12 there is no manipulation.

2.9.5.3 Pushing a Synclastic Inflatable Against Another Membrane

Within Fig. 2.58, example SI11 shows a Synclastic inflatable membrane pushed against an anticlastic mechanically prestressed membrane. Example SI12: Synclastic inflatable membrane pushed against a zeroclastic mechanically prestressed membrane. Example SI13: Synclastic inflatable membrane pushed against another synclastic inflatable membrane. Example SI14: Synclastic inflatable membrane pushed against a monoclastic inflatable membrane. Example SI15: Synclastic inflatable membrane pushed against an anticlastic inflatable membrane.

In examples SI12 and SI13 the curvature decreases. In case of SI13 it might become zeroclastic and even synclastic in the opposite direction. At the borders of a synclastic surface to its inverse synclastic curvature there will be a folding line (see Fig. 2.58) or anticlastic area (Fig. 2.59). In examples SI11, SI14 and SI15 the curvature in the surface decreases to zeroclastic and might even become anticlastic depending on the pushing force and tensile force within the surfaces.

2.9.5.4 Pushing a Monoclastic Inflatable Against Another Membrane

Within Fig. 2.58, example MI11 shows a Monoclastic inflatable membrane pushed against an anticlastic mechanically prestressed membrane. Example MI12: Monoclastic inflatable membrane pushed against a zeroclastic mechanically prestressed membrane. Example MI13: Monoclastic inflatable membrane pushed against a synclastic inflatable membrane. Example MI14: Monoclastic inflatable membrane pushed against another monoclastic inflatable membrane. Example MI15: Monoclastic inflatable membrane pushed against an anticlastic inflatable membrane.

In examples MI11 and MI15 the monoclastic surface becomes anticlastic at the contact area. In example MI13 the monoclastic surface becomes synclastic at the contact area. Adjacent to this area the surface will be a folding line (see Fig. 2.58) or anticlastic area (Fig. 2.59). In example MI14 there are two cases: In the first case the monoclastic surfaces have the same parallel direction. The monoclastic curvature decreases to zeroclastic and might become monoclastic in the opposite direction; at the boundary of the "negative" monoclastic surface to the "positive" curvature there will be a folding line or monoclastic area (Figs. 2.58 and 2.59). In the other case the monoclastic surfaces are not parallel. Then the connected area will form an anticlastic surface in equilibrium. Adjacent to this contact area there will be a folding line or anticlastic area (Figs. 2.58 and 2.59).

2.9.5.5 Pushing an Anticlastic Inflatable Against Another Membrane

Within Fig. 2.58 example AI11 shows an Anticlastic inflatable membrane pushed against an anticlastic mechanically prestressed membrane. Example AI12: Anticlastic inflatable membrane pushed against a zeroclastic mechanically prestressed membrane. Example AI13: Anticlastic inflatable membrane pushed against a synclastic inflatable membrane. Example AI14: Anticlastic inflatable membrane pushed against a monoclastic inflatable membrane. Example AI15: Anticlastic inflatable membrane pushed against another anticlastic inflatable membrane.

In example AI15 the connected area will be anticlastic. If the prestress in both membranes is equal, the surface will be a minimal surface. In examples AI11 and AI14 the connected area will be anticlastic. If the prestress in both membranes is equal, a minimal surface within the boundaries of the connected area will be formed. If not, it will differ from a minimal surface due to the overpressure at one side. In example AI13 the connected area is anticlastic and decreases to zeroclastic to become synclastic if the tension in the synclastic inflatable membrane and pushing force is strong enough.

2.9.6 *Pushing a Rigid Element into or Out of a Membrane Structure in Force Equilibrium*

The following general rules apply for surfaces in equilibrium that will be manipulated by pushing a rigid element into or out of the structure (Fig. 2.60):

If a rigid element is pushed against the surface in equilibrium, the membrane structure will adapt to the surface of the rigid element unless the boundaries/surface of the rigid element allow the flexible surface to form a new surface in force equilibrium released from the surface of the rigid element. See Fig. 2.61(E).

If there is contact between two surfaces, the angle between them is zero.

If the angle between the surfaces increases fluently from zero, there will be a gap between the surfaces. The contact of the surfaces is within a place on the two surfaces. In this case there is a smooth transition between the two surfaces into a joint surface. See Fig. 2.61(A), (C) and (F).

If the angle between the surfaces changes sharply, the release between the surfaces is at one of the boundaries of the surface. This will give a (curved) line within the other flexible surface. See Fig. 2.61(B), (C) and (E).

2.9.6.1 Pushed-Out Elements from the Inside of an Inflatable

Within Fig. 2.60, example SI16: Synclastic inflatable membrane with a pushed-out element. Example MI16: Monoclastic inflatable membrane with a pushed-out element. Example AI16: Anticlastic inflatable membrane with a pushed-out element.

If any rigid element is pushed out from the inside of an inflatable structure, an anticlastic surface at the borders of the structure develops.

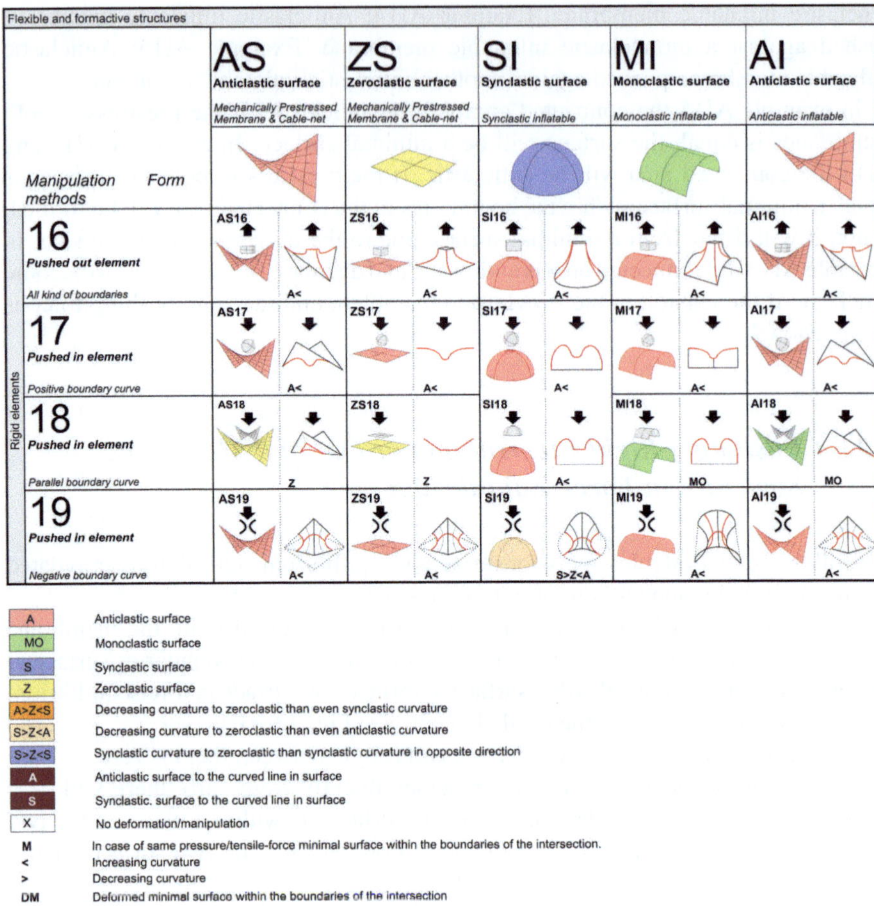

Fig. 2.60 Table: Pushing or pulling with a rigid element

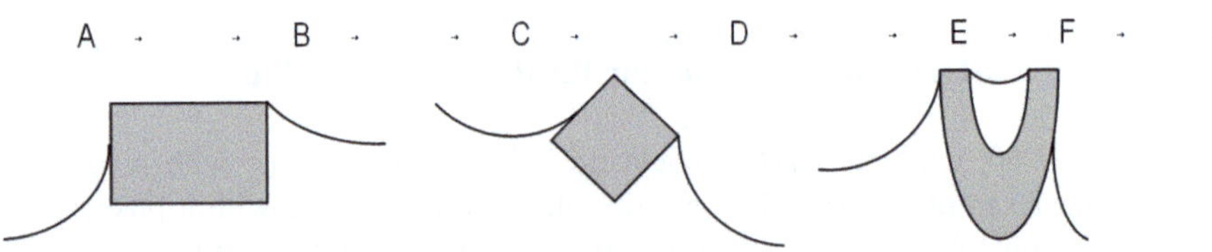

Fig. 2.61 (A, B, C, D, E and F): The sections of a surface in force equilibrium pushed against a rigid element

2.9.6.2 Pushed-in Elements Against Mechanically Prestressed Membranes

Within Fig. 2.60, example AS16: Anticlastic mechanically prestressed membrane with a pushed-in element. Example ZS16: Zeroclastic mechanically prestressed membrane with a pushed-in element. Example AS17: Anticlastic mechanically prestressed membrane with a pushed-in element with a positive boundary. Example ZS17: Zeroclastic mechanically prestressed membrane with a pushed-in element with a positive boundary. Example AS19: Anticlastic mechanically prestressed membrane with a pushed-in element with a negative boundary. Example ZS19: Zeroclastic mechanically prestressed membrane with a pushed-in element with a negative boundary.

A zeroclastic mechanically prestressed membrane and an anticlastic mechanically prestressed membrane will develop an anticlastic surface at the boundaries of any rigid element pushed in or out as long as the boundaries are not parallel.

2.9.6.3 Pushed-in Elements with a Parallel Boundary Curve

Within Fig. 2.60, example AS18: Anticlastic mechanically prestressed membrane with a pushed-in element with parallel boundaries. Example ZS18: Zeroclastic mechanically prestressed membrane with a pushed-in element with parallel boundaries. Example MI18: Monoclastic inflatable with a pushed-in element with parallel boundaries. Example AI18: Anticlastic inflatable with a pushed-in element with parallel boundaries.

If the borders of a zeroclastic mechanically prestressed membrane and an anticlastic mechanically prestressed membrane are parallel with the element pushed in, it will form a zeroclastic surface. If the direction of a monoclastic or synclastic inflatable surface is parallel to a monoclastic or zeroclastic element or straight boundary line, the surface will be monoclastic.

2.9.6.4 Pushed-in Elements with a Positive Boundary Curve Against an Inflatable

Within Fig. 2.60, example SI17: Synclastic inflatable membrane with a pushed-in element with a positive boundary. Example MI17: Monoclastic inflatable membrane with a pushed-in element with a positive boundary. Example AI17: Anticlastic inflatable membrane with a pushed-in element with a positive boundary.

If any rigid element with a positive border curvature is pushed in from the outside of an inflatable structure, an anticlastic surface is formed at the boundary of the rigid element. This also occurs if the positive boundary line is parallel to the boundary line of the membrane.

2.9.6.5 Pushed-in Elements with a Negative Boundary Curve Against an Inflatable

Within Fig. 2.60, example SI19: Synclastic inflatable membrane with a pushed-in element with a negative boundary. Example MI19: Monoclastic inflatable membrane with a pushed-in element with a negative boundary. Example AI19: Anticlastic inflatable membrane with a pushed-in element with a negative boundary.

An inflatable structure will develop a synclastic surface when a rigid element with a negatively curved boundary is pushed in from the outside.

2.9.7 The Influence of the Pattern on the Shape

Besides the forces on or in the membrane, the direction of the fibres or the structure of the cable net will influence the form of the surface. Some examples are presented in Figs. 2.15–2.18.

Figures 2.62, 2.63, 2.64, and 2.65: For equilibrium saddle shaped cable nets with equal force density proportions in cable (1) and net (1) and equal anchor points. The form of the saddle is influenced by the geometry of the net structure.

In Fig. 2.63 the diagonal (yellow) cables were added. In Figs. 2.64 and 2.65, no cables were added to the structure. In this case the fixed (red) points were rotated.

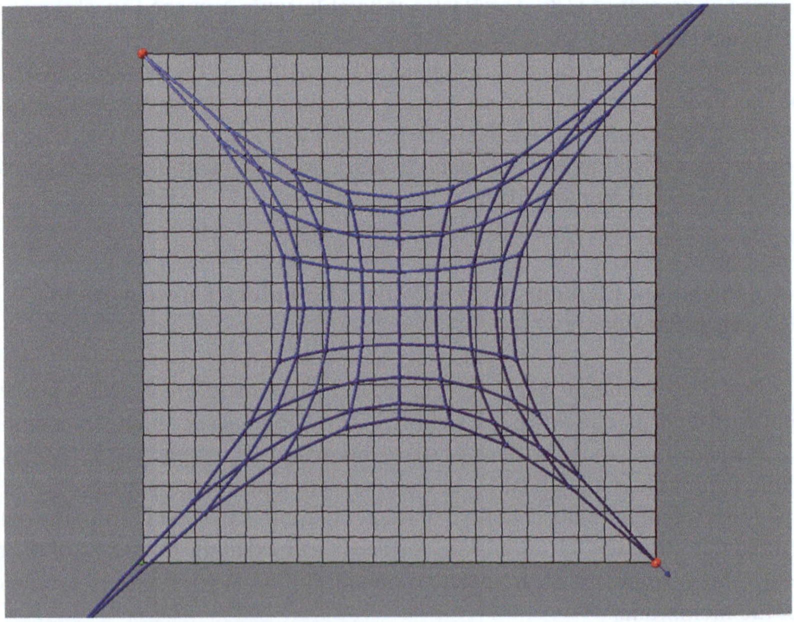

Fig. 2.62 Pattern with equal force density

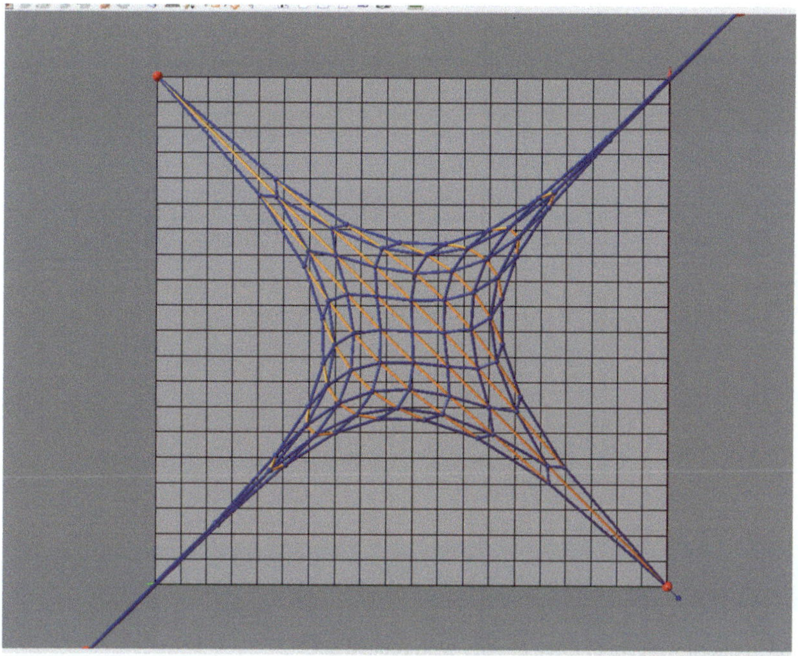

Fig. 2.63 Pattern with equal force density with extra cables

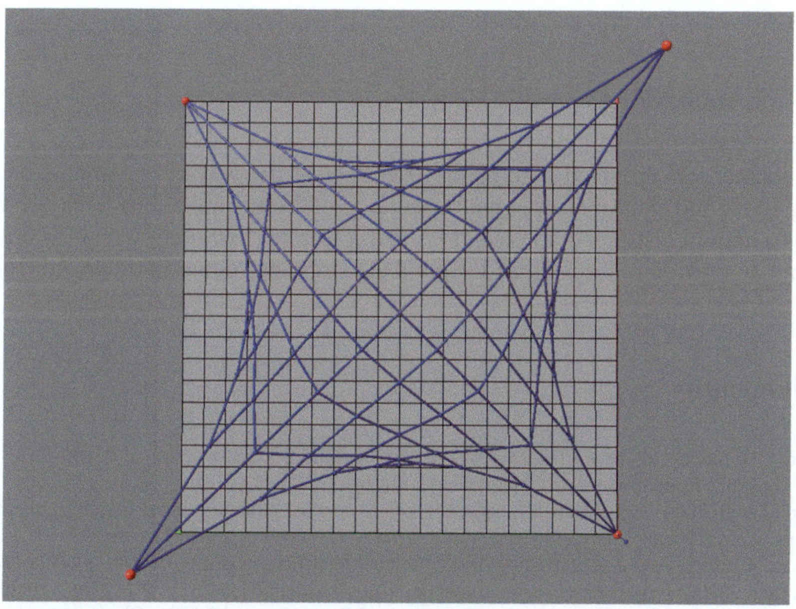

Fig. 2.64 Pattern with equal force density with rotated net

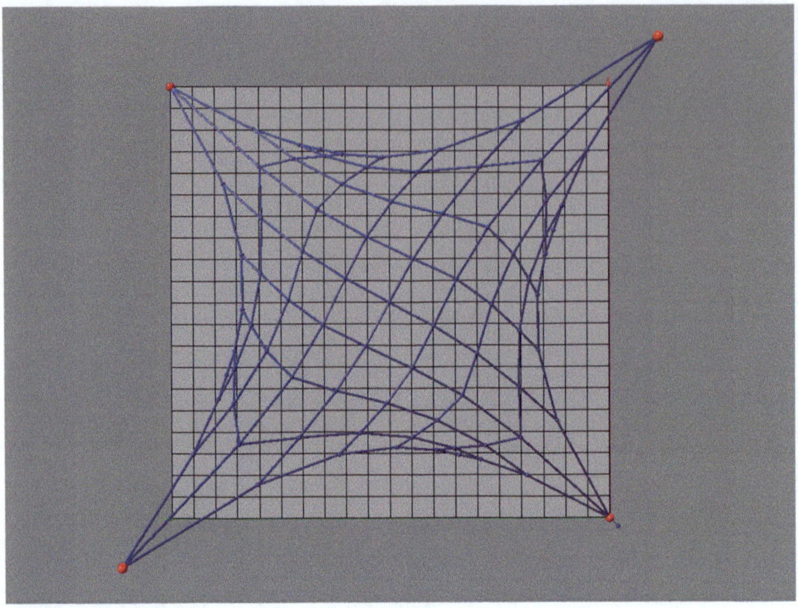

Fig. 2.65 Pattern with equal force density with rotated net

The differences between figures show the influence of the pattern on the geometry of the surface.

2.10 Conclusions

This chapter has provided a complete overview of the ways a membrane can be manipulated and of the major geometrical effects of the manipulations. This overview can be a helpful instrument for designers and researchers working with membranes. It is not an attempt to research and explain the stress behaviour in the membrane.

Bibliography

Dirnbock H, Stachel H (1997) The Development of the Oloid. J Geom Graph 1(2):105–118

Engel H (1967) Structure systems. Hatje Cantz, Stuttgart

Glymph J et al (2004) A parametric strategy for free-form glass structures using quadrilateral planar facets. Autom constr 13(2):187–202

Linkwitz K (1999) About form finding of double-curved structures. Eng Struct 21(8):709–718

Linkwitz K (2002) Experience from a course on "form finding and analysis of tension structures", held at the University of Stuttgart for 10 years. Int J Space Struct 17:159–169

Linkwitz K, Schek H (1971) Einige Bemerkungen zur Berechnung von vorgespannten Seilnetzkonstuktionen. Ingenieur-Archiv 40:145–158

Otto F, Schaur E (1979) Form, Kraft und Masse – Grundlagen. Form-Force-Mass. In: Information of the Institute for Lightweight Structures Series IL 21. George Wittenborn Inc., p 21

Pronk ADC (2012) 84 Ways to manipulate a membrane. Journal of the Int. Association for Shell and Spatial Structures

Schek HJ (1974) The force density method for form finding and computation of general network. Comput Methods Appl Mech Eng 3(1):115–134

Seffen KA, You Z, Pellegrino S (2000) Folding and deployment of curved tape springs. Int J Mech Sci 42(10):2055–2073

Chapter 3
Fluid Architecture

3.1 Introduction

This paragraph will clarify what fluid architecture is and will give reasons why the word 'fluid' was chosen instead of 'free-form' or other terms. It will also clarify the relation between fluid architecture and flexible formwork.

The word 'architecture' is defined in the oldest known architectural theory, by Vitruvius (*c.* 85 – 20 BCE) in his book *De Architectura.* The three key concepts of architecture according to Vitruvius are *utilitas* (effectiveness), *firmitas* (durability, stability, strength) and *venustas* (charm, beauty, experience). Vitruvius concludes: the integration of these three factors will lead to a special building without a matter of profusion (Heath 1989: 246–253; Hon and Goldstein 2005: 1–21; Vitruvius et al. 1914).

Although Vitruvius' theory is more than 2000 years old, it is still widely respected. High-tech architect Renzo Piano refers to the three key concepts of Vitruvius when asked to give his definition of architecture (Renzo Piano, bouwenmetwonen.net). The Vitruvian theory of architecture is one of the few theories in architecture that stands without further discussion. There is, however, much more discussion about the interpretation of architecture: the style.

Greg Lynn described smoothness in architecture in his book *Animate Form* as a kind of '*continuous variation that neutralizes those contractions inseparable from the urban fabric through constantly and fluidly incorporating these differences*' (Lynn 1998, 1999). Ito wrote in the early nineties: "*on fluid architecture in which he wanted to move beyond the history of architecture as static form, as idealized and conceptualized geometry. The mere suggestion of fluids introduces a unique perspective towards structure; architecture as an intense geometrical development of matter*" (Macapia 2006).

In the mid-nineteen-nineties, of particular interest are the post-modernist writings of Deleuze, "*from which key terms such as 'smooth space', 'folding', 'pliancy', and so on, have been transmitted into a new impetus for morphogenesis – the generation of new form- in architecture*" (Wong 2010: 237–267). Deleuze studied Leibniz

A. Pronk, *Flexible Forming for Fluid Architecture*,
https://doi.org/10.1007/978-3-030-71551-9_3

and Worringer (Worringer 1920) as an important source in his books *The Fold: Leibniz and the Baroque* (Deleuze 1993) and *Difference and Repetition* (Deleuze 1994). Deleuze proposes (citing Leibniz) that difference is better understood through the use of *dx*, the differential. A derivative, *dy/dx*, determines the structure of a curve while nonetheless existing just outside the curve itself; that is, by describing a virtual tangent. Leibniz (1646–1716) has been of great interest. Not only was he the one who described curvature mathematically, but also one who led an interesting life arguing, against people like Newton, that space, time and motion are relative, not absolute (Leibniz 1969: 271–281). Much later on, the ideas about relativity came back in the theories of Maxwell that inspired the relativity theories of Albert Einstein (1907: 371–384). Einstein proved that Leibniz' theories about relativism where right. Leibniz, living in the baroque, questioning the mainstream, advocating relativism, advocating the unification of Christian churches, stimulating the debate on research and the exchange of knowledge and showing his unconventional creativity in combining different fields of knowledge, making the first exact calculations on the form of catenaries and becoming, together with Newton, the inventor of differential calculations, can be seen as the first intellectual to take a scientific interest in fluid architecture. In a more theoretical research, it would be worthwhile to study the history, philosophies and roots of Leibniz and his importance to fluid architecture. Gottfried Semper (1803–1879), an architectural theorist, studied architecture in relation to infinitesimal calculus, tectonics and weaving. He desired to give architecture a scientific foundation through methods of systematic comparison and classification. He explained a work of architecture with infinitesimal calculus (Poerschke 2012: 119–134), suggested that the act of making is a defining factor in architectural form and expression, especially a material influence on the process of construction (Semper 1860: 107), and he postulated that enclosures (walls) have their origins in weaving (Semper 1989). Those three aspects can be linked to formative moulding referring to the doubly curved surface and the special tectonics as a result of the construction with the flexible moulding material. This flexible material is in many cases woven.

Lars Spuybroek has a limited list of realized buildings, but as an architectural theorist within the field of fluid architecture, he made a clear connection between architecture and textile. In an interview he refers to Gottfried Semper and Frei Otto as his sources of inspiration. He connects computational techniques with the tactility of textiles. By this tactility is meant the morphology of textile production techniques like weaving, bundling, interlacing and braiding. In his designs, he transfers this morphology to buildings or art objects. His design for a new World Trade Center in New York is a good example (Spuybroek 2004).

According to Deleuze, Leibniz, Semper, Spuybroek and Lynn, the objective morphology of fluid architecture has a smooth, irregular and doubly curved building envelope. Doubly curved refers to the Gaussian description of curvature. Irregular means that every element in the building envelope is unique. In combination with the double curvature, it means that the local Gaussian curvature is changing at any translation over the building envelope. This change of curvature therefore has to be nonlinear. Smoothness means a building envelope with a minimum of plasticity and detail.

Anticlastic and synclastic surfaces can be made as form-active structures like inflatables, prestressed membranes and cable structures. Form-active structures can be made irregular, smooth and doubly curved. Form-active structures have been used mostly for temporary buildings and buildings with limited conditions like canopies.

Only with the combination of more than one surface typology is it possible to increase the number of possibilities compared to the regular architecture. Synclastic and anticlastic surfaces can satisfy the morphological rules of fluid architecture but have to be combined in order to increase the limited possibilities as compared to the possibilities in "regular" architecture with zeroclastic surfaces.

Next to the morphological rules of curvature, regularity and smoothness, it is my conclusion that fluid architecture in most cases must have a combination of synclastic and anticlastic surfaces.

With this conclusion I come to a preliminary definition of the so-called fluid architecture:

1. It has to be a building with the integration of: *utilitas*, *firmitas* and *venustas.*
2. With a building envelope that is or appears to be:
    - doubly curved
    - irregular
    - smooth
    - having a combination of synclastic and anticlastic surfaces.

Later we will give a final and closer specification of the morphological aspects of fluid architecture. Within fluid architecture there are different approaches. In this research they are divided into the following categories:

- focus on Form: free-form or nonstandard architecture
- focus on Materials: fluid, elastic, liquid or blob architecture
- focus on Technology: biological, kinetic or digital architecture.

3.2 Free-Form Architecture

'Free-form' architecture is the most frequently used term. It serves the need to differentiate from other styles in architecture. The term 'free' is chosen to indicate that other styles are less free in determining the shape of a building. Only with the combination of surface typologies is it possible to increase the number of options in comparison with regular architecture. This combination of surface typologies can be seen as a liberation of the form. 'Free' is taken in the sense of 'not limited or controlled' (Cambridge Dictionary, dictionary.cambridge.org/dictionary/english/free). In the definition above, free-form architecture is irregular, smooth and doubly curved whilst the surface can be divided into elements, and every element is different. The case studies in Chaps. 4 and 5 will show that the apparent freedom in design is an illusion since the shapes of buildings are bound to rules in an optimization process necessary for the construction of the building. In design, the structural behaviour, production processes, construction techniques and material behaviour will influence the form of the building. In the following chapters those influences will be researched. The studies show that, in most cases, fluid architecture has the pretence of creating a free form, but it is not a free form at all. Marradi concluded: “the main issue is to understand the relationship between the design and the buildability of such complex envelopes” (Marradi 2013). Reading his and other papers (Veltkamp 2007) on this issue, the term ‘free-form’ is incorrect, since the building’s geometry is well-defined, and consequently not free.

The free-form vision is often captured, not in the form of real buildings, but of drawings, images, models and text. The media for the expression of visionary forms becomes a more and more important feature to create the impression of freedom. It can be concluded that free-form architecture is liberated from the domination of zeroclastic surfaces and free to combine other types of surfaces but is not, in a formal way, free at all.

3.3 Nonstandard Architecture

While a group of nonstandard architects label their architecture with names such as liquid architecture, hypersurface architecture, hybrid architecture, animated architecture, transarchitecture, performative architecture, evolutionary architecture and generative architecture, nonstandard architecture is contrasted with ‘standard’ mass production architecture (Oosterhuis and Feireiss 2006; Oosterhuis and Xis 2007).

The term ‘nonstandard architecture’ as used in the 2003 Paris exhibition is a reflection on the language of architecture and its field of application, based on the exploitation of digital elements. The notion of ‘nonstandard’ first appeared in the field of mathematics in 1961, with the work of Abraham Robinson who was strongly influenced by the work of Leibniz. Linear mathematic theory could not explain all. The introduction of the chaos theory into the electrical computer by the repeated iteration of mathematical formulas was firstly done by the experiments of Yoshisuke Ueda in 1961 and published in 1970 (Sprott 2003). Nonstandard mathematics led to the major theoretical breakthroughs that cleared the way both for fractal and catastrophe theory and for artificial intelligence. Its implications are manifold and affect every discipline to which algorithmic systems can be applied, such as artificial intelligence or morphogenesis (the development of structure). Nonstandard is obviously to be contrasted with ‘standard’. Classic modern architecture has become the standard. It claimed to be the norm for good architecture with comfort, light, clear and hygienic spaces thanks to the efforts of the second technical revolution (1880–1920) (Non-standard Architecture, centrepompidou.fr). ‘Nonstandard Architecture’ aims to assess the social, economic and political changes brought about by the widespread application of so-called ‘nonstandard’ and nonlinear production processes in design, architecture and territorial and urban policies (idem).

According to Bagneris et al. we can characterize nonstandard architecture as an architecture with doubly curved surfaces (Bagneris et al. 2008). But nonstandard architecture is a much broader concept. The term was used for an exhibition in the winter of 2003/2004 at the Centre Pompidou in Paris. This exhibition presented a historical overview and some recent work by Asymptote, dECOi architects, DR_D, Greg Lynn FORM, Kovac Architects, KOL/MAC studio, NOX, Objectile, oosterhuis.nl, R&Sie, Servo and UN studio (Non-standard Architecture, centrepompidou.fr).

The standard implies a norm that is based on agreement, either implicit or explicit, between humans on the most widely accepted or common denominator. It is what most people see as average. Considering that there is always a subjective component to a standard, the most widely accepted method depends on what the majority uses. As a result, standards cannot be described exhaustively in objective terms. If new insights are adopted by the majority, then the norm is changed and becomes the new standard. Non standard is everything that deviates from *the* standard of the majority. It might even not be recognized as part of the domain. This means that nonstandard architecture might be seen as not being architecture at all.

Fluid architecture has existed for decades, but since the 1988 MOMA exhibition on deconstructivism, curated by Mark Wigley and Philip Johnson, this approach to architecture has dominated the skyline of major cities around the world (Wong 2010: 237–267). Fluid architecture can be seen as the 'organic' version of deconstructive architecture. In 2016, after more than 25 years, fluid architecture is not such a novelty any more, but still an object of discussion and also not generally accepted (Hight 2008).

3.4 Focus on Materials

3.4.1 Fluid and Liquid Architecture

Fluid is a near-synonym for liquid. Novak defined liquid architecture as being meta morphic in time and space (Jackson 2001). We use liquid only as a visual characteristic. A liquid is without independent form. Its form depends on the surrounding surface and material into which the liquid is applied. In that case, it adopts the form of the surrounding surface. If a liquid is released into space, the molecules of the liquid are held together by the surface tension of the liquid. As a result of the surface tension, the liquid will form a minimal surface, i.e., a sphere. In another context, however, the sphere will become deformed under the influence of forces such as wind and gravity. This combination of surface tension and external forces in a gas-filled space will always lead to a doubly curved surface. So, the surface-tension leads to a smooth surface, but depending on the forces operating on the surface it will probably have an irregularly shaped (doubly curved) surface. Liquid architecture (according to Novak) is "*an architecture that breathes, pulses, leaps as one form and lands as another*." Liquid architecture is contingent on the interests of the beholder. By liquid, Novak means: "*the animistic, animated, metamorphic, as well as crossing categorical boundaries, applying the cognitively supercharged operations of poetic thinking. Metamorphosis is change in form, through time or space*" (Novak 1992).

3.5 Blob

Fluid architecture is known, especially in Holland, as BLOB architecture (or blobitecture and blobism). Blob refers to raindrops (or drops in general). "*Blobs and metal balls are isomorphic surfaces, constructed as composite assemblages of mutually inflecting parametric objects with internal forces of mass and attraction*" (Kolarevic 2000). We can postulate that a blob can react like a drop and have the same surface characteristics. In that way, it refers to liquid as described above. The free dictionary defines the word as Blob: a soft amorphous mass. A soft amorphous shape will be explained as an ill-defined or arbitrary shape.

The change in shape is used in several cartoons. The best known is *Barbapapa* developed by French architecture student Annette Tison and biology teacher Talus Taylor. One of the most important and well-known episodes of *Barbapapa* is about the building of an amorphous building. Barbapapa refuses to live in a rectangular apartment building and builds his own organic house. He is using his body as an inflatable mould to make the concrete shell structures. There is also a cartoon figure called Blob by Frederick J. Dukes, and a game *the blob* and three horror/science fiction movies "The Blob" (Fig. 3.1).

The term blob introduces a new dimension. In the animation industry, blob will also be referred to as meta-clay or as a meatball. To summarize, we can state that a blob is an object made up of isomorphic polysurfaces (Lynn 1998). The form is

Fig. 3.1 A game made by 9 students from Utrecht, The Netherlands in 2006

defined by the forces operating on the material. The interaction between shapes and forces will result in a continuous (isomorphic) polysurface.

In computer software, movies and cartoons; the blob is portrayed as something that poses a threat to the existing world. Blob architecture is also regarded as something that upsets the existing order and structure. A perception of negativity and nasty sound and image of the word blob is probably the reason why most architects prefer not to identify their buildings with the word blob. Some architects and designers have embraced the word blob, but most have decided against it (Waters 2003; Hight 2008).

3.6 Focus on Technology

The term BLOB is also an acronym used in computer software, standing for Binary Large OBject. A blob is a data type capable of storing large randomly organised data. This is contrasted with a systematic stacking of uniform information in databases. Below, a quote by one of the leading structural engineers in fluid architecture in Italy, Majowiecki: "*in the old days, we built [structures] to get us from here to there, nowadays from point A to point B and bit-by-bit*" (Vday Majowiecki, beppeg rillo.it). "*Since blobs can store binary data, they can be used to store images or other multimedia files. For example, a photo album could be stored in a database using a blob data type for the images, and a string data type for the captions*" (Sears et al. 2007). In other words, it is a large amount of information with a variable structure. Translated to geometry, this signifies a vast point cloud in which the points are related to each other in a multitude of ways (Fig. 3.2).

Greg Lynn defines a blob as a soft body affected by forces (Lynn 1999). 'Blobitecture' has become the most iconic result of the developments in three-dimensional digital technologies within the public domain. The term, first coined (in an architectural setting) by Greg Lynn in 1995, Describes "*organic, amoeba-shaped, fluid-form buildings designed using digital modelling software*" (Waters 2003; Futcher 2012). The blobitecture movement has produced dynamic and interactive architecture. Blobs, isomorphic polysurfaces, were originally developed for the study of complex molecules. The parameters that define blobs include mutual gravity (weight), extent of influence, threshold and form type (Mahesh 2001).

Free-form architecture is often called digital architecture. Kolarevic divided digital architecture into five categories on the basis of design methods: topological architecture; isomorphic architecture, animate architecture, metamorphic architecture, parametric architecture and evolutionary architecture (Kolarevic and Malkawi 2005).

Fig. 3.2 Fluid forms

3.7 Kinetic Architecture

Architecture related to waves in physical phenomena like sound and light. Architects in that field are Xenakis, Greg Lynn, Oosterhuis, Novak and Spuybroek.

3.8 Biomorphic Architecture

Biomorphic architecture is often characterized by amoeba-like shapes and is nowadays popularly referred to as 'blob-architecture' or 'blobitecture'. The biomorphic approach is based on the coherence between morphological studies of natural phenomena and the application of those morphologies in architectural structures. Open cell structures like those found in bones, foams and sponges are of special interest (Otto 1995; Banhart 2000; Yang et al. 2015: 608–615; Turner 2006) (Figs. 3.3 and 3.4).

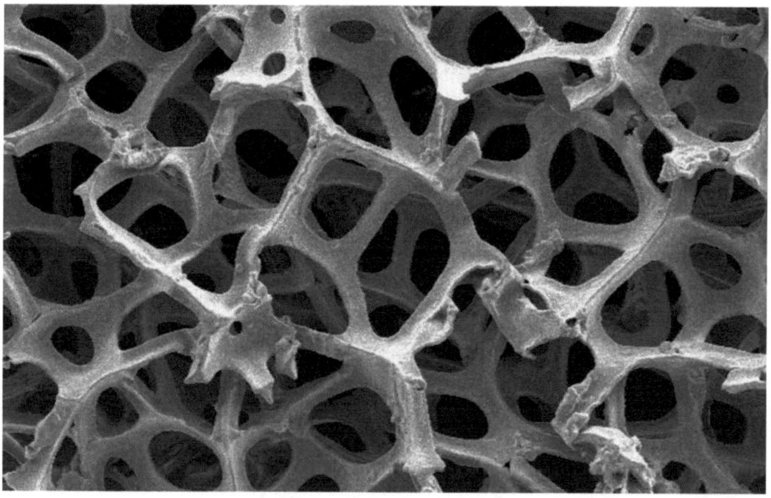

Fig. 3.3 Foam

The relation between architecture and nature, especially the human body, has had a long history and is still current. This issue is discussed extensively here because the biomorphic approach is an important motivator for many artists and architects. In addition to this motivation, there are also technical reasons for a more elaborate approach of this topic. Vitruvius, da Vinci, and Le Corbusier tried to find a relationship between human measurements and architecture, and Calatrava referred not so long ago to a twisted human torso for the design of a skyscraper in Malmö, Sweden. The first comparison between human dimensions and architecture is as old as the first buildings and written down in the earliest known book about architecture, by Vitruvius (46–30 BCE) (Vitruvius et al. 1914). Leon Battista Alberti wrote *De Re Aedificatoria* (Alberti 1988), patterned after *De Architectura* by that Roman architect and engineer. The work was the first architectural treatise of the Renaissance. It covered a wide range of subjects, from history to town planning, and from engineering to the philosophy of beauty. In 1485 it was published in full, after which it became a major reference for architects. Around 1490, Leonardo da Vinci drew the well-known Vitruvian Man. Da Vinci claimed that he created the drawing after empirical study of the human anatomy (Keele 1964: 360–370). Albrecht Dürer wrote four books on Measurement (Dürer 1525) and four books on Human Proportion (Dürer 1528). Dürer wrote his books between 1512 and 1523. He constructed five types of bodies, both male and female figures, in parts as well as in full form.

His publications were based on Vitruvius as well as on empirical observations of, 'two to three hundred living persons' (Panofsky 1955). His findings were not new but Dürer was the first to research the visual representation of the human body in a scientific way. The Vitruvian Man by Da Vinci is often used as a source for the golden section, the 'golden number' that defines the ratio between two segments. This ratio is $a/b = (a + b)/a$, the number is usually denoted by the Greek letter

Fig. 3.4 Turning Torso building, Malmo by Santiago Calatrava

phi = $(1 + \surd 5)/2 = 1.62$. The golden ratio was used more and more after 1830 in the aesthetic and architectural domain.

Between 1942 and 1955, Le Corbusier developed the "Modulor": a series of measurements of the human body based on the golden ratio (Padovan 2000: 35). He used two measurement systems, the red and the blue. For the red series he took the measure of the length of a man, 183 cm according to him, and divided this number several times by phi: 183, 113, 70, 43, 27. For the blue series, he assumed the size of a man with a raised arm, 226 cm in height, and did the same: 226, 140, 86, 54. Le Corbusier built the first Unité d'Habitation in 1947 on the basis of the Modulor. The golden ratio can be found in nature in an indirect way by the sequence of Fibonacci: 1, 1, 2, 3, 5, 8, 13, 21, 34, 55, etc. This sequence of numbers is the result of adding up the two previous numbers to get the third number. This sequence will end at infinity in the proportions of the golden section. The sequence of Fibonacci can be found in the distribution of sunflower seeds on the surface of the flower. Nowadays we use a 3D scan to measure bodies and buildings. In this way we are able to adjust products to the special measurements of a unique object. The biomorphic approach is mainly focused on the discovery of the structural principles of nature which are highly optimized by evolutionary processes. The parallel between fluid architecture and biological forms is broader and more diverse (Figs. 3.5, 3.6, 3.7, 3.8, 3.9, and 3.10).

Vertebrate animals and humans consist of a skeleton in combination with organs and muscles covered by an elastic skin. The skull can be seen as a shell structure; the remaining bones as bars under compression connected to each other by hinging joints. The skeleton is designed for compressive strength and a flexural bending. The

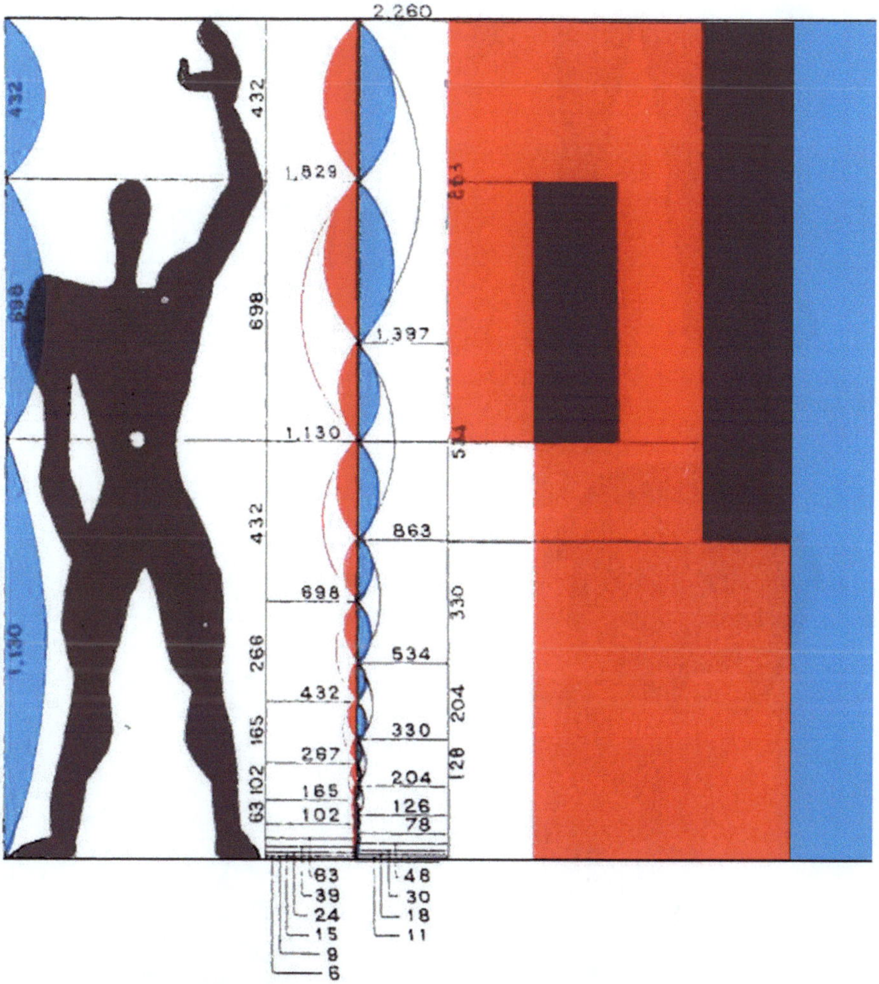

Fig. 3.5 The Modulor by Le Corbusier

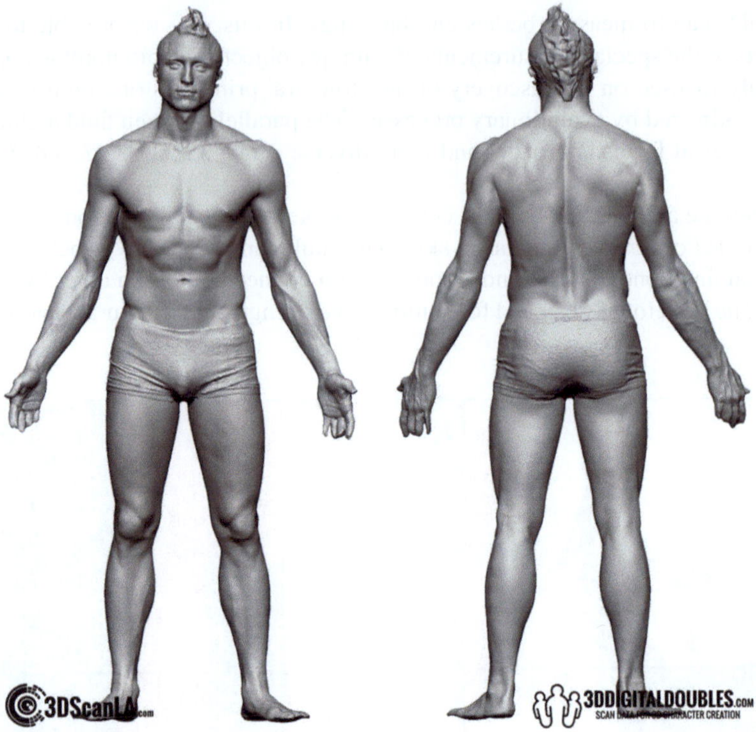

Fig. 3.6 Body scan

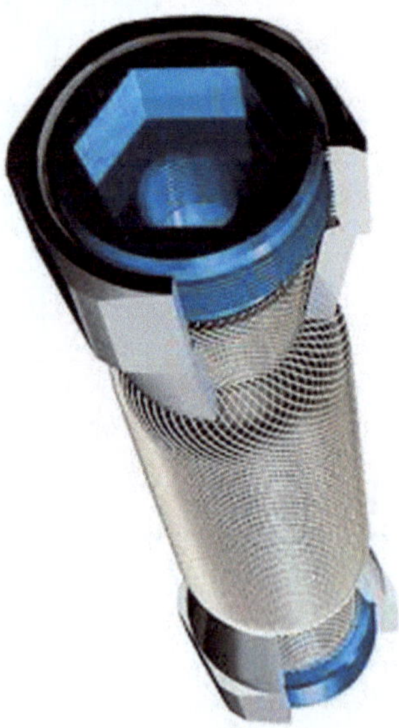

Fig. 3.7 Festo inflatable muscle

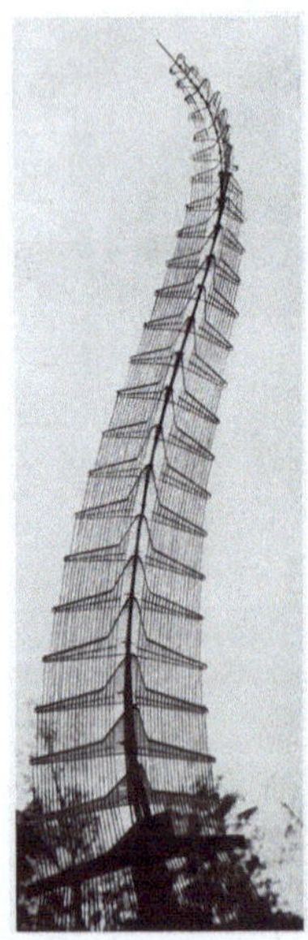

Fig. 3.8 Kinetic tower by F. Otto 1966

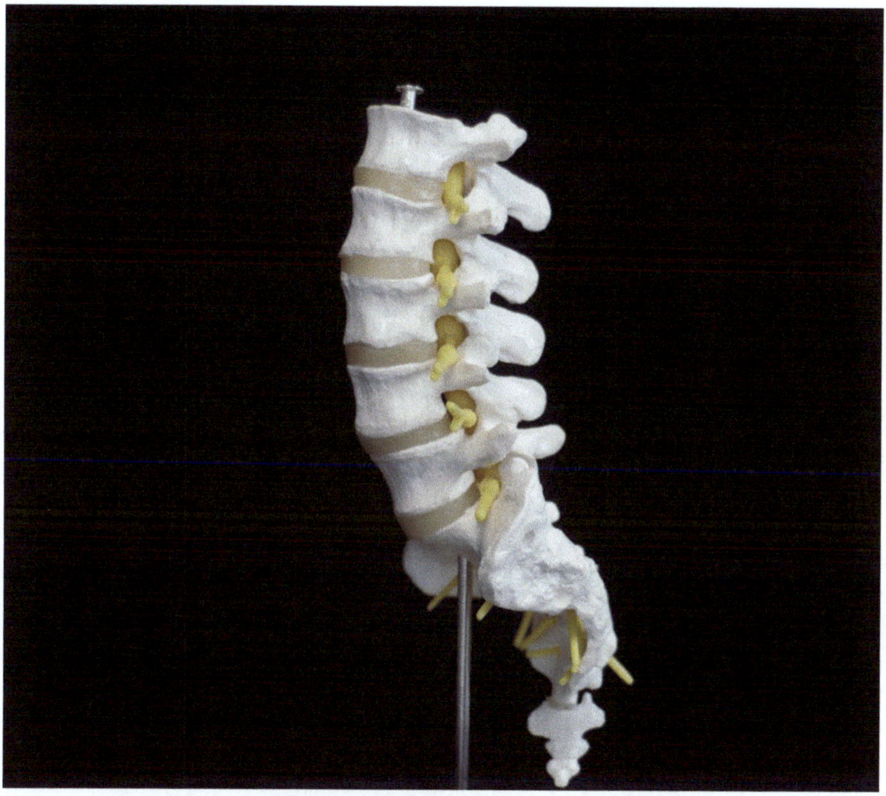

Fig. 3.9 Model of backbone

muscles can be seen as elastic ropes that will shorten by contraction. Muscles can only take tension. At both sides of the bone are muscles. In that way the muscle is able to flex or to release so that the joint can hinge back and forth. The skin is stretched around the body. The skin will wrinkle if the surface tension is lost. This will happen in case of a loss of elasticity as a result of extreme stretching or age, and if there is too much skin to cover the body inside it (Fung 1933: 568) (Fig. 3.11).

In plastic surgery it is possible to manipulate the skin, e.g., to reduce the wrinkles by applying lipids as a subcutaneous filler. This will tauten the skin. In other cases, the skin may be stretched by removing superfluous skin (face-lift). The form of the skin can also be manipulated by removing the fat layer below (liposuction) or by adding an implantation (rigid inserts or silicone).

The skin can also be manipulated by improving its condition through sport (body building), by diet or abstinence, or by suntanning, or painting with make-up or tattoos. Most people manipulate their skin surface to make themselves look more attractive. Attractive generally means: having a smooth and elastic skin that is thinner, rounder and tauter in specific places. Fat people are mainly synclastic. Thin people

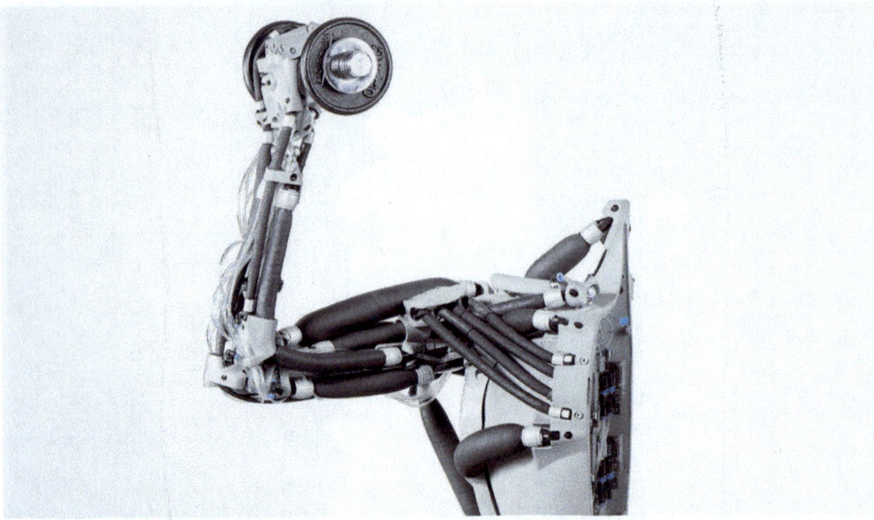

Fig. 3.10 Robot by Festo with inflatable muscles

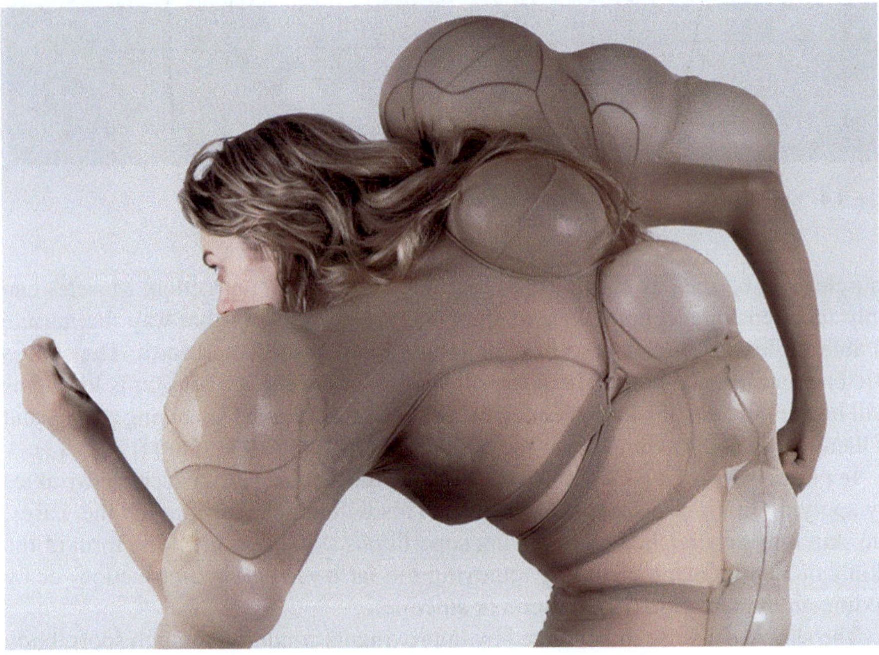

Fig. 3.11 Evaluation, 2008, Lucy McRae, www.lucymcrae.net (*Source* Lucy McRae)

are mainly anticlastic. An attractive body has a seamless combination of synclastic and anticlastic surfaces.

Fluid architecture is defined as: 1 irregular, 2 smooth, 3 doubly curved, 4 satisfying *utilitas, firmitas, venustas.* Fluid architecture in most cases has a multifaced body with a combination of synclastic and anticlastic surfaces. The same can be said of human beings. People prefer their skin to be doubly curved, irregular, and smooth. Everything in the body has a function and a strong and healthy body is highly appreciated. The amount of time people spend on their appearance and the way in which people try to impress others in nonverbal ways is indicative of this. The successful singer Dolly Parton has had extensive cosmetic surgery to achieve her idealized form. The largely cultural preference for certain physical characteristics has been linked to a mechanism for selective reproduction. A healthy person with a smooth, shiny and strong body is deemed to be more fertile. Such a person has an 'attractive' appearance, which is often wishfully associated with success in other areas of human endeavour (Reis et al. 1980: 604–617). In this sense, people seem to be perceived as a special kind of kinetic fluid architecture. In most of the arts, but also in advertising, naked bodies (Balgaradean and Babut 2015: 63–77; Lambiase and Reichert 2003) draw almost all the attention away from the surrounding environment. In fluid architecture the building is also a stand-alone entity, not or poorly connected to surrounding area, drawing all the attention and creating a world unto itself. On the one hand, this explains why, although fluid architecture is looked upon favourably for its eye-catching highbrow buildings, it is frequently controversial, too. On the other hand, the publicity generated by fluid buildings is often a reason for clients to order a stadium, museum or other public building in this style (Figs. 3.12, 3.13, and 3.14).

There is a sexual dimorphism (Gustafsson and Lindenfors 2004: 253–266) between men and women. Generally speaking, women have more fat and a smaller proportion of muscles in their upper limbs (Miller et al. 1993: 254–262), therefore they are more smoothly curved than males (Figs. 3.15 and 3.16).

To complete the analogy between people and fluid architecture, the morphology of fluid architecture may be called more masculine or more feminine depending on the smoothness of the curvature on the surface.

Maslow described the hierarchy of needs in his paper "a theory of a human motivation" (Maslow 1943: 370–396), see Fig. 3.22. Pyramid of Maslow (Fig. 3.17).

With reference to this theory, a building plays a fundamental role, providing the safety of a strong shell against the elements and other hazards—or to quote Vitruvius: *utilitas* (effectiveness) and *firmitas* (durability, stability, strength). Fluid architecture provides more than just a safe and comfortable shelter. It has the added benefit of *venustas* (charm, beauty, experience). The utilitas of a building is on a par with the physiological level of basic human functions; the firmitas comes next, providing a safe shell against outside influences, and venustas fits in with the upper levels of Maslow's pyramid of human motivation (Hershberger 1999).

The organic surfaces of fluid architecture make a strong reference to the surfaces of the human body. The attraction of the human body is used for architecture. In the hierarchy of Maslow, sexuality is another basic need. Freud sees sexuality also as an

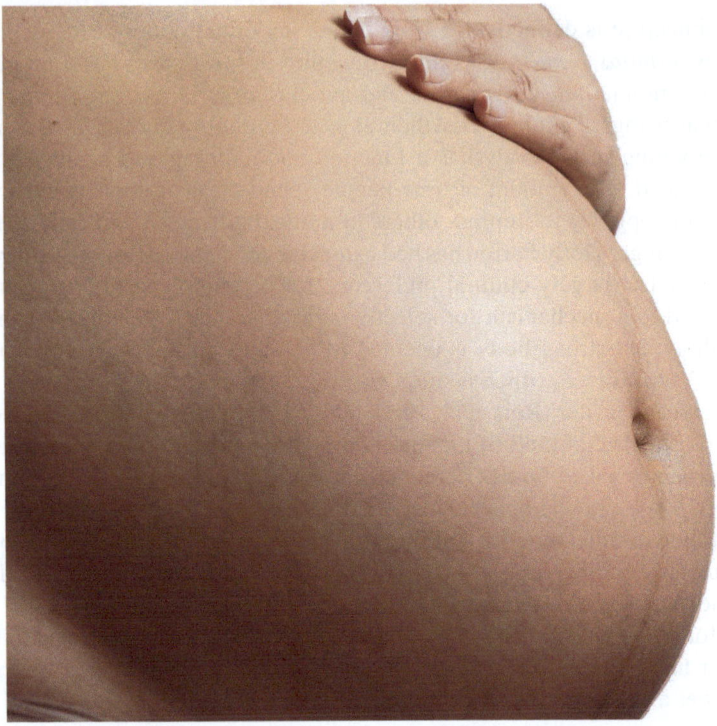

Fig. 3.12 Synclastic body

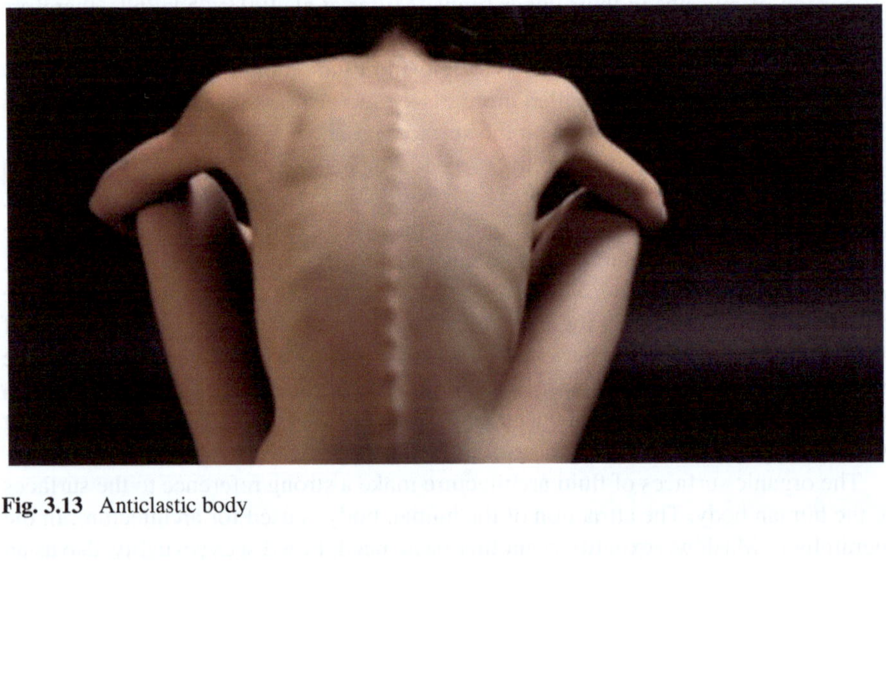

Fig. 3.13 Anticlastic body

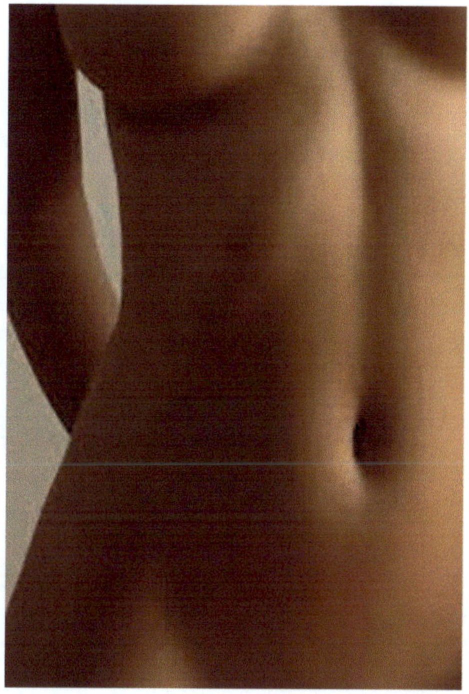

Fig. 3.14 Body with a seamless connection of synclastic and anticlastic surfaces

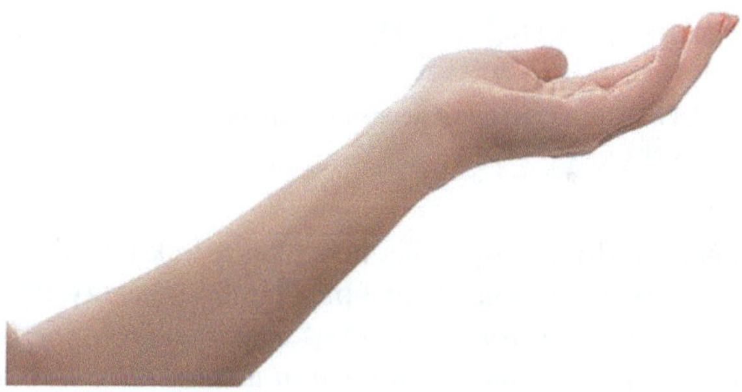

Fig. 3.15 Female versus …

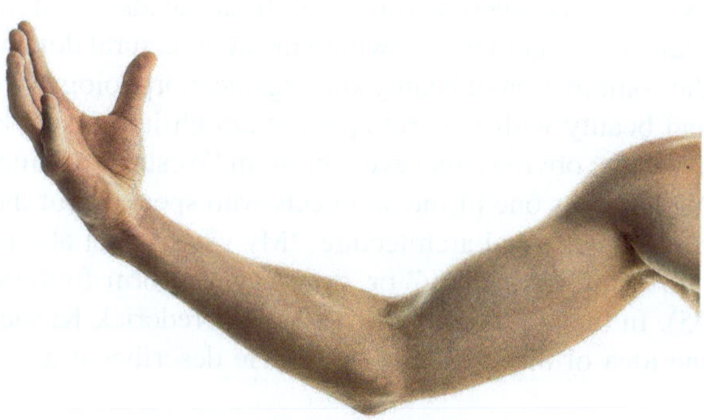

Fig. 3.16 Male arm

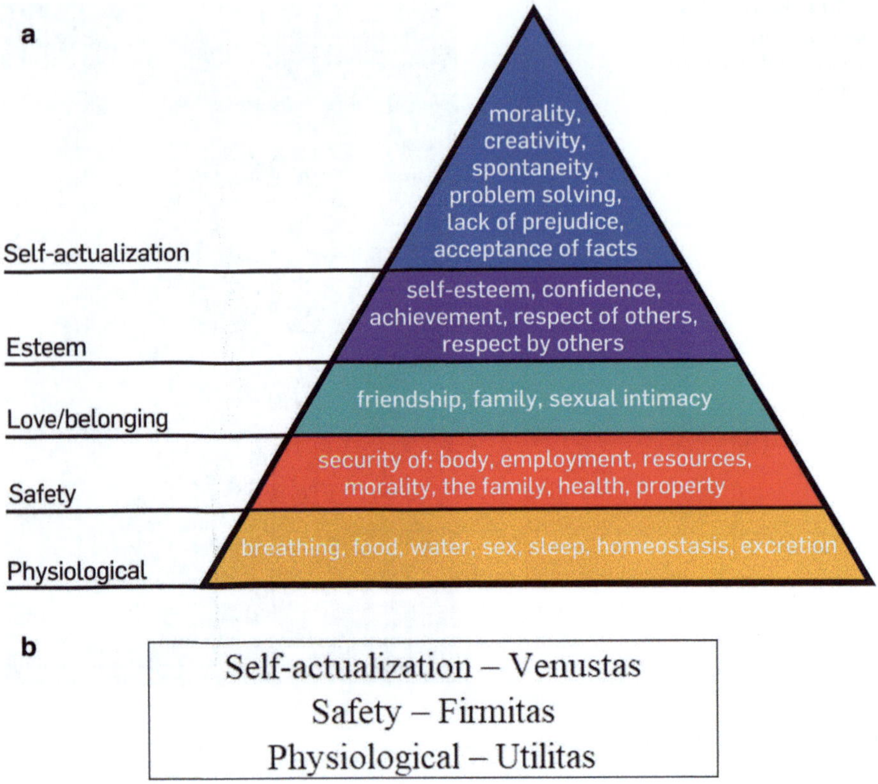

Fig. 3.17 The hierarchy of needs in the pyramid of Maslow connected to the Vitruvian aspects of architecture

important drive for people, he makes a link between beauty and sex (Freud 2002). To clarify the reason why people search for beauty, he writes: "*All that seems certain is its derivation from the field of sexual feeling. The love of beauty seems a perfect example of an impulse inhibited in its aim. 'Beauty' and 'attraction' originally are attributes of the sexual object.*" If this is truth, the architect refers to sexual motives in his search for beauty. The organic forms in fluid architecture make it even more explicit. Architecture fulfils the basic needs of humans for sheltering and safety. Fluid architecture refers within the architectural domain to the sexual preferences by the combination of beauty and organic morphology. The combination of architecture and beauty with a direct appeal through its sensual organic morphology is strong and more obvious than ever before in 'Western' architecture (Thomsen 1998). Oskar Niemeyer is one of the architects who speak about the relation between the human body, beauty and architecture: 'My work is not about "form follows function," but "form follows beauty" or, even better, "form follows the feminine"' (Metz 1997: 35). In the 1966 published vision of Frederick Kiesler (1890–1965), he introduces the idea of the "Endless house" He describes it as: "The *Endless House* is called "endless" because all ends meet continuously. It is endless like the human body. There is no beginning and end to it. The "*endless*" is rather sensuous. More like the female body in contrast to sharp-angled male architecture". He also speaks about an "elastic spatial concept" (Bogner 1988). The most explicit reference was by Toyo Ito. In a public talk about the Taichung Metropolitan Opera House on 26 October 2016, Toyo Ito mentioned the curved voids in the building. He referred explicitly to the human organs: "the voids are like the openings in a human body, like nose, ears, mouth and so on" (Ito 2016). In architecture the reference is usually implicit. The above quotes might be misconstrued as a male sexist approach. Similar features are apparent in the work of female artists such as Zaha Hadid, Niki de Saint-Phalle, Lucy McRae and Kathryn Findlay, however. Therefore, I do not subscribe to this point of view. The architect Zaha Hadid got into problems and had to defend herself when her design for the 2022 FIFA World Cup Stadium was found to look too much like a vagina. In art (Finley 1999) and primitive cultures (Rawson 1987) sexuality is more explicit and generally excepted. The influential artist Marina Abramović (1946–...) has explored her body and the limits of her body in her performances since the 1970s. Lucy McRae (see Fig. 3.16) calls herself a body architect. More erotic examples are the 1966 project by Niki de Saint-Phalle and Jean Tinguely and the work by Jeff Koons in 1991 (Figs. 3.18 and 3.19).

Fig. 3.18 Installation by Jean Tinguely, Niki de Saint-Phalle and Per Olof Ultvedt 1966

Fig. 3.19 Made in heaven by Jeff Koons 1991

The presence and availability of sexuality in society is more obvious than some time before and also the way we look on it is changing. The book *Playboy-50 years: The Photographs (2003)* (Hefner 2005) gives an overview of nude pictures of women since 1953. In this overview the pictures of the nude women become smoother over time, almost like they have a plastic skin. Pubic hair disappears in the nineties and photos are all retouched with the help of Photoshop to have "better" curves and smoother surfaces. In the way the nude women are presented there is a similarity with architecture. The introduction of Photoshop in 1990 and the introduction of computer aided design in the same period made it possible to facelift both erotic and architectural pictures. The real model or building became a representation of the photo-realistically presented perfect image in the picture. This all happened simultaneously with the introduction of fluid-architecture. And also, in fluid architecture, the representation of the smooth doubly curved building in a rendering is more seducing than the real object. Since 2011 there is a website 'render porn star' (renderpornstar.com). At this website you can look at the renderings of buildings. All the buildings presented will be nicer and better than the real thing. The gap between the image and the object has become unbridgeable: we can only endeavour to get as close as possible to the image.

Concerning the appearance of people, this gap has led to a progressively expanding cosmetic industry and the promotion of cosmetic surgery. Globally every year 0.16%

of the people have plastic surgery (ISAPS global statistics in 2015, isaps.org). In the Netherlands, the cosmetics industry represents a market of 2.8 billion euro, this is 166 euro for every inhabitant. Reliable economic figures about cosmetic surgery are difficult to find but based on a report by ISAPS there is an annual turnover of about 140 million in the Netherlands (Global Cosmetic Surgery and Service Market Report 2015–2019, Kelly Scientific Publications; ISAPS global statistics in 2015, isaps.org) and about 20 billion globally. The total budget in the Netherlands for health care in 2014 was 94.2 billion (CBS, cbs.nl). Cosmetics are 3.1% of the total budget on health care.

The building industry sees the gap between the images of the representation of fluid buildings and the way they are realized as a problem and talks about the fluid nightmare that has to be realised. The turnover for the building industry in the Netherlands over 2014 was 20.8 billion euro. If the building industry would follow the same policy as the cosmetics industry this might increase their turnover by 3.1% or 645 million euro on an annual basis only for the Netherlands.

3.9 The Digital Body

The introduction of computers in the measurement of surfaces makes it possible to make a 3D scan of an object. The ready-to-wear clothes industry developed a method to make a scan of a human body. Based on this scan, it is possible to take the personal dimensions of humans. The Lectra company (Lectra, lectra.com) has developed a full body scan. They developed software for working out the design of clothing and for generating two-dimensional patterns. It is possible to use the same method to measure buildings. The digital models can be compared, to look for similarities and differences between the surface of a designed model and the surface of the real-time object. This can be used to guide the plastering of stucco or spraying of shotcrete. Based on the scan, it is determined what amount of material needs to be added where in order to accomplish the final form (Nabulsi et al. 2010). It is possible to use a scan to digitize physical models during the design process. The digital model can be analysed for its structural behaviour, form, etc.

People prefer their skin to be doubly curved, irregular, and smooth. An attractive body has a seamless combination of synclastic and anticlastic forms covered with an elastic skin (Figs. 3.20 and 3.21).

Would it be possible and interesting to stretch a membrane or cable net for a fluid building in the same way as the skin on a body? And would it be possible to influence the surface of the building in the same way that the surface of a human body can be influenced? Mechanical principles, design techniques in fashion (fabrics), biology (tissue), computers (point clouds) and building industry can be combined. In the following chapters more about how this can be done in the design and construction of fluid architecture.

Fig. 3.20 Photo of body with projection of a regular pattern by Fernand Fonssagrives, mid-1950s (Fontana 2012, moderndesign.org)

Fig. 3.21 Photo of the inside of an inflatable by architects of air (materialist zen, materialistzen. blogspot.nl, 2011)

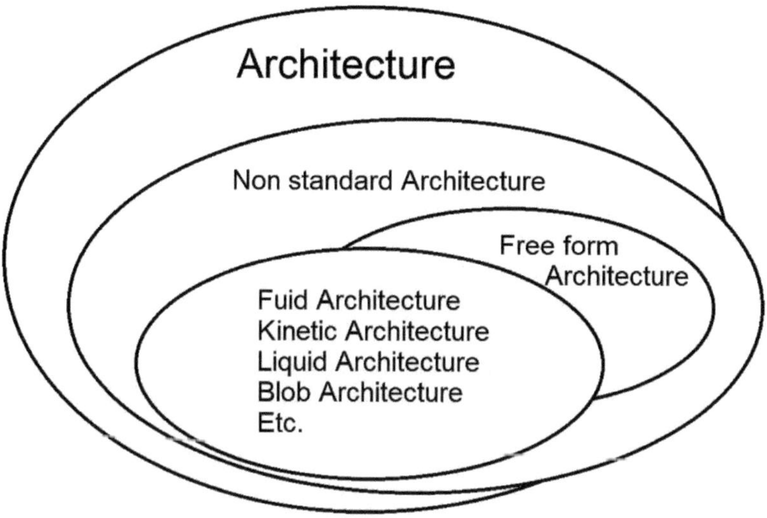

Fig. 3.22 Categorisation of fluid architecture

3.10 The Term Fluid Architecture

Above, the different categories within the so-called 'fluid' architecture were discussed. How to name this style? (Fig. 3.22).

'Liquid', 'blob', 'digital', 'kinetic', 'biomorphic', etc. are too narrow in their reference, too much focused on a material or a technique, to cover the whole field.

The name 'free-form architecture' is widely used. But in most cases, the realized building is a highly rationalized optimization of the design intention. Therefore, the word 'free' in relation to this type of architecture is not correct.

'Nonstandard' means different from other architecture. But this type of architecture is too specific to categorize it as only being different.

'Fluid architecture' will cover the whole range of styles and is specific enough. It expresses smoothness, irregularity and a possibility to have a seamless connection between doubly curved surfaces. Therefore, the words 'fluid architecture' are used in this research (Schmal 2001).

3.11 Historical Overview of Fluid Architecture

In the first part of this paragraph the historical development of "fluid architecture" is discussed. It starts with the late French Gothic style known as Flamboyant. The historical development of fluid architecture is classified in two ways. In the first classification, the relation between structure and form is the determining factor. In fluid architecture the focus is on creating fluid forms which are not only dominated by structural optimization but mainly by architecture. The different style periods are the second way in which the development of fluid architecture is classified. The following style periods have been determined: precursors, early, modern, deconstructive and individual projects. Important key buildings are the Statue of Liberty, TWA terminal, the church in Ronchamp, the Philips Pavilion and the Guggenheim Museum in Bilbao The complete separation of structure and building envelope in fluid architecture was shown for the first time in the Statue of Liberty (early fluid architecture) and reappears in the fluid architecture of the Guggenheim museum in Bilbao at the end of the twentieth century (deconstructive fluid architecture). The architectural shape of the 1958 Philips Pavilion, the 1956 TWA terminal and the 1954 church in Ronchamp are not dominated by the actual structure. The perceived structural elements are architecturally applied and are far from the most efficient structural application of doubly curved shell structures. These buildings can be seen as the first examples of "modern" fluid architecture.

It is a striking observation that every important style period (Gardner and Kleiner 2009) ends in a superfluous practice of irregular curved elements.

- For the Gothic style this is the Flamboyant style (1350–1500);
- For the Renaissance this is Rococo (1700–1800);
- For the Romantic style this is Art Nouveau (1890–1910) and
- For the modern style this is fluid deconstructivism (Since 1988).

Beside the highbrow buildings there is an interesting line of individual small-scale projects.

Fluid architecture can be defined more narrowly based on the analysis of fluid buildings in relation to the *Gestalt* Theory of Max Wertheimer and the 15 basic properties necessary to create spaces which contain a certain aspect of 'life' as determined by Christopher Alexander. This definition encloses a wide range of varieties in form and tactility of material.

Below, a historical overview of fluid buildings will be given. In addition to this overview, there will be a characterization of the periods of fluid architecture distinguished here, to place the buildings in their historical context. Domes and the most hyperbolic scales are excluded from this overview because they do not fulfil the criteria of irregularity. Where they are of technical interest, they will be discussed later. The criterion of irregularity reduces the historical overview. A particular consequence is that the work by Niemeyer, Nervi, Calatrava and most of the work by Isler will not be included in the overview because of the regularity of their morphology. 'Irregular' is meant to convey that every surface element of a building is unique, without any repetition. Some exceptions have been made for works that have been of great influence to the development of fluid architecture. Those works include the canopy of the UNESCO Building by M. Breuer and P. Nervi, Chapel Lomes de Cuernavaca by F. Candela (Futcher 2012), Sydney Opera by J. Utzon and Zürich central station by S. Calatrava. The historical overview is limited to the most important buildings of that period. Although a full overview may be desirable, it is not necessary within the focus of this research. In this research four periods are distinguished. It is generally accepted that fluid architecture emerged from the deconstructivist architecture at the end of the 1980s, but it would be interesting to go further back in history to the Gothic period, to the first time when curvature was also used in a decorative way.

The fluid architectural buildings mentioned in the historical overview will be classified in two ways: by style and by structure.

The first, stylistic, classification of fluid architecture involves categorizing by style according to: precursors, early, modern, deconstructive or individual fluid architecture as outlined below:

1. ***Precursors of fluid architecture***

Gothic and Gothic baroque.

2. ***The early fluid architecture***

The period that begins with the baroque and ends with the work of Gaudi, Steiner, Mendelsohn in 1921.

3. ***The modern fluid architecture***

The period that starts in the 1930s and runs on until the petrol crisis in the 1970s.

4. ***Deconstructive fluid architecture***

Beginning in the late 1980s with deconstructivism and later in the mid-1990s with fluid architecture until the present day.

5. ***Individual fluid architecture in private houses***

In these individual cases after the Second World War the architecture is highly influenced by personal elements.

The second way involves categorizing by the relationship between structure and architecture:

- *Type 1* a structure with vertical walls with a curved surface and decorations attached to the vertical walls. The roof structure is, from the outside, made in a different way with a different type of material.
- *Type 2* structural behaviour of the structure dictates the form of the building.
- *Type 3* structure follows the architectural designed form.

3.11.1 Precursors of Fluid Architecture

3.11.1.1 Gothic

Gothic architecture, which originally developed in France, is focused on a technical approach that shows the technical possibilities of the engineering of that time. One of the defining characteristics of Gothic architecture is the pointed or ogival arch combined with a repetitive pattern of doubly curved ceilings (Bony 1983). Arches of this type were also used in the Middle East before they were structurally employed in medieval Gothic architecture (Warren and Grabar 1991: 59–65). In England, the Gothic style led to a less technical challenge compared to France. The buildings are lower and squatter, less vertical. It also led to a more decorative style with a mesh of doubly curved lines on a doubly curved ceiling. The pattern of the ceiling is elegant but still symmetrical and repetitive (Clifton-Taylor 1967) (Fig. 3.23).

The Lady Chapel constructed by Sir Reginald Bray at the wish of Henry VII at Westminster Abbey is the last great masterpiece of English medieval architecture, and the culminating achievement of over three hundred years of development in the Gothic style, at the point where it intersects with the new movements of the Renaissance. Consider the thousands of skilled but unsung craftsmen and masons that created the thirteenth Century octagonal Chapter House, the vaulted ceiling in the Lady Chapel, and the flying buttresses that enable the roof of the nave to span around 30 m (Jenkyns 2008).

Fig. 3.23 Chapel Westminster Abbey London, Sir Reginald Bray, London, 1503 to 1519 (Westmi nster-abbey.org)

In late Gothic architecture the S-shaped arch is introduced by the decorative Flamboyant arch. The combination of S-shapes symbolises a flame. The curves come together in one sharp point. The lively effect of the S-shape is often used for decoration in windows and surfaces mostly in churches. The German art historian Wilhelm Worringer discusses the Gothic ornaments with a liberated line. He calls them "the Northern Line". This line is continually undetermined and it strives for infinity (Worringer 1920). The S-shape is structurally weak. In structures it is most of the time combined with larger and more stable arches and not at all for vaulting (Fletcher 2001) (Fig. 3.24).

Later, in Bohemia, the Gothic-baroque (Benešovska 2001) leads to an almost irregular pattern with doubly curved lines (Baroque Chapel by Santini-Aichel, 1720). Despite the decorativeness of the Flamboyant arch, the English-Gothic and the Gothic-baroque, there is still a connection with the structural behaviour of the

Fig. 3.24 A Flamboyant window Cathedral Meaux France

building. The decoration led to a diversity of doubly curved S-shaped lines as in the Flamboyant arch (Figs. 3.25 and 3.26).

The style of the St. John of Nepomuk church is absolutely unique and does not appear anywhere else in Europe. It was built in the 1720s and it's undoubtedly the most significant work by J.B. Santini-Aichel who masterfully managed to combine baroque and Gothic styles. The actual church is permeated with the number five as derived from the legend of St. John of Nepomuk. The church has a five-pointed star shape, five exits, five chapels, five altar alcoves. In the church interior there are oval-shaped peripheral chapels connected with narrower pointed arches. A considerable number of interestingly placed windows provide an abundance of light that widens the whole space in the baroque way and stresses the sculptural decoration (worldheritagesite.org).

Fig. 3.25 The Pilgrimage Church of St. John of Nepomuk, J.B. Santini-Aichel, Zelená hora, Czech Republic, 1722 (worldheritagesite.org)

Fig. 3.26 The Pilgrimage Church of St. John of Nepomuk, J.B. Santini-Aichel, Zelená hora, Czech Republic, 1722 (worldheritagesite.org)

Although the doubly curved structures are decorative and impressive, in this research they are regarded as precursors of fluid architecture because of their symmetry and regularity. In Gothic architecture the structural behaviour of the structure dictates the form of the building. It is classified as type 2.

3.11.2 The Early Fluid Architecture

3.11.2.1 Baroque and Elastic Baroque

In Gothic architecture, and even in Flamboyant curved lines there is still a link with the structural behaviour of buildings. In the baroque it is the first time that architects abandon this link and are inspired to create irregular asymmetrical forms released from the link with structure. The early fluid architecture emerges in the Baroque, with the work of Borromini, who was a leading figure in the emergence of Roman baroque architecture (Blunt 1989), and Guarini, influenced by him (Meek 1988: 6–11, 19), who used irregular curved lines in what is called the "elastic baroque" (Janson 1986: 310–312).

The word elastic refers to membranes and fluidity. The characteristics of their buildings and the reference to the word 'elastic' suggest that their buildings can be classified as early examples of more or less fluid architecture. Nowadays, fluid architecture is also qualified as 'the new baroque' and seen as a reaction to the straight lines of modernism (Willoughby 2003) (Fig. 3.27).

San Carlo alle Quattro Fontane (also called San Carlino) is an example of the elastic baroque. The church in Rome was designed by architect Francesco Borromini (1599–1667) and constructed during the period 1634–1677. The tight geometric complexity of interlocking ovals and circles creates spaciousness in the small corner church.

The façade of the church is divided into two storeys by a projecting cornice. Each storey has three bays. The entire design of the façade is a veritable example of the architect's predilection for convex and concave forms.

3.11.2.2 Rococo

The rococo is a miniature form and refinement of the elastic baroque. Rococo is a European style period from roughly 1720–1775. The name is a contraction of the French words "rocaille" and "coquille"—an asymmetrical stone and shell motive that can be found in the interior design of the Italian "barocco". Barocco meant the playful ornamentation of grottos by means of all kinds of irregular shells and stones (Fig. 3.28).

Rococo originated in France as a reaction to the strict classical style that dominated the arts during that period. The rococo embodies the sharp and refined feeling for the visual arts in a society that highly esteems elegance, artistry and light-hearted wit. A characteristic of the rococo period is the asymmetry, in other words: irregularity. The colour use is characterized by soft tints and abundant use of pastels. The rococo is the first style that strives for irregular and asymmetrical forms. Furthermore, both the rococo and the barocco are reactions to the dominant classicist style. Rococo focuses mostly on the interior, moreover there's a great contrast between the interior and the exterior design of the buildings. With the majority of the fluid architecture, a similar

Fig. 3.27 San Carlo alle Quattro Fontane, Francesco Borromini, Rome, 1638 to 1641

difference between the outside and the inside can be observed (Norberg-Schulz 1999; Varriano and John 1986) (Fig. 3.29).

In the Wies Church the rococo style has reached a point of excellence. Coming out of the darkness and loneliness of the forest around it, one is impressed to discover the apparent lightness of the rococo style of the Wies Church.

Art, culture, light and music are in perfect harmony with theology. The Wies Church is the masterpiece created by the brothers Domenikus and Johann Baptist

Fig. 3.28 Coquille (visoflora.com)

Fig. 3.29 Wies Church, Domenikus and Johann Baptist Zimmermann, Wies, Germany, 1745 to 1754 (UNESCO, whc.unesco.org)

Zimmermann (Wessobrunner School) who were stucco-workers, marbling artists as well as architects. The outside of the church is modest in comparison with the splendour of the decor inside. Gilded stucco, wood carvings and vividly coloured frescos contrast the whitewashed walls (Blunt 1988). In the baroque, the structure has vertical walls with singly curved surfaces and decorations attached to the vertical walls. The roof structure is, from the outside, made in a different way with different materials. It is classified as type 1.

3.11.2.3 Art Nouveau

In the nineteenth century, a process of industrialization was put in gear where regularity and rationalism were the highest attainable good. The buildings from this period are still decorated with ornaments from the past, with various neo-styles. Art Nouveau marks the end of the neo-styles. It can be characterized by decorations based on patterns of elegance. It was a reaction to academic art of the nineteenth century, it was inspired by natural forms and structures, not only in flowers and plants, but also in curved lines (Duncan 1994). The work of Rudolf Steiner and Erich Mendelsohn is characterized by sculptural buildings where the plastics are acquired by adding mass on the structure, walls and floors, of the building. Erich Mendelsohn (1887–1953) never sought to build in an extreme way but he aspired a fusion of the organic with strong geometric axes. He became famous with the Einsteinturm in Potsdam (Hentschel 1997) (Fig. 3.30).

The work of Gaudi's Casa Mila and Casa Battlo (Huerta 2006: 324–339) can be placed in the same tradition. Casa Mila (1906–1910), (Fig. 3.31) covers an area of more than 1000 square meters. Two courtyards, one shaped as an ellipse and the other as a circle, provide the interior with daylight. The inner walls have frescoes, reliefs and striking handrails. "La Pedrera"—'the quarry'—was the name an astounded population gave to this completely unique building. It could be compared with the steep cliff walls in which African tribes built their cave-like dwellings.

Fig. 3.30 Einsteinturm, Erich Mendelsohn, Potsdam, 1921 (3-co.com)

Fig. 3.31 Casa Milà, Antoni Gaudi, Barcelona, 1906–1910

The roof is extremely eccentric, curved and covered with fantastically abstractly formed chimneys and ventilation shafts. The building has no load-bearing walls. The pressure is distributed through supporting beams and columns which are part of a structure made of steel and concrete (Huerta 2006: 324–339; Molema 1987; Crippa and Gossel 2003). Conceived in the art nouveau style, the structure has vertical walls with a curved surface and decorations attached to the vertical walls. The roof structure is, from the outside, made in a different way with a different material. It can be classified as type 1.

The work of Gaudi goes much further than decoration. He researched the mechanical behaviour of building structures and the possibilities of construction materials. He applied this knowledge to many works like the loft of Casa Mila and his *magnum opus* the Sagrada Familia (Molema 1987; Huerta 2006: 324–339). Those works can be classified as *type 2*: the structural behaviour of the structure dictates the form of the building.

Statue of Liberty by Frederic Auguste Bartholdi (Sutherland 2003) (Figs. 3.32, 3.33, 3.34, and 3.35).

Frederic Auguste Bartholdi designed the monument (Tyler 1990). He obtained a U.S. patent for its structure. Maurice Koechlin - chief engineer of Gustave Eiffel's engineering company and designer of the Eiffel Tower, engineered the internal structure of the statue. Eugene Viollet- le-Duc was involved in the choice for the copper sheets used for the cladding of the envelope that hung on an internal frame of steel (originally puddled iron) (Moreno 2005). Although it was conceived as an enormous sculpture and not as a building, it has essentially the same structure and surface as the Bilbao Guggenheim museum. Technically it can be taken as the first example of a fluid building. By the definition of Vitruvius, it might not be architecture since the *utilitas* (effectiveness) of the structure is not housing but art and an icon for the city of New York. This 1886 building is the first with a clear separation between the building surface and the main load bearing structure, long before the 1926 *plan libre* by le Corbusier (Beton 1987) and the 1921 design for a Glass Skyscraper

Fig. 3.32 Statue of Liberty, exterior structure, Frédéric Auguste Bartholdi Liberty Island, New York, USA 1886

Fig. 3.33 Statue of Liberty, interior structure, Frédéric Auguste Bartholdi Liberty Island, New York, USA 1886

by L.M. van der Rohe (Schulze 1985). This building is classified as type 3. In 2018 they still used the same kind of structure for the 182-meter-tall Sadar Vallabhbhai Patel statue in India.

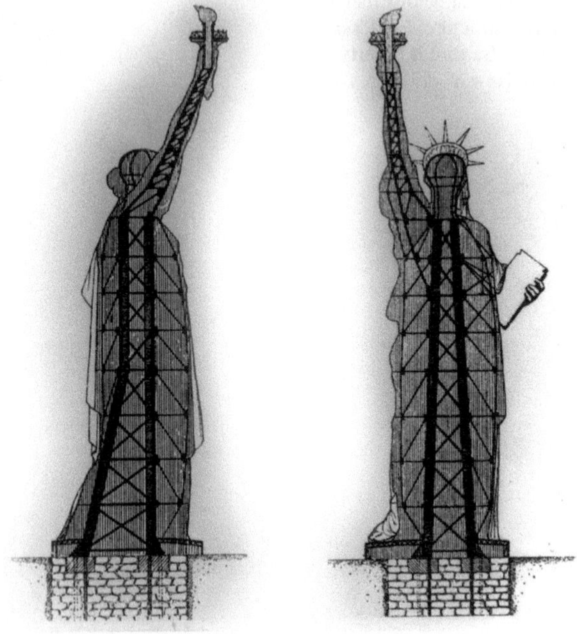

Fig. 3.34 Statue of Liberty, structural drawing

Fig. 3.35 Statue of Liberty, connection of cladding

3.11.3 The Modern Fluid Architecture 1930–1973

3.11.3.1 The Period of the Shell Structures

After the Art Nouveau period, architecture becomes more and more functional and under the influence of the Modernists, ornamentation is abandoned (Andrews 2010). Several avant-gardist structural engineers discover the possibilities of doubly curved

Fig. 3.36 Club Táchira, Eduardo Torroja, Caracas, 1956

shell structures. In the 1950s and 1960s a lot of special shell structures are built (Fig. 3.36).

The Club Táchira of Caracas, Venezuela (1956), one of the last works by Torroja (1899–1961), in cooperation with Jose Fructoso (Fruto Vivas) (1928), has the lightness and overwhelming grace of a moving sail. 'Functionality is emphasized by avoiding all strictly ornamental touches. Beauty is sought in the simple and spontaneous gracefulness of the basic outline, in the proportion of the masses, and the rhythm of the form.' It is 'a line of double curvature in space - parallel and identical directrices' (Torroja Miret 1958a UCLA; Torroja Miret 1958b Dodge) (Fig. 3.37).

The UNESCO building in Paris was achieved in cooperation by the architects Breuer, Nervi and Zehrfuss. Breuer found a functional solution: a Y-shaped structure that would give maximum light and air for the 600-odd offices. Equally successful is the butterfly-roofed conference hall. With roof and monumental façade shaped from folded concrete slabs, it attains simple dignity by the drama of its stark engineering. Says Nervi: "At last reinforced concrete has become a 'noble' architectural material" (Nervi 1958) (Fig. 3.38).

Candela, nicknamed 'The Shell Builder' due to his extensive exploration of tensile shell structures. Candela believed that strength should come from form not mass.

Fig. 3.37 Accueil de UNESCO, Marcel Breuer, Paris, 1956

Fig. 3.38 Chapelle a Cuernavaca, Felix Candela under construction, 1957

In the design of the chapel in Lomas de Cuernavaca, Mexico (1958–1959), the saddle-shaped shell fits in harmoniously with its surrounding landscape. It is a pure hyperbolic paraboloid whose slender edges seem to float free and whose roof slopes from each end down to a skylight (Moreyra Garlock and Billington 2008) (Fig. 3.39).

The factory for Sisli is a single continuous shell surface by H. Isler, with an irregular plan on seven supports. It is a hanging free-form prestressed concrete structure built in 1969 as a factory for the fire extinguisher manufacturer, Sisli SA, in Geneva. The shell has no edge beams and a general thickness of only 100 mm. The shells interact structurally. Because of the total asymmetry and complex symmetry, the forces within the surface and its buckling behaviour were determined by the use of experiments on small-scale models. The final shape of the building with the

Fig. 3.39 Factory for Sisli, Heinz Isler, Geneva, 1969

Fig. 3.40 Munich Olympic Arena, Frei Otto, Munich, 1972

Fig. 3.41 Multihalle, Frei Otto, Mannheim

double shell was developed after a series of model studies with roofs of different configuration (Chilton 2000).

Frei Otto (Otto and Rasch 1996) accomplished a complete harmony between the mechanical behaviour of a construction and the characteristic of double curves and irregularity. Frei Otto is a pioneer of lightweight tensile structures. Otto's first major structure was a saddle-shaped cable net pavilion in Stuttgart. After founding the Institute for Lightweight Structure at the University of Stuttgart in 1964, Otto continued his work with projects like the Munich Olympic Arena of 1972 pictured below (Fig. 3.40).

The Olympic stadium was considered revolutionary for its time. It included large sweeping canopies of acrylic glass stabilized by steel cables that were used on a large scale for the first time. Since designing the Munich Stadium in 1972, Otto has spent his time studying biological structures and researched grid shells that enjoy great popularity at the moment (Nerdinger 2005) (Fig. 3.41).

The grid shell in Mannheim, charming the public with its transparency, lightness and unusual shape, spans 7,400 m^2 with a maximum reach of 60 m and a maximum height of 20 m. The concept of this structure lies in the combination of form, material

Fig. 3.42 Dulles Airport Virginia USA Architect Eero Saarinen

Fig. 3.43 Interior of Dulles Airport Virginia USA Architect Eero Saarinen

and structural efficiency. The Multihalle Mannheim belongs to the family of slope forms. The simplest example is the hanging chain. The shape of the grid shell in Mannheim was designed by a form-finding method with a scale model. The load bearing structure of the shell is made of solid wood slats. On site, the slats are fixed with bolts to a wooden grid and finally covered with a PVC-coated polyester mesh. The grid shell in Mannheim is a unique building for its time because of the simplicity of the construction method used.

Eero Saarinen used the form of a catenary in the first hanging roof in complete tension for a permanent building. The terminal building of Washington DC's international airport Fig. 3.42 has a huge concrete sheet, slung between two asymmetric rows of concrete supports. Within the concrete sheet there are steel suspension-bridge cables hanging between the concrete supports. Saarinen described it as 'like a huge, continuous hammock suspended between concrete trees.' This terminal building is the first one in its structural typology. The roof can be seen as a concrete membrane and therefore of importance as a form of active moulding (Fig. 3.43).

Fig. 3.44 TWA, Eero Saarinen, New York, 1956 interior

Fig. 3.45 TWA, exterior

In these buildings the structural behaviour of the structure dictates the form of the building: they are classified as type 2.

But Eero Saarinen used the form language of structural shells also separately from the structural background. This was done in the TWA Terminal Building (Kennedy airport) in New York (Roman 2003; Pelkonen and Albrecht 2006; Merkel 2005) (Figs. 3.44 and 3.45).

"One of the most self-assured, self-confident – even self-conscious – buildings to emerge as a result of the interplay of the architectonic and engineer-inspired buildings was Saarinen's TWA Terminal Buildings at New York. It alarmed the remaining purists of modern architecture. Its bird-like symbolism, exciting forms and cavernous interior were not simply a casual reminder of the changes that had taken place in architectural thinking in the 1950s, but a demonstration of the architect's role as an originator and, in the American scene, as a 'building stylist' Clearly it represented a revival of architectural Expressionism" (Sharp 1991). Saarinen developed a remarkable range which depended on colour, form and materials. He showed a marked

Fig. 3.46 Notre Dame du Haut, Le Corbusier, Ronchamp, 1954

dependence on innovative structures and sculptural forms, but not at the cost of pragmatic considerations. He easily moved back and forth between the International Style and Expressionism, utilizing a vocabulary of curves and cantilevered forms.

At the end of his life, Le Corbusier realized some buildings with curved surfaces. "Surrealism is a key to the late works of Le Corbusier (Fox Weber 2008) most notably the church at Ronchamp, France, of 1950–1954 (Pauly 1997). *Notre-Dame-du-Haut* was a more extreme statement of Le Corbusier's late style (Fig. 3.46). Programmatically, the church is simple – an oblong nave, two side entrances, an axial main altar, and three chapels beneath towers – as is its structure, with rough masonry walls faced with whitewashed Gunite (sprayed concrete) and a roof of contrasting architectural concrete (*béton brut*). Formally and symbolically, however, this small building, which is sited atop a hillside with access from the south, is immensely powerful and complex" (Sharp 1991: 132–133). Because it was difficult to transport huge building blocks to the top of this hill, Le Corbusier used the debris of the old church to build the new chapel. After this he sprayed the wall with concrete and painted it white. A special feature is the structure of the roof which is completely hollow and made of reinforced concrete spread over a structure of beams.

The pavilion designed by Le Corbusier for the Philips Company at the 1958 Brussels World's Fair showcased a spectacle that remains a landmark in multimedia production. The structure is composed of hyperbolic-paraboloid shells which, up to now, had not been used for problems of this type. The walls are constructed of rough slabs cast in sand moulds on the ground. They are mounted in place by means of moveable scaffolding and are supported by a double network of cables, suspended along the cylindrical directrices of strongly reinforced concrete (Fig. 3.47).

Le Corbusier made, just like Saarinen, a separation of the form and the structure in his designs for Ronchamp and the Philips-pavilion (Pronk et al. 2007; Treib 1996). This separation between form and structure was shown for the first time in the Statue of Liberty and comes back in the fluid architecture at the end of the twentieth century. The form of the 1958 Philips pavilion, the 1956 TWA terminal and the 1954 church

Fig. 3.47 Philips pavilion by Le Corbusier 1958

Fig. 3.48 Olympic stadium, Kenzo Tange, Tokyo, 1964

in Ronchamp are not dominated by the structure. These buildings can be seen as the first examples of fluid architecture within the modern style period. Below, there are some more works of fluid architecture within the modern style period. All of these works are classified as type 3: the structure follows the architecturally designed form.

The 1964 Olympic Stadium in Tokyo by Kenzo Tange is formed by two sweeping curved roofs and an asymmetrical but balanced design. The plan of the stadium is in the form of two semi-circles, slightly displaced in relation to one another, with their unconnecting ends elongated into points. The entrances are located in the concave sides. The roof is supported on two reinforced concrete pillars and is made up of a system of steel cables onto which enamelled steel plates are soldered. The curved form of the roof serves to make it more resistant to wind, which can reach hurricane force in this region. The structure is engineered by Yoshikatsu Tsuboi and Mamoru Kawaguchi (Zhongije 2010) (Fig. 3.48).

The 1969 Basento bridge in Potenza (Italy) by Sergio Musmeci (Aa 2003), is supported by a doubly curved concrete slab of about 30 centimetres thickness. Sergio Musmeci (1926–1981) was professor at the faculty of architecture in Rome. He

Fig. 3.49 The 1969 Basento Bridge by Sergio Musmeci

Fig. 3.50 Sydney Opera House, Jørn Utzon, Sydney, 1973

researched lightweight structures and was influenced by aeronautical engineering, Pier Luigi Nervi, Gaudi and Frei Otto. The form finding of the slab is done by making models of hanging structures and the manipulation of neoprene foils. The doubly curved slab has a linear symmetry and repetition of the arches but is completely irregular in other respects. This bridge is the third example, next to the Philips pavilion and Dulles Airport, where the image of a mechanically stressed membrane is used for form finding of a concrete slab structure. Contrary to the other two structures this concrete slab is structurally optimized (Figs. 3.49 and 3.50).

The Sydney opera house by Jorn Utzon (1918–2008) involved one of the earliest uses of computers in structural analysis in order to understand the complex forces in the shells. The shells were created as sections from a sphere (Sharp 2002). The regular and symmetrical but doubly curved building is placed in this overview because of the importance and expressiveness of the building. It was also the last building created before the 1973 oil crisis.

After the oil crisis of 1973 the interest in doubly curved structures diminished and hit an all-time low in the 1980s. The reason can be found in the rising cost of labour. Even in Mexico (Cassinello and Torroja 2010; Burger and Billington 2006) labour is too expensive for Felix Candela to allow him to realize the complex moulds for his shell structures. In the 1970s and 1980s, the building industry is under pressure to innovate insulated building structures because of the oil crisis. This requires a drastic adaptation of new building techniques and the introduction of building physics. So, the combination of the rising costs of labour and oil in the 1970s led to rectangular buildings.

One of the exceptions is the work of Tzonis and Calatrava (2004). Although his work is symmetrical and, in many cases, involved with structural objects like bridges, he is one of the few architects who managed to design and build doubly curved objects in the 1980s and 1990s. His work is a mixture of sculpturing, architecture and structural engineering. His sketches illustrate the direct connection in design between the human form and other forms in nature. The structures are anthropomorphic or naturomorphic without blatant mimicry (Tzonis and Calatrava 2004). Below, there are two of his works: the Zürich Stadelhofen railway station and Liège-Guillemins railway station. The work of Calatrava can be seen as the connection between two periods. His work is highly sculptural and even decorative; therefore, it is no longer connected to the *modern architectural* style. But his work is also not fluid architecture because of its symmetry and regularity (Figs. 3.51 and 3.52).

3.11.4 Deconstructive Fluid Architecture

In the 1990s after a break of more than 20 years, fluid-architecture makes a comeback, stimulated by the new possibilities of computers in design and engineering combined

Fig. 3.51 Zürich Stadelhofen railway station in Switzerland by Santiago Calatrava 1990

Fig. 3.52 Liège-Guillemins railway station in Belgium by Santiago Calatrava 2009

with the morphology from other fields like biology, maths, fluidity, sound and speed (Wendland 2002; Kolarevic 2000). It begins with a style classified as "deconstructive" with reference to the deconstructivist architecture that arose at the end of the 1980s. In deconstructive fluid architecture the structure follows the architecturally designed form. The structural and architectural linkage with shell-structures is lost. This new approach to creation of form in architecture was a radical departure from simple geometry and rectilinear forms (Roudavski 2009: 345–374; Wong 2010: 237–367; Piegl and Tiller 1997). The signal for the advent of this work was the 1988 exhibition in the Museum of Modern Art (MOMA) in New York, titled 'Deconstructivist Architecture' (McLeod 1989: 22–59). Deconstructivist architecture started with a chaos of flat surfaces coming together in an irregular way. Later, these forms where smoothened to (doubly) curved surfaces (Figs. 3.53 and 3.54).

The first large building is the Bilbao Guggenheim Museum by Frank O. Gehry. The new Guggenheim Museum was probably the most frequently mentioned new building of 1998 and 1999 in architecture circles. The composition continues a curvaceous style that has become Gehry's signature. The curved components of this style were present in the early Gehry House, and a similarly sleek curvaceous cladding was displayed in the sculptural fish of the Fishdance Restaurant. It is the most important example of fluid architecture. It proved that it is possible to design and build fluid architecture. After this building, many others followed and fluid architecture has since become an important style with many architects and designs.

The Guggenheim Bilbao (Gehry and Forster 1998) was a remarkable turning point in Gehry's work and in the history of architecture because it manages to maintain the impromptu sketchiness of his drawings in the realized building. It's a sketch in real space, a sketch you can walk into. Gehry is convinced that architecture is art in the first place. Gehry states: "I approach each building as a sculptural object, a wonderful

Fig. 3.53 Jewish museum architect Daniel Libeskind

and spectacular spatial container, a space with light and air, a response to context and appropriateness of feeling and spirit. To this container, this sculpture, the user brings his baggage, his program, and interacts with it to accommodate his need. If he can't do that, I have failed" (van Bruggen 1997). In this vision the structural concept of the building is something that has to fit within the space of the sculptural envelope. Envelope and structure are separated as in the Statue of Liberty. All these buildings belong to type 3, where structure follows the architecturally designed form.

Fig. 3.54 Guggenheim Museum, Frank Gehry, Bilbao, 1997

3.11.5 Fluid Architecture in Private Houses

Beside the above outline of prestigious public or semi-public buildings it is also possible to make an outline of fluid architecture for private houses. In these buildings there is more freedom for experiments and the architecture is highly influenced by personal elements. In many cases, the building is built by the architect or the owner. Often, this architecture is ignored by architectural critics but adored by the public and/or seen as environmentally friendly or 'hippie' architecture. frequently, the fluid forms are made with ferrocement or with an adjusted application of ferrocement, most commonly a mixture of Portland cement applied over layers of woven or expanded steel mesh and closely spaced rod rebar to produce relatively thin, compound curved sheets to make hulls for boats, shell roofs, water tanks, etc. It is impossible to give a complete list. Below are some well-known, important examples since the Second World War. At best, they are structurally optimized but the architecture is predominant. They are all classified as type 3.

Ferrocement has been used from the beginning for imitation of nature. It is called "Faux Bios". Later in the 1930s in the United States, it became popular as a construction method for "Novelty Architecture" (Heimann 2001). See Figs. 3.55 and 3.56.

The Bavinger House by Bruce Goff was completed in 1955 in Norman, Oklahoma, United States. The artists Nancy and Eugene Bavinger built their house with the help

Fig. 3.55 House in ferrocement Built in 1931 by duck farmer Martin Maurer

of students, volunteers and local companies. The house consists of a 96-foot-long spiral curved wall. Apart from this wall there are no interior walls. Within the house there are "floating" platforms on different heights. Some of them have curtains for privacy. On the ground floor there are pools and plants in a kind of courtyard garden (De Long 1977, 1988).

David Greene (1937), studied at Nottingham University. His 1958 design thesis "The Mosque in Baghdad", was a design proposal to be made in shotcrete. He was one of the six members of Archigram. The nine and a half magazine editions they made between 1961 and 1974 were unconventional and had a big influence on architectural design and thinking. David Greene has taught at the AA in London, University of Westminster and Oxford Brookes University (archigram.westminster.ac.uk; Sadler 2005). The vision of Frederick Kiesler (1890–1965) published in 1966 introduces the idea of the "Endless House". There is some similarity with the design for the Mosque and also his ideas have been of great interest to contemporary architects like UNstudio.

Fig. 3.56 The Bavinger House by Bruce Goff, 1955

3.11.5.1 Hundertwasserhause Vienna 1986

The Hundertwasserhause were realised in 1986. Friedensreich Hundertwasser (1928–2000) was a painter since the early 1950s who got more and more interested in architecture. He wrote several essays on architecture and designed the "eye-slit" house and the "high-rise meadow house". The artist was invited to realize an apartment building in 1977. It took several years to find a building site. The architect Josef Karwina assisted him during the realisation of the building (Wade 2009) (Fig. 3.57).

3.11.5.2 Pierre Cardin's Bubble House (1989) 'Palais Bulles' by Antti Lovag

The architect Antti Lovag (1925) designed the Bubble House 10 km from Cannes with a view of the Mediterranean and red cliffs. For him, this house was about going back to the roots of our ancestral habitat by referring to caves, cave dwellings, organic shapes, etc. The house contains 1,200 square meters of living space, consisting of an amphitheatre (500 seats), reception room, panoramic lounge, and 10 suites. The gardens and pools are spread out across 8,500 square meters. The spheres of the bubbles were constructed with a steel mesh and rods in combination with shotcrete. The house has 28 round rooms similar to soap bubbles. Experimental materials were

Fig. 3.57 Hundertwasserhause Vienna 1986

used for the interior, like various polymers. Antti Lovag, born in Hungary, took his first architectural education in Stockholm and continued his education in Paris. In the 1960s he changed his vision on architecture and referred to forms in nature. Some quotes by him: "form of play - spontaneous, joyful, full of surprise", "Whether for economic reasons or lack of technical solutions, human beings have confined themselves to cubes full of dead ends and angles that impede our movement and break our harmony," "the straight line is an aggression against nature" "Conviviality is a circular phenomenon. The circle structures the way human life is carried out" (anttilovag.org) (Fig. 3.58).

Ushida & Findlay desired to free up form and focused on the geometry of chaos, non-linear mathematics, Mobius strip, spirals and growth within nature. In their designs interior and exterior spaces are inseparably connected to one conceptual design. Their work is influenced by Frederick Kiesler ("Endless House"), Antoni Gaudi, Bruce Goff's organic spaces, Andre Bloc's sculptures, as well as those designed by Häusermann and Chanéac in the 1960s (archilab.org). The 1993 Truss Wall House by Ushida & Findlay was designed and built in advance of the main stream in fluid architecture. The project began as a technical commission from the owner of the Truss Wall system who liked to have a method of compound curve concrete construction. The system had been in use for many years in a figurative form, made to look like dragons, curtains and the like. Ushida & Findlay used the system for creating an abstract, free and flowing form for this house. The truss wall system made it possible to achieve a totally integrated combination of special concept and structure (ushida-findlay.com) (Figs. 3.59 and 3.60).

Fig. 3.58 Pierre Cardin's Bubble House 1989

Fig. 3.59 Truss Wall House, Ushida and Findlay, Tokyo, Japan, 1993 (Short 2008)

Fig. 3.60 Truss Wall House, Ushida and Findlay, Tokyo, Japan, 1993 (Short 2008)

3.11.5.3 Steve Kornher Mexico

Steve Kornher (1950) is a designer/builder living in San Miguel de Allende, Guanajuato Mexico. He has been experimenting with lightweight aggregates like expanded clay, shale, slate, diatomaceous shale, perlite, obsidian, and vermiculite through application of heat; by expanding blast-furnace slag through special cooling processes; from natural deposits of pumice, scoria, volcanic cinders, tuff, and diatomite; and from industrial cinders. Substantial savings can be made by decreasing dead load. Lightweight ferrocement may weigh from 35 to 115 lb per cubic foot, depending on the type of lightweight aggregate used for the method of production. In this way, he uses lightweight ferrocement in the construction of underbeds for floors and roof slabs. Steve Kornher is on the technical side of design and is willing to work with different architects and able to realize curved forms in various styles (flyingconcrete.com).

3.11.6 The Categorizing of Fluid Architecture

As we have seen, the fluid architectural buildings mentioned in the historical overview have been classified in two ways (Fig. 3.61).

So far, the first way, we have classified fluid architecture according to its precursors, early, modern, deconstructive and individual projects, as described at the beginning of this section.

The second way is the classification according to the relation between structure and architecture. This classification has 3 typologies.

Type 1 a structure with vertical walls. The curved surface and decorations are "attached" to the vertical walls. At the outside the roof structure is made with a different material. The roof has a structure that is different from the walls.
Type 2 the structure dominates the form of the building.
Type 3 the structure serves the architecturally designed form.

3.12 Conclusion

3.12.1 Precursors of Fluid Architecture

In spite of the decorativeness of the English Gothic and the Gothic-baroque there is still a connection with the structural behaviour of the building. Therefore, they are classified as type 2. As a result of the repetition and regularity of the structure; these can be classified as precursors of fluid architecture.

English-Gothic Gothic-baroque	■	Precursors of fluid architecture
Elastic baroque	■	Precursors of fluid architecture
Rococo	■	Early fluid architecture
Art nouveau	■	Early fluid architecture
Modern architectural	■	Modern fluid architecture
Modern structural	■	Modern fluid architecture
Post-modern	■	Decontructive fluid architecture
Personal style	■	Individual fluid architecture

Type 1	decorative (limited structural impact)
Type 2	structure dominates architecture
Type 3	architecture dominates structure

Building / Architect	**Year**	**Type 1**	**Type 2**	**Type 3**
Chapel Westminster Abbey	1519	■		
Chapel / Santini-Aichel	1769	■		
San Carlino / Borromini	1641	■		
Weiskirche / J. Baptist Zimmermann	1754	■		
Statue of Liberty / F. Bartholdi	1886			■
Casa Mila / A. Gaudi	1910	■	■	
Maison Duldeck / R. Steiner	1920	■		
Einsteinturm / E. Mendelson	1921	■		
Ronchamp / Le Corbusier	1954			■
Bavinger house / B. Goff	1955			■
TWA / E. Saarinen	1956			■
Club Tachira / E. Torroja	1956		■	
Unesco Paris / M. Beuer - P. Nervi	1956		■	
Chapelle Cuernavaca / F. Candela	1957		■	
Philips pavilion / Le Corbusier	1958			■
Dulles Airport / E. Saarinen	1962		■	
Maison S. Baratti / V. Giorgini	1962			■
Olympic Stadium / K. Tange	1964			■
Factory Sisli / H. Isler	1969		■	
Multihalle / F. Otto	1971		■	
Munich Olympic Arena / F. Otto	1972		■	
Sydney Opera / J. Utzon	1973			■
Hundertwasserhause / F. Hundertwasser	1986			■
Bubble House / A. Lovag	1989			■
Zurich station / S. Calatrava	1990			■
Truss Wall House / Ushida & Findlay	1993			■
Bilbao Guggenheim / F. Gehry	1997			■
Art Nouveau House/ S. Kornher	2004			■

Fig. 3.61 Classification of buildings by style

3.12.2 *Early Fluid Architecture*

Early "fluid architecture" starts with rococo and ends with Art Nouveau. The work of Gaudi, Steiner and Mendelsohn can be classified as type 1. Gaudi earns a special position because his structurally inspired work also belongs to the more decorated type 2.

3.12.3 *Modern Fluid Architecture*

From the 1930s, the success period of the shell structures runs until the oil crisis in the 1970s (type 2). The "modern" fluid architecture is inspired by the shell structures and their expressive possibilities. From the 1950s, architects start designing fluid architecture with regard to the detailing and the material used in shell structures. However, they stand apart from the constructive context of the shell structures. On the basis of structure, this is classified as type 3.

3.12.4 *Individual Fluid Architecture*

In the gap between modern and deconstructive architecture there is an unbroken line of fluid architecture fuelled by the personal interest of individuals who are mostly building for themselves. In most cases this personal interest is driven by a romantic approach for a better world in harmony with nature.

3.12.5 *Deconstructive Fluid Architecture*

In the 1990s after a break of more than 20 years, fluid architecture returns to the architectural scene stimulated by the new possibilities of computers in design and engineering. The first large and prestigious building is the Bilbao Guggenheim by Gehry. In this research it is classified as post-modern fluid architecture. Post-modern fluid architecture is characterized by the structure following the architectural designed form. The structural linkage with shell-structures is lost.

Most structurally influenced styles, like Gothic architecture and the shell structures since the 1930s, have a tendency towards some kind of regularity and modularity. With the release of the structural connection in the rococo and the post-modern fluid architecture there is more space for irregularity and asymmetrical forms. Rococo and fluid architecture formed a reaction to a generally accepted architectural style with an orthogonal architecture.

The Statue of Liberty is the first building of type 3. There is separation of structure and building surface and the structure of the building has the same typology as most of the postmodern fluid architecture which began with the Guggenheim museum in Bilbao. From a technical point of view, it gives the Statue of Liberty a special position as the first technical concept for the separation between building surface and structure. In the Modern style this separation led to orthogonal buildings. It took more than a hundred years before the Statue of Liberty had a follow up with a sculptural surface on a steel frame.

3.13 The Technical Problem of Fluid Architecture

The changes in industry in the nineteenth century influenced the building industry in the beginning of the twentieth century. The most important developments were the (re-)invention of concrete and the use of metal as a structural component. The industrialization of the building industry led to a process of prefabrication of building products. Even in our time this process of shifting labour from the building site to the factory is continuing. The separation of the load bearing structure, which is harder to prefabricate, and the cladding of the building envelope, which is easier to prefabricate, is one of the main results of this industrialization process. This idea of industrialization was promoted by many architects. One of the most important forums for architects was the CIAM 1928–1946. Architects like Le Corbusier, Hugo Haring, Mart Stam, El Lissitzky and Rietveld where members of the CIAM (Mumford 2002). In the sixties, Archigram inspired architects to go even further than functionalism and challenged them to "play" with new techniques and materials. It led to the so-called high-tech architecture. The 1971 Centre Pompidou by Piano and Rogers was the first example. However, high-tech was more a baroque exposure of what can be done with technique but the core of the building stayed within the rectangular domain (Cook 1999).

After the realization of the Guggenheim museum in Bilbao, the proposals for making fluid architecture became hugely popular. Many engineers, however, faced (new) problems they could not solve.

In 2002, a Dutch engineering office ABT (2002) concluded in their report "Freedom, Design and Construction 2002": "Before fluid-architecture can be applied on a large scale, optimization on the technical level and the process level needs to be implemented. Clients view this architecture as problematic. Fluid architecture leads to problems in the area of communication and a faulty exchange of information. In order to realize fluid architecture, the integration between the different disciplines involved in building is an absolute precondition. The development in the industry is lagging behind, compared to the technical possibilities. The building sector aims at two-dimensional straightforwardness and regularity. For fluid architecture this aim would have to change to three-dimensionality" (Tessmann et al. 2009). The renowned Dutch structural consultancy Pieters Building Technology drew the same conclusion in an essay (Pieters Building Technology 2002).

Since 2002, progress has been made in the area of Building Information Modelling and computer automated production. The developments with the program Rhino and plug-ins like Grasshopper (Payne and Issa 2009) for drawing parametric designs have reached impressive results (Chiu 2010). However, the essence of the report still holds true and the problems with the realization still exist.

Bibliography

Aa VV (2003) Il ponte e la città: Sergio Musmeci a Potenza. Gangemi Editore SpA, Rome

Alberti LB (1988) On the art of building in ten books (trans: Leach N, Rykwert J, Tavenor R). The MIT Press, Cambridge MA

Andrews B (2010) Ornament and materiality in the work of Adolf Loos. Re.Building: 438–445. University of Nebraska-Lincoln, Nebraska

Anson M, Ko JM, Lam ESS (2002) Freedom, design and construction. In: Advances in building technology. ABT Velp, Velp

The Antti Lovag Project. http://www.anttilovag.org/. Accessed 11 Nov 2016

The Archigram Archival Project. Archigram Magazine Issue 1. http://archigram.westminster.ac.uk/magazine.php?id=96. Accessed 11 Nov 2016

Archilab. Ushida Findlay Partnership. http://www.archilab.org/public/2000/catalog/ushida/ushidaen.htm. Accessed 11 Nov 2016

Bagneris M, Motro R, Maurin B, Pauli N (2008) Structural morphology issues in conceptual design of double curved systems. Int J Space Struct 23:79–87

Balgaradean CM, Babut R (2015) Young people's perceptions toward sexuality in advertising. Manag Mark J: 63–77. University of Craiova, Faculty of Economics and Business Administration, Craiova

Banhart J (2000) Metallic foams: challenges and opportunities. Fraunhofer-Institute for Advanced Materials, Bremen

Benešovska K (2001) Architecture of the Gothic. Prague Castle Administration, Prague

Beton T (1987) The Villas of Le Corbusier 1920–1930. Yale University Press, New Haven

Blunt A (1988) Baroque and Rococo: architecture and decoration. Harper & Row, New York

Blunt A (1989) Borromini. Harvard University Press, Cambridge MA

Bogner D (ed) (1988) Friederich Kiesler: Architekt, Maler, Bildhauer, 1890–1965. Locker, Vienna

Bony J (1983) French Gothic architecture of the twelfth and thirteenth centuries. University of California Press, Berkeley

Bouwen met Wonen. http://www.bouwenmetwonen.net/architectuur/special/renzopiano/. Accessed 12 Apr 2011

van Bruggen C (1997) Frank O. Gehry: Guggenheim Museum Bilbao. New York, Guggenheim Museum Publications

Burger N, Billington DP (2006) Felix Candela, elegance and endurance: an examination of the Xochimilco Shell. J Int Assoc Shell Spat Struc 47. The International Association for Shell and Spatial Structures (IASS)

Cambridge Dictionary. FREE meaning in the Cambridge Dictionary. http://dictionary.cambridge.org/dictionary/english/free. Accessed 25 Feb 2016

Cassinello P, Torroja JA (2010) Felix Candela: his vocational training at the university and his subsequent relationship with the institute founded by Eduardo Torroja. J Int Assoc Shell Spat Struc 51(1). ISSN 1028-365X

CBS. https://www.cbs.nl/. Accessed 2015

Centre Pompidou. Non-standard architectures. https://www.centrepompidou.fr/en/program/calendar/event/cpBeKA. Accessed 6 Apr 2011

Chilton JC (2000) The engineer's contribution to contemporary architecture: Heinz Isler. The Engineer's Contribution to Contemporary Architecture (ECCA). Thomas Telford Ltd and Riba Publications, London
Chiu Y (2010) How to make the soft skin; a preliminary framework for the parametric design of the bionic soft skin. In: Future cities 28th eCAADe conference proceedings. eCAADe, Zurich, pp 237–242
Clifton-Taylor A (1967) The cathedrals of England. Thames and Hudson, New York
Cook P (1999) Archigram. Princeton Architectural Press, Princeton
Crippa MA, Gossel P (2003) Antoni Gaudí, 1852–1926; from nature to architecture. Taschen Gmbh, Köln
Deleuze G (1968) Différence et répétition. English edition: Deleuze G (1994) Difference and repetition (trans: Patton P). Presses Universitaires de France, Paris
Deleuze G (1988) Le Pli: Leibniz et le baroque. English edition: Deleuze G (1993) The Fold: Leibniz and the Baroque (trans: Conley T). Éditions de Minuit, New York
De Long DG (1977) The architecture of Bruce Goff: buildings and projects, 1916–1974, vol 1. Garland Publishing inc., New York
De Long DG (1988) David Bruce Goff: toward absolute architecture. MIT Press, Cambridge, MA
Duncan A (1994) Art Nouveau (the world of art). Thames and Hudson, New York
Dürer A (1525) Underweysung der Messung mit dem Zirckel und Richtscheyt. Hieronymus Andreae, Nürnberg
Dürer A (1528) Vier Bücher von Menschlicher Proportion. Hieronymus Andreae, Nürnberg
Einstein A (1907) Uber die vom Relativitätsprinzip geforderte Tragheit der Energie. Ann Phys 328(7):371–384
Finley K (1999) The nude in contemporary art. The Aldrich Museum of Contemporary Art, Ridgefield
Fletcher B (2001) A history of architecture on the comparative method. Elsevier Science and Technology, Amsterdam
FlyingConcrete. http://www.flyingconcrete.com/. 11 Nov 2016
Fontana L (2012) Fernand Fonssagrives photography: the sensual touch. via: modern design. http://www.moderndesign.org/2012/04/fernand-fonssagrives.html
Fox Weber N (2008) Le Corbusier: a life. Borzoi Books, New York, Knopf A
Freud S (2002) Civilization and its discontents. Penguin, London
Fung YC (1933) Biomechanics: mechanical properties of living tissues. Springer-Verlag, New York, p 568
Futcher T (2012) The digital age: new forms in contemporary architecture. University of Kent, Department of Architecture, Canterbury
Gardner H, Kleiner FS (2009) Gardner's art through the ages: a global history, 13th edn. Thomson Wadsworth, Belmont
Gehry FO, Forster KW (1998) Frank O. Gehry, Museo Guggenheim Bilbao (Opus 32). Edn Axel Menges, Fellbach
Gustafsson A, Lindenfors P (2004) Human size evolution: no evolutionary allometric relationship between male and female stature. J Hum Evol 47(4):253–266
Heath T (1989) Lessons from vitruvius. Des Stud 10(4):246–253. Design Research Society, Surrey
Hefner HM (2005) The playboy book: fifty years. Taschen Gmbh, Köln
Heimann J (2001) California crazy and beyond: roadside vernacular architecture. Chronicle Books, San Francisco
Hentschel K (1997) The Einstein Tower (An intertexture of dynamic construction, relativity theory, and astronomy). Stanford University Press, Palo Alto
Hershberger R (1999) Architectural programming and predesign manager. McGraw-Hill Companies, Inc, New York
Hight C (2008) Architectural principles in the age of cybernetics. Taylor and Francis, London
Hon G, Goldstein BR (2005) From proportion to balance: the background to symmetry in science. Studies in History and Philosophy of Science Part A 36:1–21

Huerta S (2006) Structural design in the work of Gaudí. Archit Sci Rev 49(4):324–339. University of Sydney, Sydney
ISAPS (International Society of Aesthetic Plastic Surgery) (2015) Plastic surgery statistics, ISAPS global statistics in 2015. http://www.isaps.org/news/isaps-global-statistics. Accessed 1 Apr 2015
Ito T (2016) Presentation for the 2016 IASS symposium at Toyo University in Japan on 26 September 2016. IASS Symposium
Jackson M (2001) Diagram of the fold the actuality of virtual architecture
Janson AF (1986) History of art, 3rd edn. Harry N. Abrams, New York, pp 310–312
Jenkyns R (2008) Westminster Abbey. Harvard University Press, Cambridge MA
Keele KD (1964) Leonardo da Vinci's influence on renaissance anatomy. Med Hist 8(4):360–370. Cambridge University Press, Cambridge
Kelly Scientific Publications (2017) Global cosmetic surgery and service market report 2015–2019. Global medical aesthetics industry: injectables, energy-based devices, cosmeceuticals, cosmetic surgery, facial aesthetics, implants, cosmetic tourism. Market Analysis & Forecast to 2022. http://www.researchandmarkets.com/reports/3133911/global-cosmetic-surgery-and-servicemarket-report. Accessed 23 Jan 2017
Kolarevic B (2000) Digital morphogenesis and computational architectures. In: Conference proceedings Sigradi, Rio de Janeiro
Kolarevic B, Malkawi A (2005) Performative architecture; beyond instrumentality. Spon Press, New York
Lambiase J, Reichert T (2003) Sex in advertising: perspectives on the erotic appeal. Lawrence Erlbaum Associates Inc, Mahwah NJ
Leibniz GW (1969) Nova Methodus pro Maximis et Minimis, itemque tangentibus, qua nec fractas, nec irrationales quantitates moratur, et singulare pro illi calculi genus in de Acta Eruditorum (1684). English edition: Leibniz GW (1969) A source book in mathematics, 1200–1800 (trans: Struik DJ). Harvard University Press, Cambridge MA
Lectra. Technology solutions for fashion, automotive and furniture. www.lectra.com. 1 Feb 2017
Lynn G (1998) Folds, bodies and blobs: collected essays. La lettre volée, Brussels
Lynn G (1999) Animate form. Princeton Architectural Press, New York
Macapia P (2006) "Emergent grid": a conversation with Toyo Ito. http://www.petermacapia.com/blog/wp-content/gallery/Publications/ItoInterview/MacapiaItoEmergentWeb.pdf. Accessed 6 Apr 2011
Mahesh S (2001) Production of digital space: On the nature of digital materiality. Conference Proceedings Architectural Information Management, Helsinki, 29–31
Marradi B (2013) Function follows form… strategies for integrated design of complex glass envelopes. In: COST action TU0905 Mid-term conference on structural glass. CRC press, London
Maslow A (1943) A theory of human motivation. Psychol Rev 50:370–396
Materialist Zen (2011) Architects of air: Amacoco. http://materialistzen.blogspot.nl/2011/07/architects-of-air-amacoco.html. Accessed 6 Apr 2011
McLeod M (1989) Architecture and politics in the reagan era: from postmodernism to deconstructivism. Assemblage 8:22–59
Meek HA (1988) Guarino Guarini and his architecture. Yale University Press, New Haven, pp 6–11, 19
Merkel J (2005) Eero Saarinen. Phaidon Press, London
Metz T (1997) 'Form follows feminine': niemeyer, 90, is still going strong. Archit Rec 185(12):35
Miller AEJ, MacDougall JD, Tarnopolsky MA, Sale DG (1993) Gender differences in strength and muscle fiber characteristics. Eur J Appl Physiol 66(3):254–262
Molema J (1987) Antoni Gaudí, een weg tot oorspronkelijkheid. TU delft, Delft
Moreno B (2005) The Statue of Liberty encyclopedia. New Line Books
Moreyra Garlock ME, Billington DP (2008) Felix Candela: engineer builder, structural artist. In: Architecture and design. Princeton University Art Museum/Yale University Press, New Haven
Mumford EP (2002) The CIAM discourse on urbanism, 1928–1960. MIT Press, Cambridge, MA

Nabulsi S, Rodriguez A, Rio O (2010) Robotic machine for high-quality shotcreting process. Aitemin, Madrid, pp 1–8
Nerdinger W (2005) Frei Otto complete works. Birkhäuser Architecture, Basel
Nervi PL (1958) Art: palace of concrete. Time Magazine, Tampa
Norberg-Schulz C (1999) The age of the late Baroque and Rococo. Rizzoli, New York
Novak M (1992) Liquid architectures of cyberspace. In: Benedikt Michael (ed) Cyberspace: first steps. The MIT Press, Cambridge, MA
Oosterhuis K, Feireiss L (2006) Game set and match II: on computer games, advanced geometries and digital technologies. Episode Publishers, Rotterdam
Oosterhuis K, Xis X (2007) Interactive architecture. Episode Publishers, Rotterdam
Otto F (1995) Pneu und Knochen. In: IL vol 35. Institut für Leichte Flachentragwerke, Krämer, Stuttgart, pp 1–236
Otto F, Rasch B (1996) Finding form: towards an architecture of the minimal. Axel Menges, Fellbach
Padovan R (2000) Proportion: Science, philosophy, architecture. Taylor & Francis, London, p 35
Panofsky E (1955) The life and art of Albrecht Dürer. Princeton University Press, Princeton
Pauly D (1997) Le Corbusier: the Chapel at Ronchamp. Springer Science & Business Media
Payne AO, Issa R (2009) The grasshopper primer. Lift Architects, Greenville
Pelkonen EL, Albrecht D (2006) Eero Saarinen: shaping the future. Yale University Press, New Haven
Piegl L, Tiller W (1997) The NURBS book, 2nd edn. Springer, New York
Pieters Bouwtechniek (2002) Detail in architecture. Pieters Bouwtechniek, Haarlem
Poerschke U (2012) Architecture as a mathematical function: reflections on Gottfried Semper. In: Williams K (ed) Architecture, systems research and computational sciences. Nexus Netw J 14(1):119–134. Birkhäuser, Basel
Pronk A, Houtman R, Afink M (2007) The reconstruction of the Philips-pavilion, vol 1. TU Einhoven, Eindhoven
Rawson PS (1987) Primitive erotic art. G.P. Putnam's Sons, New York
Reis HT, Nezlek J, Wheeler L (1980) Physical attractiveness in social interaction. J Pers Soc Psychol 38(4):604–617
Renderpornstar*. http://renderpornstar.com/. Accessed 1 Apr 2015
Roman A (2003) Eero Saarinen: an architecture of multiplicity. Princeton Architectural Press, Hudson
Roudavski S (2009) Towards morphogenesis in architecture. Int J Archit Comput 7(3):345–374
Sadler S (2005) Archigram: architecture without architecture. MIT Press, Cambridge MA
Schmal PC (2001) Blobmeister, erste gebaute Projekte. Birkhäuser Verlag, Boston
Schulze F (1985) Mies Van Der Rohe: a critical biography. University of Chicago Press Inc, Chicago
Sears R, Ingen C, Gray J (2007) To BLOB or not to BLOB: large object storage in a database or a filesystem? In: Technical report MSR-TR-2006-45. Microsoft Research, Redmond
Semper G (1860) Style in the technical and tectonic arts; or practical aesthetics (trans: Mallgrave HF). Getty Publications, Los Angeles, p 107
Semper G (1989) The four elements of architecture and other writings (trans: Mallgrave HF, Herrmann W). Cambridge University Press, Cambridge
Senagala M (2001) Production of digital space: on the nature of digital materiality. In: Conference proceedings architectural information management. Spon Press, Helsinki, pp 29–31
Sharp D (1991) The illustrated encyclopedia of architects and architecture. Quattro Publishing, New York, pp 132–133
Sharp D (2002) Twentieth century architecture: a visual history. Images Publishing, Melbourne
Short CA (2008) What is architectural design research? Build Res Inf 36(2):195–199. Taylor and Francis
Sprott JC (2003) Chaos and time-series analysis. Oxford University Press, Oxford, p 89
Spuybroek L (2004) NOX: machining architecture. Thames & Hudson, London
Sutherland CA (2003) The Statue of Liberty. Barnes & Noble Books, New York

Tessmann O, Bollinger K, Grohmann M (2009) On shells, structural landscapes and performative geometry. In: Computation: the new realm of architectural design: 27th eCAADe conference proceedings, eCAADe (Education and Research in Computer Aided Architectural Design in Europe), pp 47–52. eCAADe (Education and Research in Computer Aided Architectural Design in Europe)

The International Society of Aesthetic Plastic Surgery (2015) ISAPS international survey on aesthetic/cosmetic procedures performed in 2015. http://www.isaps.org/news/isaps-global-statistics. Accessed 1 Apr 2015

Thomsen CW (1998) Sensuous architecture: the art of erotic building. Prestel Publishing, München

Torroja Miret E (1958a) Philosophy of structures. University of California Press, Berkeley

Torroja Miret E (1958b) The structures of Eduardo Torroja: an autobiography of engineering accomplishment. Dodge, New York

Treib M (1996) Space calculated in seconds: The Philips Pavilion, Le Corbusier, Edgard Varese. Princeton University Press, Princeton

Turner CH (2006) Bone strength: current concepts. Annals New York Academy Sci 1068:429–446. Orthopaedic Research Laboratories and Biomechanics and Biomaterials Research Center, Indiana University Purdue University, Indianapolis

Tyler JF (1990) An American symbol; the Statue of Liberty. In: Americans all, a national multicultural education program. First Experience

Tzonis A, Calatrava S (2004) Santiago Calatrava: the complete works. Random House Inc., New York

UNESCO World Heritage Centre. Pilgrimage Church of Wies. http://whc.unesco.org/en/list/271. Accessed 6 Apr 2011

Ushida-Findlay. Truss Wall House. http://www.ushida-findlay.com/project/truss-wall-house/. Accessed 11 Nov 2016

Varriano J, John L (1986) Italian baroque and rococo architecture. Oxford University Press, New York

Veltkamp M (2007) Free form structural design, schemes, systems and prototypes of structures for irregular shaped buildings. In: Research in architectural engineering series, vol 6. Delft University Press, Delft. Article

Visoflora (2009) Coquille du Nautile. http://www.visoflora.com/photos-nature/coquille-du-nautile.html. Accessed 6 Apr 2011

Vitruvius P, Morgan MH, Langford Warren H (1914) The ten books on architecture. Harvard University Press, Cambridge MA

Wade O (2009) Antoni Gaudí and Friedensreich Hundertwasser. VDM Publishing, Saarbrücken

Warren J, Grabar O (eds) (1991) Creswell's use of the theory of dating by the acuteness of the pointed arches in early Muslim architecture. In: Muqarnas VIII: an annual on Islamic art and architecture. E.J. Brill, Leiden, pp 59–65

Waters JK (2003) Blobitecture waveform architecture and digital design. Rockport, Beverly

Wendland D (2002) Model-based formfinding processes: free forms in structural and architectural design

Westminster Abbey. A royal church. www.westminster-abbey.org. Accessed 5 Apr 2011

Wong JF (2010) The text of free-form architecture: qualitative study of the discourse of four architects. Des Stud 31(3):237–267. Elsevier Ltd, Amsterdam

World Heritage Site. Pilgrimage Church of St. John of Nepomuk. https://www.worldheritagesite.org/list/Pilgrimage+Church+of+St.+John+of+Nepomuk. Accessed 8 Apr 2011

Worringer W (1920) Form problems of the Gothic. G.E. Stechert & Co., New York

Yang L, Harrysson O, Cormier D, West H, Gong HJ, Stucker B (2015) Additive manufacturing of metal cellular structures: design and fabrication. JOM 67(3):608–615. Metals & Materials Society

Zhongije L (2010) Kenzo Tange and the metabolist movement. Taylor and Francis Group, London

3-co.com. Architecture. http://www.3-co.com/Public/Technical%20Information/architecture.ht. Accessed 6 Apr 2011

Chapter 4
Three Architects and Their Buildings

Below, the buildings and the working methods of three architects are described. The selection is based on their divergent approaches. In general, architects and engineers and contractors tend to present their work in the media a bit rosier than suits the real process. Therefore, it is not surprising that parties involved who are responsible for the execution take a different view of the origin of the realization and the design of the building. However, it will give an insight into their approach of the building and the way they want to deal with the problems involved with the realization of fluid-architecture.

The popularity of fluid architecture is so widespread, that I cannot give an overview of all the architects who are involved. Moreover, it will not be necessary within the framework of this book to provide a full overview of the architects that are occupied with this kind of architecture. The most authoritative architects and companies at the time of writing are listed below:

- Asymptote Architecture; Hani Rashid and Lise Anne Couture;
- dECOi Architects; Mark Goulthorpe;
- DR_D Lab; Dagmar Richter;
- Greg Lynn FORM; Greg Lynn;
- Frank O. Gehry Associates; Frank O. Gehry;
- Kovac Architects, Inc.; Marcos Kovac;
- KOL/MAC LLC; Sulan Kolatan and William MacDonald;
- Massimiliano Fuksas;
- NOX; Lars Spuybroek;
- Objectile; Bernard Cache and Patrick Beaucé;
- ONL, Kas Oosterhuis;
- Peter Cook;
- R&Sie(n); Francois Roche and Stephanie Lavaux;
- Servo; David Erdman, Marcelyn Gow, Ulrika Karlsson and Chris Perry;
- UNStudio; Ben van Berkel;
- Zaha Hadid.

A. Pronk, *Flexible Forming for Fluid Architecture*,
https://doi.org/10.1007/978-3-030-71551-9_4

Examples of work by the architects Kas Oosterhuis, Frank Gehry and Peter Cook will be reviewed. They were chosen on the basis of reputation; availability of materials and the realization of their designs.

4.1 ONL

The office ONL (Oosterhuis, www.oosterhuis.nl) established in Rotterdam is being led by architect Kas Oosterhuis and visual artist Ilona Lénard. The architects in this office work together with composers; graphic designers; visual artists and programmers; ICT aspires to bring art and architecture closer together. Renowned buildings of Oosterhuis are, amongst others: the salt water pavilion on Neeltje Jans, the waste processing station Ehors/Vloedbelt in Zendere; the WEB of North Holland and the Hessing showroom, Ekris Headlights, and the F side housing project, all in The Netherlands, and abroad under construction the CET cultural centre in Budapest and the Al Nasser HQ office tower in Abu Dhabi. The philosophy of the office is influenced by the rise of use of ICT in the same way as the ICT has nestled itself in the machines in our direct environment. According to Oosterhuis, the same development is happening in the building industry. They construct projects where the user is not only a passive consumer in his own home, but where the spaces react interactively with the user. The buildings become moving objects that react to the environment and the user. Simultaneously, the façade forms a kind of 'media-skin' through which contact can be made with the other side of the world by means of the internet. An architecture that literally starts to move—where notions such as space and time are stretched—an architecture that permanently communicates with other places, where distance no longer plays a part. This implies that the technical skin of the façade must be a system that has the ability to continuously change and where sensors have been constructed to react to its environment. The technical skin must adapt in such a manner that it will still be constructively supporting.

The office produces projects in which they try to get closer to their final aim. In the WEB of North Holland, the building envelope was combined with a secondary support structure. The skin is not a dynamic system, however, but a geometric, complex form. In his designs, Oosterhuis searches for just one detail that will prove to be 'eternally' adaptable. The parametric detail is the core of a building process that takes the architect's data and produces it directly, the so-called "file to factory" process (Boer and Oosterhuis 2004).

> Architectural bodies, too, are now the target of technological invasion. These bodies are a part of global networks, they are (linked by cables) wired. The bodies are connected to databases and their behaviour and (form) shape can be programmed. The body-specific scripts feed on data from databases that are upgrading themselves in real time. The architectural bodies can now be literally animated. Architecture no longer has a static final image; its visible form is becoming as unpredictable as the weather. (Oosterhuis 2002)

Oosterhuis would prefer to construct the building by means of CAD-CAM. He strives for a direct connection between the manufacturing method and digital design. He describes the following method: before a computer model is generated, a main model is designed out of foam. Subsequently, the form will be translated to a computer model by means of a 3D digitizer (3D-scanner). This model will serve as a reference form for the definitive constructive principle. Across this form, a 3D mesh will be generated. The final result is a model incorporating all the elements, dimensions and materials. Out of this model a supplier can generate the necessary 2D drawings. No longer does Oosterhuis want floor plans or cross-sections, but a 3D model of a building with unique elements cut and produced for that specific case with uniform connections. This will result in hundreds of different elements that can be put together on site like a jig-saw puzzle with a bolt connection. He envisages the computer as the instrument to accomplish the optimal connection of information; it should be possible to realize new dynamic and complex buildings. The architect will be playing the role of the pivot in the web. The design created by the architect will be the starting point; computers and manufacturing machines will have to cooperate in an optimal manner to realize the design. On his web site Oosterhuis published the following mission statement:

> The simultaneous design work of the architects and the engineers will inevitably lead to a better synthesis of construction and architecture, where the one is the expression of the other. We create situations where designers and engineers work directly together, while crucial data are exchanged not only from person to person, but also from laptop to laptop during the design sessions. Digital techniques are supporting the augmented design process. Working inside the common design space leads to an enormous increase in mutual understanding and effective design solutions. (Oosterhuis, www.oosterhuis.nl)

Regretfully, the complete connection of all the aspects of the building process has not yet been realized in practice. Even the latest achievements of BIM (Building Information Modelling) (Kwak et al. 2011) have their limits. The designer is not fully informed. The intended top down approach of the design in the direction of the execution can only take place if a complete information transfer is realized. And conversely, a designer can oversee and deal with all the information available.

The WEB of North Holland was exhibited for the first time at the Floriade of 2002 in the Haarlemmermeer and subsequently acquired a permanent spot on the campus of the Technical University Delft. The building constitutes a geometric, complex form. The creation appears to have descended on earth as a space-ship. The floor plan of this building is a pentagon. From this pentagon the complex form is created by a ‘space grid’, a spatial pattern of knots and staves that make a form in three dimensions. In this way the steel web forms a dynamic form and simultaneously functions as the structure for the pavilion. Oosterhuis wanted to construct the plates for the façade as mentioned above. He strives for the principle by which, despite the variety in form and size of the elements, the details are always the same. Furthermore, he strives for a direct connection between the design mechanical behaviour, detailing and production. The triangular gaps in the space frame are filled in with Hylite™

plates that are curved in one direction. The steel plates are fastened onto the steel structure by joints. In between the plates there is a big void that maintains visibility of the steel structure. The result is that the structure is affected by the weather. The exterior does not show anything of what is happening on the inside and a person will have to enter the pavilion to find out. The pavilion can be entered on two sides. The entrance deviates from the traditional arrangement for entering a building. You enter through a door that opens upwards, like the door of a Lamborghini. Inside, the space is completely different. The space inside that can be walked up and down has a surface area of 150 m^2. A visitor can interact with the 360-degree projection. The projection is made by 7 projectors. The visitor manipulates the form and content of the presentation by walking around in the five coloured sectors of the pentagonal presentation room. Every sector represents a characteristic aspect of the province of North Holland. With the help of sensors, the presentation reacts to the number of visitors in the room in the first instance. The human body calculates and re-calculates itself continuously, with fast and slow movements, but never knows a completely static moment. The light program also reacts to the movements of the continuous stream of visitors. The main carrying construction of the pavilion was constructed from the previously mentioned 'space grid'. The stiffness of the construction is gained from the stiffness of the fixed triangle. Despite the triangle as the smallest element creating the complete web, the constructive principle consists of larger parts forming a pentagon. The floor field and the roof are composed of smaller triangles within the bigger pentagon structure. The pentagon is connected by thirty trusses in the façade area. Furthermore, the roof is supported by five columns. The triangular façade elements are hiding the space between main and secondary structure. The whole pavilion measures 19 × 26 ms and is 6 m high. The outer skin that completes the steel grid consists of 360 cold curved triangles of the composite material Hylite™. This material consists of a layer of two mm polypropylene between two layers of aluminium, measuring 0.3 mm thick. Every plate has a different size and was constructed individually. The triangular surface elements are each divided into 4 triangles. The central triangle is flat and the corner triangles are singly curved (Fig. 4.1).

The suggestion of a doubly curved surface can be achieved by the combination of the triangle elements (Figs. 4.1 and 4.2). The centre triangles (purple) are flat and the corner triangles are singly curved. The two edges of the triangle in the corner have to be in the same direction to be singly curved. In Fig. 4.3, the triangle plate with the yellow corner triangles has a boundary curvature in opposite directions. The yellow triangles will be doubly curved and have to be made with elastic deformation of the material. With the triangle method as used in this pavilion it is perceived as a synclastic curvature. For anticlastic curvature the boundary lines would need to have an opposite curvature, which is impossible. The part of the WEB of North Holland that has to be perceived as having an anticlastic curvature has a low curvature and

Fig. 4.1 Triangle element divided into four triangles with a plane central triangle and singly curved triangles at the corners

Fig. 4.2 The combination of 6 triangle elements perceived as a synclastic curvature

some winkles. In fact, it is a faceted surface. Thanks to the big voids between the elements there is a suggestion of curvature (Chap. 5) (Figs. 4.4 and 4.5).

Despite the gap between vision and realization, Oosterhuis has broken new ground by his approach during the realization of this pavilion. Thanks to the introduction of computer directed production facilities it is possible to realize very complex forms. The MERO company, well known for their space grids, fully optimized their production facility. With the help of computer directed production facilities MERO was capable of making irregular frames like the new trade fair in Milan, designed by M. Fuksas. Design tools like Rhino Grasshopper are more and more able to generate production elements by parametric design (Knebel et al. 2001; Soeren et al. 2004).

Fig. 4.3 The combination of 4 triangle elements with a curvature in two directions

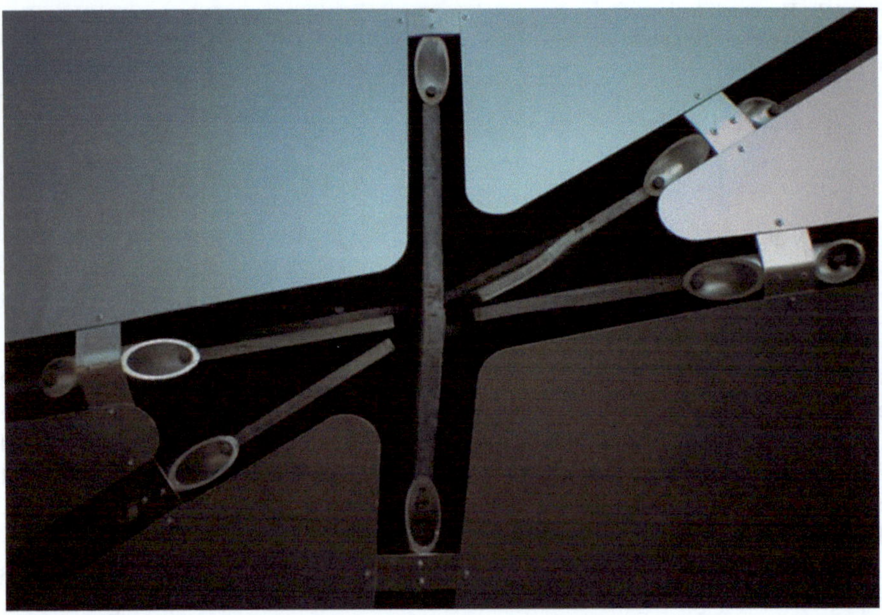

Fig. 4.4 The void between the triangle plates

Fig. 4.5 Web of North Holland by ONL

4.2 Frank O. Gehry

Instead of creating a building, Gehry creates parts of functional sculptures. His work can be characterized as a fluid version of deconstructive architecture (Chap. 3.11) (Wigley 1995). Deconstructivism is a development of postmodern architecture that began in the late 1980s. It is influenced by the theory of "Deconstruction", which is a form of semiotic analysis. It is characterized by fragmentation, an interest in manipulating a structure's surface or skin, non-rectilinear shapes which appear to distort and dislocate elements of architecture, such as structure and envelope. The finished visual appearance of buildings that exhibit deconstructivist "styles" is characterized by unpredictability and controlled chaos. Not only are his buildings divided up into different parts, so too is his working process. After an extensive iterative process, Gehry arrives at a definitive form. During this process, understanding is acquired by means of how sketches and models relate to the context, how the form relates to the function and how the form relates to the material. By connecting and stacking neutral blocks, an overview in the organization is obtained and the feeling of scale developed in relation to the environment. The models form a visual representation of the program and will be re-organized during the implication of all architectural elements. The models give the client and the team their first impressions. They are often colour-coded and developed on two scales; one team for the exterior, subdivided into urban planning scale and the larger scale, and a second team for the interior of the building. The zone in between exterior and interior is for technique (structure and installations). The exterior team and interior team are two teams working more or less independently. During this process, Frank Gehry makes the sketches that give a first impression of the form of the building and selects the sketches that will be presented to the client. The 'esquisse' and participation are the two main elements that are indissolubly connected. The 'esquisse' is a quick architectural drawing that is intended to reflect the essential ideas of the conceptual project. With the 'esquisse' the 'parti' is being set. Parti is a Latin word that is best translated with: the choice, the manner, the intention, and the way to find a solution (Gargus 1994). The sketch gives the 'big idea'. Simultaneously, the concept sketch is subjected to the interpretation of the client. It creates the possibility for the client to be involved on the architectonic form without the consequence that the architect would have to let go of certain characteristics. Subsequently, hundreds of models are made to research and translate all the characteristic elements. On a model scale insight is gained about the behaviour of the material. Even though, sometimes, this material is not exactly the same as the actual material used during the construction, it provides feedback on certain formal rules that are applicable to the material. Eventually, a definitive physical model is decided upon. The surfaces of Gehry's architecture are mainly singly curved or ruled surfaces. If there is curvature in the opposite direction, it is limited (Figs. 4.6, 4.7, and 4.8).

Fig. 4.6 Bilbao Guggenheim by Gehry with almost singly curved façade

Fig. 4.7 Main structure of the Bilbao Guggenheim museum

Fig. 4.8 The construction of the Bilbao Guggenheim museum

Next, this model is translated to a 3D digital model. This translation is made by scanning the physical model. In this manner the exact *x*, *y* and *z*-coordinates are being recorded and translated to the program CATIA (Computer-aided Three-dimensional Interactive Application), developed by Dassault Systems (3DS, www.3ds.com). In the next step, different curved areas are being separated in the program into smaller areas. The different areas are cut out by a cutting machine that is directly connected to a software model. The complete method, from the physical models to the sketch models, till the virtual models are named: 'skin in' method. The physical and virtual models function as a spatial envelope in which the construction is being developed.

The strength in Gehry's work can be found in this process, where he involves the client and others in the design. The deconstructive aspect of his architecture makes it possible to work in this way. In a more comprehensive architecture this divided process might pose problems.

> This technology provides a way for me to get closer to the craft. In the past, there were many layers between my rough sketch and the final building, and the feeling of the design could get lost before it reached the craftsman. It feels like I've been speaking a foreign language, and now, all of sudden, the craftsman understands me. In this case, the computer is not dehumanizing; it's an interpreter. Frank O. Gehry (Hadjri 2003)

In the translation from the CATIA model to the technical elements for construction it becomes clear what the true nature of the technological problems will be. The major parts are curved and every curve will be different from the other. In the Guggenheim museum, this problem was solved by reducing the curvature and limitation of the curvature to mainly single curvature and ruled surfaces with reduced curvature. The structure of the building follows a classical division in main structure, secondary structure and surface. The main structure consists of a frame connected to a steel-concrete composite floor. The frame approximates the outer surface in a faceted way and has more or less transformed squares with diagonals for stability. The vertical parts of the frame are on a slope, the other beams remain horizontal. All the elements of this frame have been prefabricated. The frame with straight beams is covered with a secondary structure of vertically curved beams (Fig. 4.7). To connect both structures, spacers are attached to the horizontals of the main frame. These spacers are adaptable in three directions and hold the vertical profiles that can absorb a light curvature. These profiles are covered with steel plates and insulation. Across these plates, a water-resistant layer is affixed. The titanium façade plates are applied on top. The plates have been fitted with a fold and overlap. The combination of overlap, small sizes and limited curvature makes it possible to finish the façade almost entirely with same-sized elements. This results in a monotonous, dynamic form that is perceived as a doubly curved surface. In order to make the surface of the façade, the tolerances in size are adapted between the main structure and the secondary structure and in the overlap of the titanium plates. In this way, all the deviations can be handled on site by the craftsmen (Figs. 4.9, 4.10, and 4.11).

The façade of the Bilbao Guggenheim comes close to the manipulation of a regular insulated steel façade as used for industrial buildings. Since the realization of the Bilbao Guggenheim museum, Kalzip (www.kalzip.com) developed a complete program to fold and bend diverging and converging strips. In this way it is possible to realize doubly curved surfaces as found in the façade and roof of the Architectuurcentrum in Amsterdam.

Fig. 4.9 Façade of the Bilbao Guggenheim museum from behind

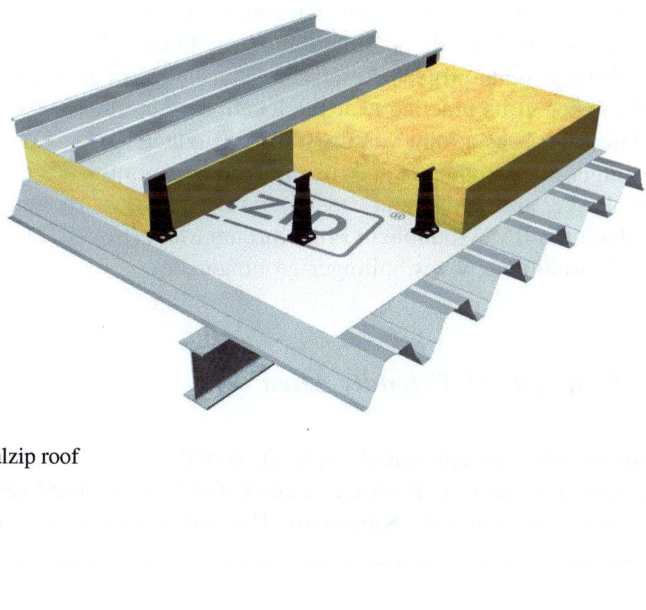

Fig. 4.10 Kalzip roof

Fig. 4.11 Architectuurcentrum Amsterdam, by Rene van Zuuk

4.3 Peter Cook and Colin Fournier

The Kunsthaus in Graz was designed by the architects Peter Cook and Colin Fournier. They both teach at Bartlett School in London. Peter Cook is one of the founding members of the legendary architectural group ARCHIGRAM in which he was active from 1961 until 1974. Fournier got his practical experience during large projects when he was working for the American Parsons Company. He also worked with Bernard Tschumi on the Parc de la Villette in Paris, France. Peter Cook had little experience with large scale projects before the Kunsthaus. In July 2001, the team of the Kunsthaus began its practice near the location where the Kunsthaus was to be realized. This team was a joint venture of three partners: Peter Cook and Colin Fournier, Bollinger and Grohmann and the Graz office Architektur Consult ZT. Of these three, Bollinger and Grohmann had experience with constructing free forms. In 1999, they built the BMW-bubble in Frankfurt am Main (Fig. 4.12).

(Bollinger + Grohmann, www.bollinger-grohmann.de).

4.3.1 The Concept: 'A Friendly Alien'

The Kunsthaus Graz had to represent the reputation of the city of Graz in the field of experimental design. Therefore, Peter Cook and Colin Fournier used new materials and forms in their proposal for the Kunsthaus. The building had to be functional for

Fig. 4.12 Kunsthaus Graz by P. Cook and C. Fournier

the users and had to fit in the old city centre of Graz without being too detached (Cook and Fournier 2004). The building can be seen as an addition to the "Eisernes Haus". It is a "suspended" volume that leaves the surface level intact by the transparent façade on the ground floor. The building is accessible to everyone who views it from the street. The upper part seems to be a translucent foil that is wrapped around the building and closes the exhibition rooms off from the outside world. Daylight enters through skylights that have been applied in the uppermost area.

4.3.2 Engineering and Construction

First, an explanation is given of how the design by Peter Cook and Colin Fournier was translated to a model on which the structure and building envelope could be based. Next, a closer look will be taken at how the building materials were applied and whether this has had any influence on the design.

4.3.3 Development of Form

The most conspicuous element of the Kunsthaus is its fluid form. This special form has played an important role in the materialization of the design. With the construction of the BMW bubble in Frankfurt am Main Bollinger and Grohmann gained experience with realizing and digitizing these kinds of forms. Compared to the BMW bubble, the Kunsthaus was in many ways the bigger challenge. It is bigger in size, 60 by 40 m compared to 24 by 16 m of the bubble, it is a permanent building as opposed to the temporary pavilion and Graz has higher wind and snow loads than Frankfurt.

The architects made their model in cast acrylic glass and a rough digital model was used for rendering. For the structural analyses the model was scanned. Gehry and Oosterhuis did the same, but in this case the close cooperation between engineers and architects changed the form to the benefit of structural optimization. By making the roof higher in the middle there were fewer bending moments. When Gehry, and Oosterhuis, too, presumably, scan their designs nothing is changed in the form of the building.

The "fluid part" of the Kunsthaus was remodelled in the CAD-program Rhinoceros. The basic form with which they started was a sphere. This sphere was subsequently deformed by means of relocating the so-called gravitational points. In this manner the digital model got its desired form, with the result that the criteria of production, lengths of the profiles, span and other constructive preconditions could be optimized.

4.3.4 The Structure

The structure of the fluid part consists of two parts: the exterior translucent surface and the grid shell at the inside. The grid shell scale is divided in regular vertical sections that have been laid across the surface every 3 m. There is a secondary structure that divides the surface into a triangular mesh. The inner structure consists of conventional floors, a number of cores and columns of steel and steel-concrete composite floors. Stability is obtained by the cores, walls and floors. The cores also serve as openings for elevators and staircases. The outer structure stands apart from the cores and has a free span. The bottom floor of the "blob" supports the complete outer structure and skin of the building (Fig. 4.13).

4.3.5 Façade

The skin consists of a doubly curved façade, although the triangle mesh structure consists of linear elements. The outer skin is wrapped around the structure underneath.

The suspended ceiling covering the steel structure is made of a prestressed metal net structure. In between, there is space for air conditioning, electricity and data cables. The thickness of the complete structure is 90 cm in total. The appearance of

Fig. 4.13 Detail of exterior of the Kunsthaus

the façade on the inside is completely different from the outside. The primary steel structure has been treated with a fire-resistant coating on the inside. The triangles have been closed by sandwich panels. The complete structure is covered with insulation and a roofing foil. Nozzles make the connection between the steel structure and the plexiglass second skin. In the space behind the Plexiglas façade an automatic sprinkler installation, a fire detection system and round neon tubes have been applied. Plexiglas was selected for its good technical characteristics. It can be easily transformed under relatively low temperatures; it is transparent and allows colour to be added to the material. 1,300 curved Plexiglas panels have been attached, which together create a joint fluid surface. Because of the form, every element is unique and has been specifically designed to fit together. The plates are attached by a point attachment by the so-called nozzles. Four different nozzles have been used. They can be turned in many directions and therefore possess the necessary adaptability. The void between the plates is present to absorb deviations of the panels, but also to transport sun-heated air outside as is usual in a second skin façade. The place of the nozzles is defined by the triangular steel structure. The deviation of the panels and voids is thus also defined by the structure below and the deviations in the steel structure will also have their effects. The tubes on the roof of the building (Figs. 4.14 and 4.15) light the exhibition room and are equipped with gratings with sun screens to regulate the light. They are a structure unto themselves and are covered with Plexiglas panels.

Fig. 4.14 Interior of the Kunsthaus

	Oosterhuis The Web	Gehry Guggenheim	Cook Kunsthaus
process of design	top-down	iterative	iterative
concept/inspiration	computer kinetics	sketches	situation/program
structure	arches and triangles	columns and beams	triangle space structures
cladding	singly curved elements	small singly curved elements	doubly curved elements
inside/outside	inside -/- outside	inside -/- outside	inside ← outside

Fig. 4.15 Table: Three works by Kas Oosterhuis, Frank Gehry and Peter Cook compared

4.4 Conclusion

The three architects reviewed here strive in their designs for integration of the façade and the structure. They continue to build with techniques that have been developed for rectangular buildings where the traditional separation between the main structure and separating elements prevails. The outer surface is in all cases not the water-resistant layer. Oosterhuis and Cook/Fournier use the façade as a doubly curved second skin façade. In all cases the use of 3D computer software makes it possible to realize the form designed by the architect. The architect is only to a certain extent capable of overseeing the technical and structural implications of the design. Frank Gehry rationalizes his design on the basis of a free sketch in the most drastic manner. In the paper "panelling architectural free-form surfaces" by Eigensatz et al. (2010). the panelling and surface of two buildings by Zaha Hadid Architects (Lilium Tower, Warsaw and National Holding Headquarters, Abu Dhabi) are analysed and geometrically optimised with the help of a panelling algorithm. In all cases, it has led to high-profile buildings, but also to problems concerning measurements of elements and the connection between the support structure and façade panels (Eigensatz et al. comes to the same conclusion). In almost all these cases the consequence is expensive buildings with technical problems. Techniques that were developed in the past to realize shell structures are hardly ever used in realizing these buildings. In the Kunsthaus, the difference between the exterior and the interior is substantial but limited. In the other two cases the difference is complete.

4.4.1 *Perceived Deviation*

The connection of complex curved elements in a building façade needs extra attention. In the three case studies above, these problems are handled differently. Eric Vastert (1992) researched the perceptibility of deviations between building elements in a dissertation. He concluded that the perceived deviation is based on three aspects; pattern of the elements, surface of the elements and the observer. Pattern is deviated according to the size of the panel; width of the join, number of joins, regularity, complexity and proportion. Surface is deviated with respect to transparency, colour, smoothness and surface structure. The influence of the observer is deviated in terms of the distance of the observer to the object, the attention, the visual capability, visual training, expectations and weather conditions. In the figure below in the first column there is a list of aspects of influence to the perceived deviations between building elements as concluded by Vastert. The second column shows how to avoid the perceived deviations.

The perceived deviation of colours is researched by James Mai (2013) (not including the aspects of transparency, smoothness and surface structure). He refers to the generally accepted colour system of Albert Munsell (1858–1919). Munsell distinguishes three colour dimensions: hue, value (lightness), and chroma (colour purity) (Kuehni and Schwarz 2008). Mai introduces the term: “Simultaneous Colour Contrast Territories”. In his research Mai concludes that, in general, a deviation in hue, value and chroma of 20% is not perceived by the observer. He calls this a territory. In his conclusion, he recommends that more research into the form and boundary of each Colour Territory needs to be done. Vastert comes to the conclusion that the deviation in dark coloured surfaces with less contrast is less perceived than in lighter colours with more contrast. The focus of the research by Vastert (1992: 30) is on the perceived deviation of voids. He concludes that this is very much related to the architectural style. Clean, simple, regular, rectangular, smooth buildings with big elements and a small width between the elements are the most sensitive to deviations in the size of the elements and voids. His research is limited to zeroclastic regular building façades. He focuses on the aspects that are easy to measure, like width and length of the join and the number of elements. He states that the distance of the observer to the element is of minor importance due to the fact that the perceptibility of deviations is mostly influenced by the proportions of the element and the join (Fig. 4.16).

Perceived deviation between building elements				
Aspects	How to avoid the perceived deviation	The Web	Guggenheim	Kunsthaus
Pattern				
Size (length of joint)	Small elements	average	small	average
Width of joint	Large	large	overlapping	small
Number of joints	High	high	high	very high
Regularity	Irregular	changing triangles	changing rectangles	mostly changing rectangles
Complexity	Complex	complex	complex	complex
Proportions	Average	average	average	average
Surface				
Transparency	Non-transparent	non-transparent	non-transparent	transparent
Colour	Dark	silver	silver	no colour
Smoothness	Bumpy	smooth	almost smooth	smooth
Surface structure	Complex	flat/monoclastic	flat/monoclastic	doubly curved
Observer				
Distance to object	High	low	low/high	average
Attention	Low	high	high	high
Visual capability	Low	x	x	x
Visual training	Low	x	x	x
Expectations	Low	high	high	high
Weather conditions	Bad	x	x	x

Fig. 4.16 Table of perceived deviation between building elements

Although fluid architecture is neither zeroclastic nor regular, but doubly curved and irregular, it is interesting to connect the research by Vastert to the three buildings in this chapter. This is done in the last three columns. The pattern aspects of the buildings allow a great deviation in the size of elements in spite of the surface and observer aspects. In the Web of North Holland, the width between the elements is relatively big. This is also the most important aspect when aiming to avoid the perceived deviations in the façade. The joins form a pattern with continuous lines that give the object a kind of kinetic gesture. In the Kunsthaus in Graz they had a much more difficult job due to the small width of the joins and the transparent smooth material. The joins where not supposed to be perceived as a pattern. In the Guggenheim museum in Bilbao there is no space between the joins, there is only a line at the overlap of the titanium elements. The size of the elements is much smaller and the lines of the joins are discontinuous. The titanium material is smooth but the surface structure is wrinkled. In the Guggenheim building it was much easier to avoid the perceptibility of deviation in the size of the elements due to the choices in material, size and pattern of the façade.

The observer aspects (for instance attention and weather conditions) will change in time but in general, fluid architecture has a high profile that calls for attention and people have high expectations. In fluid architecture the design is more sculptural and therefore the complete picture of the building will ask more attention as opposed to the attention to specific details. If the attention and distance between a certain detail and the audience are big enough, a large deviation might not be noticed. With the Guggenheim building and with the Kunsthaus, the buildings can be seen from a great distance. At the Kunsthaus, the curved part is raised and at the Guggenheim one of the most important façades is behind a mirroring pond. With the Web of North Holland, it is possible for the audience to approach the façade.

Bibliography

Boer S, Oosterhuis K (2004) Architectural parametric design and mass customization. In 5th European conference on product and process modelling in the building and construction industry–ECPPM

Bollinger + Grohmann. www.bollinger-grohmann.de. Accessed Dec 2012

Cook P, Fournier C (2004) A Friendly Alien: ein Kunsthaus fur Graz. Hatje Cantz, Stuttgart

Eigensatz M, Kilian M, Schiftner A, Mitra N, Pottmann H, Pauly M (2010) Paneling architectural freeform surfaces. ACM Trans Gr 29(4), Article 45

Gargus J (1994) Ideas of order: a formal approach to architecture. Kendall Hunt Publishing, Dubucque

Hadjri K (2003) Bridging the gap between physical and digital models in architectural design studios. Int Archives Photogramm Remote Sens Spat Inf Sci XXXIV-5/W10

Kalzip. http://www.kalzip.com. Accessed Jan 2013

Knebel K, Sanchez-Alvarez J, Zimmermann S (2001) The structural making of the Eden Domes. MERO GmbH & Co. KG, Würzburg

Kuehni R, Schwarz A (2008) Color ordered. Oxford University Press, Oxford

Kwak JM, Choi GY, Park NJ, Seo HJ, Kang LS (2011) 4D CAD application examples and directions for development in civil engineering projects. In: 2nd international conference on education and management technology IPEDR, vol 13. IACSIT Press, Singapore

Mai J (2013) Territories of color. In: Proceedings of bridges 2013. Tessellations Publishing, Phoenix

Oosterhuis K (2002) Architecture goes wild. 010 Publishers, Rotterdam

Oosterhuis K. Kas Oosterhuis. http://www.oosterhuis.nl/. Accessed 1 Dec 2012

Soeren S, Sánchez-Alvarez J, Knebel K (2004) Reticulated structures on free-form surfaces. MERO GmbH & Co. KG, Würzburg

Vastert E (1992) Visual perception of visual deviations. Eindhoven University of Technology, Eindhoven

Wigley M (1995) The architecture of deconstruction: Derrida's Haunt. MIT Press, Cambridge, MA

3DS—Dassault Systèmes. Design Engineer|Catia, Computer Aided Three-dimensional Interactive Application (CATIA). http://www.3ds.com/products/catia/. Accessed Jan 2013

Chapter 5
The Morphology of Fluid Architecture

5.1 Introduction

In Chap. 2 it was argued that the freedom in design will increase tremendously with the combination of different forms. Chapter 3 proved that fluid architecture is perceived as smooth, irregular and (doubly)curved. In this chapter, I wish to go a step further into the morphology of fluid architecture. At first it is important to be aware how people observe. Therefore, I will give a brief summary of the *Gestalt* laws. These laws influence how we perceive the form of a building. Christopher Alexander describes in his work *The Nature of Order* (Alexander 2002: 144–242) fifteen basic properties necessary to create spaces which contain a certain aspect of 'life'. In these basic properties by Christopher Alexander the *Gestalt* laws play an important role. Finally, the three buildings in Chap. 4 will be observed and compared to the fifteen properties Christopher Alexander defined for "good" architecture. In the conclusion a summary of the morphological aspects of fluid architecture.

5.2 *Gestalt* Theory

Max Wertheimer, Kurt Koffka and Wolfgang Kohler were the founders of *Gestalt* psychology. In 1912 Max Wertheimer published his first papers (Wertheimer 1912: 161–265; Wertheimer and Spillmann 2012). They researched how and what people perceive and how we interpret it. The fundamental principle of *Gestalt* perception is the law of *Prägnanz* (German for conciseness) which says that we tend to order our experience in a manner that is regular, orderly, symmetrical, and simple. The *Gestalt* effect is the form-generating capability of our senses: "The whole is other than the sum of the parts" (Koffka 1999). *Gestalt* psychologists attempt to discover refinements of the law of *Prägnanz*, and wrote the laws which hypothetically allow

A. Pronk, *Flexible Forming for Fluid Architecture*,
https://doi.org/10.1007/978-3-030-71551-9_5

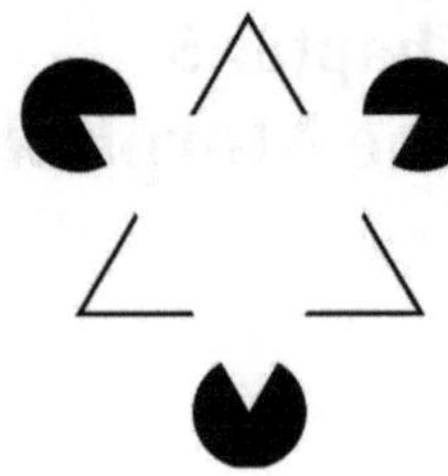

Fig. 5.1 Law of closure

us to predict the interpretation of what is there. Below, a brief summary of the *Gestalt* laws:

Law of Closure. Figure 5.1 The mind may experience elements it does not perceive through sensation, in order to complete a regular figure (i.e., to increase regularity).
Law of Similarity. Figure 5.2 The mind groups similar elements into collective entities or totalities. This similarity might depend on relationships of form, colour, size, or brightness.
Law of Proximity. Figure 5.3 Spatial or temporal proximity of elements may induce the mind to perceive a collective or totality.
Law of Symmetry. Symmetrical images are perceived collectively, even in spite of distance.
Law of Continuity. The mind continues visual, auditory, and kinetic patterns.
Law of Common Fate. Elements with the same moving direction are perceived as a collective or unit.

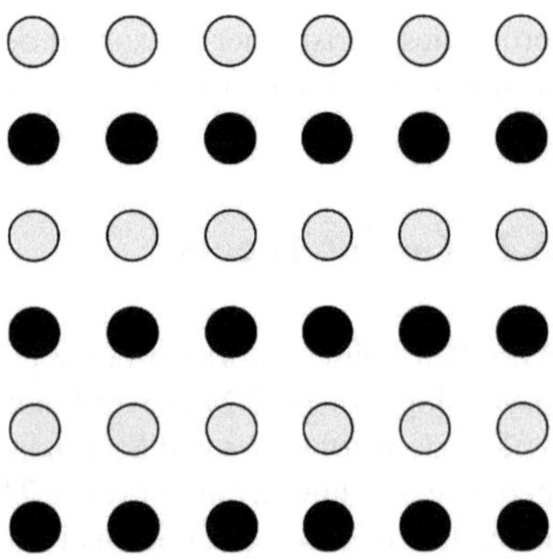

Fig. 5.2 Law of similarity

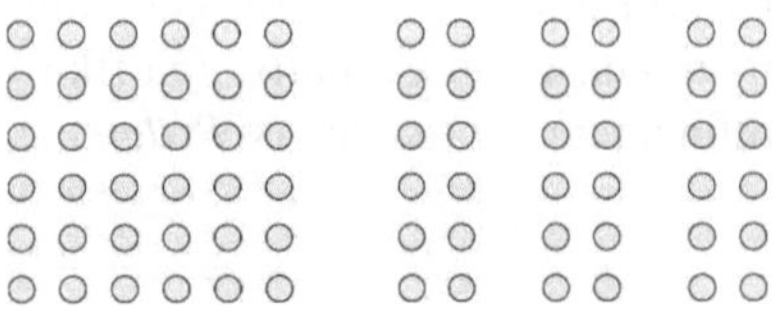

Fig. 5.3 Law of proximity

Fig. 5.4 Selfridges store in Birmingham by future systems

There are several applications where architects use the *Gestalt* laws in fluid architecture. The law of continuity is used in the Selfridges store in Birmingham by the London-based architectural office Future Systems headed by Jan Kaplicky and Amanda Levete (Fig. 5.4). The surface of this building consists of 175 mm shotcrete on a steel structure covered by insulation and 15,000 aluminium discs with a diameter of 60 cm on a black background (Hernández Hernández 1999). The large join between the elements will hide the deviation between all the elements. Everybody is aware of the deviations in the joins but no one sees it as a problem. The same happens in Fig. 5.5. In this photo by Fernand Fonssagrives (mid 1950s) of a woman's skin and shadow (Fontana 2012). The general outline of the form can still be recognized. The change in distance between the dots allows freedom of interpretation and gives a suggestion of the curved surface of the skin. The mind will fill up the gap as in the *Gestalt* law of proximity. If the form of the elements in a grid is the same but they are different in size, as in Fig. 5.6, they also will be recognized as part of the same surface in accordance with the Gestalt law of continuity. The difference in size is perceived but bridged by the mind as belonging to the same group. In the Web of North Holland, the *Gestalt* laws of proximity and continuity are used to suggest a doubly curved smooth surface. This is done by maximizing the void between the elements and by repetition in size and shape of the slightly changing triangle elements (Figs. 4.4 and 4.5).

Fig. 5.5 Photo of woman in shadow by F. Fonssagrives

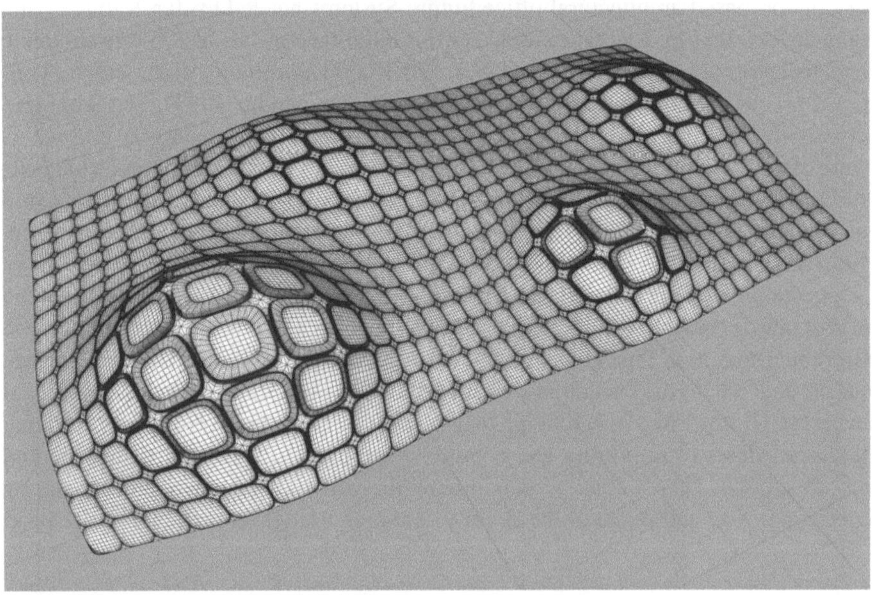

Fig. 5.6 Packaging on doubly curved surface

5.3 The Fifteen Basic Properties by Christopher Alexander

Christopher Alexander describes fifteen basic properties necessary to create spaces which contain a certain aspect of 'life'. This can be described as a feature resulting in the appreciation and beauty of spaces and matter. He states that due to the absence of those traditional design tools, modern architecture is lifeless by contrast to many historical buildings. In his essay he gives a strong opinion about the consequences for architecture in respect to his fifteen basic properties.

The fifteen properties Alexander uses to support his statement are: (1) levels of scale, (2) strong centres, (3) thick boundaries, (4) alterating repetition, (5) positive space, (6) good shape, (7) local symmetries, (8) deep interlock and ambiguity, (9) contrast, (10) gradients, (11) roughness, (12) echoes, (13) the void, (14) simplicity and inner calm, (15) not-separateness.

For this research it is interesting that he provides a way to categorize architecture by its morphology. In some cases, there is a link with the *Gestalt* laws. So, it is worthwhile to analyse the buildings in Chap. 4 in respect to the fifteen categories that Christopher described (Salingaros 1995: 638–643). In this way it is analysed where the fluid buildings are different from architecture in general. The properties will be validated in four categories and summarized in a matrix. Below the four categories:

++ this property is used and an important aspect for the appearance of this building.
+ this property is used but not essential for the appearance of this building.
– this property is not used but not essential for the appearance of this building if it was used.
–– this property is not used and will change the appearance of the building in an essential way if it will be used.

The first and last categories determine an essential aspect for this building. The last one will differentiate this building from other architecture. If this aspect is also of the essence in the other two buildings it might be essential for fluid architecture in general.

5.3.1 Level of Scale

Buildings are constructed by putting smaller and bigger parts together. All these parts have an influence on the way the building is perceived. According to Alexander each part that is used has its own centre. This topic will be elaborated in the following paragraph about strong centres. Within the total assembly the levels of scale and scale differences determine the tension between the centres and contributes to the 'life' of the building. The property of having levels of scale is not a mechanical thing which can require a wide range of different scales, but 'life' can only occur when each centre gives life to the next one. In order to create a good composition,

Fig. 5.7 The relationship between ratios 1:e (left) and 1:20 (right)

Alexander states that one jump in scale should be 1:3. So, when a larger jump of scale is necessary several steps with the ratio 1:3 should be taken, or else the relation between scale levels becomes too hard to be noticed. Too large a scale jump does not do anything to bring life to a structure.

Note that this ratio seems to be closely related to the natural number $e = 2.718$, as well as $\pi = 3.14$. The irrational number e is sometimes called Euler's number because he was the first to discover this natural logarithm. Both e and π are constants within the mathematical system. When using ratio 1:e to create one scale jump, the human brain is still able to notice and understand the relation between two levels of scale, yet the human brain fails to notice the relation and scaling when the ratio is larger, for example 1:20 as shown in Fig. 5.7.

Conclusion

In the three fluid buildings there is a direct relationship between the smallest element and the boundaries of a façade in all the cases. The elements do not change in material but they all have small differences in size and shape. In the Guggenheim, the ratio between element and façade is 1:20 or even more, in the Kunsthaus it's 1:15 and in the Web of North Holland it's 1:10. It is remarkable that there is no intermediate level of scale between the size of one element and the boundaries of the building. The surface of the building is assumed to be "fluid" and therefore the individual elements are of less importance. The result is a high ratio between the different levels. It is assumed that in architecture in general the ratio is lower than it is in fluid-architecture but it would be hard to prove that this is an essential aspect of fluid architecture.

5.3.2 *Strong Centres*

The frequent appearance of strong centres in structures is another property which brings 'life' to a structure. In his theory, Alexander states that centres play a much

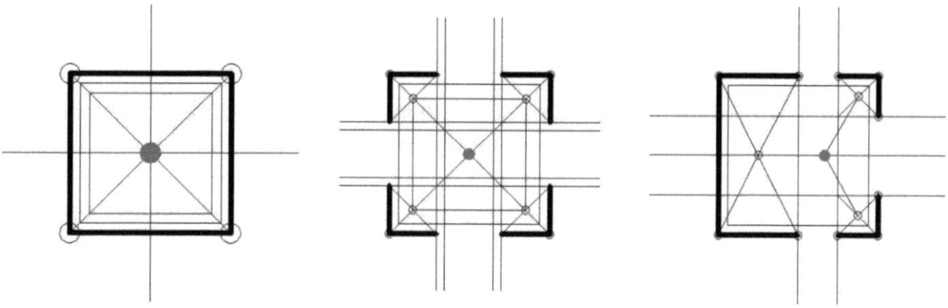

Fig. 5.8 Determination of the centroid

more fundamental role as basic elements to create architectural beauty within a composition compared to the other fourteen properties. A strong centre appears when different layers are put together to create a composition. In this way a vector field is set up. Every point within the total assembly has the property that from that point the centre of a field is in a certain direction: one direction is going to the centre and another direction is moving away from the centre. As a result, the whole visual field is oriented towards the centre and the field appears to be centred. The hierarchy of the layers creates a deep feeling and intensity of the centre. However, every layer needs to be a 'structure' which forms centres itself. The arising of a gradient of increasing intensity creates a centre in the middle. So, the field effect and the power of a centre are created by the sequence of other nearby centres leading up to it. It is this multiplicity of different centres, at different levels, that engages human beings because they feel the presence of a centre.

When concerning the mathematical definition of strong centres, this is strongly related to the determination of the centroid of a specific field. The location of this centre depends on the shape and form of the field. Figure 5.8 shows the determination of the centre of gravity of a basic geometry. When the field is divided into multiple parts, these will create new centres of gravity for each individual part. The overall centre of gravity can now be determined by the combination of these multiple individual centres. When the multiple parts form an irregular subdivision, it is possible that the centre of gravity of the total field will move away from its original point. This effect is shown in the right diagram of Fig. 5.8.

Conclusion

In the three fluid buildings, the irregular subdivision of building parts and the irregular curvature of the building surfaces makes it impossible to find a centre. In the Web of North Holland there is only one space, therefore it might be possible to find a central area, but with the two other buildings the combination of different spaces and forms is too complex. To create a centre with an increasing intensity will regularize and order a building. Fluid architecture likes to disorder and create restless "moving" spaces, therefore it is assumed that fluid architecture has to avoid elements that will structure the building. Therefore, strong centres will be avoided.

5.3.3 Boundaries

Living centres are often formed and strengthened when they are surrounded by boundaries. A boundary shapes a force field around a centre and in this way focuses the attention on the centre of a space. It helps to produce and intensify the centre which it is marking off. At the same time a boundary unites the centre which is being marked off with the world at the other side of the boundary. The boundary thus separates as well as unites a centre with the space beyond the boundary. In both ways the centre becomes more intense. According to Alexander a boundary needs to be of the same order of magnitude as the centre which is marked off by it. He also states that any centre which has 'life', like a building or a town, must have a boundary. So, without boundaries, the strong centres, which are described in the previous section, could not exist. This is shown in Fig. 5.8, where cuts in the boundaries determine the changes in centre points.

Considering mathematics, boundaries are strongly related to geometrical forms. These start with a single point. The relation between two points can be defined as a line. A line can be further developed by implementing another point so that a triangular shape is created. This triangle transforms to a rectangle which transforms into a polygon by implementing new points within the total composition. The final form can be related to the base shape of the circle. So, every bounding area can be related to this circular base form. This effect is shown in Fig. 5.9.

Conclusion

Boundaries and centres have a strong relationship. In fluid architecture one space is merged with another, therefore it is not clear where the space begins or ends. In many cases, the ceiling, wall and floor are merged. In the three fluid buildings discussed here, the irregular subdivision of building parts and the irregular curvature of the building surfaces makes it impossible to find a centre and a simple well-defined boundary. In the Web of North Holland there is only one space, therefore it might be possible to determine a central area but with the two other buildings the combination of different spaces and forms is too complex. Fluid architecture will disorder and create restless "moving" spaces, therefore it is assumed that fluid architecture has to avoid elements that will structure the building, so strong centres and clear boundaries will be avoided. The *Gestalt* law of closure ensures that even in areas that are not clearly defined the human mind imagines an enclosed space with a kind of imaginary centre. Humans have a natural inclination to look for centres or even to imagine a

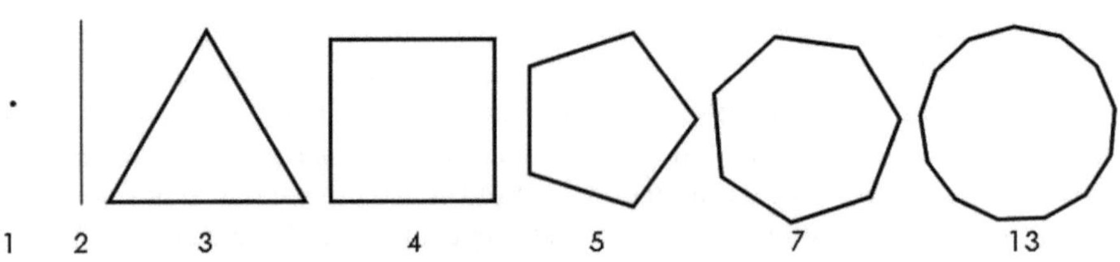

Fig. 5.9 Polygon boundaries

centre. It is assumed that if the centre is not found people will look for a new centre point. Fluid architecture is using this human characteristic to suggest the idea of movement and fluidity.

5.3.4 Alterating Repetition

The relation between centres can be intensified through the use of repetition. A certain rhythm of a repeating centre intensifies the field effect of the total composition. This effect is not, however, just a simple repetition. There are two basic characteristics when concerning repetition: the basic element and a defined action. Alexander states that when a single element is literally being repeated, a boring image appears. He thinks that the element should alterate in order to create ‘life’ in a composition. The effect of this difference is shown in Fig. 5.10. In the upper series the element is being repeated by one and the same action over and over again. The two lower series are undergoing an alterating action which results in a change of the element’s shape.

Compared to mathematics this will result in a series that alterates by undergoing the action. For example, the series of $f_n = f_{n-1} + 1$ with $f_0 = 0$, which results in the series 0, 1, 2, 3, 4, 5, ... n + 1. This is a much more interesting series than $f_n = f_{n-1}$, resulting in the series with a result equal to f_0.

Conclusion

Fluid architecture is by definition irregular. Within the three fluid buildings there will be no repetition or alterating repetition in a formal way. But on a different level, the different panels are a variation on a theme and have a strong coherence. This aspect is not of great importance to the definition of fluid architecture.

Fig. 5.10 Alteration of the element by undergoing an action, the first line shows a non-alterating repetition ($f_n = f_{n-1}$). At the lines below different actions are shown of change like lowering and narrowing ($f_n = f_{n-1}$ + action)

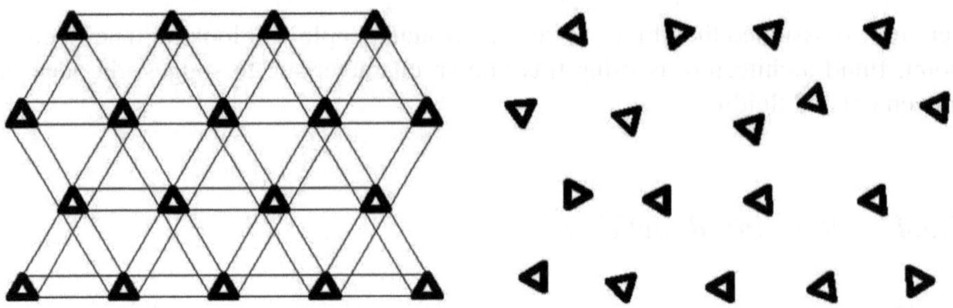

Fig. 5.11 On the left, the resulting space has a strong relation to the mass, whereas on the right it doesn't

5.3.5 Positive Space

The void which results from the positioning of mass has its own identity due to its usefulness and the relationship with the surrounding mass. Considering architectural beauty, positive space is difficult to define.

When reviewing positive space in a two-dimensional way, positive space can be the result of translation, rotation and minimal transformation of the negative space: the mass. This effect is shown in Fig. 5.11. Alexander states that the interaction between the mass and the void results in the fact that every bit of space is very intensely useful and that there is no leftover space which is not useful. So, a positive space has a very strong geometrical relation with the mass from which the space is the resultant.

Conclusion

In the three buildings discussed here, there is a loose relationship between mass and space. Not every bit of space is very intensely useful and there is enough left over. Considering Alexander, fluid architecture should be qualified as buildings with lots of negative space.

5.3.6 Good Shape

A shape can be built up from elements which all have strong centres. Every element should have its meaning for the final goal which the total composition wants to achieve. The centres of the several elements are very important in this way, because they affect, support and strengthen each other in order to create a good shape. When elements are missing or dislocated within the total arrangement, this has a negative effect on the final composition.

Mathematically, this creates a problem, because the simple sum of all the elements is not necessarily the right way to achieve an optimal result. This means that the sum of all the elements should result in a bigger outcome than the expectation. Mathematically, this is not possible ($1 + 1 \neq 3$) and it is therefore impossible to define a good shape in an objective manner.

Conclusion

The three buildings we discuss do not have very specific centres. Following the theory above it is impossible to qualify them as well shaped.

5.3.7 Local Symmetries

According to the theory of Christopher Alexander, 'life' can only occur when symmetry is present on several levels of scale. Symmetry in only one specific element is not enough to bring 'life' in general. It only appears when several elements, which all have an own symmetrical axis, are placed together to form a total composition. By using symmetry, new and stronger centres occur, which brings 'life' to the whole.

This mirror effect is a result of mathematical beauty which is very easily recognized by the human brain. The symmetry of elements attracts the human eye to the central point where the mirroring takes place, so that the new centre is being highlighted. Mirroring can be done in relation to a point, line or surface. This effect is shown in Fig. 5.12 in which a simple line is mirrored twice and creates a strong image.

Conclusion

The Web of North Holland and the Guggenheim buildings are without symmetry. The Kunsthaus has only a few symmetries in very local elements (for instance the tubes on the roof). The three pictures above show that with the mirroring of any line it is possible to create a strong and obvious centre. Fluid architecture tries to avoid strong centres (Alexander, 2nd property), therefore only very local symmetries are allowed.

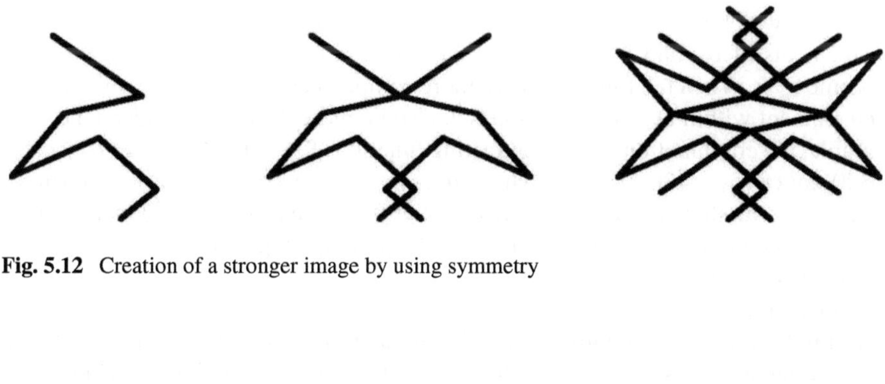

Fig. 5.12 Creation of a stronger image by using symmetry

5.3.8 *Deep Interlock and Ambiguity*

This property is concerned with the interrelation between two or more elements, which react to each other's shape to create a new unity. Space that occurs between the elements forms a new centre and gets a function as mediator between the elements of the new whole. They also react to their immediate surroundings so that the new unity is capable of being adapted by its environment. Within *Gestalt* psychology this is called the multi-stability, emergence or reification (King 2005).

Conclusion

Inside the Web of North Holland, there is just one space but in the other two buildings there is a continuous interlock between spaces. This interlock is so deep that spaces will morph into each other. The effect demonstrates that fluid architecture is without clear centres.

5.3.9 *Contrast*

When several elements of different shape, texture or colour are combined in one composition, their contrast results in a more comprehensive and unified whole. The contrast within a composition will be strengthened when the specifications of the elements differ more extremely, just like an image with only black and white surfaces is a simplified image. However, more differentiation between the elements generates a more complicated image in which each individual effect on the total image becomes harder to understand. Like a greyscale image, showing more detail, but making the image more complex. A clear contrast between the participating elements should create and allow a clear differentiation in order to create a unified image.

It can be stated that there is a relation between contrast and the mathematical subject of equations. Equations are compositions in which the participating elements all have their own function in order to create a result. When the elements of an equation have a clear relation to each other this will also result in a clear perception of the total function. For example, the combination of the contrasting functions $f_{(1)} = x^2$, $f_{(2)} = 400$, which gives $f_{(1,2)} = x^2 + 400$ as shown in Fig. 5.13. The relation between the elements of $f_{(1)}$ and $f_{(2)}$ gives a clear perception of their influence on the final outcome $f_{(1,2)}$. When the function gets a more complicated character, the influence of each element within the total function will also be more difficult to understand. For example, the combination of the contrasting functions $f_{(1)} = 1/3x^2$, $f_{(2)} = \surd(3x)$, $f_{(3)} = 300\sin(x)$, $f_{(4)} = 400$, $f_{(5)} = x^2$, which gives $f_{(1,2,3,4,5)} = 1/3x^2 - \surd(3x) - 300\sin(x) + 400 + x^2$, as shown in Fig. 5.14. Because of the more differentiated character of this function, the relation and the effect of the participating elements on the whole is much harder to understand.

Conclusion

In fluid architecture 'contrast' must be divided into, first, the contrast with the surrounding buildings and, second, the contrast between the different elements. The

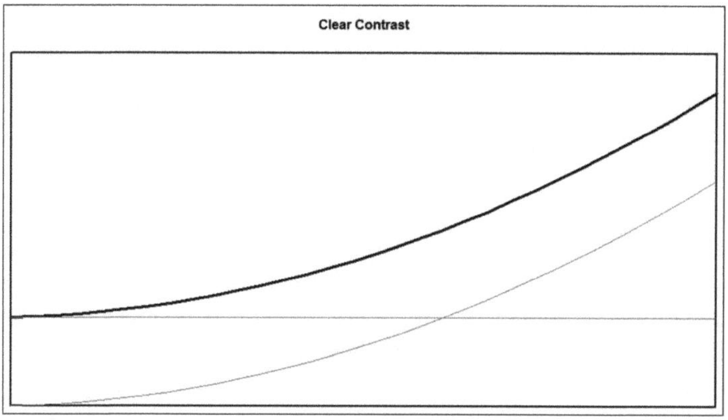

Fig. 5.13 Relation in the function: $f_{(1,2)} = x^2 + 400$ (dark line)

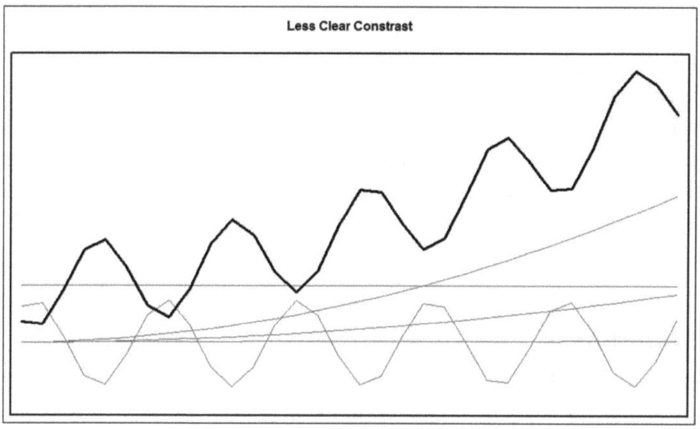

Fig. 5.14 Relationships in the function: $f_{(1,2,3,4,5)} = 1/3x^2 - \surd(3x) - 300\sin(x) + 400 + x^2$ (dark line)

three buildings we discuss here have a strong contrast in materials and shape with the surrounding buildings. For fluid architecture this contrast can be a positive element that will support the special position of the building or place in the city. The contrast between the elements is less obvious. There is a deviation in form but within a set of "rules". If the material is changed, the set of rules changes, too. If different elements are merged, there will be a less clear or almost no contrast. How to deal with the issue of contrast depends very much on the individual design of the building. In the three reference buildings the contrast with the surrounding buildings is immense. In the Web of North Holland, the contrast between different elements is limited to the kind of curvature and shape of the elements. The Kunsthaus and the Guggenheim museum Bilbao can be divided into sections with a curved surface and those with a plane surface. The curved sections use other materials than the plane sections.

Within the sections, the contrast between different elements is limited to the kind of curvature and shape of the elements.

5.3.10 Gradients

Qualities vary, slowly, subtly, gradually, across the extent of each element. This gradual change is comparable with the series described in the paragraph concerning alterating repetition. A gradient is a mediator that slowly changes the appearance in a certain direction and with a certain regularity.

Considering mathematics, a gradient shows remarkable similarity with mathematical functions. The mathematical concept of a function expresses the interdependence of two quantities; one which is given by the input which produces the output. Considering this function, it is of great importance that it must not be constant, but must change gradually. The function itself determines the rate, intensity and direction of the gradient. The tangent of the function is very important in defining these characteristics of the gradient at a certain position. When the tangent of a function is not zero, the related direction of the function describes a gradient.

Conclusion

In the three reference buildings, a gradual change is present in all elements. Of course, it depends on the design of the building but it is used in every example of fluid architecture that is divided into building elements.

5.3.11 Roughness

According to Christopher Alexander, roughness can be defined as a predefined grid which has certain imperfections which are related to the pattern. These imperfections generate a local disturbance within an image and so develop a certain tension in the overall field. In this way, the imperfections draw the attention of the spectator and bring 'life' to a composition. The effect of imperfections within a field is shown in Figs. 5.15, 5.16 and 5.17.

With reference to mathematics, this effect can be related to deviation. The deviation can be widely spread around one value or very near to a value. If the deviation has a greater spread it will be perceptible. The *Gestalt* law of *Prägnanz* is used in architecture to connect separate elements and lines to one object. *Prägnanz* is also used to adapt deviations in size and order. Not all visible deviations will be consciously perceived.

Alexander sees perceived deviation as a positive thing that will add "life" as long as the imperfections are limited and placed within a grid. Vastert (Chap. 4) sees it as something that can be tolerated as long as it is not perceived. The pattern aspects within fluid architecture (a high irregularity, number of elements and width of join) will allow great deviations without these being perceived. The surface aspects

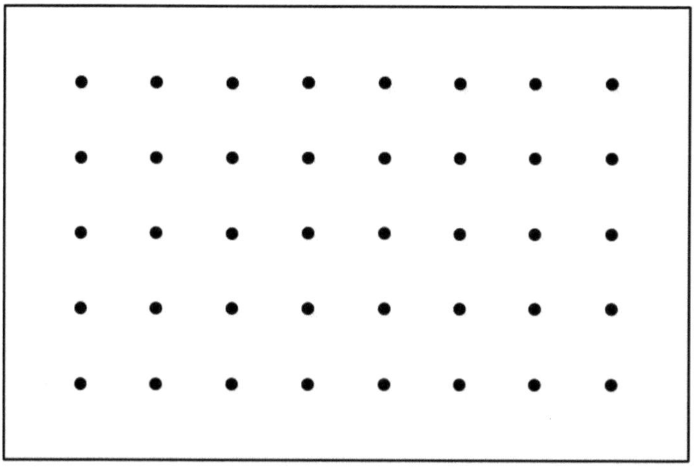

Fig. 5.15 Points in a regular grid

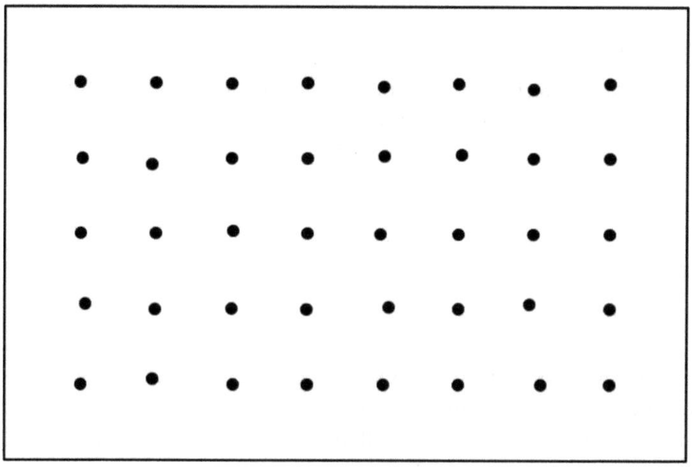

Fig. 5.16 Disturbance of the regular grid of Fig. 5.15 by shifting certain points, resulting in a livelier composition

(transparency, smoothness etc.) within fluid architecture have an opposite effect. Within fluid-architecture all the elements are different in size and curvature (cf. 10. Gradients). The grid and surface aspects of the elements are so strong that all the elements are still recognized as being part of a grid. A gradual change in size and curvature is accepted as long as the deviation in size and shape will not distort the grid Roughness, defined by Alexander as a grid which has certain imperfections which are related to the pattern as part of the grid, and will not contribute to the fluid character of fluid-architecture if it is perceived as being an imperfection. The roughness according

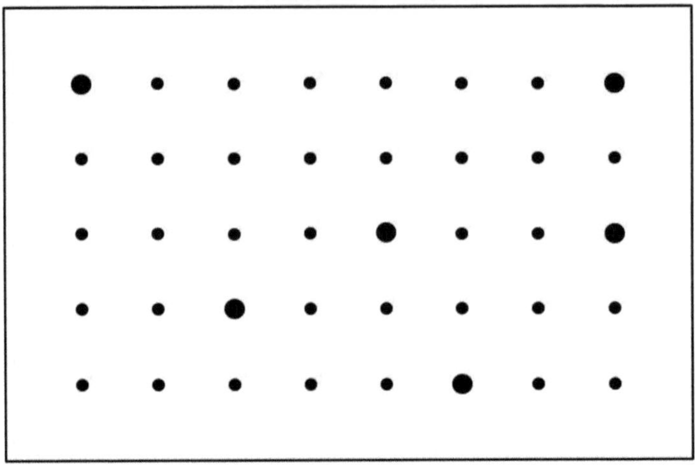

Fig. 5.17 Disturbance of the regular grid of Fig. 5.15 by scaling of certain points, resulting in a livelier composition

to Alexander will disturb the dynamic character of fluid architecture. The research by Vastert gives rules to prevent roughness. Those rules must be followed to avoid perceived deviations although the pattern aspects of fluid architecture allow great deviations before they are perceived as such.

5.3.12 Echoes

An echo can be defined as a repetition of a base object with a possible alteration. An important aspect considering echoes is the geometrical reference to a base shape. Alexander states that echoes depend on angles and families of angles which are prevalent in a design. In general terms, there is a deep underlying similarity among the elements of a composition which generates the perception that everything seems to be related.

The relation between echoing elements in a larger whole can be the result of a transformation by mathematical aspects like translation, scaling and rotation or a combination of these as shown in Figs. 5.18, 5.19, and 5.20. Note that every transformation results in the same geometrical shapes as present in the previous one.

Conclusion

In the three buildings there are no echoes. In the Kunsthaus there is a repetition of light tubes. In the other two buildings there is no repetition at all. In all buildings there is a *gradient*. The Selfridges building by Future Systems in Birmingham proves that repetition is possible. Although I have not come across an example, echoes must be possible within fluid architecture. This is proved by Fig. 5.8 with packaging on a doubly curved surface.

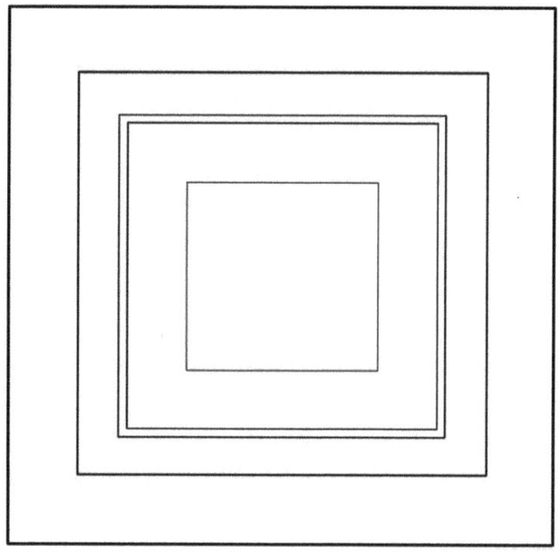

Fig. 5.18 Repetition by scaling

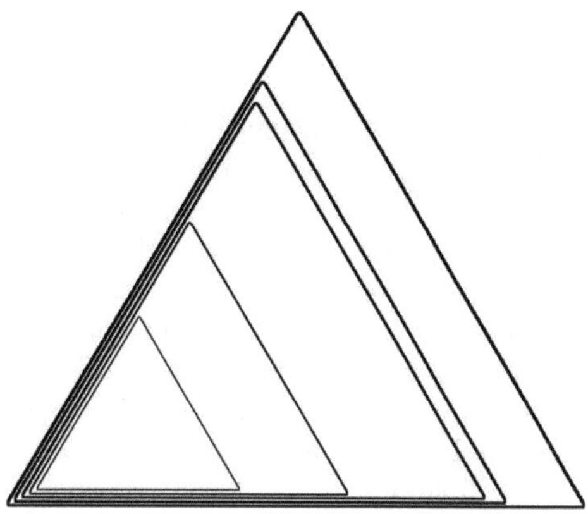

Fig. 5.19 Repetition by scaling and translation

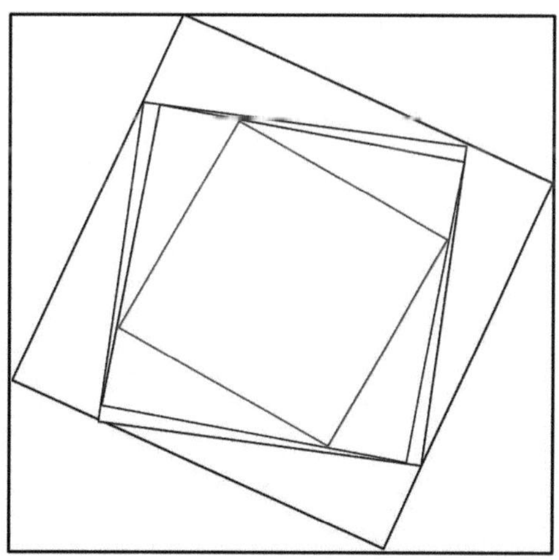

Fig. 5.20 Repetition by scaling and rotation

5.3.13 *The Void*

The void is used to invite dynamism (activity) through the use of empty space. Surrounding structures form inner spaces and the absence of objects in these results in a space with the opportunity to develop activities. A void is necessary in every building, it results in a place of rest and calm. The surrounding fabric determines the feeling of the void, yet it is important that the void has enough space to exist and to be used. It is impossible to relate this property to a mathematical aspect.

Conclusion

Fluid architecture will use the void to connect the different parts of the building and to create interlock and ambiguity. The Web of North Holland has only one room but in the other two buildings the spaces are connected by a void. (Fluid) architecture without a void is possible but always less interesting.

5.3.14 *Simplicity and Inner Calm*

In order to create understandable architecture, the use of simple geometrical shapes is necessary. The proportions of these geometrical shapes need to be unusual. The shapes will be the result of an unusual parameter like length, size, width, etc. It is also important that there are few imperfections within the total composition, although they are necessary to create a living whole as explained in the paragraph about roughness.

Mathematically speaking, this means the use of understandable shapes, like squares and triangles that must have one unusual parameter which will determine the composition. An example which shows the result of simplicity and inner calm is shown in Figs. 5.21 and 5.22.

Conclusion

Fluid architecture likes to suggest movement. It is by definition complex and without inner calm.

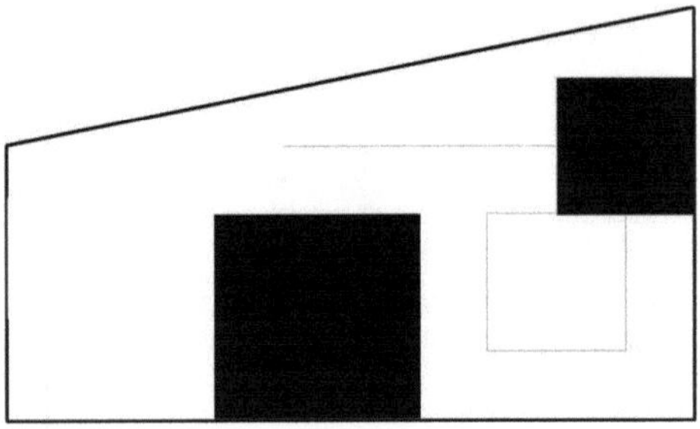

Fig. 5.21 Without inner calm and

Fig. 5.22 With inner calm

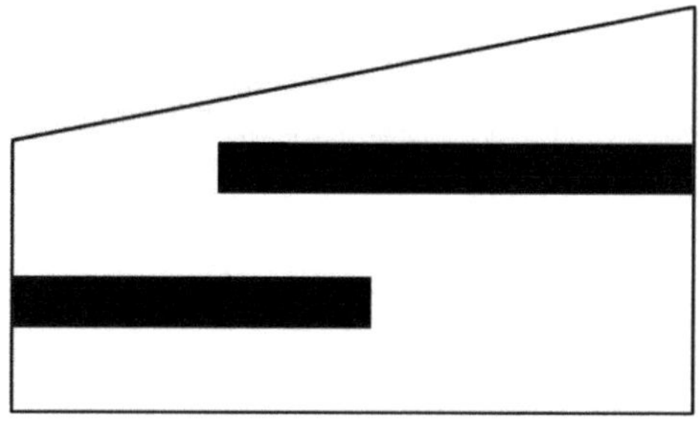

5.3.15 Not-Separateness

A shape can be considered not-separate, when it is experienced as one with the world and not separated from it. This means that any centre which has 'deep life' is connected, in feeling, to what surrounds it, and is not cut off, isolated, or separated. According to Alexander, this last property about not-separateness is the most important property of all the properties described before. He states that a shape cannot have 'life' when it is not unified with its surroundings, even though it should contain the other fourteen properties within its design. Considering not-separateness in mathematical terms, this property cannot be defined in an objective way.

Conclusion

Fluid architecture is by definition rebellious and therefore separate from the surrounding environment. It needs its surroundings to be unique. A clear contrast with its surroundings will make the building even better.

5.4 Conclusion

Concerning the statement of Alexander, the fifteen basic properties above that are necessary to create lively spaces should be positive for architecture in general. For fluid architecture 8 of them are negative and 3 of them are positive. This big difference gives a clear profile to fluid architecture and may explain the strong opinions held by the public towards fluid architecture (Fig. 5.23).

Fluid architecture uses the *Gestalt* laws for the interpretation of elements by the viewer to predict what will be perceived. Fluid architecture leads to dynamic spaces related to each other by voids. It uses differences in height, size, scale and shape, generally built with materials that will be perceived as smooth. Smaller elements are collectively interpreted as one bigger surface.

this poperty is used and important for the appearance of this building. ++				
this property is used but not essential for the appearance of this building. +				
this property is not used but not essential for apearance of the building if it was used -				
this property is not used and will change the appearance of the building in an essential way if it was used --				
	fluid			
Fifteen basic properties by C. Alexander	The Web	Gugenheim	Kunsthaus	fluid architecture
1. level of scale,.	+	-	-	+-
2. strong centers,	--	--	--	--
3. boundaries,	--	--	--	--
4. alternating repetition,	+	+	+	+-
5. positive space,	--	--	--	--
6. good shape,	--	--	--	--
7. local symmetries,	--	--	--	--
8. deep interlock and ambiguity,	++	++	++	++
9. contrast,	-	+	+	+-
10. gradients,	++	++	++	++
11. roughness,	--	--	--	--
12. echoes,	--	--	--	+-
13. the void,	+	++	++	++
14. simplicity and inner calm,	--	--	--	--
15. not-separateness	--	--	--	--

Fig. 5.23 Table: fifteen basic properties of architecture related to fluid architecture

5.5 Defining Fluid Architecture

At the end of this chapter, it is possible to combine the conclusions of Chaps. 3, 4 and 5 and to come to a final definition of fluid architecture.

In this research fluid architecture is qualified as:

1. Irregular (Sect. 3.1).
2. Being perceived as smooth (Sect. 4.4).
3. Being perceived as doubly curved (Sect. 5.2).
4. A building with Utilitas, Firmitas and Venustas (Sects. 3.1 and 3.8).
5. A multifaced body with a combination of perceived synclastic and anticlastic surfaces (Sect. 2.6).
6. Avoiding strong centres.
7. Having no clear boundaries.
8. Enclosing an undefined space.
9. Having an undefined shape.

10. Having main boundaries without symmetry. If symmetry is found it is a local symmetry.
11. Merging different volumes. If the size of the building is big enough the different spaces within the building are connected with a void.
12. Having a complex geometry.
13. Presenting a unique object, there is no relation or a very limited relation with other buildings on that site.
14. Having smaller building elements, gradually changing in size and shape, that do not draw attention and that contribute to a building surface of a bigger size (Sect. 5.3.11).
15. Having a perceived fluidity of the surface that refers to organic forms (Sect. 3.8).

This more precise definition of fluid architecture shows where this type of architecture is different from other types of architecture. The approach of Alexander is only used to categorize the different aspects of architecture in a framework and to clarify the differences. In the chapters above form-active material and fluid architecture have been researched and defined. The technique and material of the building elements made with flexible moulding will be researched in Chaps. 6–11.

Bibliography

Alexander C (2002) The nature of order, book I: the phenomenon of life. The Centre for Environmental Structure, Berkeley, pp 144–242

Fontana L (2012) Fernand Fonssagrives photography: the sensual touch, Modern Design. http://www.moderndesign.org/2012/04/fernand-fonssagrives.html. Accessed Dec 2012

Hernández Hernández JM (1999) Selfridges Birmingham. Future Systems, UK, pp 1999–2003

King BD (2005) Max Wertheimer and Gestalt theory. Transaction Publishers, Piscataway

Koffka K (1999) Principles of Gestalt psychology. Psychology Press, Hove

Salingaros NA (1995) The laws of architecture form a physicist's perspective. Phys Essays 8(4):638–643

Wertheimer M (1912) Experimentelle Studien über das Sehen von Bewegung. Zeitschrift fur Psychologie 61:161–265

Wertheimer M, Spillmann L (eds) (2012) On perceived motion and figural organization. MIT Press, Cambridge, MA

Chapter 6
Structure

6.1 Structure Systems

In the classification of structural elements according to Büttner and Hampe (1985) and Schodek (Schodek 2000) the criteria of rigid vs. non-rigid and line elements vs. surface elements are used. Heino Engel uses the criteria of structural behaviour. Engel offers 4 criteria:

Form-active (Fig. 6.1),
Vector-active (Fig. 6.2),
Section-active (Fig. 6.3) and
Surface-active (Fig. 6.4).

From structural behaviour Engel moves to typology and from there to the integration of a structural typology into an architectural concept. A comparison of structural systems starting with the criteria of structural behaviour seems the purest; therefore, I will use this classification by Heino Engel. Nevertheless, I like to make some comments on the classification of Engel in relation to the aspect of rigid vs. non-rigid, the capability to resist bending moments and the criteria of height. Height-active is a combination of surface, section and sometimes vector-active behaviour and not a new typology in itself. Therefore, only the first four criteria of Engel are used in this research. Moreover, the fifth criterion (height) is not mentioned in (Veltkamp 2007: 19–21), which confirms my assumption that this criterion is of secondary importance. The aspect of rigid vs. non-rigid and the capacity to resist bending moments will be discussed below.

The distinctions between structural typologies by Heino Engel are used in this research as a theoretical basis to compare different structures (Fig. 6.5).

It is possible to compare the 4 typologies by H. Engel on different aspects like (Fig. 6.6):

1. Forces (bending moments, tension, and compression).
2. The ability to construct the structure on site or to prefabricate.
3. The ability to make two-dimensional and three-dimensional structures.

A. Pronk, *Flexible Forming for Fluid Architecture*,
https://doi.org/10.1007/978-3-030-71551-9_6

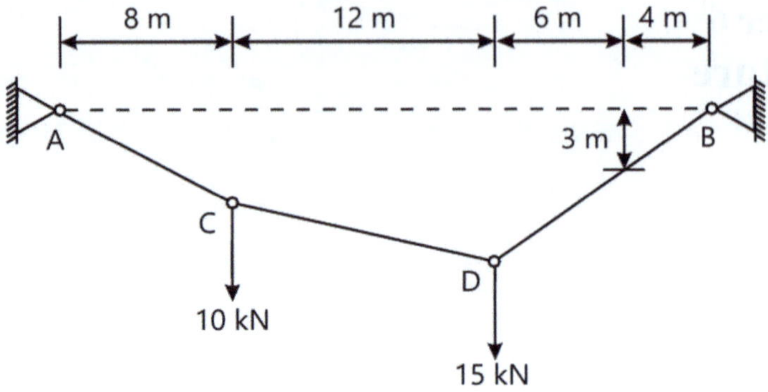

Fig. 6.1 Catenary (form-active)

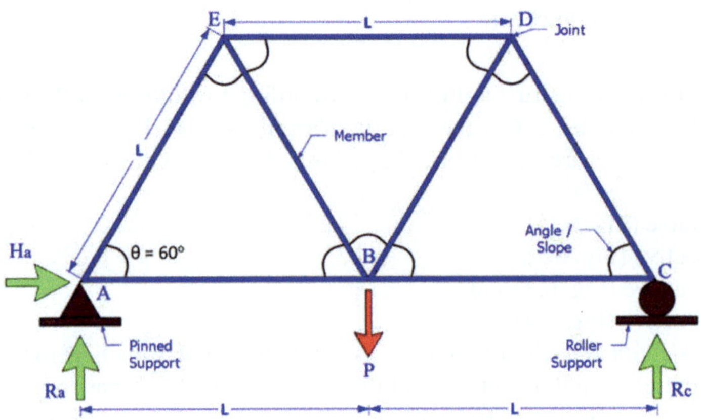

Fig. 6.2 Truss (vector-active)

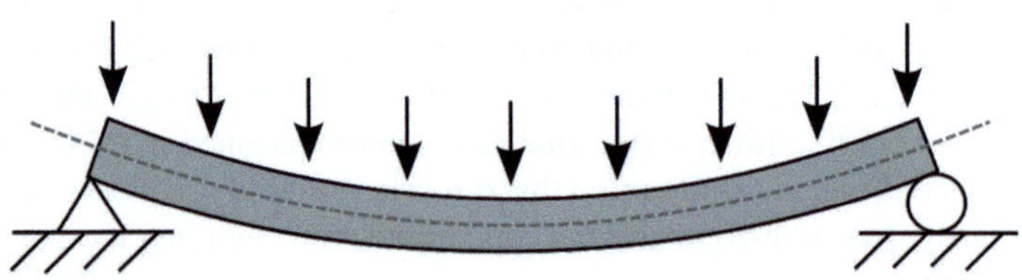

Fig. 6.3 Beam (section-active)

4. The form of the structure: 2D—3D and the form of the surface: zeroclastic, monoclastic, anticlastic, synclastic.
5. Applications.

The diagram above follows the typological deviation by H. Engel as published in his book Structure Systems (Engel 1999). According to Engel, vault structures, thrust

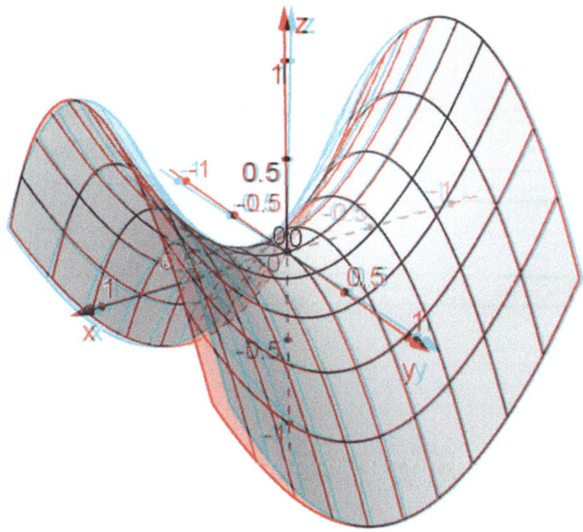

Fig. 6.4 Shell (surface-active)

Structure family		Definition	Structure type
1	Form Active	are systems of flexible, non-rigid matter in which the redirection of forces is effected by a self-found FORM design and characteristic FORM stabilization.	1.1 cable structures 1.2 tent structures 1.3 pneumatic structures 1.4 arch structures
2	Vector Active	are systems of short, solid, straight, linear members (bars), in which the redirection of forces is effected by VECTOR partition, i.e. by multi-directional splitting of single forces.	2.1 flat trusses 2.2 transmitted flat trusses 2.3 curved trusses 2.4 space trusses
3	Section Active	are systems of rigid, solid, linear elements in which the redirection of forces is effected by mobilization of SECTIONAL (inner) forces.	3.1 beam structures 3.2 frame structures 3.3 beam grid structures 3.4 slab structures
4	Surface Active	are systems of flexible, but otherwise rigid planes resistant to compression, tension, shear, in which the redirection of forces is effected by surface resistance and particular SURFACE form.	4.1 plate structures 4.2 folded plate structures 4.3 shell structures

Fig. 6.5 Table: Classification of structure systems by Engel

lattices and arches are form-active and shell structures are surface-active. So not only membrane structures and cable structures but also funicular structures like arches. Engel describes vault structures and thrust lattices as three-dimensional funicular arches (Engel 1999: 120–131). If the joins behave like flexible hinges the structure can be seen as form-active. In practice the joins are not flexible and will be able to resist moments. The large number of non-flexible hinges in the direction of the surface will make them behave like a shell structure resisting high moments in the direction of the surface of the structure. Surface-active structures have the capacity to resist high moments in the direction of the surface of the structure. Vault structures

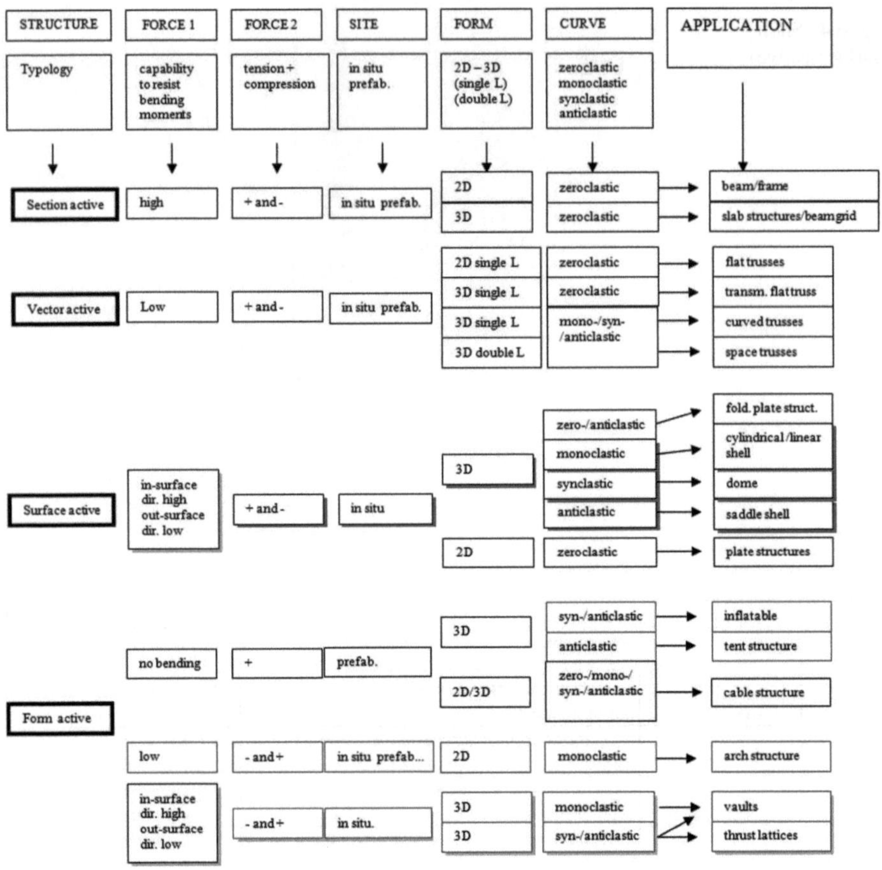

Fig. 6.6 Comparing structural typologies of Heino Engel

and thrust lattices have the same capacity and have the structural behaviour of a shell-structure. Therefore, it is better to characterize vault structures and thrust lattices as surface-active. Pure funicular arches (Engel 1999: 112–117) stay within the form-active typology. Contrary to the other form-active structures, they are made of rigid materials capable of resisting compression where the other structures can only resist tension and have a kind of flexibility within the material.

Within this research the focus is on the transfer from non-rigid structures to rigid structures. Rigid arches will fall outside this focus and flexible only is not enough. The word flexible is used in a broad way for all kinds of adjustable building systems, for instance kinetic structures, like scissor structures, which behave in a totally different way. To be precise, the word flexible is combined with a form-active material. All flexible materials will behave in a form-active way. Therefore, the word form-active can be dropped and it is enough to speak of flexible moulding as was done in the title of this research.

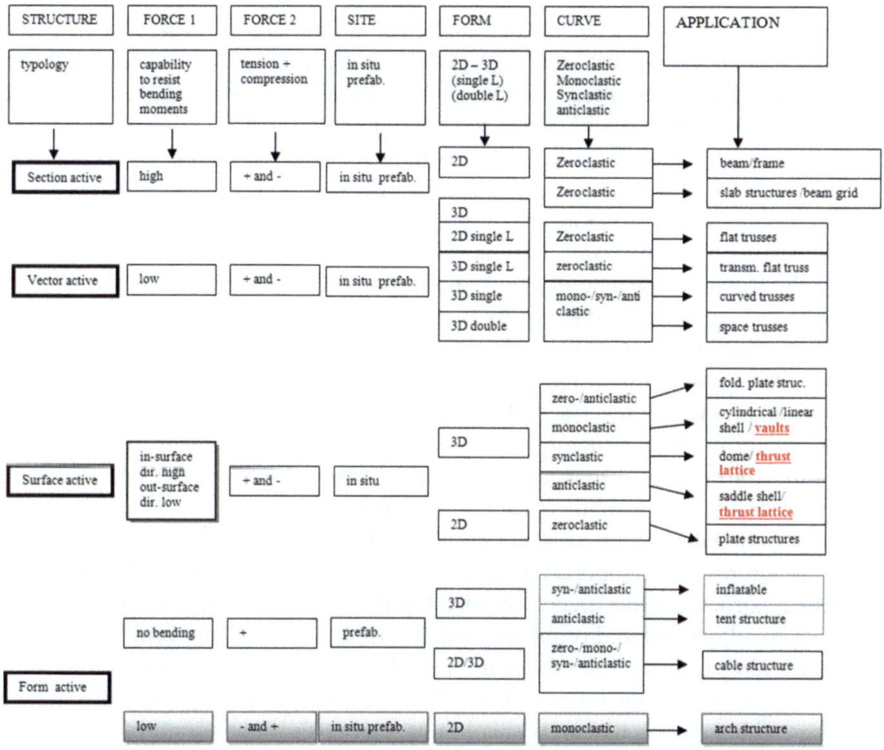

Fig. 6.7 Diagram based on Structure Systems by Heino Engel, adjusted for vault and thrust lattice structures because they have been categorized as surface-active. Arch structures are form active but form a special category as a result of the rigidity of the material and the capacity to resist compression

As a result of the conclusion above, the diagram according to the typological deviation by H. Engel is adjusted to the diagram below (Fig. 6.7).

6.2 The Relation Between Structure and Fluid Architecture

Section-active and vector-active structures are the most common structures. It is possible to curve a section-active structure for architectural reasons. In that case there is no relation between form and structural behaviour. Section-active structures with a curvature are material, labour and machine intensive. To produce a curved structure is within the western economic proportions between labour, material and machine cost almost always more expensive. Most of the manufacturing methods such as extrusion, casting, rolling, etc. for making section-active elements are focused on the production of straight linear elements. To curve these products during the manufacturing process will always require extra efforts and investments. To curve straight elements afterwards is always an extra effort. In the structural nature of section-active structures there is no reason to curve a structure and for economic reasons it is cheaper to avoid curvature. Section-active structures also have a low material efficiency. Fluid architecture is by definition curved therefore it is better to avoid section-active structures in fluid architecture (Sects. 1.3, 3.12, 5.5).

With vector-active structures it is possible to approach curved structures in an efficient way. For example, this is done in the construction of the Statue of Liberty, the Bilbao Guggenheim (Sects. 3.11.2 and 4.2) and projects like the "FrankfurtHochVier" by Helbig and Knippers (Knippers and Helbig 2009: 111–126). Structurally speaking, it is no problem to make fluid architecture within this structural typology. The problem is in the gap between the non-fluid vector-active structure and the fluid architectural surface of the building (Sect. 3.12).

Flexible form-active structures are very efficient in their use of material. The form of these structures is dictated by the structural behaviour of the structure. The membrane structures are highly optimized in a structural way but have problems with for instance the building physics (sound, heat). Therefore, they are mostly used for special structures like temporary structures, and outdoor structures, like canopies. "Funicular" structures like arches are also highly optimized in a structural way but are easier to adjust to avoid problems with building physics. The form of all form-active structures is dictated by their structural behaviour, therefore the architectural possibilities in shaping the form are limited. The combination of flexible form-active structures with supports like arches, columns, etc. gives more possibilities to influence the form of the structure (Sect. 2.9). The more supports, the more possibilities there are.

Surface-active structures have better possibilities to solve building physics than flexible form-active structures. Surface-active structures have a doubly curved skin and are easy to combine with section-active and vector-active structures. There is more freedom in shape than with form-active structures. Surface-active structures are usually made on site, but it is possible to divide the shell up into pieces that can be prefabricated. The use of prefabricated elements in shell construction goes back to 1940, when Pier Luigi Nervi used prefabricated lattice ribs for airplane hangars in Italy. In 1958, Nicolas Esquillan used prefabricated internal shear walls for shells in Paris, Grenoble and Turin. In 1957, Ilia Doganoff constructed a fully prefab shell in Bulgaria. The shell by Doganoff consisted of a series of conoid shells, each of them formed out of four prefab elements which rest on prefabricated beams. A second example is the 1961 shell by Heinz Hossdorf in Switzerland. Other works in the 1950s and 1960s using prefabricated elements are the 1958 Philips Pavilion by Le Corbusier and Xenakis (Pronk et al. 2007), the 1963 Cinerama Dome in Hollywood California, the roof of which comprises 316 mostly hexagonal precast concrete panels in 16 patterns, each weighing some 1,500 kg, and a Russian system for the production of cupolas (Sanchez-Arcas 1961).

Recent examples are a cupola in Australia built with the experimental Binix system (Bini Systems, www.binisystems.com/binix.html). This system has triangular prefabricated elements on an inflatable mould to bring them into position. Another project is the Duxford Imperial War Museum (1997) by Ove Arup and Partners.

If several shells with the same shape will be made, the mould can be made next to the building site. In that case the shell will only be lifted to its final place like the cylindrical shells for a sports hall in Daly California USA (Sanchez-Arcas 1961) and the Stuttgart Federal Garden Fair shell (1977) by German engineer Jörg Schlaich. This example of very thin large precast shell elements consisted of glass reinforced concrete with a thickness of only 12 mm and combined, they spanned a circular area with a diameter of 26 m. The (2005) shell near Millau, France consists of prefabricated curved elements of self-compacting high strength fibre-reinforced concrete with a total length of 28 m and a maximum height of 850 mm, but with a hollow core. The skin is only 100 mm thick. The elements are put side-by-side on a temporary framework and then the elements are pressed together by longitudinal prestressing ties. They are brought into position by a crane and placed on temporary scaffolding. The elements are made on the building site due to their sizes.

With respect to prefabrication, the question is how to apply prefabrication: as a whole in large parts, or in very small easily transportable parts. In general, the aim is to have a maximum of repetition with a minimum of joins and work on site. The work on site is based on transportation, erection, and the measurements of the tolerances and the joins to be used. Obviously, they are closely interrelated. Except for the few examples mentioned, surface-active structures are mostly made in one piece at the building site on a complex three-dimensional mould.

In fluid architecture the surface is by definition irregular, therefore all the elements within a building concept are unique. There are two ways to tackle this potential problem:

- to find an irregular formwork that is affordable on the building site.
- to find a way to prefabricate and assemble unique building elements.

Considering the behaviour of materials with flexible and form-active properties, it is assumed that those materials can be used to make irregular formwork on the building site or to prefabricate and assemble unique building elements. The research below and in the following chapters will prove that materials with flexible and form-active behaviour are used for both ways.

The possibilities for using flexible and form-active structures as a mould to make other typologies are illustrated in the diagrams below. Form-active structures and surface-active structures have a strong similarity with one exception, the difference in construction: in situ vs. prefabricated off site. The ability of form-active structures to be the prefabricated mould for surface-active structures is known and practiced in the fabrication of domes and shells. Wallace Neff (1942) received a patent in 1941 for a system by which the form was inflated to the shape of the structure, and the reinforcing bars and shotcrete were placed on the exterior of the inflatable form. Many domes with inflatable formwork have been built all over the world by companies like Binishell (Binishells, www.binishells.com) (Italy), YSM for Building (Y.S.M. For Building, www.domecrete.com) (Israel), Pirs SA (Domepirs, www.domepirs.com)

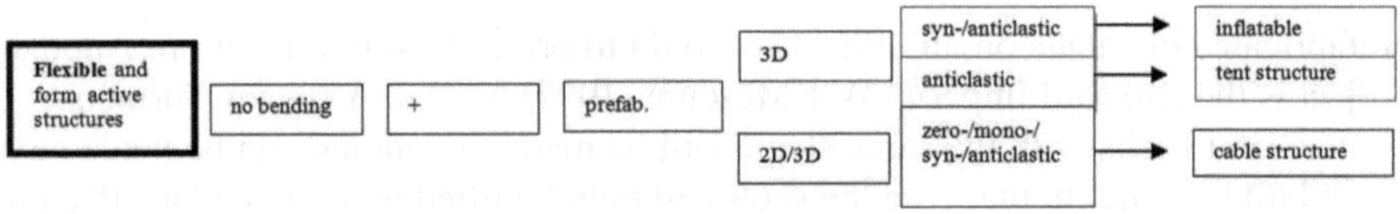

Fig. 6.8 Flexible and form-active structures

(France) and Monolithic Dome (Monolithic, www.monolithic.org; Hennik 2005; West 2016). Also mechanically stressed membranes have been used for shell structures. James Hardress de Warenne Waller (1884–1968) developed the corrugated concrete shell structure known as the Ctesiphon construction in 1941 for the first time. The experience in making shell structures with a membrane mould is the starting point for this research to the construction of fluid architecture. A complete overview of the innovations for rigidizing flexible and form-active structures can be found in Chapters 7 Induction, 8 Composites, 9 Concrete, 10 Glass and 11 Ice.

The high cost of the mould is the most important reason for architects, clients and structural engineers to avoid this typology. S. Bhooshan and M. El Sayed of Zaha Hadid Architects Computation and Design Group compared the construction of a shell structure with fabric formwork in India with a shell made with a "waffle" formwork in Mexico (Bhooshan and El Sayed 2012). In their conclusion they preferred the fabric formwork because "it offers more possibilities for such an intuitive, physically-based architectural design framework than other, rigid, formworks". In the construction of a shell designed by M. Fuksas in Eindhoven, fabric formwork was compared with a wooden mould for economical optimization. In general, a membrane mould turns out to be 35% cheaper (van Dessel et al. 2003: 1–10). On the construction of a staircase in Vergennes, Vermont, A. Lawton concluded: "the adaptability and lightness of the fabric and forming system were ideal for this steep, difficult to access site. It required less energy and cost to construct than a conventional form system, with the added advantage of a unique aesthetic" (Lawton and Miller-Johnson 2012) (Fig. 6.8).

Flexible and form-active structures can be used as a mould for all kinds of structures. The most important characteristic of this moulding technique is that it can be prefabricated and easily transported to the building site.

The diagrams below show all the theoretical options of flexible and form-active structures to be used as a mould. In those cases, the typology of the structure is changed into another typology with one exception, of the inflatable becoming an arch, in which case the typology stays within the form-active domain. In the chapters about materials there will be examples of all the known combinations.

The diagram below shows the different possibilities of cable structures (form-active) to be the mould for all kinds of trusses and (vector-active structures) and thrust lattices (surface-active).

The diagram below gives the different possibilities of tensile structures (form-active) to be the mould for all kinds of shell structures like folded plate structures and saddle shell structures (surface-active structures) and slab structures and beam grids (section-active).

The diagram below gives the different possibilities of inflatable structures (form-active) to be the mould of all kind of arch structures (also form-active structures) and dome structures, and shell structures (surface-active) and beam, and frame, and slab, and beam grid structures (section-active) and all kinds of trusses (vector-active) (Figs. 6.9, 6.10 and 6.11).

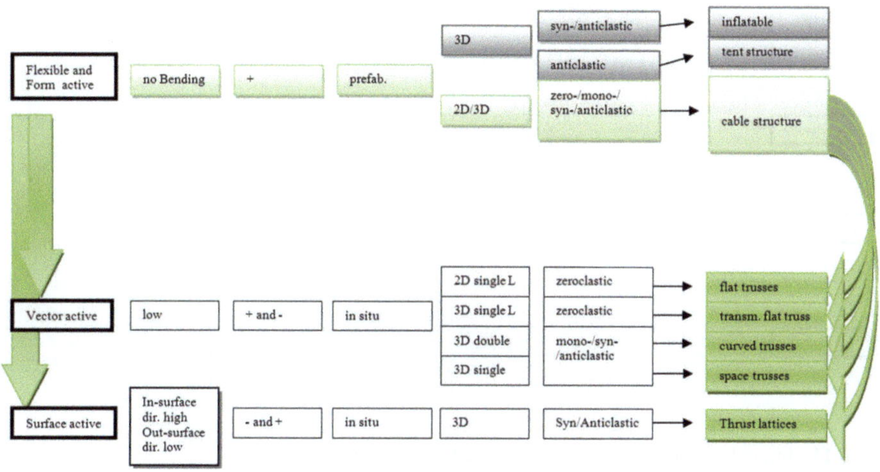

Fig. 6.9 Cable structures to be the mould for all kinds of structures

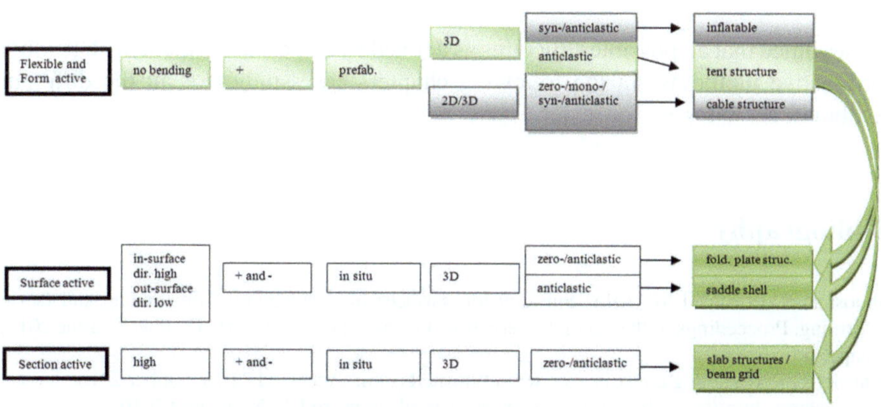

Fig. 6.10 Tensile structures to be the mould for shell structures

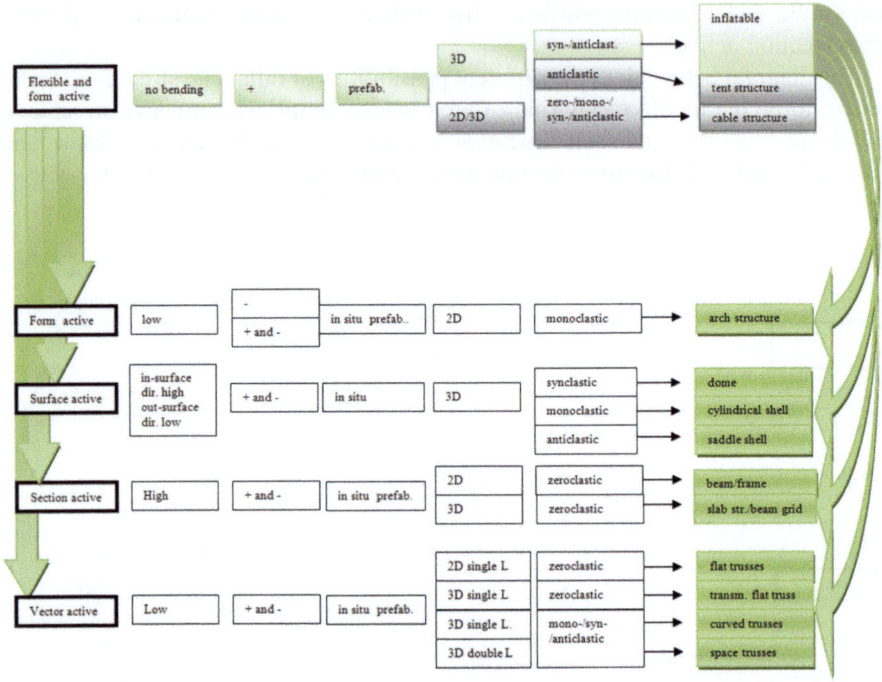

Fig. 6.11 Inflatable structures to be the mould for other structures

6.3 Conclusion

It is assumed that all possible structures as defined by Engel can be made with flexible form-active moulding. In the following chapters case studies of all the different combinations will pass for review.

Bibliography

Bhooshan S, El Sayed M (2012) Sub-division surfaces in architectural form finding and fabric forming. Proceedings of the Second International Conference on Flexible Formwork, June 2012, paper 6

Binishells | Redesigning construction. www.binishells.com. Accessed 21 November 2016

Bini Systems. http://www.binisystems.com/binix.html. Accessed 21 November 2016

Büttner O, Hampe E (1985) Bauwerk: Tragwerk; Tragstruktur. Ernst und Sohn, Berlin

Domepirs. www.domepirs.com. Accessed 21 November 2016

Engel H (1999) Structure systems, 2nd edn. Verlag Gert Hatje, Stuttgart, pp 112–117, 120–132

Hennik P (2005) Pneumatic formwork for irregular curved thin shells, shape vs. force, shell vs. pneumatic structure, Master thesis. Delft University of Technology, Delft

Knippers J, Helbig T (2009) Recent developments in the design of glazed grid shells. International Journal of Space Structures 24(2):111–126

Lawton A, Miller-Johnson R (2012) Case study—Fabric formed stair. Proceedings of the Second International Conference on Flexible Formwork, June 2012, paper 19

Monolithic. Monolithic Dome homes, schools, churches, storages, gyms and more. https://www.monolithic.org/. Accessed 21 November 2016

Neff W (1942) Building construction. US patent 2,270,229

Pronk ADC, Houtman R, Afink M (2007) The reconstruction of the Philips-pavilion (V1). Hussain Mousa Dashti (ed), Proceedings of the conference Sources of architectural form Kuwait. Technische Universiteit Eindhoven, Eindhoven

Sanchez-Arcas M (1961) Form und Bauweise der Schalen. M. Sanchez-Arcas VfB. Verlag für Bauwesen, Berlin, pp 151, 165, 168, 184

Schodek DL (2000) Structures, 4th edn. Prentice-Hall, Upper Saddle River

Van Dessel S, Chini AR, Messac A (2003) Feasibility of rigidified inflatable structures for housing. ASCE Journal of Architectural Engineering 9(1):1–10

Veltkamp M, (2007) Free form structural design. Delft University Press, Delft

West M, (2016) The fabric formwork book: methods for building new architectural and structural forms in concrete. Routledge, New York

Y.S.M. For building. www.domecrete.com (redirects to http://ysmforbuilding.com/en/ysm-for-building-ltd/). Accessed 21 November 2016

Chapter 7
Induction

7.1 Conclusion on Structure

From a structural point of view, form-active structures are the most favourable typology. In form-active structures, the load cases on the structure determine the form of the structure. Surface-active structures offer more possibilities in the form finding of the structure and the combination of surface-active structures with vector-active and section-active structures will lead to a hybrid structure offering endless possibilities for the design of the form. From a structural point of view this is less optimal; most of the time it will lead to a structure with high bending moments not in the direction of the surface. This can be resolved with extra columns, beams or truss structures to support the surface. True three-dimensional surface-active structures are fluid architecture if they are irregular. For most designers/architects the structural optimization is not the leading consideration for their designs. In that case a hybrid structure with a continuous surface supported by arches, section-active and vector-active structures is the best option. The structural engineer has the task of introducing as much surface-activeness as possible and avoiding high bending moments perpendicular to the surface.

The dissertation of Martijn Veltkamp arrived at the same conclusion about the use of surface-active structures. The system he developed was a curved structure to support and follow the surface of the architectural envelope. The system is a curved load bearing structure for fluid architecture made from circular elements with a triangular section in combination with monoclastic curved plates. Although this solution is economically optimized for fluid architecture in "first-world countries" it is far more expensive when compared to straight structures. Up till now it has not been used in the construction of any fluid architectural building (Veltkamp 2007). The roof of the Kunsthaus in Graz (Sect. 4.3) was made higher to have more surface-activeness within the single layer triangular load bearing structure. From these and other examples like the work by Kilian and Ochsendorf (2005) one may conclude that there is consensus about the benefits of striving for maximum surface-activeness with the structural systems to use in fluid architecture.

A. Pronk, *Flexible Forming for Fluid Architecture*,
https://doi.org/10.1007/978-3-030-71551-9_7

7.2 Conclusion on Form

There is an architectural demand to make smooth irregular and doubly curved architecture. This architectural desire can be explained by several motivations:

- Better tools emerging for designing and visualizing curved structures (Sects. 3.6 and 3.12).
- The search to distinguish from other architecture, cities, museums, etc. (Sect. 3.1).
- The architectural drive to create beauty using doubly curved surfaces by reference to nature (Sect. 3.8).

The mainstream of architecture and even the avant-garde have until the beginning of the nineteen nineties concentrated on zeroclastic forms (Sect. 3.12). With zeroclastic forms it was possible to make an endless amount of combinations. The ultimate step in this form typology was the deconstructivism of the late nineteen eighties and early nineties (Sect. 3.11.4). To push the limits further, fluid architecture was the next step after deconstructivism. Using curved surfaces within one domain (for instance only synclastic) is much more complex compared to the use of zeroclastic surfaces. It also does not bring a larger amount of possibilities. Only with the combination of different form typologies is it possible to increase the amount of design possibilities dramatically (Sect. 2.6). Therefore, designers leaving the domain of the zeroclastic surface are almost obliged to design surfaces with a combination of anticlastic and synclastic curvatures. The endless way anticlastic and synclastic surfaces can be combined gives the most freedom (free form) in design (Sect. 3.10). The drive for beauty leads to a wrinkleless smooth connection between the surfaces. The form aspects in relation to the *Gestalt* laws and the fifteen basic properties proposed by Christopher Alexander (Sect. 5.5) give a closer specification of the form aspects in fluid architecture.

Fluid architecture can be qualified as being:

1. Irregular.
2. Smooth.
3. Doubly curved.
4. A multifaceted body, in many cases with a combination of synclastic and anticlastic surfaces.
5. It has to be a building with: Utilitas, Firmitas and Venustas.

7.3 Conclusion on Building Technology

The problems in constructing the structure and the building envelope of fluid architecture are still unsolved. Surface-active structures are the most suitable structural systems but they come with excessive moulding costs. In 2004, Anish Kapoor made

Fig. 7.1 Cloud Gate by A. Kapoor (2004)

"Cloud Gate", an immense polished "bean" out of stainless steel. This work of art is of great importance to fluid architecture because it shows that it is possible to make big smooth doubly curved surfaces on the scale of a building if there is enough money and effort to realize the artistic objective. This shiny art object is unique and Kapoor used membranes for his biggest works. For "Marsyas" in the Turbine Hall at Tate Modern in London (2002), he stretched a red PVC membrane between three massive rings. For the inflatable he created at the Grand Palais in Paris (2011) he also used red PVC membrane. These pieces of art show the possibilities of membranes in forming extreme and complicated sculptural surfaces (Figs. 7.1, 7.2 and 7.3).

The building envelope deals with the same moulding problem. The ability of flexible form-active structures to be the prefabricated mould for surface-active structures is well known and still successfully practiced in the fabrication of domes. The theoretical range of options to make curved surfaces with membranes is large. There are 84 different combinations to manipulate surfaces in force equilibrium (Sect. 2.9.1). Each of the 84 methods will lead to new possibilities, the result is endless but not all imaginable forms can be made in this way or should be made in this way. The possibilities are not limited to the change from flexible form-active structures into a surface-active shell. It is possible to make all of the four structural typologies as defined by Engel (Sect. 6.1). Flexible form-active structures can have a fluid connection from one form to another as well as a strong curved line.

Fig. 7.2 Marsyas by A. Kapoor (2002)

Fig. 7.3 Leviathan by A. Kapoor (2011)

The conclusion on structure, form and technique leads to the following findings.
Findings:

1. The possibility to make a fluid connection between doubly curved and other surfaces is important in fluid architecture.
2. The ability of flexible and form-active structures to be the prefabricated mould for other structures is proven by many projects and companies. Most of the projects are focused on the making of dome or dome-like structures.
3. The experiments by Linkwitz (Sect. 2.9.2) (Linkwitz and Schek 1971: 145–158) and Otto (1973) show the basic principles and endless possibilities for the manipulation of prestressed membrane structures, catenaries and inflatable structures.
4. Surface-active structures are the best typology to make fluid architecture. If a true surface-active structure is not possible, a hybrid combination with section-active and vector-active structures needs to be considered.
5. The ability to manipulate flexible and form-active structures and adapt other structural typologies leads to an endless number of forms.

Based on the findings above, I assume that in most cases flexible and form-active structures are able to be the mould for the construction of fluid architecture in an efficient way.

To prove this, several experiments and projects by other researchers and by myself were executed using different materials. But before these experiments and projects were executed it was necessary to know if designers can work within the limits of the flexible moulding system. To examine this, three professional artists were asked to work with form-active moulds. They were assigned to design a pavilion that would fit in their artistic oeuvre and they were assisted by architecture students of TU/e. The artists had no structural background and were free to work any way they wanted. Finally, they had to present a scale model with an elastic membrane or foil. In that way they proved that their designs can be executed with a flexible form-active mould. The artists were Jurgen Bey, Ronald Cornelissen and Gabriel Lester and the project was supported and mainly funded by SKOR *Stichting Kunst en Openbare Ruimte* and NBKS *Noord-Brabantse Kunst Stichting*. Only the most interesting design proposal was actually built. All artists worked in their own way to arrive at a concept.

The work by Ronald Cornelissen is a specific, personal expression. After finishing his design, he used the B.B.S. method to simplify the realisation of it.

The work by Gabriel Lester is inspired by all kinds of influences like nature, technique, movies, etc. The main themes in his work are light and projections. The method we showed him was placed in his palette of inspiration.

Jurgen Bey likes to start from a theoretical point of view before he makes his design. He wants to know all the ins and outs of this new method. It seems that in his design he wants to show the ultimate expression of the method in his art.

All three artists had in common that only at the end of their conceptual phase did they prove that their proposal could be built with a flexible form-active mould. A jury of architects and artists commented on the design proposals. They judged that the proposal by Gabriel Lester was too much an example of "classic" fluid architecture and did not really fit in his oeuvre. The proposals of Ronald Cornelissen and Jurgen Bey fitted in their oeuvres perfectly. From a technical point of view, the proposal by Jurgen Bey was more challenging. It was also more 'design' whereas the proposal by Ronald Cornelissen was more highbrow 'art'. Therefore, the jury and TU/e decided to build the pavilion designed by Jurgen Bey (Sect. 10.12). Afterwards, the three artists where asked if they had felt any artistic or conceptual impediments on account of the technical moulding method they had to use. None of them felt this had presented any problems (Figs. 7.4, 7.5 and 7.6).

There is a strong relation between the technology used and the tectonics of an object. Flexible form-active moulding will have very specific tectonics. Therefore, it was of great significance that none of the artists had any problems with that. The Jury came to the same conclusion as the artist because they found in two of the three cases that the design proposal fitted perfectly in their oeuvres. The reason why the third proposal did not fit in the oeuvre of Gabriel Lester was not so much in the technical limitations of the moulding technique as it was in his effort to please the client with a nice design.

This design experiment shows that it is likely that flexible form-active moulding can be used by any designer and a deep understanding of structural systems by the designer is not necessary.

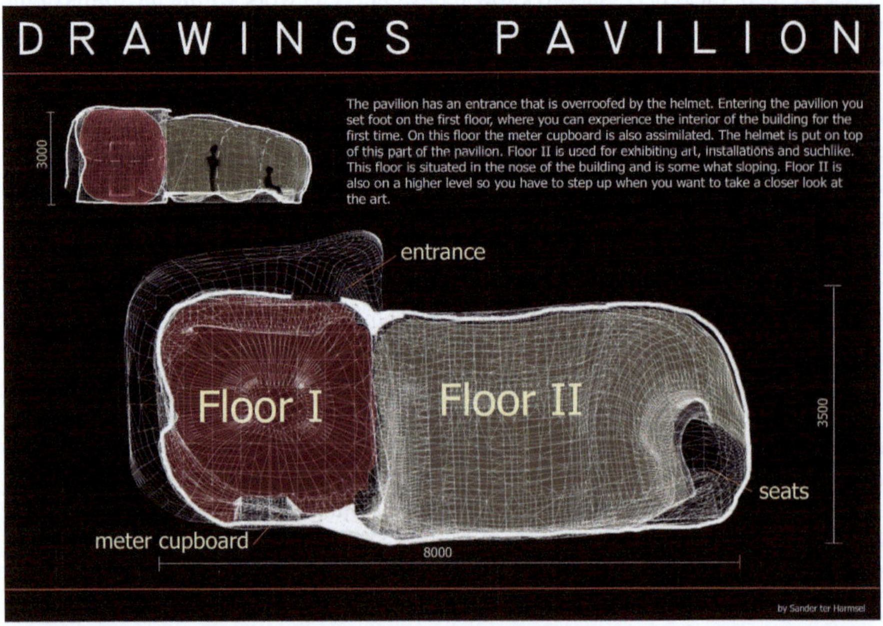

Fig. 7.4 Design proposal for TU/e blob pavilion by Ronald Cornelissen

Fig. 7.5 Design proposal for TU/e blob pavilion by Gabriel Lester

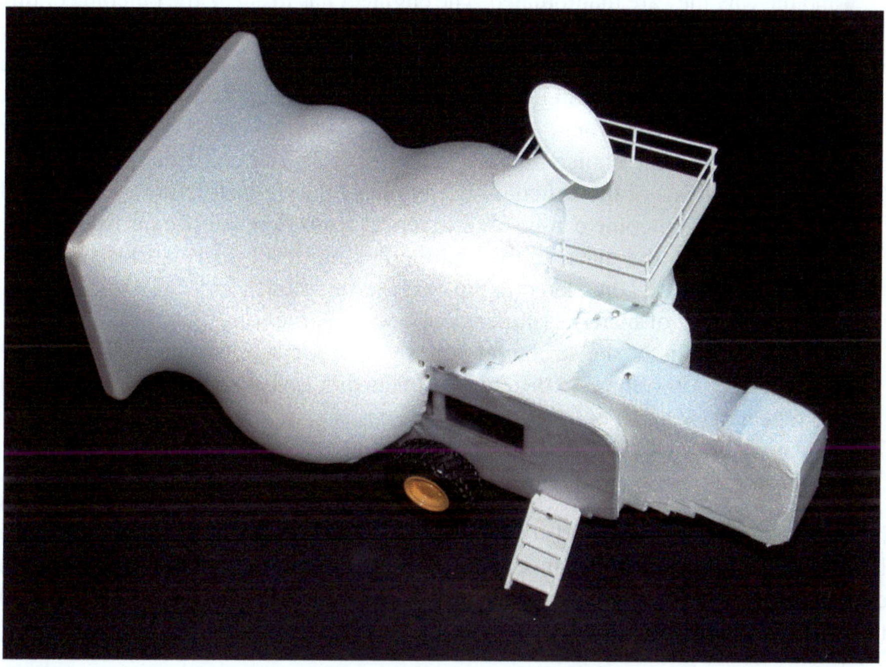

Fig. 7.6 Design proposal for TU/e blob pavilion by Jurgen Bey

In the experiment above, the use of a flexible form-active mould was the starting point. In daily life, designing a building is a cyclic process involving many different stakeholders (Lichtenberg 2005). Functional and architectural reasons dominate in a normal building design process and engineering and construction are not the items to start with. The sooner the designer integrates the idea of using a flexible form-active mould in his design process the better and easier the integration process between architecture and structure will be.

When designer and engineer decide to use a flexible form-active mould, the following steps have to be pursued to arrive at an integrated optimum process and ditto product:

1. At the end of his design process the designer makes a physical or digital model in force equilibrium.
2. By rigidizing the physical model, or by 3D printing the digital model, the form can be analysed.
3. In case of a physical model it has to be scanned and made digital.
4. With this step the engineering process begins. Like any design process this is also a cyclic process. It starts with analysing the surfaces in force equilibrium and the way they are manipulated. The engineer has to check the outcome of this step with the designer. If the engineer has a good understanding of the intentions of the designer, he may come up with some alternatives that will simplify the construction of the building. At this stage the chemistry between designer and engineer is very important.
5. The next step is the choice of material, dimensions and construction.
6. The first analysis of the structural behaviour of the rigid structure as developed in step 5. The outcome can be used to repeat from step 4 above, as follows.
   4B After step 6 the material load on the flexible form-active surface is better known and can be considered in the engineering model.
   5B Fine tuning of material, dimensions and construction.
   6B Full analysis of the structural behaviour of the rigid structure. If the final result of steps 6–6B is very different from the first analysis, then the cycle should be repeated (4C, 5C, 6C).
7. The construction of the building.

The engineering in this process is complex and advanced. Also, the engineers have to combine different fields of knowledge about the engineering of surfaces in force equilibrium and, for instance, shell structures. The architectural demand for a certain form will make it even more complex. An approach with chemistry between the different specialisms is important in this process (Fig. 7.7).

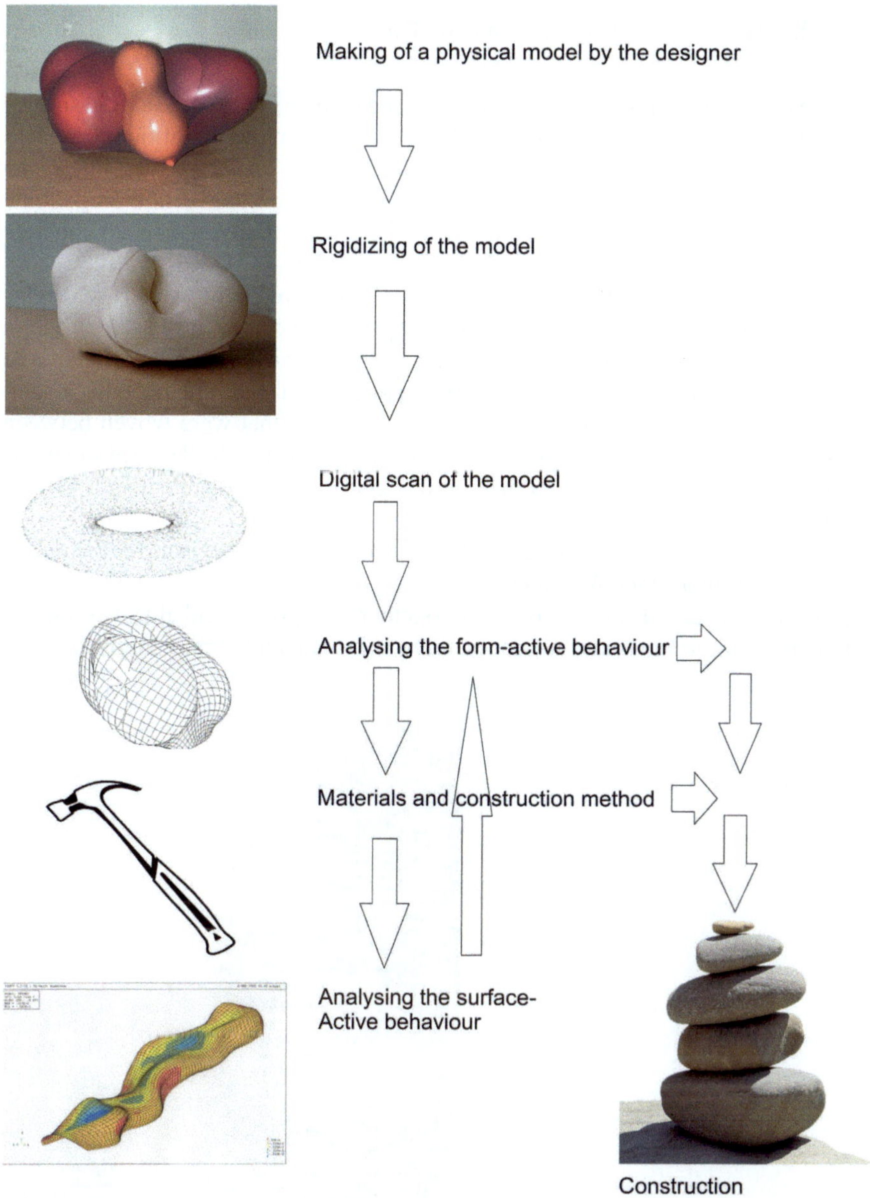

Fig. 7.7 Method for design and construction

7.4 The History of Flexible Moulding

There are many types of form-active structures and many ways to make a rigid surface with the help of a form-active structure. There has been a long tradition in research on the combination of form-active structures and concrete. The historical line of the techniques developed in the past with a special focus on form-active moulding. Some of the techniques may also be interesting for other materials. We start with concrete because that is the biggest category (West 2016; Hawkins et al. 2016).

11,000–6,000 BCE Wattle and daub began with the weaving of twigs. The origins of architectural weaving were probably developed more or less at the same time in several places presently situated in Peru, China, Mexico and the Middle East. Around 6,000 BCE the technique was imported and applied in primitive early buildings in Europe. The walls were made of twigs or thin branches that were woven between upright stakes, more or less like the weaving of rattan baskets. The daub (plaster) was strengthened with fibres like straw, hay, or flax chopped to workable length (Scarre 2005; Graham 2003) (Fig. 7.8).

2,500 BCE The Egyptians used a combination of canvas and plaster for the masks of mummies (Millar 1899: 343–344).

1,500 BCE The use of linen and canvas, together with plaster and glue, was known and practised by the Egyptians for decorating coffins and embalming human bodies.

Fig. 7.8 The wood frame of wattle & daub walling, established by the Iron Age

70 BCE In his book *De Architectura*, Marcus Vitruvius Pollio refers to "fabric" formwork several times. In his chapter about stucco, he mentions wattle & daub. To avoid cracks in the stucco he advises to tie a layer of bound reed to this structure before applying the final stucco to the surface of the wall. For vaults, Vitruvius recommends the use of reeds tied together to form the lower surface on which a sand mortar is applied (Vitruvius et al. 1960: 205). The Villa Medici, Rome 1576 CE was built on Roman vaults. They were made with reed formwork. It is assumed that fresh cut reed was bent to a semi-circular shape. As reed has a limited strength it is assumed that the initial thin layer of concrete was cured before proceeding with additional construction work. The reed vault may have been supported with struts (Bronski 2010). Also the vaults of the roman aqueduct in Mauretania have impressions of reeds that are still visible (Roman Aqueducts, www.romanaqueducts.info/aquasite/cherchell/index.html). For making a dam, Vitruvius mentions the use of two cofferdams formed of oaken stakes with ties between them to be filled with "concrete" (Vitruvius et al. 1960: 162).

1437 The history of tapestry for wall covering (wall hangings) is long and probably goes back to the Egyptians (Encyclopaedia Britannica, www.theodora.com/encyclopedia/t/tapestry.html). Vitruvius doesn't mention tapestry but if a wall is in a state of dampness all over, Vitruvius advises "to construct a thin wall a little away from it on the inside" (Vitruvius et al. 1960: 209). The Italian painter Cennino Cennini (1370—c. 1440) writes that fine linen soaked in glue and plaster and stretched on a wooden frame was used for forming grounds for painting (Cennini 1954). The Encyclopaedia Britannica mentions that textile wall coverings (with paintings on it) have been known for a long time, canvas and mortar were used in Great Britain up to the 1850s (Encyclopaedia Britannica, www.theodora.com/encyclopedia/p2/plasterwork.html). Another application was canvas used as a binding power to prevent the mortar around washhouse coppers from cracking or expanding when subjected to heat. The mortar was composed of equal parts of haired lime and gritty road scrapings (Millar 1899: 344).

1852 Francois Coignet (1814, 1888) started to experiment with iron and concrete. In 1853 he realized the first steel reinforced building. The building is four floors high and is located at 72 rue Charles Michels in St Denis, Paris (Collins 2004).

1855 Joseph-Louis Lambot (1814, 1887), is the inventor of what was later called Ferrocement. Ferrocement is a cement-based composite with a reinforcement of small- diameter steel rods. The mortar is a mixture of sand, cement and water, it is applied in layers on a woven mesh of small-diameter steel rods. It is used for thin curved sheets and shells. It can be applied to boats, water-tanks and shell roofs. In 1841 he applied himself to agriculture and started constructing tanks using cement and iron reinforcement. In 1848 he made his first ferrocement boat. He patented the boat in 1855 and exhibited it at the world fair in Paris (Exposition Universelle—1855) (Newby 2016). The original prototype is preserved at the Museum of Brignoles.

1856 Alexander Desachy, a French modeller, patented fibrous plaster. A way to produce mouldings, ornaments and other works of art, with surfaces of plaster, "with the aid of plaster, glue, wood, wire, and canvas or other woven fabric" (Millar 1899: 344). The Encyclopaedia Britannica mentions: "fibrous plaster is given by plasterers the suggestive name 'stick and rag', and this is a rough description of the material, for it is composed of plaster laid upon a backing of canvas stretched on wood. It is much used for mouldings, circular and enriched casings to columns and girders and ornamental work, which, being worked in the shop and then nailed or otherwise fixed in position, saves the delay often attendant upon the working of ornament in position" (Encyclopaedia Britannica, www.theodora.com/encyclopedia/p2/plasterwork.html) (Figs. 7.9 and 7.10).

1867 Joseph Monier (1823–1906) started his experiments with reinforced concrete in 1849. He was not the only one in his time experimenting with the reinforcement of concrete (see Joseph-Louis Lambot and the work of W. Wilkinson), but he is generally seen as the person who made the best effort. He had several patents on this reinforcement for horticulture (1867), pipes and basins (1868), building façades (1869), bridges (1873) and beams (1878). In 1875 he made the first ever reinforced bridge at the castle of Chazelet (Bosc 2001: 180). Because the initial idea of ferro-cement quickly led to conventional reinforced concrete of thicker form and whilst reinforced concrete became very popular, the use of ferrocement and related research fell relatively dormant shortly after its introduction.

1899 Gustav Lilienthal (1849–1933) was a social reformer and a pioneer in building and construction technology. He was also involved in aviation. His brother Otto Lilienthal was the first 'flying man'. In 1899 he patented a fabric formwork floor system U.S. Pat. 619,769 "The invention consists in spreading some pliable but sufficiently impermeable fabric or paper over the beams intended to carry the ceiling

Fig. 7.9 First ferrocement boat by Joseph-Louis Lambot (1848)

Fig. 7.10 Ferrocement fence

that is to be built by covering the fabric with wire netting and pouring concrete on top of the covering thus formed" (Fig. 7.11).

1911 Patent by C.C. Condie, Revetment-Mattress. US Pat. 984,121. 1916 patent by A.C. Chenoweht Protective Reinforced Concrete Construction. US Pat. 1,187,502 and 1922 Patent by J. Store Method of Constructing Subaqueous Concrete Structures. US Pat. 1,421,857. Several patents using permeable fabrics (geotextiles) padded with soil or concrete were found for civil engineering applications for the protection of dams, dunes, canals, etc. The textile is brought in tension to strengthen (reinforce) the soil or concrete. These techniques are still explored (Fig. 7.12).

1926 Nose, T. Process of Constructing Culverts or Pipes of Concrete, US Pat. 1,600,353, This is the first publication concerning inflatable moulds. The patent describes inflatable rubbers used as mould for the casting of sewers. The hoses with a length of 90 meters were laid on a concrete base and inflated. When the concrete was cured the hoses were deflated and removed (Sobek 1987). In a 1934 US patent 1,964,386 he describes the effect of combining inflated tubes to influence the form of the flexible mould. This technique was still in use in 1989 by the firm NOE-Schaltechnik (NOE, www.noe.de). Variations on the patent by Nose for the application of a hollow floor system where made in 1949 by Mathews, C. B. et al.: Inflatable Core for Use in Casting Hollow Concrete Units, US Pat. 2,485,898, The 1951 patent by Ford, C.J.: Wingfoot Corporation. Retractable Form Structure, U.S.

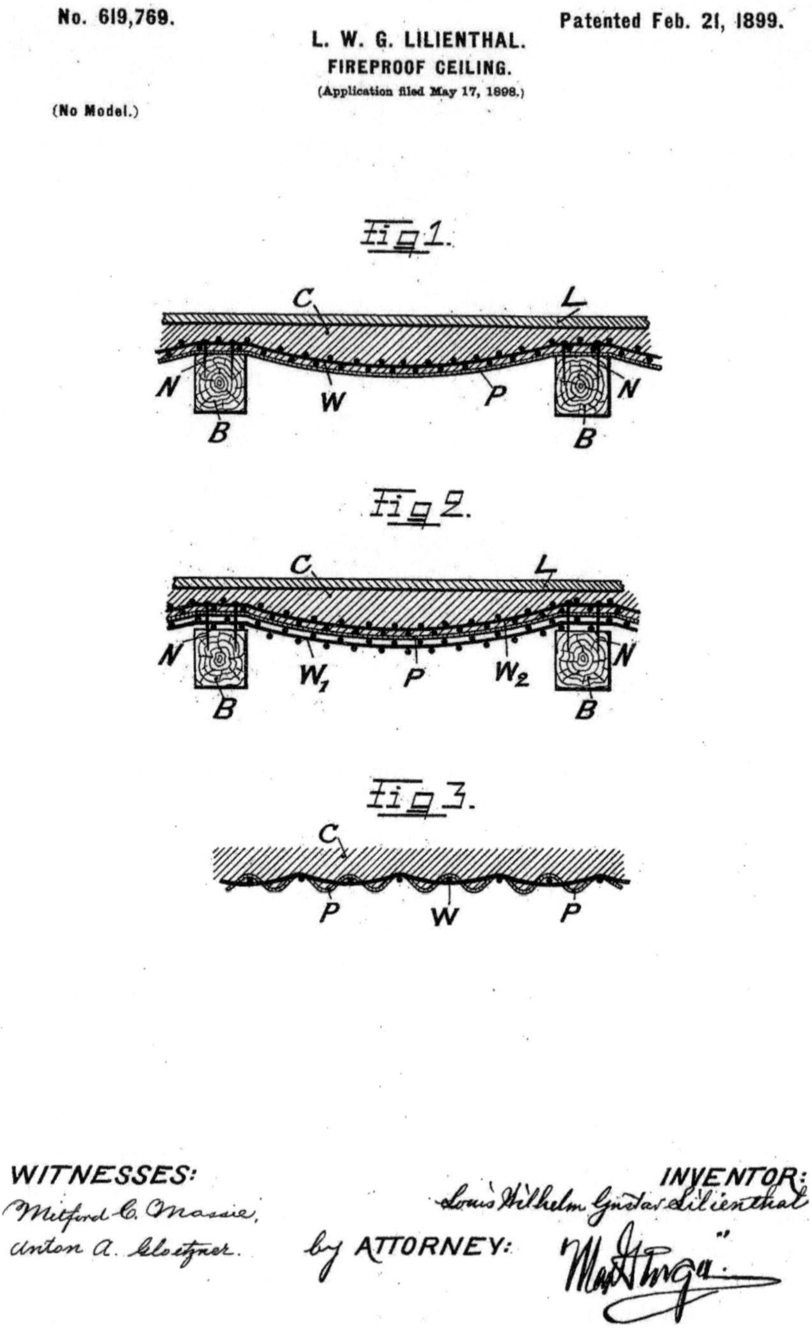

Fig. 7.11 1899 US patent 619,769 by G. Lilienthal

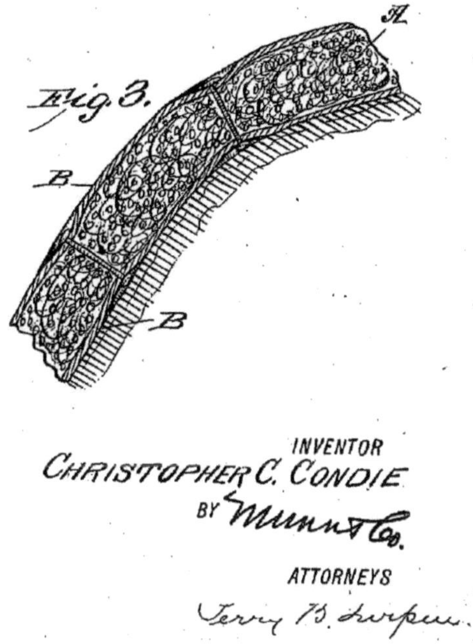

Fig. 7.12 1911 US patent 984,121 by C.C. Condie

Pat. 2,579,423, consists of an inflatable corner to join two moulding elements and seems to me to be inappropriate (Fig. 7.13).

1934 During the Great War, James Hardress de Warenne Waller 1884–1968 invented the rigidizing of military tents by covering tensile structures with cement slurry and water. Later he would call this method Nofrango. He developed this technique in the early 1930s. In 1934 and 1935, he obtained US Pat. 1,955,716 and US

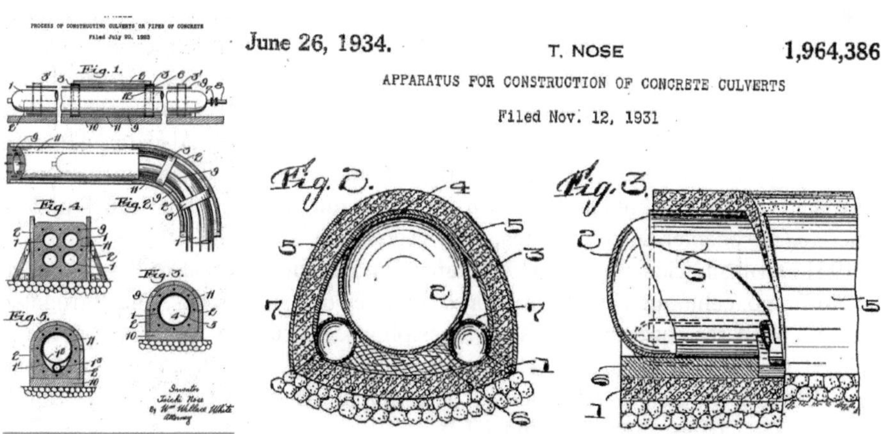

Fig. 7.13 1934 US patent 1,964,386 by T. Nose

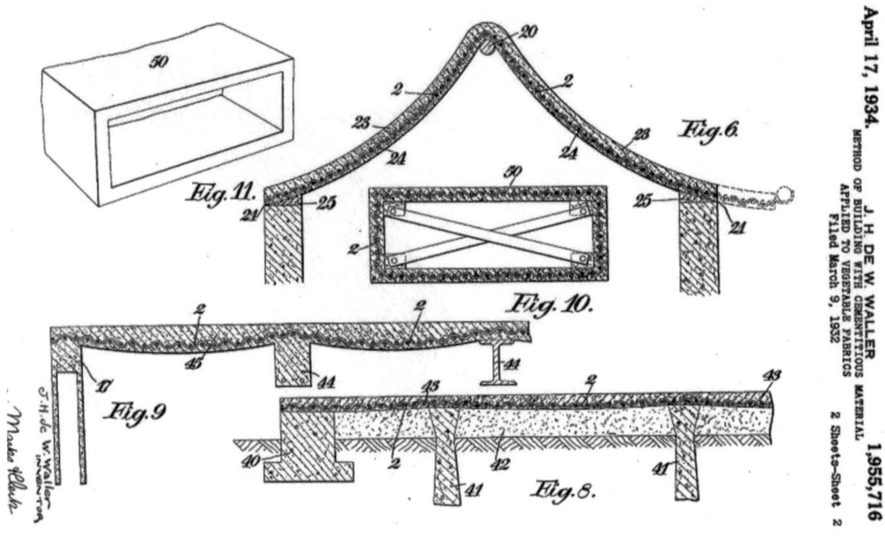

Fig. 7.14 Patent by Waller

Pat 2,015,771, which show various examples of the Nofrango building method. In 1937, D. Farrar et al. and in 1971 S.A. Parker used the flooring systems by Waller and Lilienthal in their patented floor and building system for a complete house US Pat. 2,096,629 and US Pat. 3,619,959 (Fig. 7.14).

1936 Karl Billner, developed "Method of and Apparatus for Treating Concrete, US Pat. 2,046,867". He described a way to extract water after casting concrete by sucking the water away with a vacuum machine. This will lower the water-cement ratio, reduce air bubbles and blowholes. Later patents by Arthur brooks and Byron Bender (1943 US Pat. 2,310,391) and Yokota et al. (1988 US Pat. 4,787,597), the master thesis of Bindhoff (1968) (Delijani 2010; Al Awwadi Ghaib and Górski 2001: 1459–1465) and the research by J. Orr et al. (2012). Show that fabric liners have a positive effect on the durability and strength of the concrete surface. Fabric liners are commonly used to lower the water- cement ratio after casting of concrete. The vacuum technique is used in other applications to transport the liquid material like the vacuum infusion of composites. Due to the inhomogeneous substance of concrete, there is no successful application known of the vacuum transportation (infusion) with concrete (see chapter on composites; also: vacuumatic research 1970).

1941 During the Second World War, James Hardress de Warenne Waller combined the Nofrango technique with the support of inverted catenary arches. The corrugated concrete shell structure known as the Ctesiphon construction was developed and realized for the first time in 1941. He also developed a double layered structure (Fig. 7.16). In 1949, Candela used the Ctesiphon construction for his first shells (Conlon 2012) (Figs. 7.15, 7.16 and 7.17).

1941 Wallace Neff (1895–1982) patented the first concrete dome with an inflatable formwork, U.S. Pat. 2,270,229. He developed a low-cost air form house made by

Fig. 7.15 Ctesiphon construction by James Waller (1942)

Fig. 7.16 Double layered Ctesiphon construction by James Waller (1950)

Fig. 7.17 Shell by Felix Candela (1951)

reinforced concrete cast on site over an inflatable balloon. Openings such as doors and windows were made by frames placed against the inflatable form. In 1942 he made a 45 m diameter dome with a height of 3.35 meters as a result of experiments between 1934 and 1941. The system was used in the 1940s and 1950s. He built over 1200 domes, mostly in West Africa, Egypt and Brazil (Head 2012). He reported improvements in his construction method in patents obtained in 1948 (GB 603,655) and in 1955 (GB 737,140) (Fig. 7.18).

1943 Pierre L. Nervi (1891–1979) recognizes the possibilities of ferrocement and patents new insights in this technique. He patents a technique with mats of iron mesh formed over a mould or falsework and doused with cement rich concrete. He calls it the official birth of ferrocement although the term is not new. This technique was tested by him for thin-shell structures and mass production of thin concrete elements. From 1943 to 1948, P.L. Nervi experiments with ferrocement by building boats. The first applications of this method were a 400-ton motor vessel and three motorboats for the Italian Navy. The large motor vessel was built but never launched, due to the start of World War II. The advantages of a well-built ferrocement shell construction are low weight, low maintenance costs, long lifetime and absence of need for a mould. The disadvantage of ferroconcrete construction is its labour-intensive nature (see also Chap. 3). Ferrocement was revived in the 1960s through the interest of amateur boat builders and small fisheries. The total volume fraction of steel mesh reinforcement in ferrocement generally ranges from about 2 to 8%. It is physically difficult to put more than 8% into the composite. Other potential reinforcements were also available including meshes made from natural fibres (jute or sisal) and polymeric meshes (or textiles or fabrics) of various kinds, such as nylon, polypropylene, and polyester. For

Fig. 7.18 Scaffolding for Tetto Palazzetto dello Sport Roma 1958 by Pier Luigi Nervi

instance, the floor of the 1967 Australia Square high-rise in Sydney was built with precast ferrocement elements by Nervi (Solomita 2015) (Figs. 7.19, 7.20, 7.21, 7.22, 7.23, 7.24 and 7.25).

1959 Heinz Isler (1926–2009) presented a paper under the title "New Shapes for Shells" at the first IASS conference in Madrid in 1959. He introduced three form finding methods, two of which were based on fabric: (1) the inflatable membrane, and (2) hanging cloth (the third based on freely shaped hills) (Isler 1961). He considered that the (reverse) hanging cloth had the most potential and showed experiments with wet hanging cloth on a freezing winter evening, as illustrated in his paper. He used the hanging cloth method for form finding in many works like the roofs of the N1 Autobahn service station at Deitingen Süd (1968) and the Sicli SA factory in Geneva (1969) (Chilton 2009: 51–62). Isler researched inflatables for ice domes and a 1977 low-cost housing project in Maghan, Iran and he realized several other "bubble" projects in Ponthierry, Riyadh, Aarwangen and Langenthal. The most architectural project is the 1980 Haus Balz in Stetten auf den Fildern, Leinfelden-Echterdingen, Esslingen. This house was insulated on the inside with foam (Schlaich and Schüller 1999).

1962 Frei Otto built an experimental dome for a building exhibition in Essen (DEU BAU). In the late 1950s the building industry produced some polymer building

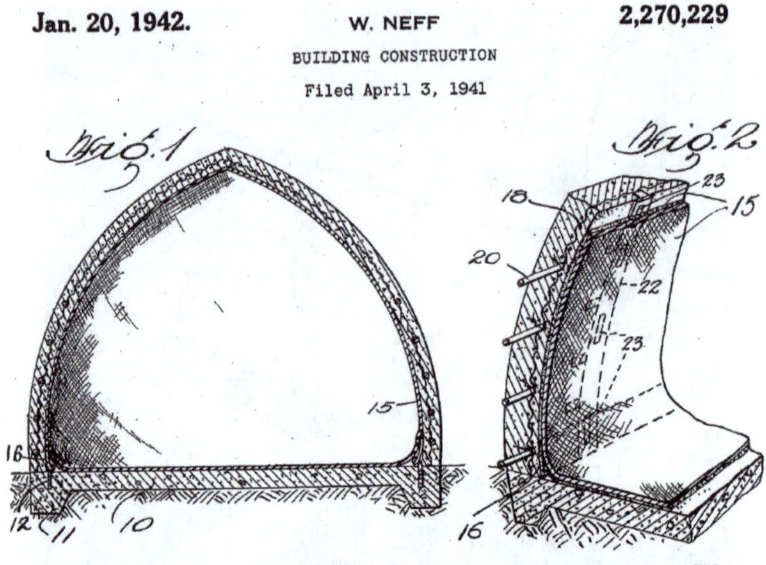

Fig. 7.19 W. Neff patent 1941

Fig. 7.20 Sicli, concrete shell by Heinz Isler, Geneva 1970 (*Source* ETH Zurich)

components like sandwich panels and tough coatings. They raised the idea of spraying foam over an inflatable form (Allison 1959). Otto studied air supported structures from 1957 until 1961 (Otto 1962a: vol 1; 1962b: vol 2). Over this period, he researched several proposals for the rigidizing of shell structures with polymers. In 1962 he realized an 8 m diameter 3,5 m high dome. The inflatable was fastened to a metal ring anchored to the ground. At the inside first a 2–3 mm of glass fibre reinforced polyester layer was sprayed and then a 1.5-meter-high 20 mm thick layer

Fig. 7.21 Sicli, structure of concrete shell by Heinz Isler, Geneva 1970 (*Source* ETH Zurich)

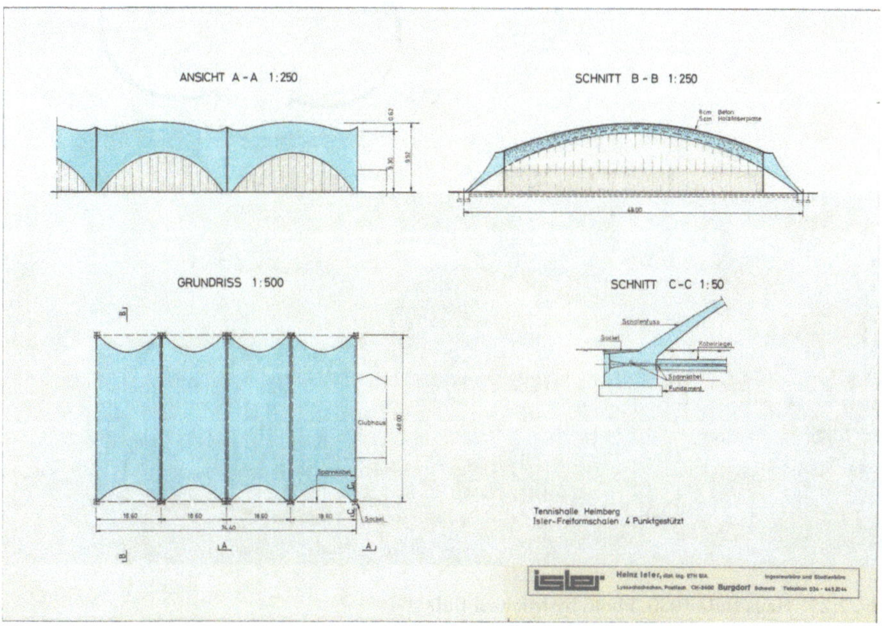

Fig. 7.22 Schematic drawing for tennis hall in Heimberg by Heinz Isler, 1978 (*Source* ETH Zurich)

Fig. 7.23 Tennis hall in Heimberg by Heinz Isler, 1978 (*Source* ETH Zurich)

Fig. 7.24 Haus Balz 1980. Drawing of ground plan by Michael Balz

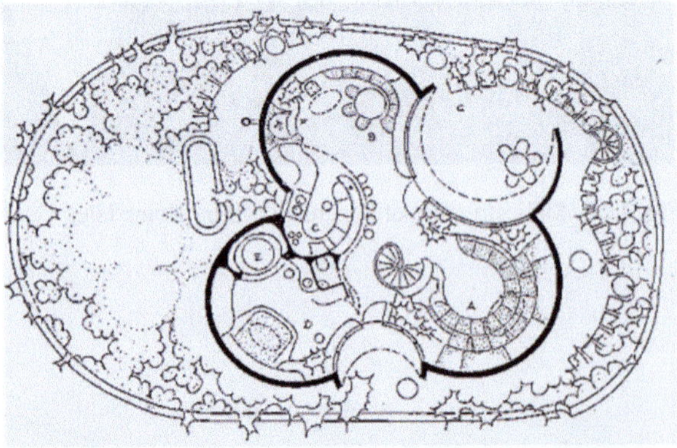

Fig. 7.25 Haus Balz 1980. Photo by Michael Balz

Fig. 7.26 Form-finding by Frei Otto

of foam was sprayed (Perlite-Palatal) (Roland 1965). This dome was the first that was sprayed from the inside (Fig. 7.26).

In 1964, the Italian architect Dante Bini (b. 1932) initiated experimental work including the Bini Dome. In 1965 he obtained an Italian patent and in 1968 a US patent: Method for Erecting Structures, U.S. Pat. 3,462,521, In his method, he erects flat cast concrete surfaces to domes by inflation. The reinforcement and concrete of the domes are placed on the flat surface of a deflated dome. A thin PVC sheet encloses the concrete and reinforcements. During inflation, the reinforcement is able to slide into the right position, the concrete is held by the reinforcement and membranes on both sides. The concrete is deviated over the surface to maintain thickness. The outer membrane can be left in place for weatherproofing of the dome, the inner membrane can be reused. For insulation a second dome is made inside. The space in between the two shells is filled with insulation material. Only rotational shapes have been made with this system. Later Bini used inflatables to erect prefabricated elements (Binix) and 3D truss-systems (Binistar). Binisystems built over 1,500 buildings in 23 countries. (See also 2005 patent by J. Kollegger). Bini also designed the Cupola house with an inflatable mould for famous Italian director Michelangelo Antonioni and his lover, actress Monica Vitti. Antonioni was the director of the movie Blow-up, among others (Figs. 7.27 and 7.28).

1964 Hiam Heifetz (1922–1993), Israel patent IL 21340. He studied the problems concerning the construction of inflatable domes thoroughly from 1960 to 1965. His research was based on the construction by W. Neff (inflatable mould with shotcrete dome). His improvements consist in the use of higher pressure and he improved the anchoring and pressure control system. He realized the first house in 1967 and

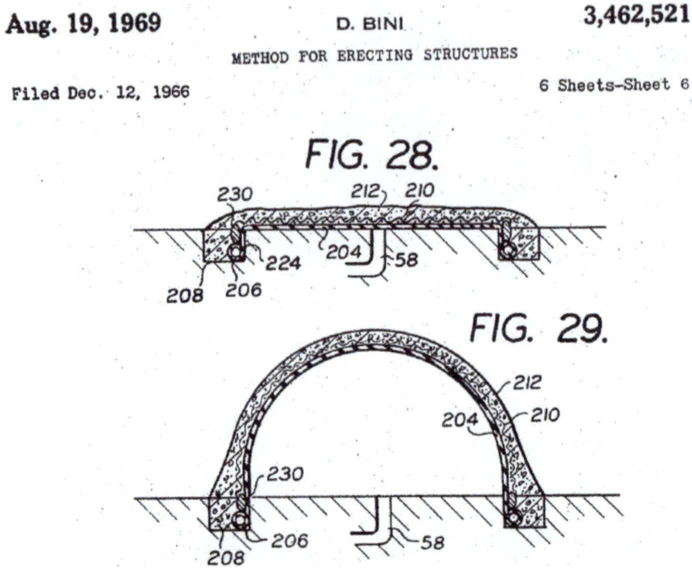

Fig. 7.27 Patent by Bini 1968

Fig. 7.28 Cupola House for Michelangelo Antonioni and Monica Vitti

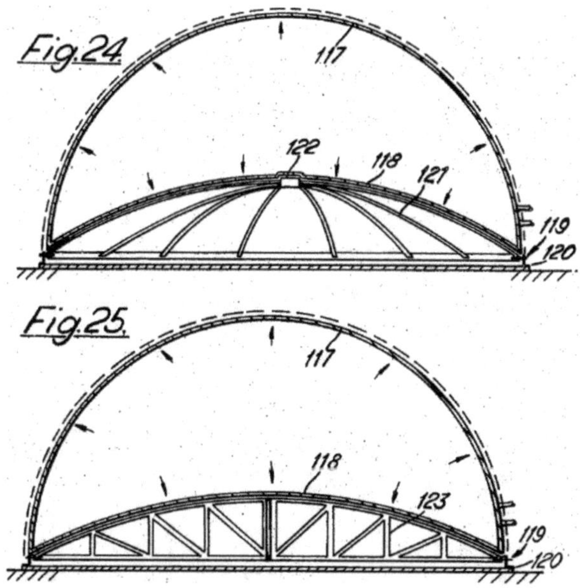

Fig. 7.29 Patent by Hiam Heifetz

established his own company (Y.S.M.) in 1969. He realized over 40,000 domecrete structures. Besides domes, he also realized joined and semi cylindrical ("banana shaped") structures (Fig. 7.29).

1964 Horrall Harrington had US patent 3,118,010: Method and Apparatus for Constructing a Concrete Structure, and later several other patents in this field. Like Otto, he sprayed from the inside. He was the first to make patterns with a wire mesh on the inside surface of the inflatable. This made it possible to influence the form of the inflatable and to divide the shell in ribs and thinner parts. Kokawa and R. van Hove used the principle of manipulating an inflatable with a rope net 40 years later (See below) (Fig. 7.30).

1960s Inflatable structures have been researched for aerospace engineering since the 1960s (Ault 1965). The research has focused on deployable systems for large inflatable antennas and sunshields. The advantages for space engineering are in the small volume and mass when compared to conventional methods. The inflatables can only be used in space if they are rigidized after inflation because of the small meteoroids that will perforate the inflatable. Seven different ways to rigidize membranes with fibre polymer composites have been developed for space engineering over the last 50 years. Four of them ([1] Thermally cured thermoset composites, [2] Ultra-violet cured composites, [3] Plasticizer or solvent boil-off composites, [4] Foam rigidization) are used for civil applications (Cadogan and Scarborough 2001) (Figs. 7.31, 7.32 and 7.33).

1966 Lloyd Turner had US Pat. 3,277,219: Method of Moulding a Building Structure By Spraying a Foamed Plastic On the Inside Of an Inflatable Form. The Californian-based architect used PU-foam for rigidizing the inflatable dome from the inside. This foam can be reinforced by spraying glass fibres. The Tecton Corp. used

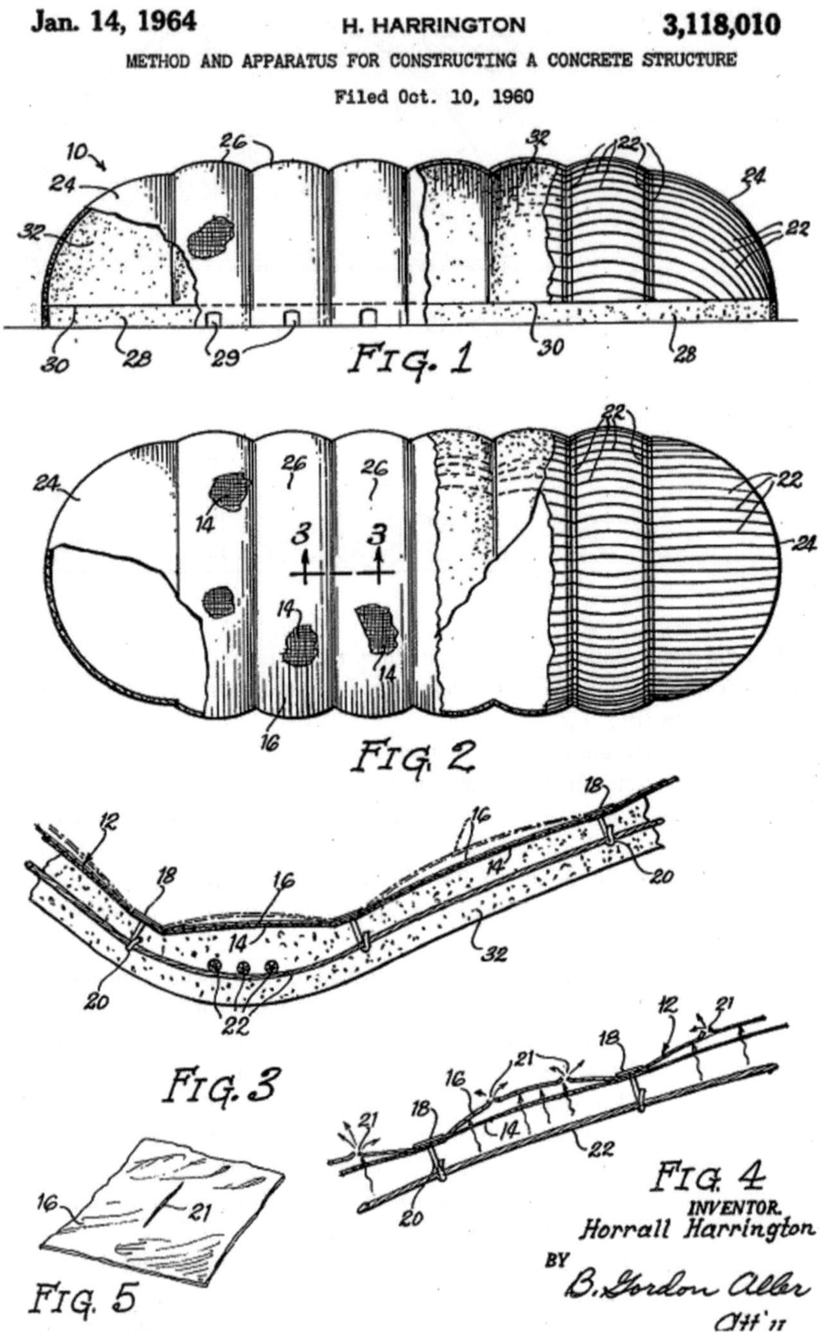

Fig. 7.30 Patent by Horral Harrington

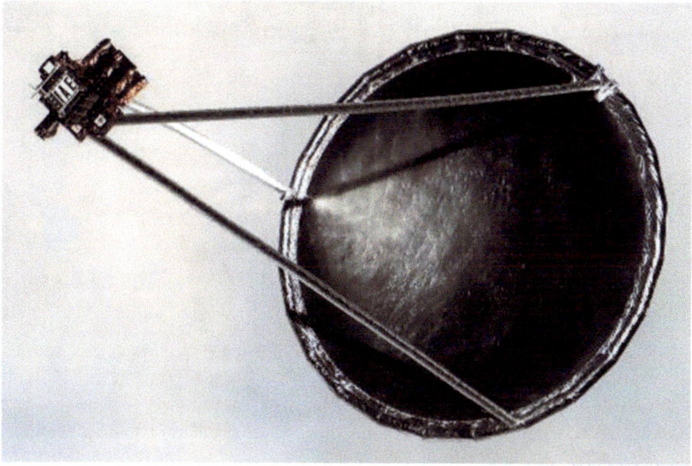

Fig. 7.31 The space shuttle-launched inflatable antenna experiment in orbit

Fig. 7.32 Inflatable re-entry vehicle

Fig. 7.33 Inflatable space habitat patent by Bigelow

an elastic membrane, polyurethane foam, steel fibre concrete with reinforcement rods and an exterior finish of synthetic on elastomer basis. Tecton Corp built more than 1,000 domes between 1968 and 1987. Monolithic Dome Institute by Barry, David and Randy South patented more or less the same system in 1975 (spraying foam and concrete on the inside of an inflatable dome). Since 1975, they have constructed over 500 domes. The largest span was 90 m in diameter. The French company i-PIRS built about 75 domes mainly in France since 1988 using the same system (Figs. 7.34, 7.35 and 7.36).

1969 The improvement of materials in the 1950s provided a new impetus to the ideas on civil engineering with geotextiles (*cf.* 1911 above). It led to new and better applications and was for Richard Fearn, Kenzo Unno and Mark West a source of inspiration for architectural applications. The technique is still used by the Fabri-form Company. Patents: Lamberton, B. A.: Method and Arrangements for Protecting

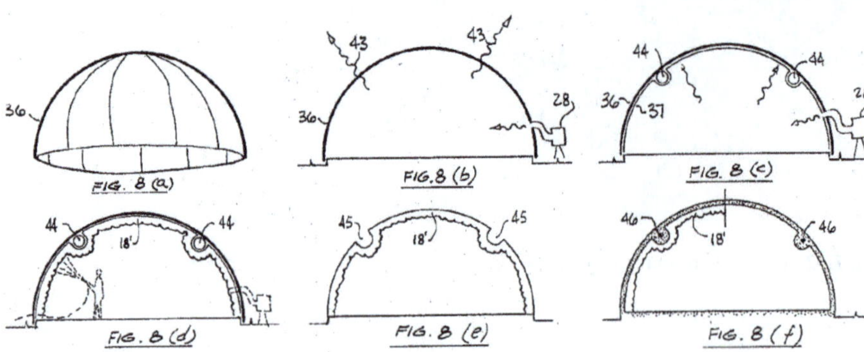

Fig. 7.34 Patent by Turner 1966

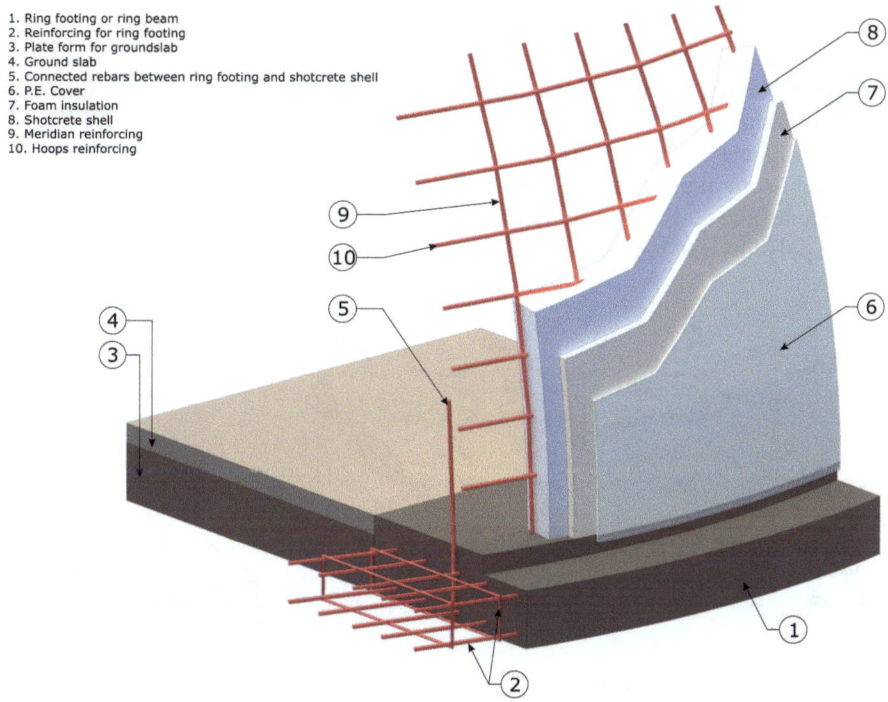

Fig. 7.35 Construction of dome by i-PIRS

Fig. 7.36 Domes by i-PIRS

Shorelines. U.S. Pat. 3,396,542 and in 1969 Patent on Fabric Forms for Concrete Structures. U.S. Pat. 3,425,228, 1969. Mora R. L.: Inflatable Construction Panels and Method of Making Same, U.S. Pat. 3,388,509, Hillen, H. F. J.: Form for Constructing a Slab for Talus or Bottom Protection. U.S. Pat. 3,425,227 (Figs. 7.37 and 7.38).

1969 Miguel Fisac (1913–2006) was the first to use the tactility of fabric formwork as an architectural concept. In 1969, he patented his new ideas and realized them in the Centro de Rehabilitación para la MUPAG in Madrid. He used smooth and flexible polyethylene lamina hanging from a rigid structure that acts as a formwork.

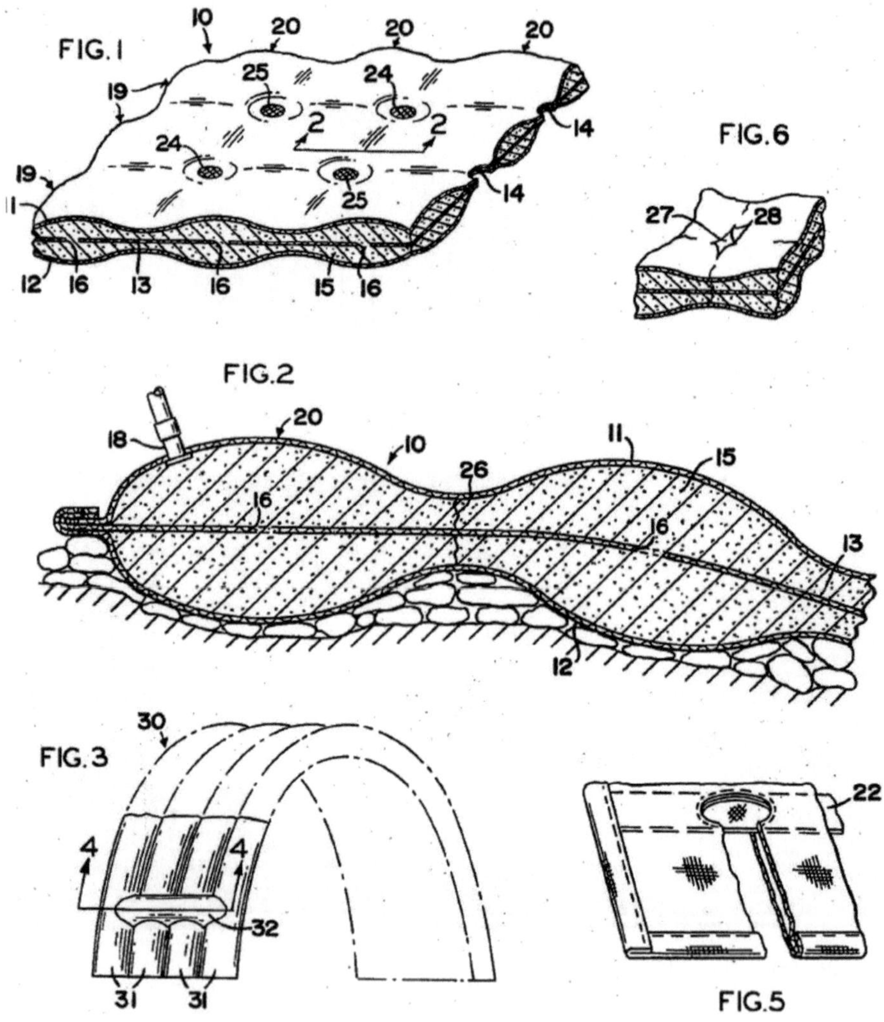

Fig. 7.37 Patent by Lamberton

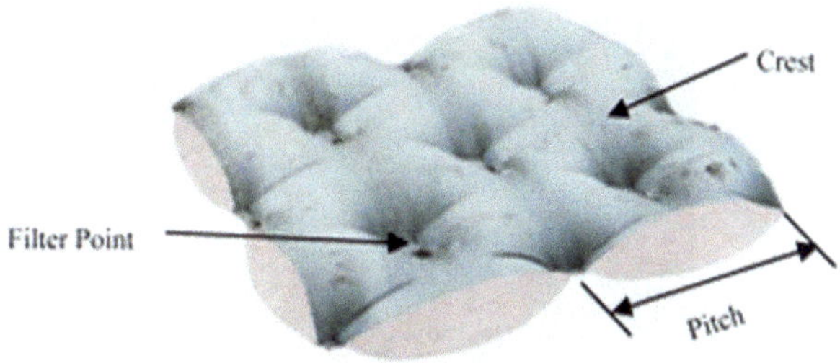

Fig. 7.38 Fabriform system (*Source* Fabriform)

"The result that the weight of this soft material gives to the concrete when poured is real and effective; the concrete takes on the texture of the material in a tactile way".

1970 Vacuum technique can be used as a liner to remove water (see 1936) but also to hold the material in place or to rigidize the fabric mould. At Queen's University Belfast several students in 1970 were the first to experiment with vacuum structures for building applications. They made a vault for a carport. Other researchers experienced in vacuum structures (or deflated structures or structures in underpressure) are Werner Sobek (See 1987) for a façade and pavilion in 2007 (Haase and Sobek 2007: 1148–1159), the students of the TU Delft by making a bridge in 2006 (Pronk et al. 2006), for transport of liquid between two membrane surfaces in 2006 (Chap. 8), and Rob Verhaegh (2010) and Frank Huijben 2009 (Huijben et al. 2012) in their research on fabric formwork (Huijben et al. 2009) (Figs. 7.39, 7.40, 7.41 and 7.42).

1970 Bayer AG introduced 5 m diameter foam domes for refugee dwellings. They were used in Turkey, Peru and Nicaragua. Polyurethane Foam (®Hartmoltopren) was sprayed on the outside of an inflatable on site. With one inflatable mould it

Fig. 7.39 The 'Pagoda', Office tower, Madrid 1955 by Miguel Fisac

Fig. 7.40 Construction of vacuum arch by F. Huijben (*Source* F. Huijben)

was possible to make 15 domes a day. Over 1,000 domes were made. In the same year, Frankenstein and Sons Ltd (UK) developed a system for extreme climates. The inflatable is sprayed with 20 mm thick layer at the outside and a 100 mm thick layer at the inside. The system can be erected in 3–4 hours and has been used in polar regions. In Wageningen in 1967, B. Hangelbroek (Institute Agricultural Factory Buildings) did two experimental structures of polyurethane foam, by spraying on the exterior of a PVC pneumatic hall. And in 1972 Pieter Huybers realized a temporary exhibition (Iglonium) in Tilburg with the Bayer technology (Huybers 1999) (Fig. 7.43).

1966 The Insituform company developed a system for the repair of sewers with the addition of a polymer composite on the inside of the sewer. After cleaning and repair of the sewer the membrane hose is brought inside and inflated. Then the polymer composite is cured and after rigidizing, the hose is released of overpressure and removed. The 1971 system is still in use and successfully distributed in many countries (Insituform, www.insituform.com).

Fig. 7.41 Final arch made with vacuum forming by F. Huijben (*Source* F. Huijben)

Fig. 7.42 Test of inflatable igloo with deflated surface by Pronk et al. (2006)

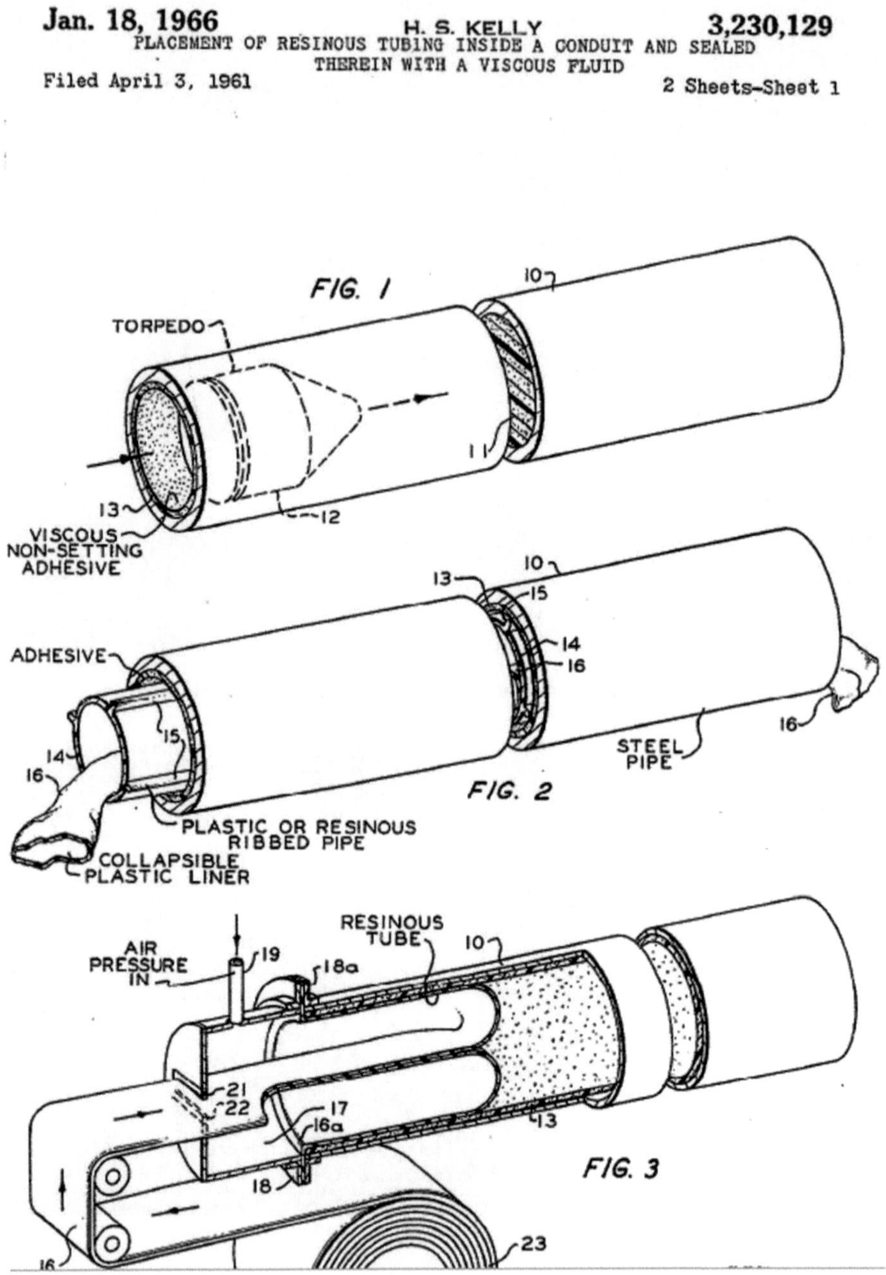

Fig. 7.43 1966 patent by Insituform Foam

Fig. 7.44 Very thin (12 mm) temporary glass fibre reinforced concrete shell designed by Schlaich, Stuttgart, Germany, 197 (*Source* J. Schlaich)

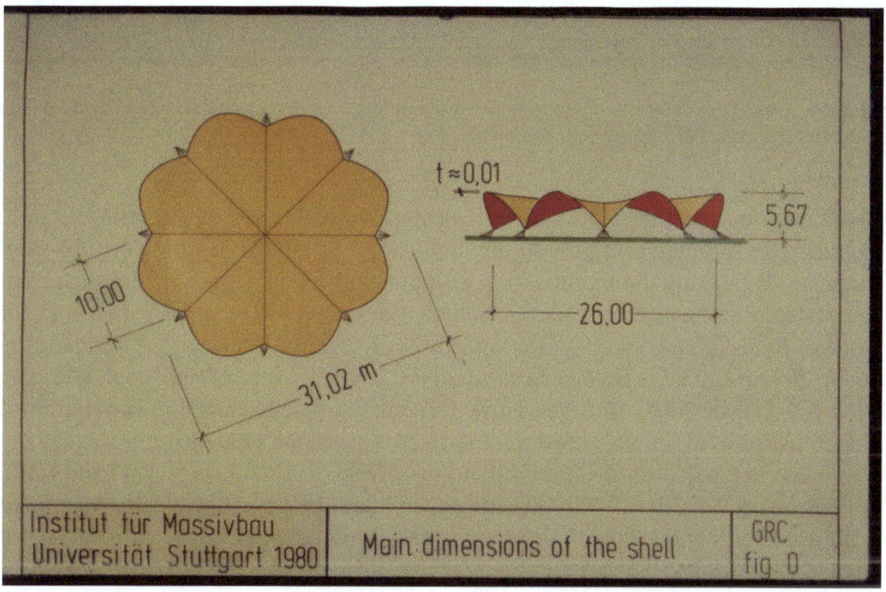

Fig. 7.45 Very thin (12 mm) temporary glass fibre reinforced concrete shell designed by Schlaich, Stuttgart, Germany, 1977 (*Source* J. Schlaich)

1977 Jörg Schlaich Built a shell for the Federal Garden Exhibition in Stuttgart in 1977 which he made of glass fibre reinforced concrete (Figs. 7.44, 7.45 and 7.46). At the end of the 1960s the complex formwork of shells became a problem. Therefore Schlaich, who had built several shells, initiated several research initiatives. He researched the developments by the Pilkington Brothers of alkali-resistant glass fibres. In collaboration with the Heidelberger Cement Company, who did research on GRC (Professor Meyer), they realized the temporary shell structure in GRC for the

Fig. 7.46 Very thin (12 mm) temporary glass fibre reinforced concrete shell designed by Schlaich, Stuttgart, Germany, 1977 (*Source* J. Schlaich)

garden exhibition in Stuttgart. The shell spans 26 m with a thickness of only 12 mm. Schlaich also initiated research on the application of inflatable formwork for shell structures. Especially the modification of inflatables by cables across the surface of the inflatable was a topic of interest. Schlaich designed a 100 m span shell roof for the Max-Eyth-See ice skating rink and in 1983 realized a prototype designed by Werner Sobek and Fritz Bacher for a small water reservoir near Heilbronn (Holgate 1997: 52–55). In 2005, Rob van Hove (NL) had a patent on the manipulation of inflatable formwork for the construction of a cross vault WO2007061299 He realized a prototype in Eindhoven 2008 and a building in Veghel in 2011 (Figs. 7.47 and 7.48).

1984 Nader Khalili (1936–2008) designed a building system of Sandbags filled with onsite earth and arranged in layers or long coils (compression) with strands of

Fig. 7.47 Inflatable mould for cross vault by R. van Hove, Eindhoven, The Netherlands 2008 (*Source* R. van Hove)

Fig. 7.48 Prototype cross vault by R. van Hove, Eindhoven The Netherlands 2008 (*Source* R. van Hove)

barbed wire placed between them to act as both mortar and reinforcement (tension) (1999 U.S. patent 5,934,027). This technique with the filled sandbags looks like the technique used in civil engineering where soil-filled geotextiles are used to protect dams, canals, earth walls, etc. E.T. Dicker (Fig. 7.52) with his 1975 patent was the first to publish an architectural application of this technique. In 1991, Nader Khalili founded (CalEarth) to realize his ideas for cheap dwellings. Between 1991 and 1993 the first dome shaped prototypes were built and tested (CalEarth, www.calearth.org). Since 1995 many buildings were raised with the support of Cal-Earth (Stohr 2006: 106) (Figs. 7.49, 7.50 and 7.51).

1986 Mark West began his first studies in 1986 and has continued to the present. His work and efforts are mainly realized in combination with teaching, workshops and exhibitions. In the beginning his work was more artistic and focused on columns and the tactile surface of the fabric formwork. In the 1990s he founded CAST (Centre for Architectural Structures and Technology) and made contact with Kenzo Unno and Rick Fearn (Veenendaal et al. 2011: 164–177). In 2001 he realized his own workshop. Since then, the work of CAST has become more structural, realizing

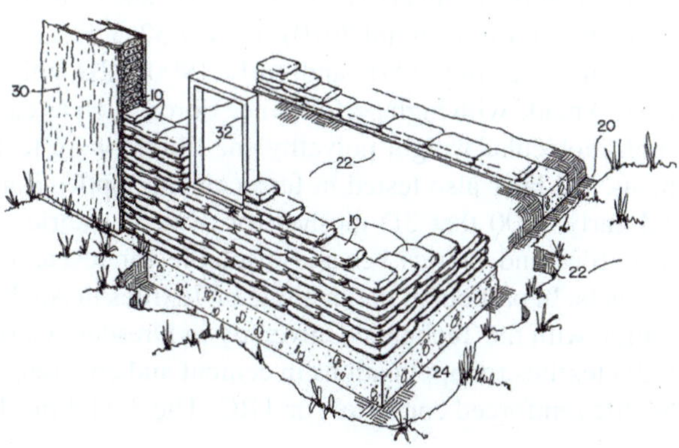

Fig. 7.49 Patent by Dicker, E. T. on the construction method of assembling bagged, settable modules. U.S. Pat. 3,922,832, 1975

Fig. 7.50 Construction of Eco dome by Cal Earth

Fig. 7.51 Realized Eco dome by Cal Earth

form optimized beam structures, with finally the "bone" beam as the most optimized beam. After 2006 the work of CAST has focused mainly on mechanically stressed shell structures (University of Manitoba, www.umanitoba.ca/cast_building/people/mark_west.html; Delijani 2010) (Figs. 7.52, 7.53 and 7.54).

During the mid-1980s and early 1990s, 2D polymeric meshes (or textiles or fabrics) made with high performance fibres, such as carbon, glass, Kevlar, or Spectra (high molecular weight polyethylene fibres) were tested for Polymeric Reinforcements and later also tested in ferrocement applications. It is only in the late 1990s and early 2000 that 3D meshes (or distance fabric) derived from the technology of textiles and fabrics became available for research studies in ferrocement type products. In particular, the Institute of Textiles in Aachen (ITA), Germany, in collaboration with the Technical University in Dresden, Germany, is pioneering a number of 3D textiles for applications in cement and concrete composites; they use the term "textile reinforced concrete", or TRC. The 3D fabrics have the advantage of placing

Fig. 7.52 Work by CAST, West and Araya (2012) (*Source* M. West and R. Araya)

Fig. 7.53 Work by CAST, West and Araya (2012) (*Source* M. West and R. Araya)

Fig. 7.54 Textile mould by CAST, West and Araya (2012) (*Source* M. West and R. Araya)

the reinforcement exactly where it is needed and tailoring its properties for particular applications. They also offer a tremendous advantage in simplifying the construction process and saving on labour cost and such 3D meshes can be readily produced in thicknesses from about 10 to 50 mm, a range perfectly suitable for ferrocement and thin cement composites applications (Figs. 7.55 and 7.56).

Fig. 7.55 Concrete Canvas detail (*Source* Concrete canvas)

Fig. 7.56 Concrete Canvas shelter 2003 (*Source* Concrete canvas)

1987 Ph.D. by Werner Sobek about structural calculations and form finding of inflatable formwork: *Auf Pneumatisch Gestützten Schalungen Hergestellte Betonschalen.* Werner Sobek researched inflatable formwork at the office of Schlaich and Bergmann and in 1983 realized a prototype for a larger shell in Heilbronn. After his Ph.D., he worked some years for Schlaich and Bergmann and became a professor, first in Hannover and later in Stuttgart. One of his research topics was inflatable structures (with overpressure but also with underpressure).

Early 1990s Kenzo Unno, Richard Fearn and Mark West have known each other since the early 1990s. Kenzo Unno (Japan) discovered the strength of plastic netting (woven geotextiles) in his search for low-cost construction methods. He developed an insulated load-bearing wall made with fabric formwork (URC "Unno Reinforced Concrete"). The padding surface of the concrete wall is used as an extra design aspect. It has been used to realize some houses in Japan (Unno 2008). The URC system is used by American architect-builder Sandy Lawton of Arro Design, Vermont, USA (Lawton and Miller-Johnson 2012). In 1993, Richard Fearn obtained a patent: Building Foundation and Floor Assembly. U.S. Pat. 5,224,321. Fearn (Fab-Form company) has introduced and realized many projects with his geotextile formwork systems for columns, foundations and walls (FAB-FORM, www.fab-form.com) (Figs. 7.57, 7.58 and 7.59).

Fig. 7.57 Wall system by Kenzo Unno (*Source* K. Unno)

Fig. 7.58 Fabric form Foundation by Richard Fearn (*Source* R. Fearn)

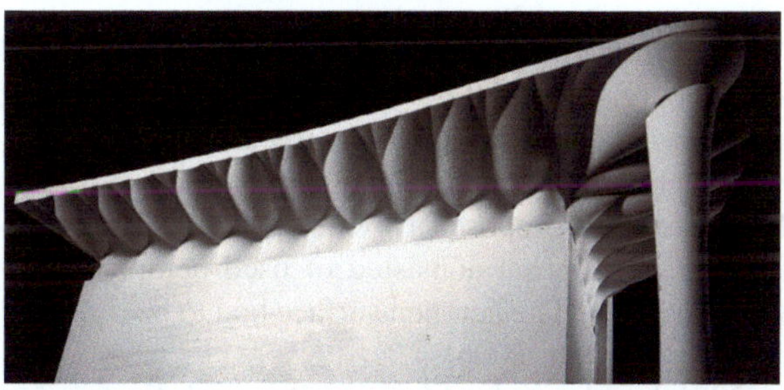

Fig. 7.59 Slab-column by A Lawton and R. Araya (*Source* A. Lawton and R. Araya)

1995 Ernesto Neto (b. 1964) began exhibiting in Brazil in 1988 and has had solo exhibitions abroad since 1995. With Vik Muniz, he represented their country at the 2001 Venice Biennale, Neto's installations are large, soft, biomorphic sculptures that fill an exhibition space, that viewers can touch, poke, and even sometimes walk on or through. These are made of white, stretchy, stocking-like material—amorphous forms stuffed with Styrofoam pellets or, on occasion, aromatic spices (Fig. 7.60).

Fig. 7.60 Suspended structure by Ernesto Neto

1997 Tsutomu Kokawa has created several ice domes in the north of Japan up to a 25-meter span. These igloos are lower and have a wider span compared to other domes. Kokawa manipulated the inflatable on the outside with cables like Schlaich did in his research. In this way he realized a geodetic dome with a grid of ice beams with optimized shell elements in between the surface of the grid. Matti Orpana made a 15 m span igloo with an inflatable mould and Arno Pronk an artificially cooled igloo with an inflatable mould in 2004. Lancelot Coar realized a wobbly structure with fibreglass bars and hanging fabric as a mould for an ice vault (Coar et al. 2016: 144). In 2014, 2015 and 2018 Arno Pronk et al. realised three world records. In 2014 a 30-meter span igloo, in 2015 a 21-meter-high ice shell and in 2018 they improved the 2015 record with the realisation of a 30,5-meter-high ice shell, the flamenco tower. In 2018 he also applied for a patent on the 3D printing of structural reinforced ice.

1996 In the late nineteen-nineties, several artists like Marcel Wanders, François Azambourg, Jurgen Bey, Rem Koolhaas and Ernesto Neto were inspired by the rigidizing of funicular polymer structures.

1999 Four Japanese architects, Ichiro Katase, Takashi Kawano, Takeshi Chiba and Ken Takeyama, proposed a low-tech balloon system for a transitional housing competition. The proposal consisted of used hempen sacks sewed together and inflated with a balloon. The hempen surface had to be rigidized with mortar. They made four prototypes (Stohr 2006).

2001 Instahome by Mankind Research Foundation Inc. in Silver Spring, Maryland USA. They introduced a building system with an inflatable roof rigidized with foamcrete. The foamcrete has a thickness of 10 cm and has a high proportion of bubbles and is sprayed and kept in place by chemicals. There is limited information about this system; probably the Instahome was never built.

2001 Research on Prada foam walls by a team from OMA in cooperation with Vincent de Rijk. The first experiments with inflatables were with a gypsum cast.

The thorough research into open cell structures with inflatable moulds has led to the building of several walls in Prada stores in the US and Italy.

2001 Arno Pronk started research on and teaching about the realization of Fluid Architecture using flexible moulding material. With the following results: 2002 realization of an optimized composite beam with inflatable mould for a stage covering in Soest. 2004 igloo in Amsterdam. 2005 a combination of synclastic and anticlastic surfaces in the realization of a composite pavilion (designed by Jurgen Bey et al.). 2006 patent for a construction method for the production of irregular shell structures. 2006 research realization of Philips pavilion. 2007 research for the construction of "Parkgarage". 2008 consultant in shell designed by Fuksas in Eindhoven. 2008 Open-cell or Bone façade and realization of smooth doubly curved composite façade element. 2009 adjustable mould for composite elements and water-filled Tensairity beam. 2010 Bone dome. 2011 adjustable mould for heat-soaking glass. 2012 tensegrity glass façade.

2002 Alan Chandler and Remo Pedreschi's students Julia Hartmann at the University of East London and Francesca Crosby at Edinburgh University both began experimenting with fabric formwork, drawing inspiration from the work of Mark West at the Centre for Architectural and Structural Technology (CAST). Inspired by this work, many projects, workshops and the Ph.D. research by Daniel Sang-Hoon Lee followed. In 2007 Alan Chandler (UEL) and Remo Pedreschi (Edinburgh) published a book about their research on Fabric Formwork (Chandler and Pedreschi 2007) (Figs. 7.61, 7.62, 7.63, 7.64 and 7.65).

2003 Concrete Canvas, UK development of a concrete impregnated textile. The concrete is held between the two layers of a distance fabric. The concrete textile was initially developed in combination with an inflatable for emergency sheltering like the 1999 Japan project (see above) but is mainly used in civil engineering and some design (art) applications (Stohr 2006: 106; Concrete Canvas, http://www.concretecanvas.co.uk/).

Fig. 7.61 The Norfolk Wall by A. Chandler of the University of East London (*Source* A. Lawton)

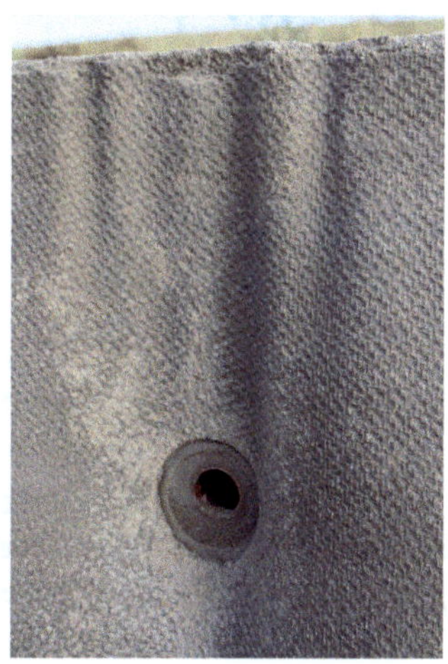

Fig. 7.66 The 2003 "Crumpled Cloth bridge" (*Source* W. Jack)

Fig. 7.67 The 2006 composite slide (*Source* W. Jack)

Fig. 7.68 The 2012 crushed wall in Cornwall (*Source* W. Jack, photo by Simon Burt)

shell. During inflation, the elements are lifted into their final position. Simultaneously, winches tighten the structure in the direction of the circumference to ensure the correct position of the elements. After inflation, the interfaces are filled with "grouting" material. Patent WO 2005/068740.

2006 Patent by Arno Pronk on a system for the production of irregular shell structures with synclastic and anticlastic surfaces. Several irregular shell structures can be made with standardized inflatables in combination with a wire mesh. The mesh can be transformed by inflatables and/or other elements as required. The mesh needs to be rigidized with mortar in order to become a shell structure. Patent NL 1032665.

2006 Jean Nouvel used a polymer foil in a rigid formwork for the decoration of concrete elements in the foyer of the Danish Broadcasting Corporation's Concert Hall.

2007 Ph.D. by G. Zimmermann, University of Kassel (Zimmermann 2007), an inflatable dome with a kind of open-cell structure. The inflatable mould consists of a two-layer membrane, connected via direct welding or flap like in formwork mattresses for marine application (see also 1960 above). The space between those weldings will act as concrete chambers. The mesh can vary on the surface and generate different patterns.

2007 Ellen de Bolster finished her Ph.D. about Conceptual Design Methodology for Modular Lightweight Structures in Cement Matrix Composites at the VUB. Tine Tysmans continued this research at the start of her Ph.D. (see 2010 below).

2008–2012 Since the first International Conference On Fabric Formwork organized by CAST at the University of Manitoba in 2008, the field has grown tremendously. The International Society Of Fabric Formwork (ISOFF) (Fabric Forming, http://fabricforming.org/) was established (In 2012 the name was changed to International Society Of Flexible Formwork). Also, the amount of research institutes and the diversity in research topics has grown. Therefore, it may happen that the list below may have some gaps.

2009 Hanil Guest House, Seoul, South Korea by Byoung Soo Cho architects (BCHO). In their research on production technology BCHO was supported by CAST. They realized this fabric formwork façade by draping a fabric over PVC pipes. The concave curves were formed by the pipes below and the convex curves by the free hanging deflections of the membrane between the supports. The elements were made horizontally and put into position afterwards (Figs. 7.69, 7.70, 7.71 and 7.72).

2009 Dani Marti made a façade for the entrance of the Westfield Centerpoint in Sydney. He made a weaving of thick ropes and made rubber/fibreglass moulds of them. The rubber moulds were used as an inlay on the rigid mould. The fibreglass concrete façade elements with woven rope pattern were cast on this rubber inlay (Orr and Marti 2012) (Figs. 7.73 and 7.74).

Fig. 7.69 "Doughnut" with inflatable mould by Zimmermann (*Source* G. Zimmermann)

Fig. 7.70 Double layered membrane by Zimmerman (*Source* G. Zimmermann)

Fig. 7.71 Hanil Guest House by BCHO 2009 (*Source* BCHO)

Fig. 7.72 Part of the façade made with fabric mould of Hanil Guest House (*Source* BCHO)

Fig. 7.73 Weaving mould for façade by Dani Marti 2009 (*Source* D. Marti)

2010 Daniel Sang-Hoon Lee did his Ph.D. Study on construction methodology and structural behaviour entitled "Fabric-formed Form-efficient Reinforced Concrete Beams". This Ph.D. was done under the supervision of Remo Pedreschi of the University of Edinburgh, Scotland, UK 2010. The aim of this research is to find an optimum

Fig. 7.74 Façade for Westfield Centerpoint in Sydney by Dani Marti 2009 (*Source* D. Marti)

form for design and construction of concrete beams made with fabric formwork. Daniel Lee has continued his work at the Royal Danish Academy of Fine Arts.

2010 FattyShell. Like the 2008 bone project at TU/e in Eindhoven, students of the University of Michigan used rubber and concrete to make an open cell structure. Grompies was the outcome of a student workshop at the AA using the elastic properties of nylon in combination with plaster, like in the bone dome by students of the TU/e.

2011 Remy & Veenhuizen designed several soft concrete benches based on the elasticity of rubber and tactility of the fabric (Figs. 7.75 and 7.76).

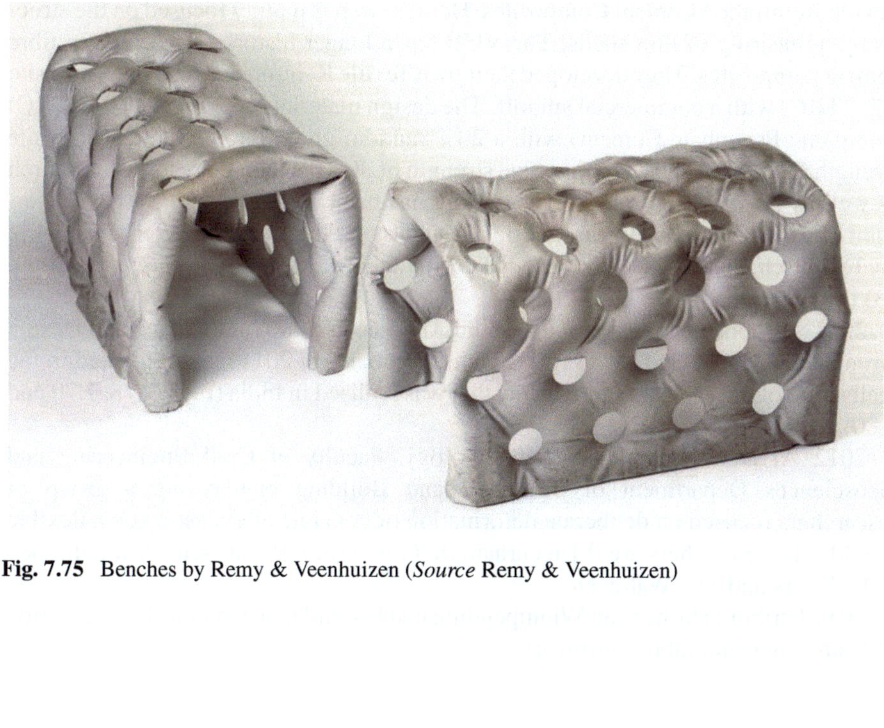

Fig. 7.75 Benches by Remy & Veenhuizen (*Source* Remy & Veenhuizen)

Fig. 7.76 Chair by Remy & Veenhuizen (*Source* Remy & Veenhuizen)

2010 T. Tysmans wrote her Ph.D. on Design of Anticlastic Shells In Innovative Textile Reinforced Cement Composites. Her research is mainly focused on the structural engineering of thin shells. The VUB has a longer history of research in fibre cement composites. They developed their own Textile Reinforced Cement Composite (IPC-TRC) with a commercial spinoff. The design material properties of (IPC-TRC) (Inorganic Phosphate Cement) with a 20% random glass fibre textile has a tensile strength of 20 MPa and a compressive strength of 40 MPa. More theoretical research as well as practical feasibility of moulding has been done by VUB (Vrije Universiteit Brussel) Marijke Moleart and Jan Wastiels and the WTCB (Wetenschappelijk en Technisch Centrum voor het Bouwbedrijf) by N. Cauberg (Cauberg et al. 2009: 53) (Fig. 7.77).

2011 S. Bhooshan works at Zaha Hadid Architects in the Computation and Design Group and is connected to the ETH as Ph.D. student. In 2011 he participated in the realisation of the shell structure below that was realised in India (Figs. 7.78, 7.79 and 7.80).

2012 At Delft University of Technology, Faculty of Civil Engineering and Geosciences Department of Structural and Building engineering a group of researchers focused on deliberate deformation of concrete after casting on a flexible mould. The researchers are P. Eigenraam, S. Grünewald, B. Janssen, H.R. Schipper, K.J. Vollers and J.C. Walraven.

2012 Fariborz Hashemian Winnipeg finished his Ph.D. on structural optimization of beams by using fabric formwork.

Fig. 7.77 Fabric mould for shell by T. Tysmans et al. (*Source* T. Tysmans)

Fig. 7.78 The waffle shell in India by S. Bhooshan et al. (*Source* S. Bhooshan)

Fig. 7.79 Support structure for waffle shell in India by S. Bhooshan et al. (*Source* S. Bhooshan)

2012 Anne Mette Manelius finished her Ph.D. "Fabric Formwork for Concrete - Investigations into Formwork Tectonics and Stereogeneity in Architectural Constructions" at the Royal Danish Academy of Fine Arts Schools of Architecture, Design and Conservation. In her research, she made different prototypes to test different ways of production. In the period 2009–2011, she realized several projects making columns, walls and furniture (Figs. 7.81 and 7.82).

2012 Tim J. Ibell, professor at the University of Bath at the time is responsible for the research into fabric-formed concrete structures. John J. Orr organized the second international conference on flexible formwork in Bath 2012. Other researchers on this topic connected to the University of Bath are A. Bak, A.P. Darby, M.C. Evernden, R. Foster J. Gartbett, H. W. Lee, T. Lewis, P. Shepherd, P. Richens, K.Z. Kostova, and Prof. C.J.K. Wiliams. In general terms, their research is focused on the structural engineering aspect of flexible formwork. Tim Ibell and John Orr are currently attached to the University of Cambridge (Figs. 7.83 and 7.84).

Since 2012, the research on flexible formwork has increased (Hawkins et al. 2016). Below, there is a list of important others in this field or research: D. Veenendaal (ETH)

Fig. 7.80 The waffle shell in India by S. Bhooshan et al. (*Source* S. Bhooshan)

Fig. 7.81 The Ambiguous Chair by A M. Manelius (*Source* A.M. Manelius)

Fig. 7.82 Demoulding of Ambiguous Chair by A M. Manelius (*Source* A.M. Manelius)

Fig. 7.83 Testing of fabric cast double T beams by J.J. Orr, A.P. Darby, T.J. Ibell and M.C. Evernden (*Source* J.J. Orr)

Fig. 7.84 Fabric cast Double T beams by J.J. Orr, A.P. Darby, T.J. Ibell and M.C. Evernden (*Source* J. Johrr)

(Veenendaal 2008; Veenendaal et al. 2011: 164–177; West 2016; Veenendaal and Block 2014: 39–50; Veenendaal and West 2016), J.J. Orr (2012; Spadea et al. 2015; Orr et al. 2013: 1236–1245), D.S.H. Lee (2011) K. Kostova (2016) W.J. Hawkins et al. (2016) and P. Shepherd (Bath), G. Tang (Nottingham), A. Chandler and Pedreschi (2007), R. Pedreschi (2013; Pedreschi and Lee 2014; Pedreschi 2015; Milne et al. 2015) (Edinburgh), T. Tysmans et al. (2012), E. Verwimp et al. (2012, 2016) (VUB), F. Huijben (2016: 1–8) (TU/e) B. Kromoser and Kollegger (2015a: 161–171; Kromoser 2015; Kromoser and Kollegger 2015b: 187–198) and F. Hashemian (2012; West and Araya 2012; West 2013) (Fig. 7.85).

7.5 Fibre Reinforcement

Most of the work in flexible moulding has been done in concrete. The developments in fibre reinforced concrete end shotcrete have been of importance to flexible moulding techniques. A brief historical summary of the developments is given below.

3,000 BCE Egyptians used mud mixed with straw to bind dried bricks. They used lime and gypsum as a binding mortar.

3,000 BCE Chinese used cementitious materials to hold bamboo in their boats and in the great wall.

1st C. BCE Fibres had been used as reinforcement since ancient times. Horsehair and straw were found in mud bricks (Graham 2003). Vitruvius mentions fibres as an additive to mortar (Vitruvius et al. 1960: 45).

Fig. 7.85 Model of shell structure at ETH by D. Veenendaal and P. Block 2012 (*Source* D. Veenendaal & P. Block Photo by Michail Lyrenmann)

c.1900 Asbestos cement used to be the most widely used fibre-reinforced concrete composite. Because of the health risk involved, there was a need to find a replacement for the asbestos used in concrete.

1950s The concept of fibre reinforced materials was researched for composites and became a topic of interest for the application in concrete.

1960s Steel, glass (GFRC), and synthetic fibres such as polypropylene fibres were used in concrete, and research into new fibre-reinforced concretes continues today.

Shotcrete

1907 Shotcrete ('gunite') was invented by American Carl Akeley, who used it to fill plaster models of animals. In this technique, the dry mortar is blasted out of the hose with compressed air. Water is injected in the nozzle with the release of the mortar (dry mix process). In 1911, Carl Akeley was granted a patent for his inventions, notably the "cement gun" (Bernard 2004: 97).

1919 Shotcrete was introduced in Europe by Carl Weber and the Torkret Company in Germany to repair damaged concrete.

1925 For the Zeiss Planetarium in Jena, Dishinger, Finsterwalder and Bauersfeld realized a reinforced dome with a span of 25 m and 6 cm thickness by spraying shotcrete against a timber formwork.

1950s The wet-mix process was invented and in the 1960s, the rotary gun, with an open hopper that could be fed continuously was introduced. The nozzle man is the person controlling the nozzle, on larger work the nozzle can be held by a mechanical arm. Small fibres are usually used to control cracking due to both plastic shrinkage and drying shrinkage.

1960s Addition of structural (steel, glass) fibres into the shotcrete improves the crack resistance (or ductility) capacity of the concrete. To process Steel Fibre Reinforced Shotcrete (SFRS) a piston pump is needed (instead of rotary gun) and control of a homogeneous distribution of the steel fibres in the concrete is important. Since the 1990s, Steel Fibre Reinforced Shotcrete (SFRS) has been researched more systematically by introducing testing methods and standards (Shah 1974: 257–262; Swammy 1992).

7.6 Pinbed Moulding—Close-Packed Matrix

Pinbed moulding uses a reconfigurable tool, it can be used for shaping mechanical parts (Buchholz 2007). In this research, the pinbed is used indirectly; to create a casting surface. The tip of the pins can describe points on a three-dimensional surface. There are two types, close- and non-close-packed pins. The pin actuation method can be mechanical, pneumatic, hydraulic or magnetic. The pin surface has to be smoothened. This can be achieved through the stiffness of the material in the void between the pins or by covering the pins with a deformable interpolating layer. Pinbed moulding can be static, or varied during the curing. Pinbed moulding can be more precise than membrane moulding. Over a period of 140 years (1863–2004), 35 patents on pin bed moulding were granted and more than half of these were granted in the past 15 years (1990–2004) (Munro and Walczyk 2007). Below, we discuss the close-packed pins and later the non-close-packed pins.

1863 The first known U.S. Patent on pinbed moulding is by J. Cochrane. U.S. Patent No. 39,886 (Fig. 7.86).

1967 Renzo Piano proposed a shell structure for the outer pavilion of the XIV Milan Triennale. This shell covered an area of about 2,500 m^2 with a rigid continuous membrane. The structure was made of a double layer of rigid polyester and expanded polyurethane, with areas of restricted light absorption. The design was never built. In his research, Piano also proposed an adjustable mould to transform his model into a one-to-one mould for the production of building elements (Massimo 1984; Piano 1969: 38–49).

2002 Based on the close-packed "free-floating" pins as researched by Hess (1931), Humphrey (1971), Nakajima (1969: 1546–1554), David Hardt et al. (1981: 140–144; 1982: 165–169), Haas and Kesselman (1996), Sullivan et al. (2000) and Walczyk et al. (1998: 287–296). The Cyril Bath Company (Cyril Bath Company, www.cyrilbath.com) realized their own pinbed moulding machine. The so-called stretch forming method will stretch plane plates over a pinbed mould. Ph.D. student Sebastian Boers (TU/e) developed an alternative system in 2006 with zigzag steel sheets to separate the parallel pins and established his company Optimal Forming (Boers, www.optimalforming.com) (Fig. 7.87).

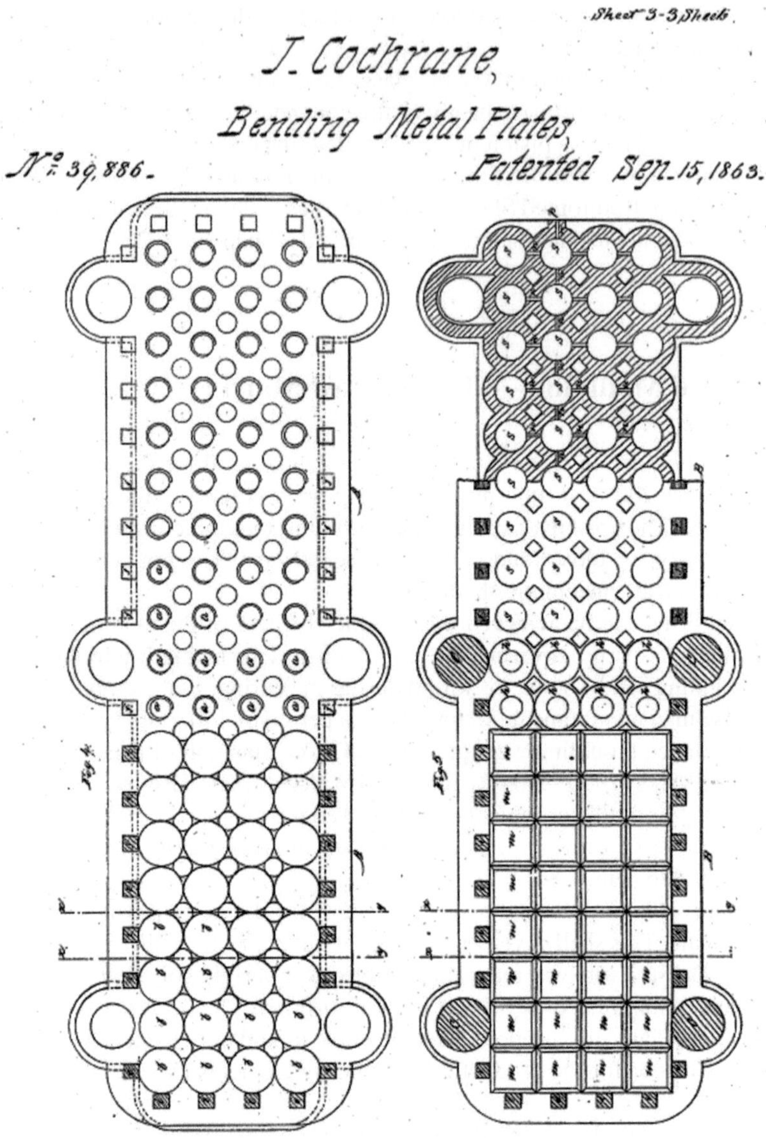

Fig. 7.86 1863 patent by J. Cochrane "Bending Metal Plates"

2007 The only other known commercial applications of reconfigurable tooling based on close-packing of a pinbed is by Huntsman Advanced Materials. The mould used for this process is 268 cm long and 210 cm wide and looks like a large needle bed. The needle bed is covered with a smooth flexible foil and vacuum adjusted to the pins.

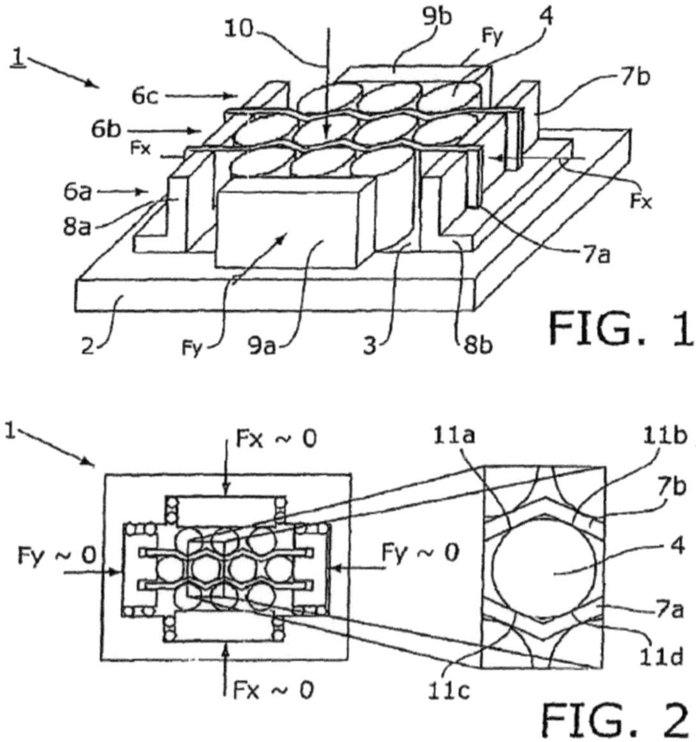

Fig. 7.87 Adjustable mould by Sebastiaan Boers 2006 Pat. WO 2008/004858

7.7 Pinbed Moulding—Non-Close-Packed

In general, the cost of this moulding method is related to the number of pins used and the way they are actuated. The wider the voids between the pins, the cheaper the method. Increasing the gap between the pins means that the deformable interpolating layer will be more important for the end result. If the gap between the pins and the interpolating layer becomes more flexible, the surface will behave like a membrane (Chap. 2). The interpolating layer can be mechanically deformed by stretching the material or by inducing a thermal change, e.g., in the case of wax or glass. The pins pushing the interpolating layer are still packed in a steel frame. The frame separates the pins and makes it possible to operate them individually.

1992 North Sails has been using 3DL™ technology, which consists of an articulating moulding surface actuated by motorized straight-line linkages located underneath. Although the tool that North Sails uses is not a pure pin-type tool, it demonstrates how a computer-based reconfigurable tool can be combined with membranes for the production of curved surfaces (North Sails, http://na.northsails.com/).

2003 Students of Arno Pronk (TU/e Eindhoven): Jeroen Helvort 2003 and Ivo van Rooy and Pieter Schinkel 2009 did their master thesis on adjustable moulding. Based on this research, Christiaan Raun and Mathias Kristensen (Aalborg University) developed their automated adjustable mould with a multilayer flexible mat in 2010 and established a company for flexible moulding (ADAPA) (Jepsen et al. 2010: 2726–2737; Kristensen and Raun 2010) (Figs. 7.88 and 7.89).

In 2007, Ph.D. student Daan Rietbergen (TU Delft) and Professor Karel Vollers obtained a patent on a technique for forming doubly curved (glass) panels by transforming a flat (glass) panel by pushing up parallel slides with curved edges. In 2010, they obtained a second patent based on individual adjusters (Vollers and Rietbergen

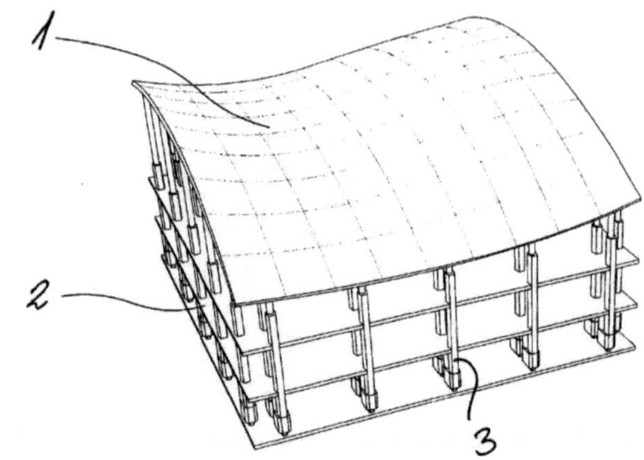

Fig. 7.88 Adjustable mould by Kristensen and Raun (2010), Pat. WO 2012/065614

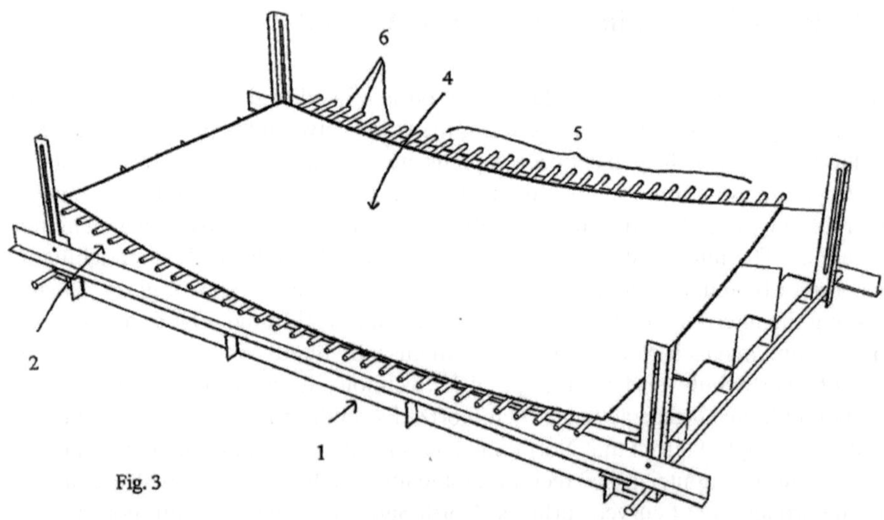

Fig. 7.89 Adjustable mould by Daan Rietbergen and Karel Vollers 2007 WO 2009/002158

2009). Karel Vollers has focused on glass and Roel Schippers on concrete (Schipper 2015; Vollers and Rietbergen 2010; Eigenraam 2013). In 2011, Pronk et al. developed an adjustable mould for the transformation of plane glass panels by manipulation of a woven steel mesh. In 2012, Karel Vollers developed a fully automated pinbed for the production of doubly curved glass panels (FD Technologies, http://fdtechnologies.eu/). The glass itself bridges the gap between the pins, removing the need for an interpolating layer. Controlling the curing of the material leads to a smooth surface without imprints (Fig. 7.90).

2011 S. Oesterle et al. (ETH Zürich) developed a method based on re-usable wax on site. The wax is poured on a flexible actuated mould. After curing, the wax elements can be assembled on-site as an infill element within standard scaffolding (Oesterle et al. 2012) (Fig. 7.91).

Fig. 7.90 Adjustable mould by Karel Vollers Free-D (*Source* K. Vollers)

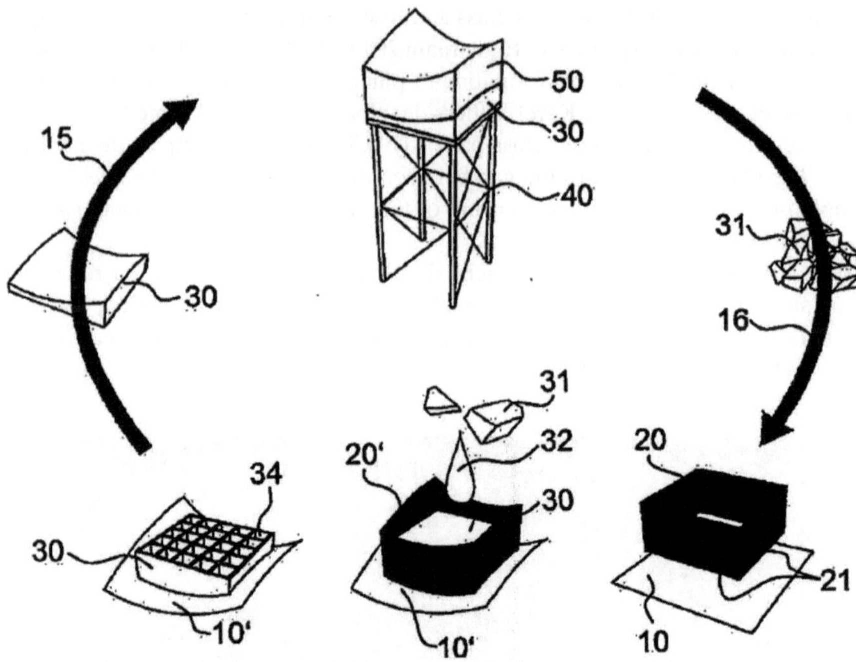

Fig. 7.91 Patent for an adjustable mould by S. Oesterle et al. wax element cast on adjustable mould

7.8 Diagram with Technical Possibilities

The list below presents an overview of the possibilities known for the materials: concrete, water/ice, polymer composites and glass. The possibilities are characterized according to 6 different aspects. This overview provides an opportunity to look for new combinations. The product and function of the product are of minor importance in this overview. Case studies are mentioned in the right-most column. Of course, this is not a definitive list. As new techniques are developed, the list can be extended and new combinations can be made.

This is an overview: the technical possibilities are not clarified further in this chapter. In the chapters about the materials concrete, ice, polymer composites and glass, all combinations and aspects will be clarified. Below you will find a diagram with an overview of the list above.

7.9 Concrete, Glass, Ice and Composites

Summarizing, form-active structures are the most favourable typology for flexible moulding. Within fluid architecture engineers should introduce surface-active

Fig. 7.92 Diagram of materials and techniques to make fluid architecture by using a flexible moulding material.

systems as much as possible. Flexible mouldings are in most cases the most efficient mould for the construction of fluid architecture. Three artists have proven that they can work with flexible moulding systems to realize their art works without a profound understanding of the structural theory behind this method. An overview of the historic development of flexible moulding and the most important combinations of flexible moulding, technique and material has been given here. The diagram in Fig. 7.92 gives an overview of all the technical possibilities. The following chapters present a summary of all the case studies in this research. The case studies are grouped by material and technique.

Bibliography

AEGION. Insituform. www.insituform.com (redirects to https://www.aegion.com/about/our-brands/insituform). Accessed 23 November 2016

Al Awwadi Ghaib M, Górski J (2001) Mechanical properties of concrete cast in fabric formworks. Cem Concr Res 31:1459–1465

Allison D (1959) Plastics move up in building. Architectural Forum 111(3)

Ault GR (1965) Self rigidizing expandable sandwich aerospace shelters and solar collectors. Archer Daniels Midland Co. and Goodyear Aerospace Corp, Akron

Bernard ES (2004) Shotcrete: More Engineering developments. Proceedings of the 2nd International Conference on Engineering Developments in Shotcrete. Taylor & Francis, London, p 97

Boers S. Optimal forming. http://www.optimalforming.com. Accessed 26 August 2012

Bosc JL (2001) Joseph Monier et la naissance du ciment armé. Editions du Linteau, Paris, p 180

Bronski M (2010) Week 39 and beyond—"Rome… by all means Rome" (And back in the U.S.A.). https://vitruviusfootsteps.wordpress.com/2010/10/01/week-39-and-beyond-%e2%80%93-%e2%80%9crome%e2%80%a6by-all-means-rome%e2%80%9d-and-back-in-the-u-s-a/. Accessed 23 November 2016
Buchholz U (2007) Casting technique enables faster manufacture of automotive components. http://www.materialstoday.com/composite-parts/features/casting-technique-enables-faster-manufacture-of/. Accessed 23 January 2017
Cadogan DP, Scarborough S (2001) Rigidizable materials for use in gossamer space inflatable structures. AIAA gossamer spacecraft form, Seattle
California Institute of Earth Architecture. CalEarth. http://calearth.org. 23 November 2016
Cauberg N, Parmentier B, Vanneste M, Mollaert M (2009) Shell elements of architectural concrete using fabric formwork—Part 1: concept. DJ Oehlers, MC Griffith, R Seracino eds, Proceedings International Symposium on Fibre Reinforced Polymer Reinforcement for Concrete Structures, FRFRCS-9, paper ID 00188 on CD-Rom, p 53
Cennini C (1954) The craftsman's handbook, vol 2. Courier Corp, North Chelmsford
Chandler A, Pedreschi R (2007) Fabric formwork. Riba Publishing, London
Chilton JC (2009) 39 etc.: Heinz Isler's infinite spectrum of new shapes for shells. Proceedings of the International Association for Shell and Spatial Structures, pp 51–62
Coar L, Mueller C, Laet L de, Hare J, Wiese K, Oberlin S (2016) Fabrigami: design and fabrication of an origami-inspired ice and fabric shell. Proceedings of the International Society of Shell and Space Structures, University of Tokyo, Tokyo, p 144
Collins P et al (2004) Concrete: the vision of a new architecture. McGill-Queen's Press, Montreal
Concrete Canvas. Concrete canvas: erosion control & containment. http://www.concretecanvas.co.uk/. Accessed 28 February 2017
Conlon C (2012) James Waller's contribution to the advancement of shell roof construction and fabric formed concrete. Paper 10, Proceedings of the second international conference on Flexible Formwork
Cyril Bath Company (2005). http://www.cyrilbath.com. 23 November 2016
Delijani F (2010) The evaluation of changes in concrete properties due to fabric formwork. University of Manitoba, Winnipeg
Eigenraam P (2013) Flexible mould for production of double-curved concrete elements. Delft University of Technology, Delft
Encyclopaedia Britannica. Plaster-work. https://www.theodora.com/encyclopedia/p2/plasterwork.html. Accessed 23 November 2016
Encyclopaedia Britannica. Tapestry. http://www.theodora.com/encyclopedia/t/tapestry.html. Accessed 23 November 2016
FAB-FORM. Fab-Form homepage. http://www.fab-form.com/. Accessed 23 November 2016
Fabric Forming. http://fabricforming.org/. Accessed 23 November 2016
FD Technologies. http://fdtechnologies.eu/. Accessed 23 November 2016
Graham T (2003) Wattle and daub: craft, conservation and Wiltshire case study. University of Bath, Dept of Architecture and Civil Engineering, Bath, Dissertation. http://www.tonygraham.co.uk/house_repair/wattle_daub/WD.html
Haas EG, Kesselman M (1996) Adjustable form die. US Patent 5,546,784
Haase W, Sobek W (2007) Vacuumatics—Deflated forms of construction. Detail 10:1148–1159
Hardt DE, Olson BA, Allison BT, Pasch K (1981) Sheet metal forming with discrete die surfaces. Ninth North American Manufacturing Research Conference Proceedings, pp 140–144
Hardt DE, Webb RD (1982) Sheet metal die forming using closed-loop shape control. CIRP Ann 31:165–169
Hashemian F (2012) Structural behaviour and optimization of moment-shaped reinforced concrete beams. University of Manitoba, Winnipeg
Hawkins WJ, Herrmann M, Ibell TJ, Kromoser B, Michaelski A, Orr JJ, Pedreschi R, Pronk ADC, Schipper HR, Shepherd P, Veenendaal D, Wansdronk R, West M (2016) Flexible formwork

technologies: a state of the art review. Structural Concrete. Ernst & Sohn Verlag für Architektur und technische Wissenschaften GmbH & Co. KG, Berlin
Head J (2012) No Nails, No Lumber: The Bubble Houses of Wallace Neff. Chronicle Books, San Francisco
Hess F (1931) Process and apparatus for manufacturing automatically accurate individual foot supports for shoes. US Patent 1,826,783
Holgate A (1997) The Art of Structural Engineering: The work of Jörg Schalich and his team. Axel Menges, Stuttgart, pp 52–55
Huijben F (2016) Vacuumatic formwork: a novel granular manufacturing technique for producing topology-optimised structures in concrete. Granular Matter 18(2):1–8
Huijben F, Herwijnen F van, Nijsse R (2009) Vacuumatics 3D formwork Systems: Customized Free-Form Solidification. Proceedings of International Conference on Textile Composites and Inflatable Structures, Stuttgart
Huijben F, Herwijnen F van, Nijsse R (2012) Structural Morphology of VACUUMATICS 3D Formwork Systems: Constructing Thin Concrete Shells with 'Nothing'. ICFF Conference proceeding paper 14
Humphrey DH (1971) Mold forming device. US Patent 3,596,869
Huybers P (1999) Draagconstructies van gebouwen, dl 10: Pneumatische constructies. TU Delft, Faculteit Civiele Techniek vakgroep Utiliteitsbouw, Delft
Isler H (1961) New Shapes for Shells. Bulletin of the International Association for Shell Structures, 8, c-3
Jepsen CR, Kristensen MK, Kirkegaard PH (2010) Flexible mould for precast concrete elements. Proceedings of International Symposium of the International Association for Shell and Spatial Structures (IASS): Spatial Structures—Temporary and permanent: Shanghai, China, November 8–12–2010. China Architecture and Building Press, Shanghai, pp 2726–2737
Kilian A, Ochsendorf J (2005) Particle-springh systems for structural form-finding. IAAS Journal 46(147)
Kostova K (2016) Design and constructability of fabric formed concrete: physical modelling for assessment of digital form finding methods. Proceedings of 11th FIB International PhD Symposium in Civil Engineering. University of Tokyo, Tokyo
Kristensen MK, Raun C (2010) A flexible mat for providing a dynamically reconfigurable double-curved moulding surface in a mould. US Patent 9,168,678 B2
Kromoser B (2015) Pneumatisches Verformen von ausgehärtetem Beton—die Errichtung von Betonschalen aus ursprünglich ebenen Platten. TU Wien, Vienna
Kromoser B, Kollegger J (2015a) Application areas for pneumatic forming of hardened concrete. Journal of the International Association for Shell and Spatial Structures (IASS) 56(3):187–198
Kromoser B, Kollegger J (2015b) Pneumatic forming of hardened concrete—Building shells in the 21st century. Structural Concrete 16(2):161–171
Lawton A, Miller-Johnson R (2012) Case study fabric formed stair. ICFF 2012 Conference Proceedings Paper 19
Lee DSH (2011) Study of construction methodology and structural behaviour of fabric formed form-efficient reinforced concrete beams. University of Edinburgh, Edinburgh
Lichtenberg JJN (2005) Slimbouwen®. Aeneas Publishers, Boxtel
Linkwitz K, Schek HJ (1971) Einige Bemerkungen zur Berechnung von vorgespannten Seilnetzkonstruktionen. Ingenieur-archiv, pp 145–158
Massimo D (1984) Renzo Piano, projects and buildings 1964–1983. Architectural Press, London
Millar W (1899) Plastering plain and decorative. London: B.T. Batsford; New York: Truslove, Hanson & Comba ltd., London & New York, pp 343–344
Milne KR, Pedreschi R, Richardson L (2015) Tailoring fabric formwork. Proceedings of the International Society of Flexible Formwork (ISOFF), Amsterdam
Munro C, Walczyk D (2007) Reconfigurable pin-type tooling: a survey of prior art and reduction to practice. Journal of Manufacturing Science and Engineering 129(551)

Nakajima N (1969) A newly developed technique to fabricate complicated dies and electrodes with wires. Bulletin of the JSME 12(54):1546–1554

Newby F (2016) Early reinforced concrete. Routledge, New York

NOE. NOE-Schaltechnik. http://www.noe.de/. Accessed 23 November 2016

North Sails. North Sails—The worldwide leader in sailmaking. http://na.northsails.com/. Accessed 23 November 2016

Oesterle S, Vansteenkiste A, Mirjan A (2012) Zero waste free-form formwork. Proceedings of the ICFF 2012, paper 24

Orr JJ (2012) Flexible formwork for concrete structures. University of Bath, Bath

Orr JJ, Darby AP, Ibell TJ, Evernden MC (2012) Optimization and durability in fabric cast 'Double T' beams. ICFF Conference Proceedings Paper 25

Orr JJ, Darby AP, Ibell TJ, Evernden MC (2013) Durability enhancements using fabric formwork. Magazine of Concrete Research 65(20):1236–1245

Orr JJ, Marti D (2012) Sydney rope wall. ICFF 2012 Conference proceedings, poster 15

Otto F (1962a) Zugbeanspruchte Konstruktionen, vol 1. Ullstein, Frankfurt am Main

Otto F (1962b) Zugbeanspruchte Konstruktionen, vol 2. Ullstein, Frankfurt am Main

Otto F (1973) Tensile structures; design, structure, and calculation of buildings of cables, nets, and membranes, vol 1–2. The MIT Press, Cambridge, MA

Pedreschi R (2013) Fabric formed concrete structures and architectural elements. Proceedings of Structures and Architecture: New concepts, applications and challenges, ICSA 2013 Second International Conference on Structures and Architecture, Guimarães

Pedreschi R (2015) A preliminary study of the strength of non-prismatic columns using fabric formwork. Proceedings of the International Society of Flexible Formwork, Amsterdam

Pedreschi R, Lee DSH (2014) Structure, form and construction. Fabric Formwork for Concrete. Proceedings of Across: Architectural Research through to Practice: 48th International Conference of the Architectural Science Association 2014

Piano R (1969) Progettazione sperimentale per strutture a guscio—Experimental project of shell structures. Casabella 335:38–49

Pronk ADC et al (2006) TU Delft. http://bk.tudelft.nl. Accessed 13 November 2016

Roland C (1965) Frei Otto-Spannweiten; Ideen und Versuche zum Leichtbau. Ein Werkstattbericht. Ullstein, Berlin

Roman Aqueducts. Cherchell (Algeria). http://www.romanaqueducts.info/aquasite/cherchell/index.html. Accessed 23 November 2016

Scarre C (2005) The human past: World prehistory and the development of human societies. Thames & Hudson, London

Schipper HR (2015) Double-curved precast concrete elements: Research into technical viability of the flexible mould method. Delft University of Technology, Delft

Schlaich J, Schüller M (1999) Ingenieurbau Führer Baden-Württemberg. BauwerkVerlag GmbH, Berlin

Shah SP (1974) New reinforcing materials in concrete. Journal of the American Concrete Institute, 257–262

Sobek W (1987) Auf Pneumatisch Gestützten Schalungen Hergestellte Betonschalen, Stuttgart

Solomita P (2015) Pier Luigi Nervi vaulted architecture: towards new structures. University Press, Bologna

Spadea S, Orr J, Yang Y (2015) Bespoke reinforcement for optimised concrete structures. Proceedings of IASS Annual Symposium: Future Visions—IASS/ISOFF

Stohr K (2006) Design like you give a damn: Architectural responses to humanitarian crises. Metropolis Books, New York, p 106

Sullivan EV, Haas EG, Schwarz RC, Kesselman M, Peck AN, Papazian JM (2000) Individual motor pin module. US Patent 6,012,314

Swammy RN (1992) Fibre reinforced cement and concrete. Proceedings of the 4th International Symposium held by E. RILEM & F.N. Spon 17

Tysmans T, Wastiels J, Adriaenssens S, Verwimp E (2012) Flexible reinforcement systems for spatially curved concrete structures. Proceedings of the 2nd International Conference on Flexible Formwork, 2012. University of Bath, Bath

University of Manitoba. Mark West. http://www.umanitoba.ca/cast_building/people/mark_west.html. Accessed 23 November 2016

Unno K (2008) Lecture by Kenzo Unno. Fabric Formwork Conference Speakers Presented at the First International Conference on Fabric Formwork

Veenendaal D (2008) Evolutionary optimisation of fabric formed structural elements. Master's thesis

Veenendaal D, Block P (2014) Design process for prototype concrete shells using a hybrid cable net and fabric formwork. Eng Struct 75:39–50

Veenendaal D, West M (eds) (2016) The fabric formwork book: methods for building new architectural and structural forms in concrete. Routledge, New York

Veenendaal D, West M, Block P (2011) History and overview of fabric formwork: using fabrics for concrete casting. Structural Concrete 12(3):164–177

Veltkamp M (2007) Free form structural design, schemes, systems & prototypes of structures for irregular shaped buildings. Research in Architectural Engineering series vol 6. Delft University Press, Delft

Verhaegh RWA (2010) Free forms in concrete: the fabrication of free-form concrete segments using fabric formwork. TU Eindhoven, Eindhoven

Verwimp E, Remy O, Wastiels J, Tysmans T (2012) Structural stay-in-place formwork in textile reinforced cement composites for very slender concrete columns. Proceedings of the 2nd International Conference on Flexible Formwork, 2012. University of Bath, Bath

Verwimp E, Tysmans T, Mollaert M (2016) Numerical evaluation of structural stay-in-place formwork in textile reinforced cement composite for concrete shells. Advances in Structural Engineering 19(4)

Vitruvius P, Morgan MH, Langford Warren H (1960) Vitruvius the ten books on architecture. Dover Publications, Mineola, p 45

Vollers KJ, Rietbergen D (2009) A method and apparatus for forming a double-curved panel from a flat panel. US Patent 12/633,055

Vollers KJ, Rietbergen D (2010) Werkwijze en Mal voor het Vervaardigen van een Gebogen Paneel. Delft University of Technology, Delft

Walczyk DF, Lakshmikanthan J, Kirk DR (1998) Development of a reconfigurable tool for forming of aircraft panels. J. Manufacturing Systems 17(4):287–296

West M (2013) Fabric-formed pre-cast panels—Simple, economical formwork for natural forms and structural efficiency. Proceedings of Façade Tectonics: Precast Concrete, USA

West M (2016) The fabric formwork book: methods for building new architectural and structural forms in concrete. Routledge, London

West M, Araya R (2012) Recent fabric formwork construction projects. Proceedings of the Second International Conference on Flexible Formwork, University of Bath

Zimmermann G (2007) Membran Beton Gitterschalen Tragwerke—Entwicklung und Vorbemessung. Books on Demand

Chapter 8
All the Possibilities Researched for Ice

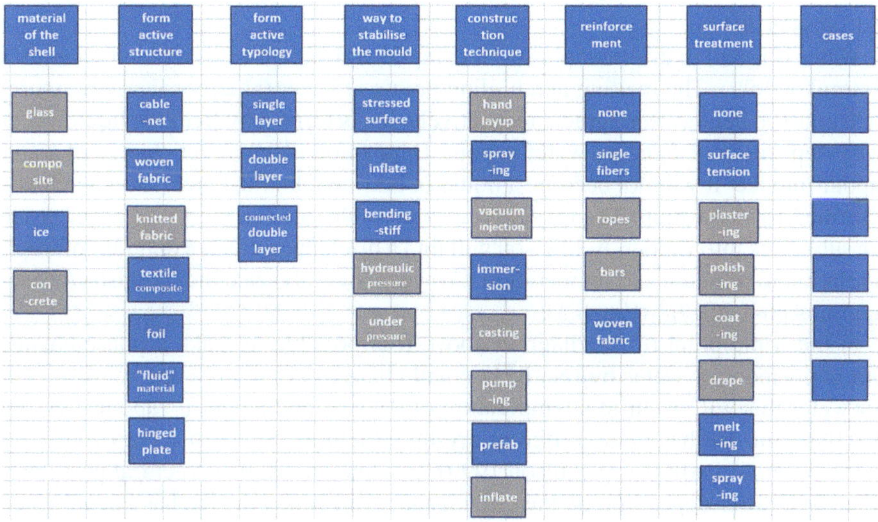

Fig. 8.1 Diagram for ice

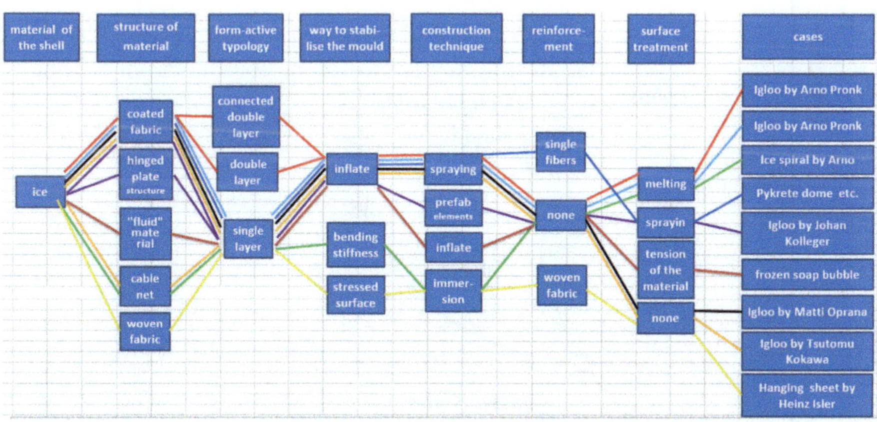

Fig. 8.2 Case studies for ice and the techniques to make fluid architecture by rigidizing form-active structures

A. Pronk, *Flexible Forming for Fluid Architecture*,
https://doi.org/10.1007/978-3-030-71551-9_8

8.1 Introduction

This introduction is a brief overview for the material ice. The focus will be on ice as a construction material. The paragraphs with different applications of case studies will show non-civil structures (Figs. 8.1 and 8.2). This is just an introduction with a focus on the application of ice. Please look at the 2013 master thesis of Janssen and Houben for a more specific introduction to the material properties of ice (Houben and Janssen 2013).

Water is found in different phases: ice, liquid and vapour in significant quantities (Petrenko 1999). The melting and boiling points of water (at atmospheric pressure) are 0 and 100 degrees centigrade, respectively. Ice is formed by freezing liquid water or by deposition of freezing water vapour. Deposition is the direct change of water vapour into ice. It will not become a liquid in between. For instance, snow in clouds is formed by deposition. The reverse is also possible. Ice can melt into water or ice can sublimate into vapour. Below you will find a brief phase change diagram of water in relation to pressure and temperature (Alyvea, www.alyvea.com/chemistry/phase-diagrams.ph-1-2017p) (Fig. 8.3).

In the triple point of the diagram above it is possible to make a phase change into ice, vapour or water by relatively small changes in pressure and temperature. For water this will occur at a temperature of 0.01 °C and a partial vapour pressure of 0.006 atm. Ice has many different crystal structures. Some of them occur only under extreme conditions (Schulson 1999). It is known that the structural properties of ice will differ depending on the circumstances in which the ice has been formed

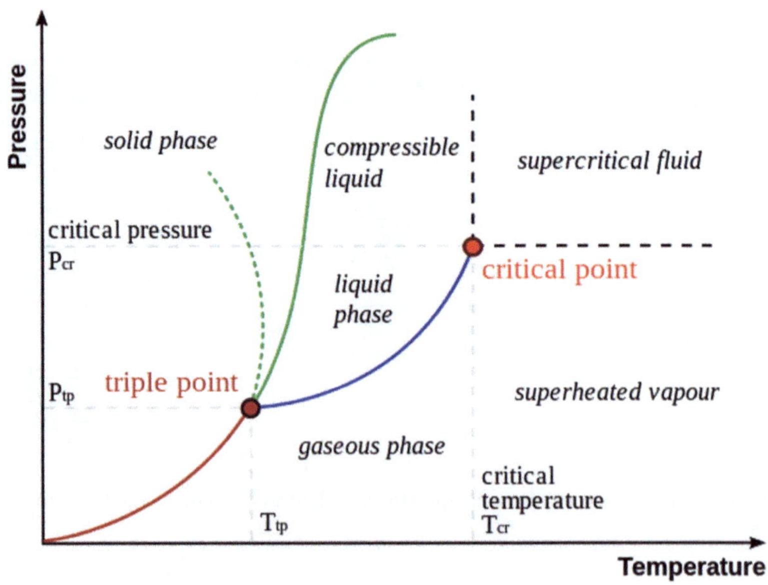

Fig. 8.3 Phase change of water

(Hooke et al. 1980). Nevertheless, we will focus on ordinary ice formed under regular conditions. Snow is a combination of ice and air formed at a temperature close to the melting point of water. Snow can be classified by its density, grain size and water content (Petrovic 2003). Both ice and snow have been used to make structures. The oldest known structure is the igloo. Nowadays the engineering of ice has many applications, for instance: offshore, mining, recreation, bridges, roads, polar missions and even space missions (Makkonen 1994; Doran et al. 2010). Ice is a cheap and omnipresent material in cold regions. The use of local resources is a great benefit because of the high transportation cost of other building materials in those areas.

The crystal structure and orientation affect the mechanical behaviour of ice as does the mixture of ice with soil and fibres (Schwartz 1980). Pure ice can be compared to brittle ceramics although the strength, elasticity and toughness are significantly lower. Snow is much weaker than ice. Under load, Ice can vary from brittle to ductile (Schroedel 2002). This depends on the temperature, density and crystal structure.

Ice has about 1/10th of its compressive strength in tension. For snow this is even lower. High-density snow-ice is usually considered to be half as strong as clear blue ice as to tensile strength. The tensile strength can vary from 0.7 MPa to 3.1 MPa. The usual tensile strength at a temperature range of −10 to −20 °C is 1.43 MPa. The flexural strength, determined by a three- (or four-)point bending test, has a range of 0.5 to 3 MPa, with an average of 1.73 MPa, at −5 °C or lower (Bureau Veritas 2010). The crushing strength of ice in a temperature range of −10 to −20 °C is 8 to 10 MPa. The compressive strength of ice in the same temperature range is between 5 and 25 MPa. The strength of ice increases in both tension and compression with decreasing temperature. The effect of temperature is much stronger in compression than in tension. Both the strength of ice and the mode of failure depend on the loading rate and the history of loading. Even the slope of the stress-strain diagram, conventionally used for determining the elastic modulus, depends on the loading rate. Therefore, it is important to report strength and modulus in terms of strain rate (Makkonen 1994).

The ductility of ice depends on the following factors: strain-rate, stress, temperature, ice type and grain size. In tension, the ductile/brittle transition is relatively abrupt and strain rate insensitive whereas in compression this is more gradual and strain rate sensitive (Gold 1977: 197–212).

The elastic modulus (relation between stress and strain) of ice depends on: temperature, crystal structure and stress application. The elastic modulus can vary 30%, and even more when close to the melting point. The average young's modulus (parallel to the c-axis of the crystal) is 12 GPa and 8.6 GPa in the most compliant direction. The young's modulus and Poisson ratio for random orientated polycrystals is 9.0–11.2 MPa and 0.29–0.33 MPa.

All mechanical properties of ice are sensitive to temperature changes. At temperatures near the melting point ice becomes very ductile and prone to creep. Creep will occur very soon, especially at higher levels of stress. The creep behaviour of ice is influenced by the crystal structure, load, grain size, impurity content and temperature. Ice has an anisotropic creep behaviour due to the hexagonal structure of the ice crystals. The creep behaviour of a (snow-) ice dome is described by T. Kokawa. It is notable that creep deflection has a linear function at the beginning, but the deflection rate increases with time until collapse. Collapse occurs after temperatures above freezing persist for 2–3 days. Because of the creep, a large deformation (accelerating creep) will be visible before collapse, and this therefore predicts the collapse (Kokawa and Watanabe 2011).

Stationary water freezes from the outside in and water expands when it becomes ice. Therefore, the pressure of water trapped in ice can get very high and will be big enough to break the surrounding surface of ice. Methods to prevent the surface from freezing faster than the core are to insulate parts of the surrounding surface, to freeze thin layers of ice or to spray water on a surface of ice.

Ice, like most materials, becomes denser at lower temperatures. The changes are equal in different directions. Cooling or heating the ice by more than 15 °C will create stresses leading to fracture, i.e. thermal shock. Experiments by J.J. Petrovic (Petrovic 2003) showed that a shell of 2–3 centimetres has a 50% chance that fractures will occur at a temperature difference of 15 °C and 100% at 20 °C.

The traditional way to build an ice structure is by cutting bricklike elements out of snow or ice. These have to be stacked together and the cracks and spaces between the elements have to be filled with snow. The inside of the building will generally be smoothed to avoid localized dripping of meltwater. The fluctuation of the outside and inside temperatures will freeze and melt the inner surface of the structure. After some time, the structure will form a continuous shell (Anderes 1983). Alternative ways to make an ice/snow structure have all been developed by or since the work of Heinz Isler in the late 1950s. Those methods are based on the inflation or pre-tension of membranes and cables for moulding (Steurer-Systems, www.steurer-systems.com/de/produkte/schrank-systeme/igluformer/der-igluformer) (Figs. 8.4 and 8.5).

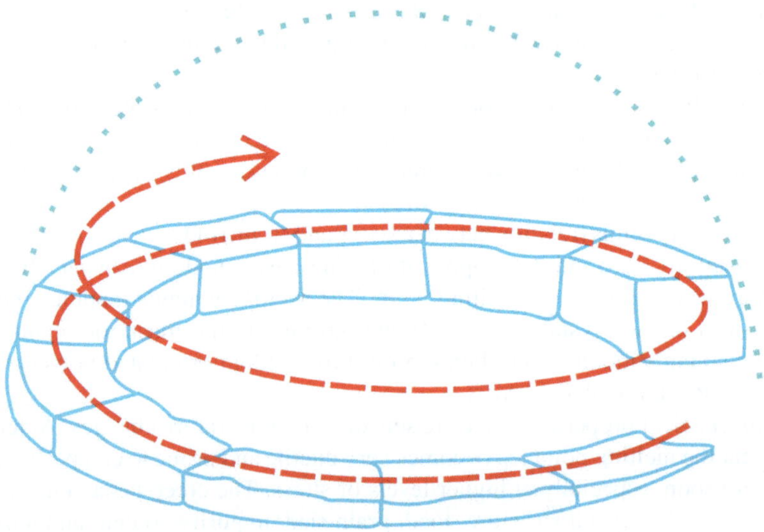

Fig. 8.4 Construction of an igloo

Fig. 8.5 Realization of igloo

Case Studies

The cases below are split into two groups:

- ice structures in an atmospheric temperature below 0 °C.
- ice structures in an atmospheric temperature above 0 °C.

8.2 Ice Structures in an Atmospheric Temperature Below 0 °C

8.2.1 Heinz Isler's Hanging Sheet

Flexible Mould
The form-active structure: woven fabric (Fig. 8.6).
Form-active typology: single layer.
Way to stabilize the mould: stressed surface.

Composite Material
Technique to handle the rigidizing material: submersion/sprayed.
Reinforcement of the material: woven fabric.
Surface treatment: none.

Project
Name and location: Hanging Sheet, Switzerland.
Design and Engineering: Heinz Isler (1926–2009).

Fig. 8.6 Isler's experiments with ice structures (*Source* ETH Zurich)

Contractor: None.
Year: ca. 1957.

Introduction

Heinz Isler (1926–2009) worked with fabric formwork to design and produce shell-structures made of ice. Isler was known as an innovative and out-of-the-box thinking man. He used gravity in his form finding for funicular and stable forms for shell structures. He started experimenting with physical models around 1957. One of his working methods involved the use of suspended stretched sheets to be frozen in the open air. Every winter he did some experiments by freezing fabric or spraying water over objects like plants, inflated balloons, hanging ropes and fine garden netting. Isler showed a variety of thin doubly curved shell structures in ice. Isler, like Gaudi, concluded that the reversed shape of hanging cloth will be the optimized shape for shell-structures. Based on his models, he considered the constructive, economic and aesthetic aspects of his shell structures (Chilton 2000; Abel and Chilton 2011; Prerost 1990; Billington and Jameson 2003; Isler 1972; Stanley and Glockner 1975: 447–462; Otto 1977: 118–120).

Material

Heinz Isler lived in Switzerland. In winter, the average temperature was below 0 °C. These climatic conditions were ideal for Isler to experiment with flexible models in his own garden. He used anything that was within his direct reach, such as branches, bushes, nets, towels, sheets, balloons, etc. Fabric is elastic, it can handle tensile forces and will wrinkle in a direction that is roughly perpendicular to the tensile force. Moreover, it is relatively cheap. The combination of ice and fabric will convert into a self-supporting shell. These properties make frozen fabrics outstanding for form-finding experiments for shell structures.

Form

In his modelling, Isler used the pull of gravity on the (wet) fabric as well as external forces. By adding water to the fabric, the fabric sags due to the forces of gravity. The hanging sheet has one or more mounting points. The zeroclastic surface has become a synclastic dome shape. There might be some wrinkles in the direction of the mounting points and the edges of the surface will become anticlastic. The anticlastic edges are typical for the shells developed by Isler. The form-finding method is demonstrated in Fig. 8.7, where the fabric is mounted between four points.

In case there is one mounting point in the middle, the fabric will be pushed upwards by the column and will hang like a tent-structure (Fig. 8.6). The sheet will hang down with strong wrinkles from the mounting point in the middle. The strong curvature of the wrinkles has a positive influence on the stability of the shell structure. Strong local curvature has a positive effect on the prevention of buckling (Ramm 2011: 143–154; Chilton 2012; Pronk and Dominicus 2012).

Fig. 8.7 The shape of the wet hanging sheet, a negative dome-shape (*Source* J. Chilton)

8.2.2 Frozen Soap Bubble

Flexible Mould
The form-active structure: the material of the shell in fluid condition (Fig. 8.8).
Form-active typology: single layer.
Way to stabilize the mould: inflate.

Composite Material
Technique to handle the rigidizing material: inflate.
Reinforcement of the material: none.
Surface treatment: surface tension of the material.

Project
Name and location: Frozen Soap Bubble.
Design and Engineering: None.
Contractor: None.
Year: ?

Soap film and soap bubble structures are important in form finding of flexible and form-active structures. A soap bubble consists of a liquid layer with a certain surface tension. In the case of water, the surface is not stable. Soap will decrease the surface tension to a third of that of pure water and will stabilize the surface though the Marangoni effect: As the soap film stretches, the surface concentration of soap decreases, which in turn causes the surface tension to increase. The soap strengthens

Fig. 8.8 Frozen soap bubble

the weakest parts of the surface and prevents the other parts from stretching excessively. The shape of a soap film or soap bubble is caused by the surface tension in combination with external forces. Without any external forces there would be a perfect minimal surface. In case of an inflatable bubble this would be a sphere. The sphere will be deformed due to gravity and the movement of air. Many other forms are possible when bubbles and soap films are combined with other bubbles, cables, points or boundary elements.

Freezing

In air below −25 °C soap bubbles will form a sphere and freeze while they are floating. The surface is very thin and when they hit an object they will shatter. When the bubble is filled with warm air it will form small almost perfect spheres. When the air is cooled there will be a partial collapse of the surface due to the decrease in relative pressure of the air inside the bubble. In 2016 students of the KU Leuven did some experiments with the freezing of combined bubbles and soap films between boundaries. The polyhedral soap film so formed was frozen successfully. The limited size and thickness of the frozen soap-structures are problematic for practical applications. If this limiting factor can be overcome, it might be possible

Fig. 8.9 Multiple soap bubbles

to find useful applications for this technique (Hutchings et al. 2000: 45–49; Taylor 1976: 489–539; Oprea 2000; Boys 1890; Isenberg 1992; Stein 2005) (Fig. 8.9).

8.2.3 Ice Shells with Inflatable Formwork

Flexible Mould
The form-active structure: coated fabric/foil (Figs. 8.10 and 8.11).
Form-active typology: single layer.
Way to stabilize the mould: inflate.

Composite Material
Technique to handle the rigidizing material: spraying.
Reinforcement of the material: none.
Surface treatment: none.

Project
Name and location: Several projects by Heinz Isler, Matti Orpana, Steurer, Carl Koepcke and Marshall Prado.
Year: 1957, 2002, 2005, 2012.

Beside the winter experiments with hanging materials, Heinz Isler also did some experiments with balloons sprayed with water while it was freezing (Pronk et al. 2016a). This resulted in an ice palace. The firm Steurer Systems

Fig. 8.10 Inflatable mould for igloo (Photo by Bart van Overbeeke)

Fig. 8.11 Inflation of mould (Photo Bart van Overbeeke)

have made a commercial application, using inflatables to make igloos for pleasure (Steurer-Systems, www.steurer-systems.com/de/produkte/schrank-systeme/igl uformer/der-igluformer). Matti Orpana of Tensotech Consulting did some extraordinary things. In 2005 he made a very big and possibly one of the biggest single surface inflatables till now in Finland. The inflatable building is 180 meters long, 90 meters wide and 23 meters high. It is designed for Finnish weather conditions and snow load. The inner surface of the membrane is covered with 50 mm of insulation (Fig. 8.12).

In the north of Finland, Matti Orpana made igloos with a 15-meter span and a 15-meter height. They are the biggest single-surface igloos made with an inflatable mould. The inflatable is covered with ice or snow. The vertical section of an igloo has the form of a catenary. In ice the walls are 90 cm thick and in snow the walls are 3 meters thick. Due to the thickness of the surface, the dead load of the igloo is dominant compared to the variable load and there will be compression only. The igloos are made at −15 °C; They remain for one season: 5 months from November to April. The igloos are used as ice hotels with various types of accommodations (Figs. 8.13, 8.14 and 8.15).

Carl Koepcke and Marshall Prado from the Harvard Graduate School of Design realized an ice shelter based on the experiments by Heinz Isler with inflatables and ice (Fig. 8.4). The project combines the use of ice and spraying water with the parametric modelling and digital fabrication of cutting patterns. An ice shell of approximately ½ inch thickness was capable of supporting itself without further bracing.

Technique

A structural form was developed to create a self-supporting, thin walled ice shell. The form for the mould was created by connecting 92 plastic panels. Such panels supply a lightweight and packable guideline for the shelter. These plastic panels define the inflatable framework for the shelter (Cilento 2012; Futures Plus, futuresplus.net/2012/04/06/fabricating-grounds-ice-shelter-ins tallation-koepcke-prado; GSD News, www.gsd.harvard.edu/2012/03/carl-koepcke-march-12-and-marshall-prado-march-11-featured-in).

Fig. 8.12 Inflatable sports hall of 90 by 180 m by Matti Orpana (*Source* M. Orpana)

Fig. 8.13 Covering of inflatable with snow (*Source* M. Orpana)

8.2.4 Ice Domes by Tsutomu Kokawa

Flexible Mould
The form-active structure: Coated fabric and cable net (Fig. 8.16).
Form-active typology: Single layer.
Way to stabilize the mould: Inflate.

Composite Material
Technique to handle the rigidizing of the material: Spraying.
Reinforcement of the material: None.
Surface treatment: None.

Project
Name and location: Ice Shell by Tsutomu Kokawa, Hokkaido, Japan.
Design and Engineering: Tsutomu Kokawa.
Contractor: Tokai University.
Year: 1997–2012.

Introduction

Tsutomu Kokawa was a professor at Hokkaido Tokai University in Japan. Hokkaido is the northernmost island of the 4 main islands of Japan. It has a continental climate

Fig. 8.14 Inflatable covered with ice by spraying water (*Source* M. Orpana)

with rain and snow (100 mm/month) spread over the year. The average temperature in the period from December to March is below 0 °C, which is relatively moderate when compared to the north of Finland, where the average temperature is −20 °C. Hokkaido has the same north latitude as Monaco while the north of Finland is above the polar circle. In Hokkaido, the winter period is shorter and less cold but with more snow when compared to the north of Finland. Since 1997, Tsutomu Kokawa has created several ice shells in Japan. His igloos are lower and can have a wider span than the ones created in Finland by M. Orpana. Therefore, he had to optimize the structure with a grid of ice beams. Because of these supports, the rest of the structure can be thinner. He made several test builds with 20-meter span ice domes. In 2012 he made a dome with a diameter of 25 meters (Kokawa et al. 2012; Kokawa 2002: 93–100).

Kokawa's ice shells are constructed by inflating a two-dimensional membrane covered with ropes anchored to a snow-ice foundation. The form of the inflatable is controlled by changing the length and geometric pattern of the ropes. The membrane

Fig. 8.15 Inside of snow igloo with ice bar (*Source* M. Orpana)

Fig. 8.16 Ice structures by Tsutomu Kokawa (*Source* T. Kokawa)

does not require three-dimensional cutting patterns. Most of the forces are handled by the ropes. The limited demands on the membrane make it an easy, flexible and cheap system (Kokawa 1984: 122–138; Kokawa et al., 2000: 23–29) (Fig. 8.17).

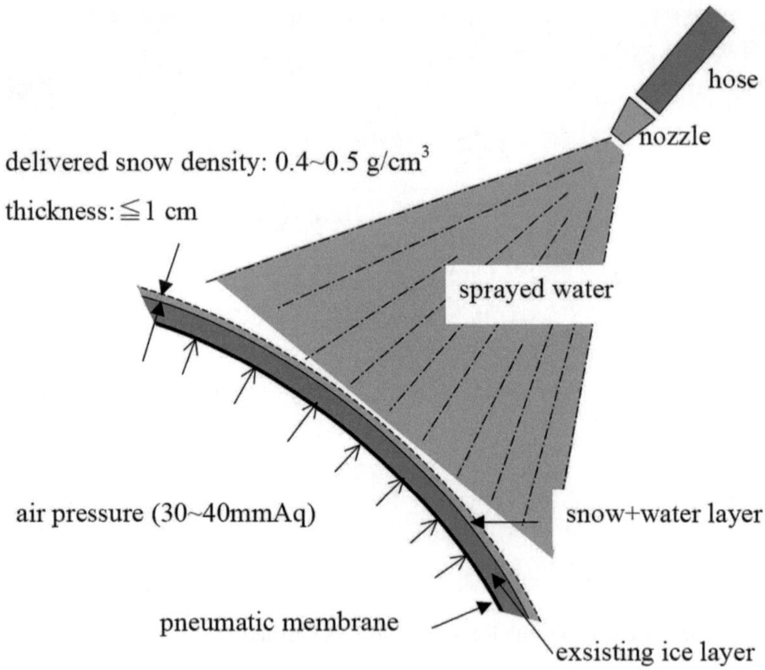

Fig. 8.17 Application of snow and water, Snow is sprayed on the membrane together with sprayed water for creating the skin of the ice shell (*Source* T. Kokawa)

The inflatable is covered with a 1 cm snow layer by blowing milled snow with a rotary snow blower. A fog of water is sprayed on top of the snow with a high-pressure adjustable nozzle. The water is held by the snow and will freeze to the surface of the layer below. The structure can be realized at temperatures below −10 °C. It is necessary to build up the igloo by thin layers of about 1 cm each to avoid deflections of the inflatable and to prevent geometrical imperfections or a heterogeneously built surface. The enclosure of a layer of dry snow might cause delamination of the shell structure. The application of snow and water is repeated until the desired shell thickness (1/100th of the span) is reached. It normally takes 1.5 h to attain 1 cm thickness.

After finishing the shell structure, the inflatable membrane and ropes can be removed for reuse. The geometric pattern of the ropes will make a grid of ice beams at the inside of the shell structure. In the Netherlands the firm Beton Ballon (BetonBallon Projecten, www.betonballon.nl) has used almost the same principle for making concrete structures. In 2005 they applied for a Dutch patent and in 2007 a world patent (van Hove 2007). In 2015 they applied for a new patent, also based on an inflatable (van Hove 2015). In 1999 David South applied for a patent were he used an inflatable with a rope net for making large dome structures in concrete (South 1999). Tsutomu Kokawa applied for a Japanese patent and has practised the combination of inflatables with tensioned cables and a curing material since the beginning of the 1980s (Kokawa 2005: 973–982; 2006: 8).

Technique

Below, the construction of an ice dome with an inflatable dome is explained (Kokawa 1987) (Figs. 8.18, 8.19, 8.20, 8.21, 8.22 and 8.23).

Foundation

The end rope, which is connected to the cover ropes covering the membrane bag, is anchored to a snow-ice foundation. This foundation is constructed by using plenty

Fig. 8.18 Construction of snow-ice foundation ring (*Source* T. Kokawa)

Fig. 8.19 Two-dimensional PVC membrane bag is placed on the snow ground. The membrane bag is connected along the outskirt of two plane pieces (*Source* T. Kokawa)

Fig. 8.20 Three-dimensional formwork by inflating a two-dimensional membrane bag covered with reticular ropes anchored to the snow-ice foundation ring. The bag is inflated by a portable air blower which also automatically keeps a constant air pressure under construction (*Source* T. Kokawa)

Fig. 8.21 Covering the formwork with a thin snow-ice layer (less than 1 cm) by blowing milled snow with a rotary snow blower, spraying water and letting it freeze naturally at temperatures below −10 °C (*Source* T. Kokawa)

Fig. 8.22 Repeating the application of snow and water until the desired shell thickness is reached, then removing the bag and ropes for reuse (Photo by B. v. Overbeeke)

Fig. 8.23 After the surface of the ice shell is cured, the membrane is deflated. The boards, ropes and membrane are removed and the ice shell is finished (*Source* T. Kokawa)

of snow in combination with water before the inflation of the formwork. The tension force in the rope which is formed by its curvature and the load of the membrane bag has to be smaller than the tension strength of the rope. The positioning of the anchor in the ring and its sectional area should be carefully designed. If the anchor is not strong enough, the formwork becomes unstable and runs the risk of breaking down. Besides the position of the anchor, it is also important to smoothly connect the lower part of the shell to the ring. If this is not done carefully and if the lower part of the formwork does not touch the ring, it will take very long to fill up the clearance with blown snow. In addition, the width of the foundation has to be carefully designed. Working space for spraying water on the ring has to be included in the dimensions of the ring. With a large dome, a heavy construction machine such as a large snow blower for applying snow on the foundation ring has to be taken into consideration when dimensioning the ring (Kokawa 2012) (Fig. 8.24).

Spraying Technique to Handle the Rigidizing Material

After the three-dimensional formwork of a two-dimensional membrane bag with ropes is inflated, the skin of the ice shell can be made. To quickly produce a high-quality ice to cover the membrane bag, several devices are needed. A rotary snow blower blows about 0.4–0.5 g/cm^3 of crushed snow onto the membrane bag with an interval of 45–90 min. Each application takes from 10 to 30 min, depending on the size of the shell. During this application, a high-pressure adjustable nozzle sprays tap water onto the snow. The amount of water sprayed on the snow depends on the size of the dome. For a dome of 10 m it has a broad range from 14 to 40 l/min. The snow absorbs the water thus preventing water from running down the sides. Snow which has been sprayed with tap water is called “milled snow”. An outside temperature of

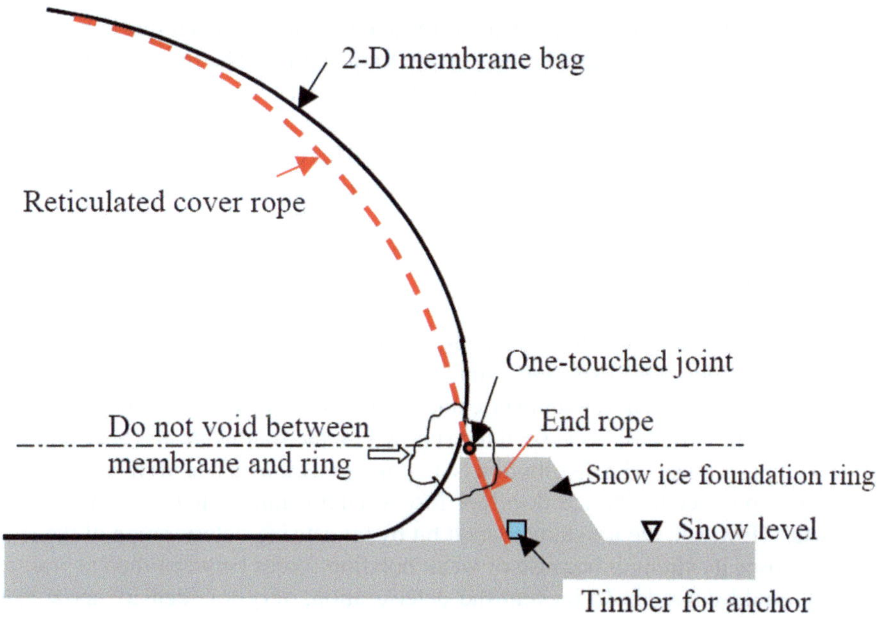

Fig. 8.24 Foundation detail (*Source* T. Kokawa)

−10 °C ensures that the snow-water mixture freezes solid. A mix of snow and water or snow-ice solidifies more quickly than if only applying water (Kokawa 2011).

While spraying milled snow onto the membrane, it is important to keep the snow depth to a thickness of less than 1 cm. Otherwise, only the upper layer of the snow will be frozen, which leads to a sandwich structure with snow between the ice surfaces. This "sandwich snow" leads to imperfections in the completed ice shell. It is important to complete this step with meticulous care. It takes about 1.5 h to obtain a thickness of 1 cm with a density of about 0.83–0.88 g/cm^3. After reaching a certain value, it can support its own weight and the weight of the new snow-water layer. The spraying of water and snow is repeated until the right thickness is reached: this is normally about 1/100th of the span.

Material

Snice is a wet kind of snow made with a snow cannon. In addition to other uses, snow and ice can be used for creating ice shells (Kokawa 2006). Because an ice shell is only used for about three months, it needs to be constructed rapidly, easily and economically. The method of creating an ice shell was developed by looking at the method used for existing reinforced concrete shells.

In the past, circular membrane bags with different reticular patterns of ropes were used. Normally, PVC is used as the material for the membrane bag. However, nowadays the large commercially available blue-sheet which is made of polyethylene is used. This material is inexpensive and is resistant to low temperatures. It is easy to make because there is no three-dimensional cutting applied. The ropes are made out of polypropylene because of the low cost and the good mechanical properties of the material. They play an important role, not only in the form-finding of the shell, but also giving local curvature by cutting into the bag. This curvature, a humped surface, appears to increase the structural capacity by forming a combination of a grid shell with a regular shell.

Maintenance

When using ice and snow, the influences of meteorological conditions have to be taken into consideration. Especially sunlight on one side of the ice shell can have devastating effects. The north-south orientation of the dome creates an unequal deformation. In north latitudes, because of the sunlight on the south side, the deformation on the south is biggest. The distribution is no longer ideal, which leads to an increase in deformation speed. The ice does not lose strength immediately because of its latent heat. However, the ice starts to melt bit by bit causing deterioration of the ice. If the ice loses its strength because of weak bonding forces between the ice grains, the shell could degrade. To prevent the deterioration, very low-density snow can be blown onto the ice shell, which functions as thermal insulation (Kokawa 1985: 155–170). Currently, large creep deformation, melting of the ice, ice thickness less than 6 cm, and weather conditions are the main reasons for discontinuing the use of the shell (Fig. 8.25).

Problems

Problems often occur during the construction process and during maintenance. One of these problems occurs when there is snowfall of more than 10 cm. When the snow falls on the thin ice layer in the beginning of the construction process, workers have

Fig. 8.25 Large deflection before collapse (*Source* T. Kokawa)

to remove it. This needs to be done before the application of snow and water in order to avoid an inhomogeneous build-up of the shell structure ('sandwich snow'). However, approaching the dome from above is not possible because the dome can collapse under its own load if the ice thickness is less than 6 centimetres for a 15 m dome. Similar problems occur when the ice starts to melt due to weather conditions. Workers have to blow snow onto the ice; however, they are unable to blow the snow everywhere. The success of an ice shell construction depends strongly on the weather conditions in the initial stage of the creation of the dome.

Form

A spherical cap is used to construct the shape of the ice dome. The spherical cap is able to support its own weight through the form of the cap alone, without using other construction applications. The form is inspired, among other things, by the Pantheon in Rome; by applying arches in that dome, the force of the stone above entryways, windows, and passages could be redirected to the walls and piers. By applying the same principle in the ice shell, using a membrane and ropes, it is possible to create a dome that is able to carry its own weight.

The construction of free form ice shells is possible by influencing the length and the geometric pattern of the ropes, the shape of the inflated membrane will follow the form of the net. A so-called synclastic surface can be manipulated by the rope net as described in '84 ways to manipulate a membrane'. In this way it is possible to make a free shape formwork model with a bag of two-dimensional membranes and a net (Fig. 8.26).

Dimensions

Diameter (m)	Inner diameter (m)	Height (m)
10	8.6	3
15	12.6	5
20	17.6	6.3
30	25.2	9.2

Fig. 8.26 Table: dimensions of a spherical ice dome with a circular membrane bag (*Source* Vasiliev and Gladkov)

40 m Span Dome

Theoretically, a dome with a span of 20 to 30 m is feasible (Kokawa 2005). Several experiments were done in the past resulting in a feasible theory. However, the question of whether it is also possible to create a dome with a span of 40 m still remains unanswered. It would be possible to create a dome of 40 m if the equipment to make it were available. The Pantheon in Rome is the biggest unreinforced concrete dome with a base diameter of 43 m. Ice is easier to manufacture than stone but not as strong. Theoretically it should be possible to realize an ice dome with a diameter of 40 m. Based on experiments done with domes of up to 25 meters, the construction of a 40 m ice dome has a trial prediction of:

- Geometry: 40 m base diameter, 16 m height, and 30 cm thickness.
- Construction period: 10 days in total (3 days for foundation ring, 1 day for inflation, 5 days for application of snow and water, 1 day for deflation).
- Creep displacement: 1.7 cm/day (51 cm/month)

8.2.5 *Pykrete Dome*

Flexible Mould
The form-active structure: Coated Fabric and Cable Net (Fig. 8.27).
Form-active typology: Single layer.
Way to stabilize the mould: Inflate.

Composite Material
Technique to handle the rigidizing material: Spraying.
Reinforcement of the material: Sawdust.
Surface treatment: None.

Project
Name and location: Pykrete Dome, Juuka, Finland.
Design and Engineering: Arno Pronk, Jorrit Hijl, Remy Houben, Frank Janssen, Roel Pluijmen, et al.
Contractor: Eindhoven University of Technology, Municipality of Juuka.
Year: 2013–2014.

Fig. 8.27 Pykrete samples

Test Material	Temperature °C	Dynamic Elastic Modulus GPa	Compression strength MPa	Relative compression strength	Flexural strength MPa	Relative flexural strength
"plain" ice	-5	7.0	1.5	1.0	1.8	1.0
	-20	8.0	2.5	1.0	2.5	1.0
Ice with 2% fibreglass net	-5	8.0	2.9	2.0	5.1	2.9
	-20	8.8	5.0	2.0	8.5	3.4
Ice with 4% fibreglass cloth	-5	8.2	3.2	2.1	5.2	3.0
	-20	9.2	6.0	2.4	10.0	4.0

The results of tests by N.K. Vasiliev and M.F. Gladkov for Ice specimens reinforced with fibreglass net or cloth uniformly distributed through the thickness of the ice.

Fig. 8.28 Table with test results by Vasiliev and Gladkov

Introduction

Ice can be mixed with 14% (by weight) of sawdust or wood pulp. This composite is called Pykrete after Geoffrey Pyke of the British Royal Navy who proposed the material in 1942 for the construction of a huge unsinkable aircraft carrier (Gold 1989). Compared to ice it is much tougher. Below, the properties of pykrete, ice and concrete are compared in the table in Fig. 8.28 by N.K. Vasiliev (Vasiliev and Gladkov 2003). If this material is used, it might be possible to make domes with a diameter far exceeding 30 ms in span. Inspired by the work of Tsutomu Kokawa and the potential of fibre reinforced ice we have realised a 30-meter dome in Juuka Finland.

Pykrete

Ice composites is a relatively new subject and deserves more attention. A brief historic overview on the research of ice composites is given below (Houben and Janssen 2013; Vasiliev et al. 2015: 56–63; Pronk et al. 2016d). It all started with the Eskimos. They used lichen to strengthen their igloos. During WWII, Geoffrey Pyke came up with the idea of building an aircraft carrier out of reinforced ice. Max Perutz, who in 1962 won the Nobel prize for chemistry, researched pykrete for this application but in the end the ice ship was never built. After WWII not much research was done. Coble and Kingery did some tests adding materials like wood, fibreglass and asbestos to the water. In 1976, small reinforced ice domes were built by Stanley and Glockner. After 1976 no reinforced ice structures were built and all tests took place at laboratory level. Several tests are described by L.W. Gold (1977: 197–212). In the winter of 2013, a group of students at the University of Alaska made an arch out of ice reinforced with wood shavings (Fig. 8.29).

Fig. 8.29 Sample of ice and sawdust (Photo by B. v Overbeeke)

1900 Inhabitancy of northern regions traditionally used lichen to strengthen their igloos.

1942 Project Habakkuk by G. Pyke, J.D. Bernal end M. Perutz design proposal for an aircraft carrier of reinforced ice (Perutz 1946).

1943 During WWII ice in Lake Ladoga (Fin. Russia) was reinforced with logs etc. for heavy military transport.

1943 L.W. Gold project Habakkuk testing of reinforced ice beams (Gold 1989).

1944 Runway for aircraft B52 through ice reinforced with fibreglass.

1957 Dunaev investigated sea ice strengthened by hygroscopic agents (sawdust and slag) (Dunaev 1957: 1–26).

1963 Coble and Kingery looked at ice reinforced with wood products, fibreglass and asbestos (Coble and Kingery 1963).

1965 Kagan and Wuori Research on the properties of ice and snow reinforced with sawdust.

1976 Stanley and Glockner built small reinforced domes (Stanley and Glockner 1976).

1980 Jarret and Biggar used geotechnical fabrics to reinforce ice.

1986 Nixon and Smith researched the toughness and bending strength of wood-reinforced ice (Nixon and Smith 1986).

1988 Kuehn and Nixon researched toughness and bending strength of wood reinforced ice along with some simple economic aspects (Kuehn and Nixon 1988).

1992, 2008, 2009 Haynes and Sirotyuk did several studies on composite ice.

2003 N.V. Vasiliev research on dispersion of forces in fibre-reinforced ice (Vasiliev and Gladkov 2003).

2006 K.L. Altunina, study of ice strengthened by cryogels (Altunina 2006: 1669–1672).
2010 A. Bastian Emulsifier in Water (xanthan) as reinforcement (Bastian 2010).
2013 University of Fairbanks, Pykrete arch, experiment by students.
2014–2015 Pronk et al. Largest dome (30 meter) and highest dome (21 m) with sawdust-reinforced ice (Pronk et al. 2015; Pronk and Borgart 2014).
2016 Pronk et al. research and application of cellulose reinforced ice (Pronk et al. 2016b; 2016c: 363–375).

Pykrete Dome

The goal of this research was to develop a building method for the application of ice composites in reinforced ice structures by realising a dome with a 30-meter span. To reach this goal, literature was studied and an overview was made of the material properties of ice and reinforced ice. Existing building methods were studied. This desk research was verified by tests. The evaluation of this research led to the design and making of the 30-meter span pykrete dome in Juuka.

Construction Technique
The method of spraying/blowing turned out to be the most suitable construction technique to make large span structures. With the method of spraying/blowing, thin layers of material can be applied on a mould from a distance. Possibilities for homogeneous adding of fibres during the spraying are essential.

Base Material
As a base material, the combination of water and snow, which forms a slush, is the most suitable. The snow makes the mixture "slushy" and therefore stickier. It also lowers the temperature of the water and increases the maximum layer thickness per spray session. The function of the water is to partially melt the snow to a denser snow-ice. Experiments with pumping a slush of snow and water proved that the slush turns into solid ice inside the pump when pressure is applied to the material. Therefore, the snow and ice have to be mixed on the surface of the shell. It is possible to use only water, but this method is much slower. The mixture of fibres and water has to be applied separately from the snow. The snow will be blown by a rotary snow blower onto the surface of the mould. The sawdust and water have to be sprayed onto the snow in thin watery layers.

Scale, Type and Material of Reinforcement
A homogeneously distributed microscopic fibre reinforcement creates the most effective composite for making shell structures. Microscopic reinforcement is preferred, because of its isotropic properties. Fibrous reinforcement scores well in both mechanical properties and processability. Of all reinforcement materials available, the natural materials perform well in environmental footprint and costs. Sawdust, wood shavings and wood chips are used in the experiments. Tests showed that adding 10% of sawdust is the most effective cellulose fibre reinforcement.

The compressive strength of plain ice is 3.18 N/mm^2, compared to layered ice 6.55 N/mm^2 and 10% sawdust reinforced ice 12.45 N/mm^2.

The flexural strength of plain ice is 1.24 N/mm^2, compared to layered ice 1.80 N/mm^2 and 10% sawdust reinforced ice 3.74 N/mm^2.

A higher structural efficiency means less material and therefore also a shorter freezing/construction time. Dome structures make the best use of snow/ice which has a relatively high compression strength compared to its tensile strength. The load transfer in these kinds of structures is uniformly distributed and therefore the type of reinforcement is preferred to be homogenously distributed and randomly orientated.

Type, Rigidity and Reuse of the Mould

Inflatables perform best when it comes to criteria like formability, reuse and weight, compared to other moulding systems. The shape of the membrane can be made with a 3D cutting pattern. A cheaper and more elementary way is to use a simple 2D membrane bladder and use a rope cover to determine the shape of the inflatable. The mould can be reused when the ice structure is finished.

Sawdust will give a brown colour to the ice. Paper dust has a white colour. For aesthetic reasons, the sawdust reinforcement can be replaced by paper dust, which will have the same mechanical properties.

The final Pykrete Dome design is the result of a combination of various studies. The research on ice shells by T. Kokawa provides a simple and efficient construction method. The construction method is analysed and improved for an optimal design of the Pykrete Dome. The relatively weak and brittle construction material ice is reinforced and replaced with the stronger ductile material pykrete. By using an almost similar construction material, the construction process can remain much the same and is only adjusted where necessary. The Pykrete Dome was made by a team of 50 people from the Netherlands and two from Belgium and it was supported by the local community of Juuka in Finland. The construction time on site was limited to 3 weeks. Juuka Finland is one of the coldest regions in the European Union with temperatures down to −15 °C and −30 °C and a long-term average temperature of −11 °C in January. Despite these statistics, there can be large fluctuations in temperature. The Winter of 2013/2014 was the warmest in 150 years and our building time with a temperature below −8 °C was reduced to 7 days. To speed up the construction process and to be less dependent on the climate conditions, the foundation system of Kokawa was replaced by the use of ground anchors. The method used was as follows. First the inflatable and rope cover were put into position and fixed to the anchor points. Next, after inflation, loose snow and ice was applied to the foundation. This process was repeated until a solid, thick snow-ice foundation ring was created. Due to the possibility to use heavy equipment to construct the foundation ring, the construction time of the foundation ring was greatly reduced. The foundation requires a certain height, width and slope to realize the actual shell structure. To minimize the curvature of the inflatable from the ground to the net structure the inflatable had a 10% oversize. In this way, the oversized inflatable filled the corner between the ground and the rope net and moreover, the bulges between the net and the ribs in the ice will become bigger. The anchors were fixed in the ground. This can be done in any temperature, but in contrast to the method used by Kokawa, the anchors cannot be reused at another location. The change in foundation made the dome lower at the same span. For environmental reasons we worked with a PE foil since this is easier to recycle than PVC.

Because we were not expecting to reuse the PE mould after deflation anyway, we cut a hole at the back of the inflatable to speed up de deflation process. After deflation, a large amount of water was found inside the ice dome. To bring the temperature down we pumped the water away and made 4 holes in the shell for ventilation. After deflation of the inflatable, the ice dome had 10 h of freezing with an outside temperature of −25 °C and an inside temperature of −8 °C before it was tested to prove the safety of the structure. To guarantee the safety during the opening ceremony, the pykrete dome was tested with sandbags with a dead load of 1850 kg. There was no measurable deformation (Figs. 8.30, 8.31 and 8.32).

The shape of the Pykrete Dome was spherical. The internal dimensions were measured with a laser device. In addition, the location of the pykrete layers was analysed and documented, as follows.

- Span 29,060 mm.
- Height [Dome] 9750 mm.
- Height [Entrance] 3100 mm.
- Height [Pykrete] 5680–6650 mm.
- Shell Thickness 150–400 mm.

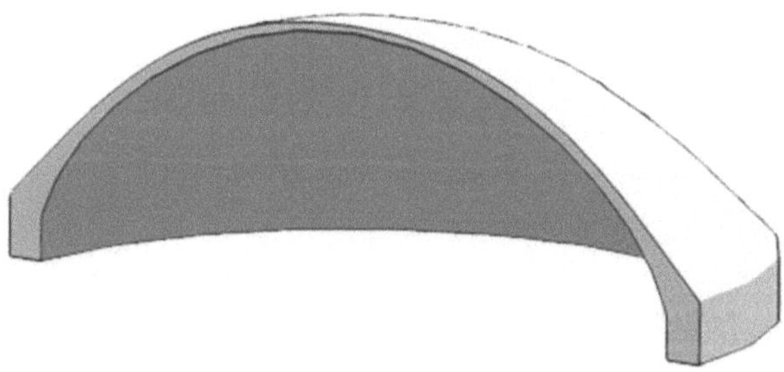

Fig. 8.30 Cross-section of dome with lifted foundation ring of snow-ice

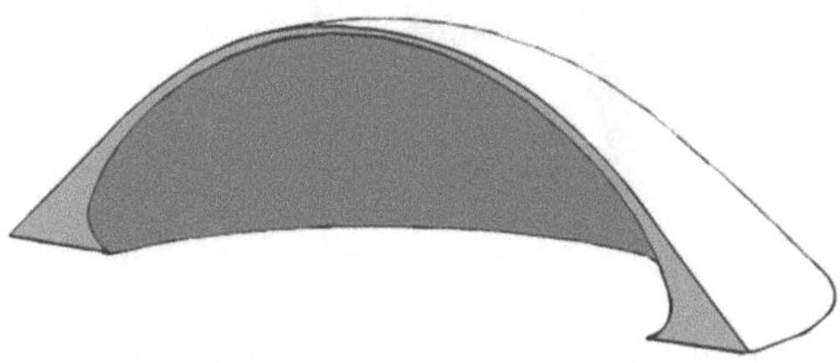

Fig. 8.31 Cross-section of dome with unlifted foundation ring

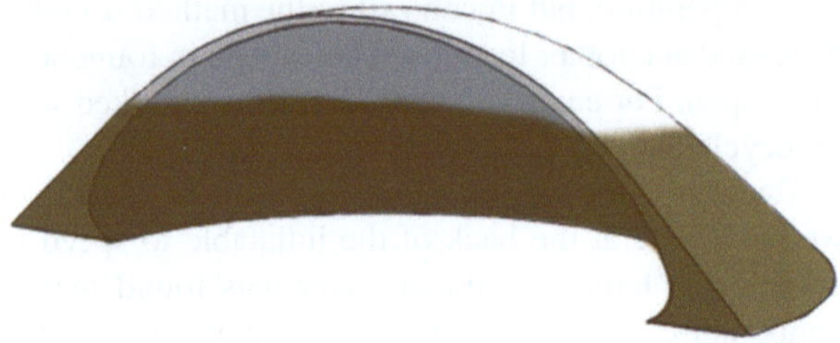

Fig. 8.32 Cross-section of dome with pykrete foundation ring

The thickness of the shell was measured at different locations in the structure. The thickness could be measured by drilling holes in the shell. The sections revealed the gradation in the shell dimensions from approximately 2 meters at the base to just 150 mm at the top of the shell (Figs. 8.33, 8.34 and 8.35).

Structural Calculations

The numerical model was calculated in the program Diana with the same properties (geometry, material, load) as the analytical model. Therefore, the output data could be compared to each other and validated. The numerical model was designed with the software program Rhinoceros (Rhino). To calculate the model, a mesh of 1000 mm

Fig. 8.33 Professor Nicolai Vasiliev and his wife at the opening of the Pykrete Dome (Photo by B. van Overbeeke)

Fig. 8.34 Spraying pykrete on inflatable mould (Photo by B. van Overbeeke)

Fig. 8.35 The exterior of the Pykrete Dome during construction (Photo by J. Rutgers)

was chosen. The model was defined as an ice shell structure with a circular foundation ring that is fixed by freezing to the ground surface.

Dimensions

- Type Curved Shell.
- Span 30,000 mm.
- Shell Thickness 300 mm [t].
- Constrained Hinged.
- Radius 16,800 mm [r/a].
- Top 0 [φ].
- Base 63.40 [φ].

Model

- Mesh 500 mm.
- Axis Spherical Orientation [0, 0, −7561].

Material

- Settled Snow 300 kg/m^3 [V.F. Petrenko].
- Snow-Ice 850 kg/m^3.
- Viscosity 3500 N/mm^2 [E].
- Poisson's ratio 0.3 [v].
- Shear Strength 2.1 N/mm^2 [W. Weeks].

Stress Measured Compression/Tensile

- Strength compression/tensile
- N/mm^2 Average [Lowest] Average [Lowest].
- Snice, Kokawa (Kokawa 1985) −3.95 [−3.75] + 1.2 [+1.0].
- Plain Ice, TU/e −3.18 [−2.00] + 1.24 [+1.02].
- Layered Ice, TU/e −6.55 [−4.38] + 1.80 [+1.39].
- Pykrete, TU/e −12.45 [−12.45] + 3.73 [+3.59].

The structural behaviour of ice and pykrete differs from source to source. Because the exact value is not defined, the lowest value found in previous research is used in the calculation model of the Pykrete Dome. To create a safe structural environment, the model is calculated under extreme conditions. The calculation includes body load, extreme snow load [1.5 m] and extreme wind load. Furthermore, a safety factor is added to cover extreme situations.

Body load

- 850 kg/m^3 × 0.3 m 0.00255 N/mm^2.
- 0.00255 × 1.2 [Safety Factor] 0.00306 N/mm^2.

Snow load

- Snow Density [Settled] 300 kg/m^3.
- 300 kg/m^3 × 1.5 m 0.0044 N/mm^2.
- 0.0044 × 1.5 [Safety Factor] 0.0066 N/mm^2.

Wind load

- Standard wind load 0.001 N/mm^2.
- 0.001 × 1.5 [Safety Factor] 0.0015 N/mm^2.

- Windward side +0.0015 N/mm^2.
- Leeward side -0.0015 N/mm^2.

The most important stresses and forces are the maximum absorbable force/stress that might occur in the structure. The proportions between both will indicate the safety factor in the structure.

Foundation

Compression forces

- Maximum allowable force: 300 kN.
- Current maximum force: 84.2 kN.
- Factor: 4.6.

Tensile forces

- Maximum allowable force: 153 kN.
- Current maximum force: 17.7 kN.
- Factor: 8.6.

Shear stresses

- Maximum allowable force: 315 kN.
- Current maximum force: 41 kN.
- Factor: 7.7.

Shell

Compression forces

- Maximum allowable stress: −2.00 N/mm^2.
- Current maximum stress: −1.08 N/mm^2.
- Factor: 1.9

Tensile forces

- Maximum allowable stress: +1.02 N/mm^2.
- Current maximum stress: +1.02 N/mm^2.
- Factor: 1

The tensile force in the shell is a very local peak at the entrance. Therefore, also this value is assumed to be safe (Fig. 8.36).

Testing the Pykrete Dome

After establishing these parameters, dome samples were made to test the material properties for compression. Because it was not possible to drill cylinders out of the Pykrete Dome, cubic samples of 90 × 90 × 90 mm were cut out with a chainsaw. The pykrete samples were much stronger and therefore harder to cut out compared to pure ice. The samples were cut at five different locations and heights. At each

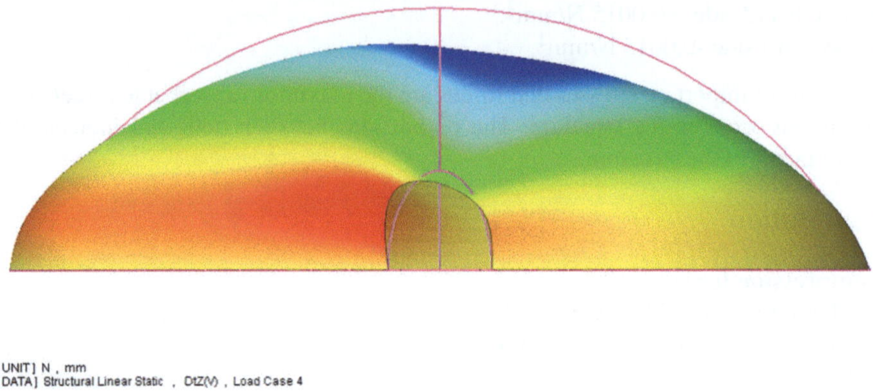

Fig. 8.36 Structural behaviour of ice dome in asymmetric load case

location, five different samples were taken. None of the samples were homogeneous and each location provided a unique composition of the construction material. The samples were a combination of layers of pykrete, snow and ice. It was clear that pykrete layers within the samples could resist higher forces than regular ice. The ice shattered after high compression, while the pykrete layers remained in shape and slightly deformed over time. The compressive strength of the ice layers in the top of the Pykrete Dome was high. The compressive strength values were even higher than the theoretical strength values. It can be concluded that the spraying method creates a high quality of ice. The pykrete samples [B] were 21% stronger than the ice samples [E]. However, the samples where inhomogeneous and only an average of 42% consisted of fibre reinforced ice. The pykrete dome was realized in unusual climate conditions.

Conclusion

This dome was the first application of sprayed pykrete and the largest span made with an ice structure. Theoretically it is possible to design a dome with a span of 100 m. The analyses in Diana calculate the following maximum stresses to be allowable:
Loads

- Body Load 0.00816 N/mm^2.

Geometry

Type Curved shell
Span 100,000 mm.
Shell Thickness 800 mm.

- Constrained Fixed.
- Radius 56,000 mm [r/a].

- Top 0 [φ].
- Base 63.40 [φ].
- Opening No.

Model

- Mesh Size 2000 mm.
- Orientation Spherical [0, 0, −27,300].

Results

Finally, the stress values were calculated at the bottom, middle and top sections of the shell structure. The values were compared to the previously calculated models and to the allowable stress in the construction material.

Sxx

- Location Bottom Middle Top.
- Base [N/mm^2] −0.07 −0.157 −0.151.
- Middle [N/mm^2] −0.003 +0.0065 +0.038.
- Top [N/mm^2] −0.262 −0.275 −0.288.

Syy

- Location Bottom Middle Top.
- Base [N/mm^2] −0.215 −0.273 −0.552.
- Middle [N/mm^2] −0.468 −0.332 −0.222.
- Top [N/mm^2] −0.265 −0.273 −0.288.

Location Top

- Displacement [mm] −8.47.

Conclusion

The stress values show that the maximum absorbable stress values of the construction material are not exceeded:

- Maximum compressive stress −0.552 [N/mm^2].
- Maximum tensile stress +0.038 [N/mm^2].

Because the maximum stress values are not exceeded, the shell thickness can be thinner than in previous designs to reduce the dead load of the structure. Application of pykrete allows for even thinner shell thickness, resulting in lower dead load of the structure. The analyses show that a theoretical approach of an ice shell with an extreme span [100 m] is possible. But it also shows that with the application of pykrete much more is possible. To research and prove just what is possible, several other projects have been realized like the Sagrada Familia in Ice in 2015, and the world record of highest shell structure in ice, the 2018 flamenco ice tower with a

Fig. 8.37 Laura Schreuder at the opening of the Sagrada Familia in Ice Juuka

hight of 30,5 meters and an average thickness of the shell of 25 cm (7 cm at the top and 40 cm at the bottom) (Figs. 8.37, 8.38 and 8.39).

Acknowledgement

These projects were realized with de support of various companies in the Netherlands, Finland and China and the professors and students of Eindhoven University of Technology and the Harbin Institute of Technology.

8.2.6 An Ice Dome of Hinging Plates Constructed with an Inflatable Formwork

Flexible Mould
The form-active structure: hinged plate structure (Fig. 8.40).
Form-active typology: single layer.
Way to stabilize the mould: inflate.

Ice
Technique to handle the rigidizing material: prefab elements.
Reinforcement of the material: ropes/cables.

Fig. 8.38 The 30.5-meter-high Flamenco ice tower in Harbin (Photo by Maple Village [Harbin])

Fig. 8.39 A bird's eye view of the Flamenco ice tower (Photo by Guanyi)

Fig. 8.40 Principle of the creation of an ice dome (*Source* S. Dallinger and J. Kollegger)

Surface treatment: none/plastering.

Project
Name and location: Ice Shell Experiment, Obergurgl, Tyrol.
Design and Engineering: Sonja Dallinger, Benjamin Kromoser and Johann Kollegger, Institute
for Structural Engineering, Vienna University of Technology.
Contractor: Vienna University of Technology.
Year: December 2008/January 2009.

Introduction

This construction method for ice domes was developed at the Institute for Structural Engineering at Vienna University of Technology. With this method, concrete as well as ice shells can be created by transforming plane plates into a shell by means of lifting hinged plates with an inflatable formwork. In December 2005, an ice dome was constructed with a span of 13 m (Kollegger et al. 2010).

The plane elements, which are placed on a planar working surface are assembled by means of tendons. The tendons connect the elements in a radial direction, forming a hinging plate structure. In order to transform the hinging plate structure to its final shape, an inflatable formwork (polyester-PVC membrane) is placed below the hinging plates. The inflatable formwork lifts the hinged plates into their final position. Tendons along the circumference carry the hoop forces after the removal of the inflatable formwork. The interfaces are filled with grouting material and post-tensioning can be applied.

Initially, 2 circular ice plates with diameters of 6 m and 13 m were built, and were cut into 96 individual elements that were held together by means of tendons. Due to the inflating of the pneumatic formwork, the ice plate with the diameter of 6 m became a hemisphere with a diameter of 4 m and a height of 2 m. Figures 8.41, 8.42 and 8.43 shows part of the shaping process of the ice shell with the original diameter of 13 m.

The method can work with prefabricated elements and will form a facetted structure approximating the real shape of the shell. The use of prefabricated elements is remarkable for the construction of shell structures. The researchers applied this method also to the erection of concrete dome structures. The maximum required air pressure in that case, inside de inflatable formwork, amounted to 37 mbar. There are some rare examples with prefabricated elements (Chap. 6) like the Philips pavilion by le Corbusier and Xenakis (Sect. 11.18) and with a Russian system for realizing domes (Dallinger and Kollegger 2009).

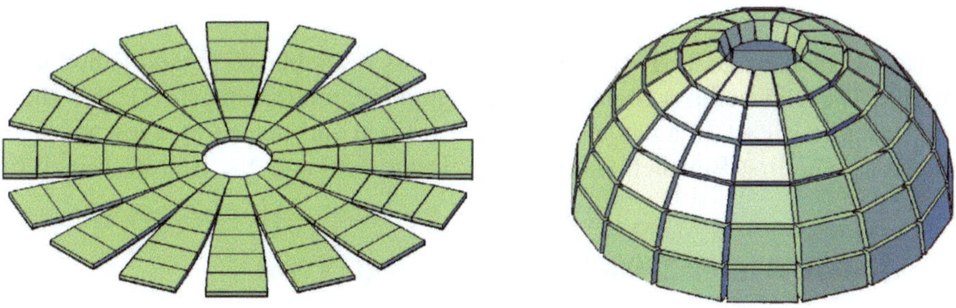

Fig. 8.41 Polyhedron dome with the inflatable formwork method (Source S. Dallinger and J. Kollegger)

Fig. 8.42 Inflation of the smaller ice model (*Source* S. Dallinger and J. Kollegger)

Fig. 8.43 Inflation of the smaller ice model (Source S. Dallinger and J. Kollegger)

This method by Dallinger and Kollegger bears resemblance to the method used by Bini. The Binishell is a circular based, monolithic, reinforced concrete thin shell. The structure is cast on a flat surface and also lifted and shaped by a low-pressure inflatable. The first Binishell was made by Dante Bini and Nic Bini back in 1964 (Sect. 11.13).

Ice Models Winter 2008/2009, Obergurgl, Tyrol

In the winter of 2008/2009, after building the wooden model, two models where created from ice. At the beginning, one flat plate was 6 m and the other flat plate was 13 m. The process in both cases was the same. A frame is placed around the circumference of the flat circular plate. Once this is ready, water is put into the frame and is frozen so that the flat plate is pure ice. This circular flat plate is cut into the right 96 parts and the rest of the ice is removed. The ducts are created by cutting the paths, placing the ducts and filling the gaps. After this, water is put on top of the structure to ensure that the ducts and the elements are held together by means of steel tendons and the cable winches are anchored in the right place. The inflatable formwork is filled with air and the elements lift and may be put into their final position. Figures 8.42, 8.43 and 8.44 show the building process of the smaller shell (Dallinger and Kollegger 2009, 2012; Kollegger et al. 2004: 274–276; 2012; BetonBau TU Wien, www.betonbau.tuwien.ac.at/print/forschung/aktuell/ice-shells.html; MatJack, matjack.thomasnet.com).

Concrete Model

Also, a concrete model is created out of flat plate with a diameter of 13 m. At the end of the transformation a hemispherical concrete shell with a diameter of 8.4 meters was created. This flat plate, too, was divided into 96 elements. Only 6 different formworks

Fig. 8.44 Inflation of the smaller ice model (Source S. Dallinger and J. Kollegger)

were needed for the production of these 96 elements. This made the process cheaper than if there had been 96 different formworks.

Figures 8.45 and 8.46 show a real project in which they used this technique. The project was realised by Benjamin Kromoser in 2017. Figure 8.45 shows the inside and Fig. 8.46 the realised shell from the outside (Kromoser and Kollegger 2015a: 161–171; 2015b: 187–198; Croft 2012).

Conclusion

With this new building method, a lot of different materials can be used besides ice, concrete or wood to transform flat plates into a shell. In this paragraph, the method

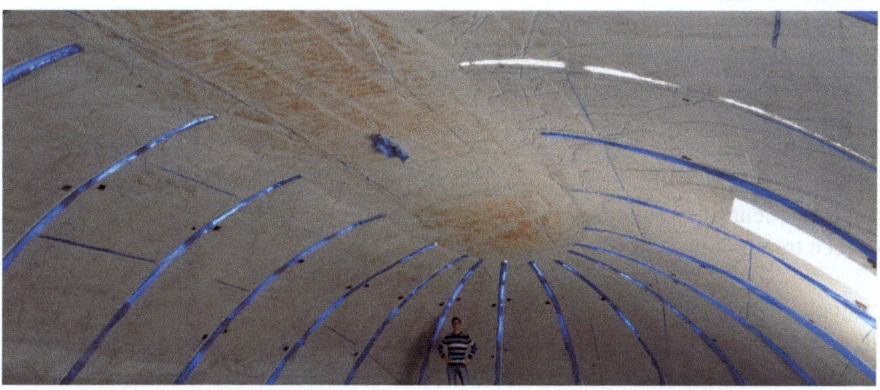

Fig. 8.45 Interior of the shell structure (Source B. Kromower and J. Kollegger)

Fig. 8.46 Exterior of the shell structure (Source B. Kromower and J. Kollegger)

with ice is the main topic but it is interesting to see the possibilities with other materials and other shell-shaped structures since the inflatable does not have to have a sphere-like form.

8.3 Ice Sculptures with an Air Temperature Above 0 °C

8.3.1 Igloo Connected to a Cooling Machine

Flexible Mould
The form-active structure: membrane (Figs. 8.47 and 8.48).
Form-active typology: single layer.
Way to stabilize the mould: inflate.

Composite Material
Technique to handle the rigidizing material: spraying.
Reinforcement of the material: none.
Surface treatment: melting.

Project
Name and location: Igloo Amsterdam the Netherlands.
Design and Engineering: Arno Pronk, Dirk Osinga.
Contractor: Easy Cool.
Year: 2004.

Fig. 8.47 Spraying of inflatable igloo; and final result

Fig. 8.48 Spraying of inflatable igloo; and final result

Introduction

Although the Netherlands are located around the 52nd parallel of north latitude, about 1300 km north of Monaco, Boston and Sapporo (Japan), the country has a moderate climate. The average temperature in February is 1 °C and in September 18 °C. In order to make ice sculptures, the temperature must be manipulated with a cooling machine. This can be done by cooling the air inside an insulated room or by direct cooling of a surface. The first way is the most common one. But the disadvantage of this method is that the complete space has to be insulated. Air is a good insulator with a low heat capacity; therefore, it will require a lot of cold air and a long time to freeze a substantial volume of ice in a cooled space. The other way is by making a matrix of ducts filled with water-glycol on a surface and directly connect the ducts to a cooling device. To make ice for a skating rink, the water-glycol filled ducts with a temperature of −12 °C are sprayed with a fog of water. Comparing the two different methods, Fig. 8.49 shows the heat capacity and heat transportation of water, ice and air (Figs. 8.50 and 8.51).

80 times more energy is necessary in order to transform 1 kg of water into ice, than is needed to increase the temperature of 1 kg water by 1 K. Water has a 4 times higher heat capacity than air. Air has a 100 times better capacity for insulating.

	Heat capacity J/kg.K	Heat transportation W/(m.K)
Ice	2.2 * 1000	2.1
Air	1.0 * 1000	0.024
Water	4.18 * 1000	0.6
Ice into Water	345 * 1000 J/kg	

Fig. 8.49 Table: heat capacity and transportation in ice air and water

Fig. 8.50 Matrix of cooling ducts for ice rink

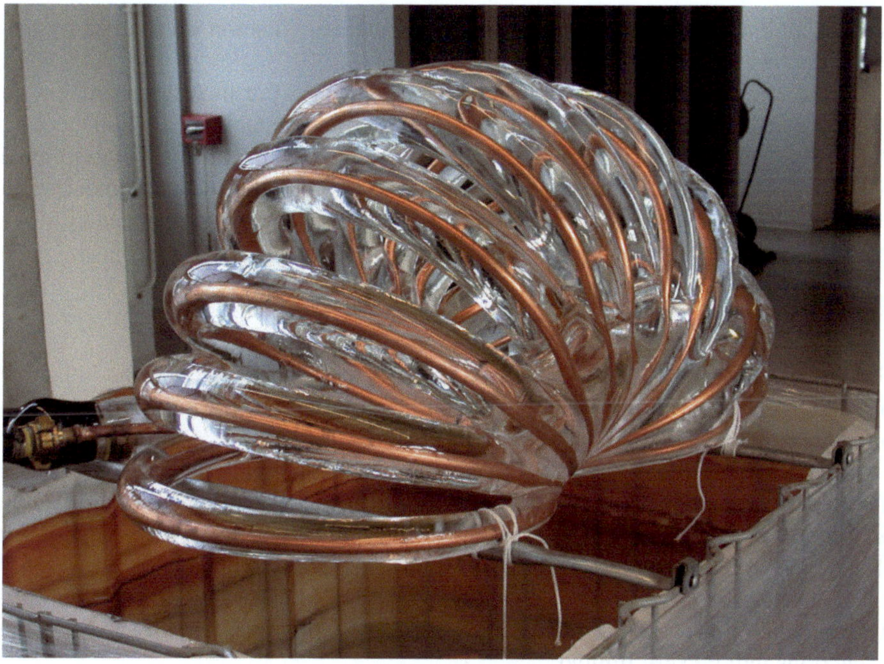

Fig. 8.51 Spiral with ice

The low heat capacity of stationary air insulates the ice object, so relatively little energy is needed to cool down the igloo when it is erected and the temperature of the air can be high as long as there is limited convection. The high heat capacity of water and the much higher density of water makes cooling with a water glycol mixture more efficient and results in much smaller volumes as compared to cooling with cold air. The combination of direct cooling with water-glycol on ice with the insulating capacities of air makes it possible to create ice sculptures with an ambient air temperature above 0 °C. Based on this knowledge, several experiments were carried out and a complete igloo was realized with an ambient air temperature of + 18 °C.

In 2004, an ice spiral was realized for the Dutch Design Week in Eindhoven. The ice spiral was made by connecting copper tubes to a cooling machine with a water-glycol solution. The tubes where submerged in a water basin, forming completely transparent ice. Due to the high air temperature, the surface of the ice was melting and freezing sequentially as the cooling machine was turned off and on. The same principle was used by Heineken in May 2006 for the launch of their new beer: Heineken extra cold. The new product was introduced by connecting the iron of the beer tap to a cooling device. In this way the condensation water on the tap froze to a column of ice on a bar. The spiral in this experiment is not form-active but it is possible to make a form-active matrix of ducts floating in water or hanging in the air. It proves that not only can flat ice surfaces, as for skating, be made in this way but also all kinds of other structures. In 2006 a patent application was lodged for this fabric cooling system (Pronk and Klepper 2006).

In September 2004, an igloo was made for a business fair in Amsterdam. First a small experimental prototype was made by winding an inflatable mould with cooling ducts (Fig. 8.51). The ducts were connected to the cooling machine and then sprayed with water. After one hour, a two mm layer of ice had formed. The experiment was a success and gave enough confidence to make the full-scale igloo (Pronk and Osinga 2005: 8).

Execution

For the execution of the igloo, 2,000 m of ducts were wound around the inflatable mould to create a grid of ducts with a spacing of 5 cm. This work took 3 days with 5 people (Fig. 8.53).

The ducts where connected to a cooling device filled with water-glycol of −12 °C. The ducts were sprayed from the outside with a fog of water for 3 days. The inflatable was removed and the ducts where sprayed on the inside of the igloo for one more day (Fig. 8.54). To collect more information about the structural behaviour, we roughly calculated the shape of the igloo with the help of the program Marc/Mentat (Fig. 8.55). The calculation proved that for ice shell structures with limited measurements the deflections will be very limited.

The cooling capacity is most critical during the making of the Igloo. The cooling capacity was calculated with the formulas below:

Qw = cooling of water to a temperature of 0 °C = M*C*dT
Qr = rigidizing process of water into Ice = M*C
Qi = cooling the ice from 0 °C to −12 °C = M*C*dT
M = the total mass = surface of the inflatable * thickness of the ice = 100 m^2 * 0.1 M = 10 m^3 = 9,170 kg
C = the heat capacity (Fig. 4)
dTw = 18 K dTi = 12 K
Qtot = Qw + Qr + Qi = 4,095.74 MJ = 1,137 kW/h

In this indicative calculation, the energy lost in transmission and by radiation during the process has been taken out. We had a chiller (cooling device) with an effective cooling capacity of 100 kW/h. The theoretical minimum time (without transmission and radiation) to make an igloo of this size is 11 h and 20 min. It took us 4 days. It is important to fine tune the cooling capacity and the amount of sprayed water to minimize the construction time and energy consumption (Figs. 8.52, 8.53, 8.54 and 8.55).

Fig. 8.52 Connection of the spiral to a cooling device

Fig. 8.53 Spiral with ice on inflatable

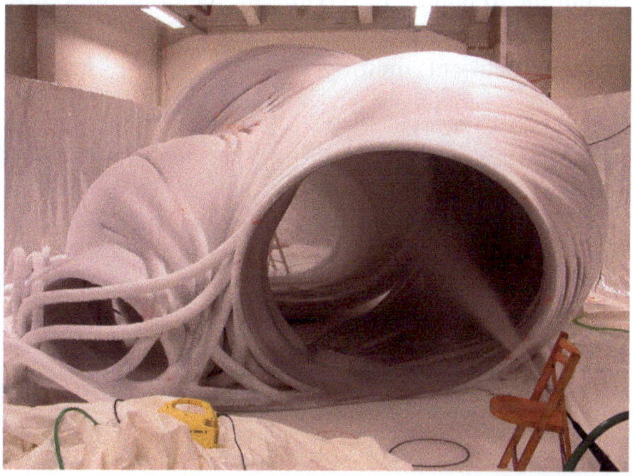

Fig. 8.54 Spraying of the inside

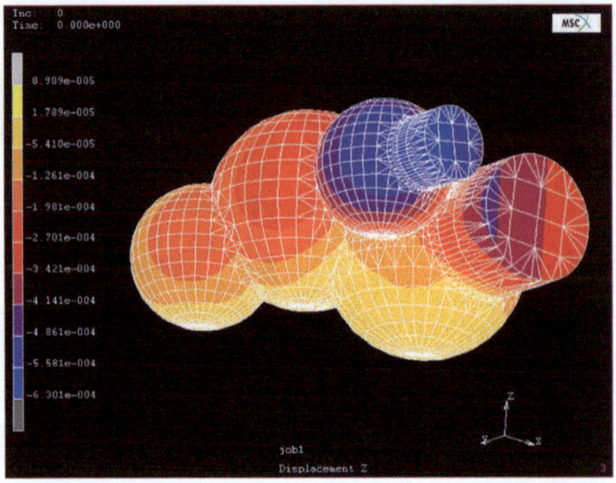

Fig. 8.55 Calculation in Mark/Mentat

Conclusion

The benefit of this method is that an igloo can be constructed with an ambient air temperature of +18 °C with the help of a cooling machine and an inflatable mould but without the need to build an expensive insulated room.

8.4 Water-Filled Membranes

Flexible Mould
The form-active structure: membrane (Figs. 8.56, 8.57 and 8.58).
Form-active typology: double layer.
Way to stabilize the mould: inflation/underpressure.

Composite Material
Technique to handle the rigidizing material: spraying.
Reinforcement of the material: none.
Surface treatment: melting.

Fig. 8.56 Distance fabric with foam

Fig. 8.57 Inflatable with sandwich membrane (*Source* Buitink Technology)

Fig. 8.58 Water-filled second skin membrane

Project

Name and location: Experiments With Heat- and Sound-adapting Membrane, Netherlands.

Design and Engineering: Arno Pronk, Tim de Haas, Roberto Maffei, Heiko Martin.

Contractor: Easy Cool Buitink Technology.

Year: 2006–2009.

Introduction

In 2004, the igloo project in Amsterdam proved that it is possible to make an igloo by covering an inflatable surface with a matrix of ducts connected to a cooling device (Sect. 11.6). To make ice, the ducts are filled with water-glycol with a temperature of −12 °C and spraying the mould with a fog of water. Multilayer membranes filled with a heat-transmitting substance/material make it possible to heat up or cool down a space by radiation and convection. The experiments with distance fabric revealed the problems associated with making a curved surface with a distance fabric. When an underpressure is used, the technique of vacuum infusion can be used to make a heat-transmitting membrane with a capacity of 500 W/m^2K. The working of the heat-transmitting membrane can be improved through proper use of fluid dynamics. The addition of extra membranes may filter the radiation of the sun and further insulate the construction. Applications of this heat-transmitting membrane can be found in the climate control of membrane structures and buildings.

Igloo With a Heat-transmitting Membrane

The 2004 igloo project was a prototype just for one occasion. To make it commercially feasible it would help if the time for making the igloo could be shortened as much as possible. The construction time at a fair is 3 days maximum. As calculated, the theoretical time for making an igloo of 10 m^3 with a chiller of 100 kW/h is 11 h and 20 min. In practice, this will be at least 48 h due to energy loss in transmission. After the igloo is finished, there has to be some time for cleaning and setting up an exhibition, building a bar, etc. If this takes 12 h it leaves 12 h for the installation of the mould, ducts and cooling device. It is possible to weave a curved distance fabric, but there is no manufacturer that will produce custom made small amounts. The other option is to hold two surfaces together by means of underpressure. This technique is used in the vacuum infusion process and discussed in the research by Frank Huijben (Sect. 11.7) (Huijben 2014).

Vacuum Infusion

Vacuum infusion is a method in which a (thermoset) resin is injected between two surfaces. A distance fabric is placed in between to separate the two surfaces and to reinforce the polymer after the infusion. After enclosing the surfaces in a membrane, a pressure differential is applied that impregnates the fabric with resin. The pressure differential is obtained by means of an underpressure or near vacuum. The infusion has to take place within the curing time of the resin used. By using a vacuum infusion technique, it is possible to have an evenly spread surface with a high density of fibres and a low emission of styrene. This technique is used for objects made of (fibre-reinforced) polyester, like (sailing) ships, airplanes and cars. The different pressures operating on the mould are limited by the atmospheric pressure of approx. 1 bar on the membrane. The pressure difference within the sandwich (water-glycol) and the outside (air) depends on the flow capacity of the liquid within the drainage material.

This is determined by the cross-section, the position (distance), permeability of the material and the viscosity and velocity of the fluid. As the fluid acts on the bottom surface by gravitational forces we can see that the height of the fluid column and the density of the fluid play a part in the force balance acting on the surface of the membrane. By using the direction of the fluid flow and the average under- or overpressure of the system relative to the atmospheric pressure it is possible to decrease or increase the pressure on the liquid within the limits of the atmospheric pressure of 1 bar (Figs. 8.59, 8.60, 8.61, 8.62 and 8.63).

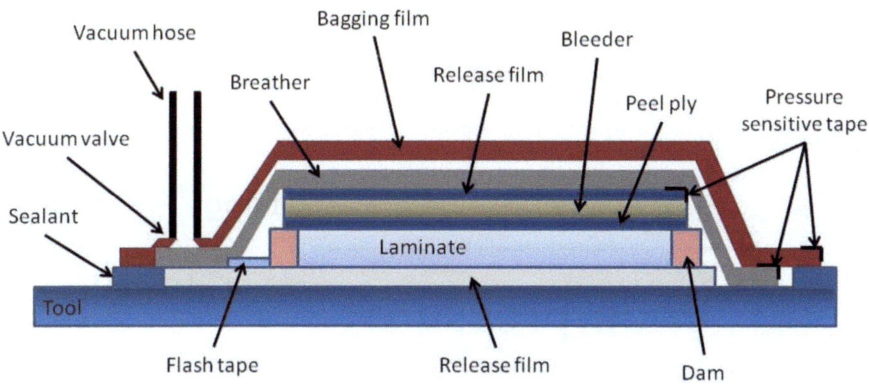

Fig. 8.59 Vacuum bagging

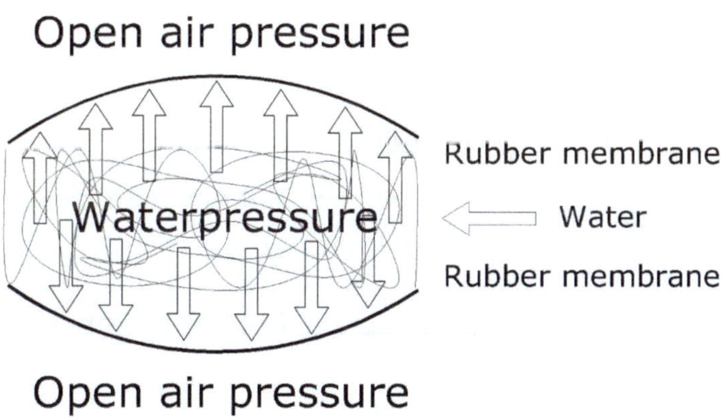

Fig. 8.60 Sandwich with overpressure

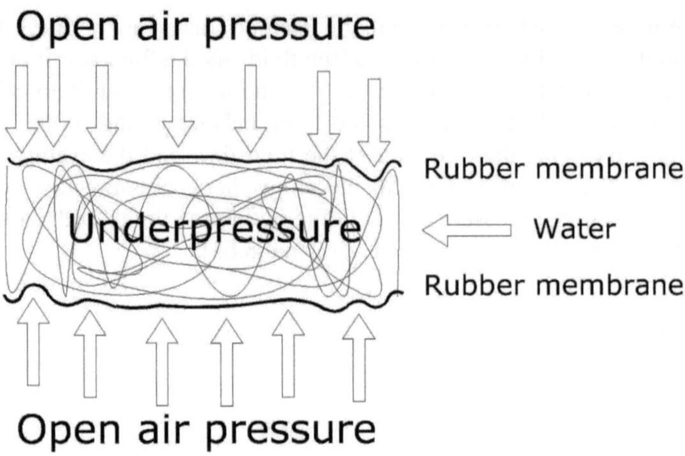

Fig. 8.61 Sandwich with underpressure

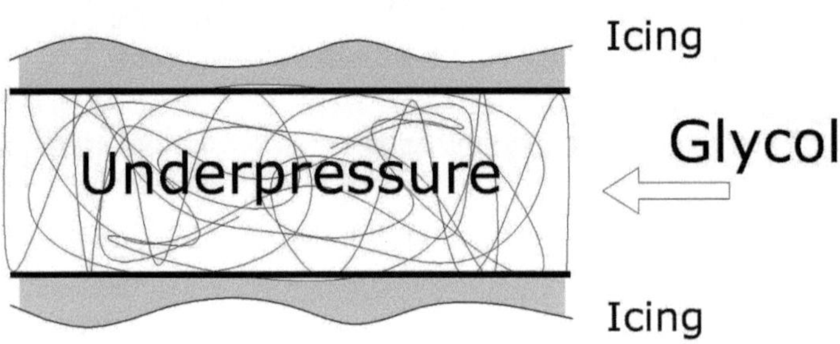

Fig. 8.62 Sandwich with glycol in underpressure and ice

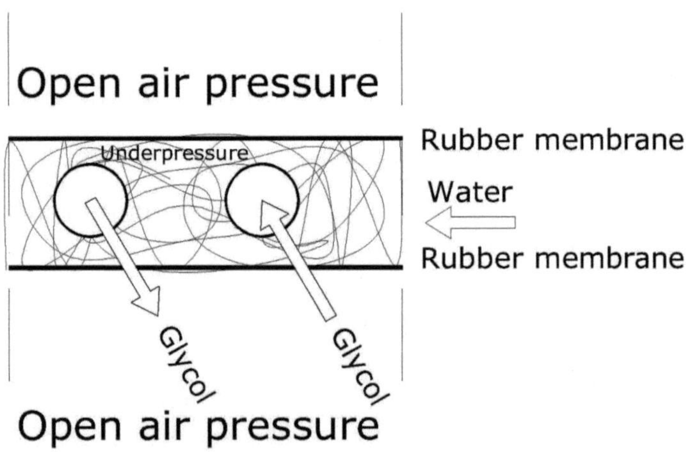

Fig. 8.63 Sandwich with glycol and tubes

Heat-adapting Membrane

To make a heat-adapting membrane, the resin is replaced by water-glycol. The volume between the two layers is put into underpressure. To keep the spacing between the layers, a permeable drainage material is used. It is possible to transport water through the cross-section of this material. When glycol is added to the water, it becomes possible to cool the water below 0 °C. In this way, it is possible to make ice on the surface of the membrane. With warm water, the membrane structure can be used for heating. Also important for cooling down and heating up fluids are the hydrodynamics. When a fluid is pumped through a surface, there is a certain amount of flow. The Reynolds number (Schramek and Recknagel 2007) is the most important quantity in hydrodynamics. It determines whether the flow is turbulent ($Re > 3500$) or laminar ($Re < 2300$). A turbulent flow is able to transfer more thermal energy to a surface compared to a laminar flow. In a turbulent flow, the mixing of the fluid makes it possible to have a transition of approximately 1500–2000 W/m^2K. In a laminar flow this is 500 W/m^2K.

To determine if the previously calculated design works, some tests were performed (Pronk et al. 2007). Four different sample pieces were made:

A distance fabric with a rough surface connected by short rope—every square centimetre was used. Thickness approx. 13 mm; material PVC-coated distance fabric.

Two membranes filled with a fibre (drainage mat). Thickness approx. 13.5 mm, material PVC foil and PVC drainage fibre (Fig. 8.64).

A bigger and wider view of piece nr. 2. Thickness approx. 13.5 mm, material PVC foil and PVC drainage fibre.

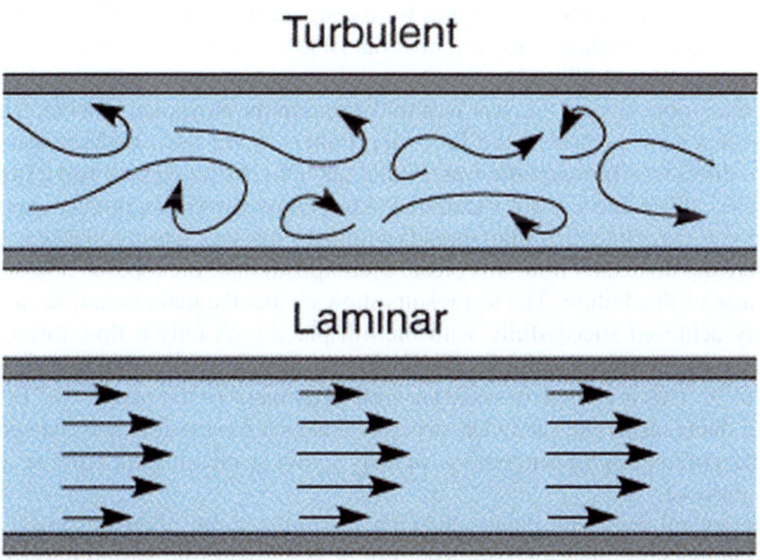

Fig. 8.64 Laminar and turbulent flow

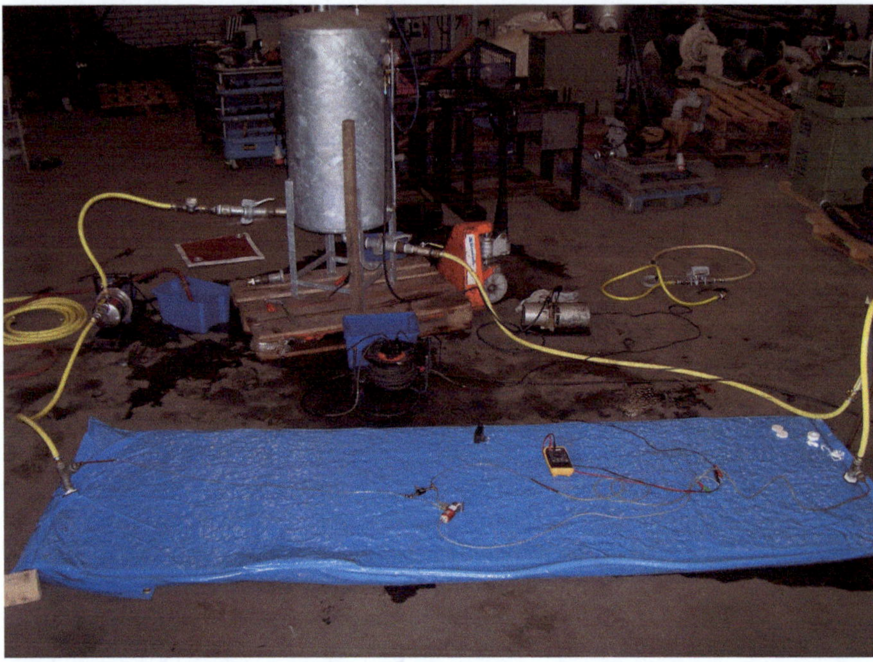

Fig. 8.65 Testing of sample

A double membrane filled with tubes and fibre, effectively divided into two different compartments: one, the tubes for example filled with glycol, two, the fibre filled with water. Thickness 30 mm, material EPDM foil, fibre reinforced tubes and PVC drainage fibre (Fig. 8.65).

The main objective of this test was to determine the Reynolds number. The test setup/arrangement measured the following things: water flow, underpressure and pressure difference between the water input and exit. Since piece 1 had connected membranes, this piece was not vacuumed. After two test cycles, a problem appeared with piece 1, the connecting short threads within the distance fabric collapsed. Therefore, there are no results from this piece, decomposition of the ropes is probably the main cause of this failure. The test results show us that the main target; Re > 3,500, was only achieved successfully with the 4th piece with only a flow through the tube in the membrane. Therefore, the maximum cooling capacity of this surface is 500 W/m^{2}K. This is enough to make ice with an ambient air temperature of 18 °C. A frame of ducts inside the sandwich structure makes it hard to fold and transport the inflatable structure. A higher cooling capacity would be pleasing but is not necessary for this purpose.

After several tests with plane sandwich structures, a full igloo was made based on the sandwich in under-pressure (see Fig. 8.57). The result was an igloo-shaped double membrane within a 30 mm fibre layer to transport the cooling liquid. It is hard to make a 100% fluid proof closed envelope out of a PVC-coated polyester reinforced membrane in a complex form with a lot of seams. In this model, the number of seams was high. The centrifugal pumps used had not been designed for conditions between 0 and 1 bar. If there is any air in the system, the centrifugal fluid pumps will shut down. It is possible to evacuate leaking air with a vacuum pump at the expansion buffer. Still, it is better to avoid leakage. With this inflatable igloo we did not make an ice surface because of the problems mentioned. The system was not stable enough due to air in the centrifugal water pump. The problems with this prototype can be solved by using an inflatable with fewer seams, without leakage, and a self-venting pump. A second prototype has not been built (Fig. 8.66).

Sound-adapting Membrane

In the realisation of second skin façades for buildings, heat and sound adaption are the two functional reasons to apply water-filled sandwiches. It is even possible in the renovation of existing buildings to modernise the appearance of an old office building in combination with the improvement of the climate control just by adding a second skin to the existing façade. The purpose of this second skin façade is to catch the heat of the sun and to store this heat in a solar boiler. The dead load of the

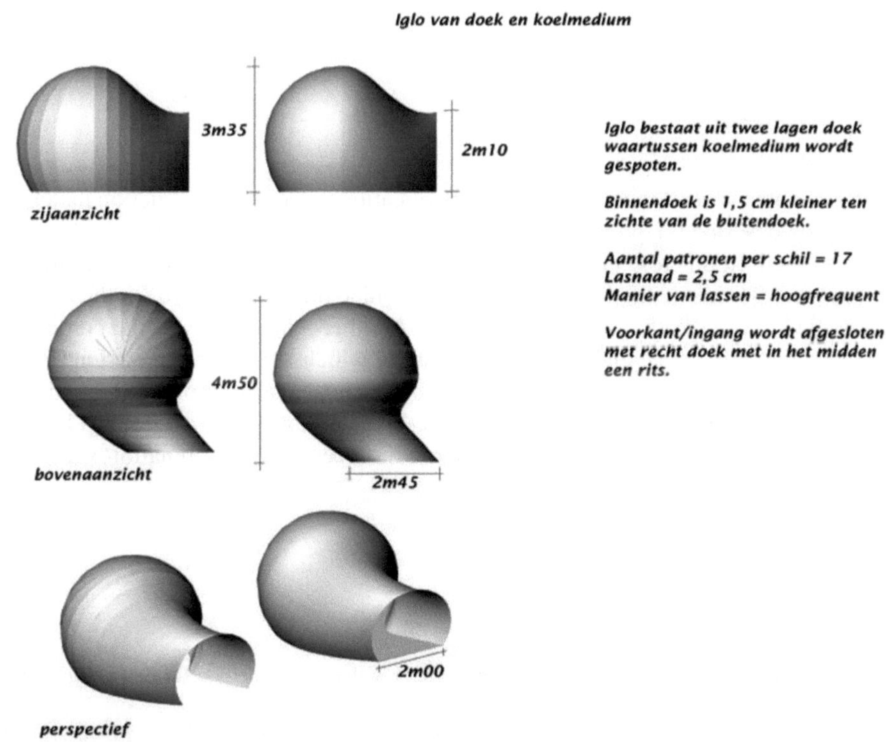

Fig. 8.66 Design of igloo with sandwich membrane

water may affect the sound insulation of the building. The layup of the sandwich is similar to the igloo, only the materials have been changed to two layers of ETFE foil and a translucent PVC 15 mm layer. The sandwich is in underpressure and fits into a frame that is fixed to the existing façade. In the prototype (Fig. 8.57), an inflatable sleeve is applied to stretch the membrane within the frame.

To measure the sound insulation, a test was done in the acoustics laboratory of the TU/e (Pronk et al. 2009). This research concerns a 200 mm thick panel filled with water. The airborne sound insulation of this panel was measured in accordance with standard ISO 140-3. Results show that the single number rating for the airborne sound insulation Rw for an average water layer thickness of 200 mm is 48 dB. The single number rating measured in this test is higher than the one expected from extrapolation of previous tests. An Rw value of 45 dB was expected and 48 dB was the measured result. The increase of 3 dB could be a consequence of the presence of two layers of the distance fabric instead of the single layer in the previous test. This value of sound insulation for a 200 mm thick water panel (surface density of 200 kg/m^2) is comparable to the sound insulation of a 100 mm thick concrete wall (surface mass of 230 kg/m^2) or a 150 mm thick brick wall (surface mass of 250 kg/m^2). However, in practice, a thinner layer of water (around 80–100 mm) should be preferred to avoid problems of stability. In 2005. a patent was applied on this fabric cooling system (Pronk and Klepper 2007).

8.5 Summary

In this chapter, several innovations on the application of ice as a construction material have been described. Up to 2013 the reinforcement of ice with cellulose had rarely been applied. In this research we combined reinforced ice with inflatable moulding techniques. By this method it was possible to realize the world record for the largest ice dome (2014) and tallest ice shell (2018). Another innovation was the combination of a cooling device and an inflatable mould which led to the realization of ice structures with an ambient air temperature of +20 °C. Furthermore, a method has been developed by which to combine double-layered fabrics and a (cooling) liquid, which made it possible to cool or heat large surfaces. This combination was also tested for its sound-absorbing qualities.

Bibliography

Abel JF, Chilton JC (2011) Heinz Isler—50 years of new shapes for shells. J Int Assoc Shell Spat Struct 52(3169):131–134

Altunina KL (2006) Mechanical and thermal properties of cryogels and foamed cryogels produced from aqueous solutions of poly(vinyl alcohol). Russ J Appl Chem 79(10):1669–1672

Alyvea. Chemistry—phase diagrams. http://www.alyvea.com/chemistry/phase-diagrams.ph-1-2017p. Accessed 24 Jan 2017

Anderes F (1983) Ice palaces. Macmillan of Canada

Bastian A (2010) The mechanical properties of select formulations of Pykrete augmented with an emulsifier in uniaxial compression

BetonBallon Projecten. http://www.betonballon.nl/. Accessed 27 Jan 2017

BetonBau TU Wien. Ice Shells. http://www.betonbau.tuwien.ac.at/print/forschung/aktuell/ice-shells.html. Accessed 27 Jan 2013

Billington DP, Jameson WD (2003) The art of structural design: a Swiss Legacy. Yale University Press, New Haven

Boys CV (1890) Soap bubbles and the forces that mould them. Dover Reprint, Mineola

Chilton JC (2000) The engineer's contribution to contemporary architecture: Heinz Isler. In: The engineer's contribution to contemporary architecture (ECCA). Thomas Telford Ltd and Riba Publications, London

Chilton JC (2012) Form-finding and fabric forming in the work of Heinz Isler. In: ICFF 2012 second international conference on flexible formwork, University of Bath, Bath

Cilento K (2012) Fabricating grounds: Ice Shelter installation / Koepcke + Prado. Via Arch Daily. http://www.archdaily.com/219734/fabricating-grounds-ice-shelter-installation-koepcke-prado/. Accessed 4 June 2013

Coble RL, Kingery WD (ed) (1963) Chapter 12 Ice reinforcement. In: Ice and snow. MIT Press, Cambridge, MA

Croft C (2012) Inflatable concrete… Italians in London. Via Twentieth Century Society. http://www.c20society.org.uk/2012/19/inflatable-concrete. Accessed 27 Jan 2017

Dallinger S, Kollegger J (2009) Pneumatic formwork for concrete and ice shells. In: International Conference on Textile Composites and Inflatable Structures, CIMNE, Barcelona. Institute for Structural Engineering, Vienna University of Technology, Vienna, AT

Dallinger S, Kollegger J (2012) New construction method for ice domes. Vienna University of technology, Vienna, AT, Institute for Structural engineering

Doran PT, Lyons WB, McKnight DM (2010) Life in Antarctic Deserts and other cold dry environments: astrobiological analogs. Cambridge University Press, Cambridge

Dunaev ES (1957) Sea water ice as a construction material. Proc. D.V.P.I. 46(11):1–26

Futures Plus. Fabricating Grounds Ice Shelter Installation. http://futuresplus.net/2012/04/06/fabricating-grounds-ice-shelter-installation-koepcke-prado/. Accessed 4 June 2013

Gold LW (1977) Engineering properties of fresh-water ice. J Glaciol 19(81):197–212

Gold LW (1989) The Habakkuk project—building ships from ice. National Research Council of Canada, Ottawa

GSD News (2012) Carl Koepcke (March '12) and Marshall Prado (March '11) featured in ArchDaily. https://www.gsd.harvard.edu/2012/03/carl-koepcke-march-12-and-marshall-prado-march-11-featured-in/. Accessed 4 June 2013

Hooke RL, Mellor M, Jones SJ, Martin RT, Meier MF, Weertman J (1980) Mechanical properties of polycrystalline ice: an assessment of current knowledge and priorities for research: report prepared for the International Commission on Snow and Ice, with support from the U.S. National Science Foundation. Cold Regions Science and Technology 3(4):263–275

Houben RWG, Janssen FHME (2013) Reinforced ice structures. Department of Built Environment, Eindhoven University of Technology, Eindhoven

Huijben F (2014) Vacuumatics 3D Formwork Systems. Eindhoven University of Technology, Department of Built Environment, Unit Structural Design, Eindhoven

Hutchings M, Morgan F, Ritore M, Ros A (2000) Proof of the double bubble conjecture. Electronic Research Announcements (American Mathematical Society) 6(6):45–49

Isenberg C (1992) The science of soap films and soap bubbles. Dover, Mineola

Isler H (1972) Eis-Versuche. 2te Internationalen Symp. "Weitgespannte Flachentragwerke". Universität Stuttgart, Stuttgart

ISO (1995) Laboratory measurements of airborne sound insulation of building elements. https://www.iso.org/standard/3958.html. Accessed 27 Jan 2017

ISO (1996) Acoustics—rating of sound insulation in buildings and of building elements—Part 1: Airborne sound insulation. https://www.iso.org/standard/4940.html. Accessed 27 Jan 2017
Kokawa T (1984) Field study on ice shell. In: Proceedings of 25th IASS Symposium, Dortmund 2, pp 122–138
Kokawa T (1985) Experimental study on ice shells in asahikawa. Cold Reg Sci Technol 11:155–170
Kokawa T (1987) A field study on 10 m-span ice shell. Hokkaido Tokai University Department of Art and Technology, Bull
Kokawa T (2002) Field study of a 30-m span ice dome. J Int Assoc Shell Spat Struct IASS 43(2):93–100
Kokawa T (2005) State of the art developments in ice shell construction. In: CD-ROM Proceedings of 17th Canadian Hydrotechnical Conference, pp 973–982
Kokawa T (2006) Current study-subjects on ice shell construction. CD-ROM of IASS in Beijing 2006:8
Kokawa T (2011) ICE shell construction in Hokkaido. In: Structural Engineers World Congress 2011
Kokawa T (2012) Building techniques for ice shell as temporary structure. School of Design, Tokai University, Asahikawa
Kokawa T, Watanabe K (2011) Shape and creep deflection analysis of ice dome. Department of Architecture, Hokkaido Tokai University, Asahikawa
Kokawa T, Itoh O, Watanabe K (2000) Ice-shell review and recent application. J IASS 41(132):23–29
Kokawa T, Watanabe K, Watanabe T (2012) Ice shell—contemporary 'Kamakura'. In: Archi-cultural translations through the Silk Road, 2nd International Conference. Mukogawa Women's University, Nishinomiya
Kollegger J, Preisinger C, Kaulfus M (2004) Ice shells for temporary event architecture. Struct Eng Int 4:274–276
Kollegger J, Eng M, Dallinger S (2010) Construction method to build ice shells with pneumatic formwork. Master Thesis by J. Perez Latorre, Vienna
Kollegger J, Kromoser B, Dallinger S (2012) The building of ice and concrete shells by employing pneumatic formwork. Institute for Structural Engineering, Vienna University of Technology, Vienna
Kromoser B, Kollegger J (2015a) Pneumatic forming of hardened concrete—building shells in the 21st century. Struct Concr 16(2):161–171
Kromoser B, Kollegger J (2015b) Application areas for pneumatic forming of hardened concrete. J Int Assoc Shell Spat Struct (IASS) 56(3):187–198
Kuehn GA, Nixon WA (1988) Reinforced ice: mechanical properties and cost analysis for its use in platforms and roads. Thayer School of Engineering, Dartmouth College, Hanover, NH
Makkonen L (ed) (1994) Ice and construction. Technical Research Centre of Finland, Espoo
MatJack. All categories on MatJack. http://matjack.thomasnet.com. Accessed 27 Jan 2017
Nixon WA, Smith RA (1986) The fracture toughness of some wood-ice composites. Cambridge University, Engineering Department, Cambridge
Oprea J (2000) The Mathematics of soap films: explorations with Maple®. Student Mathematical Library, vol 10, 1st edn. American Mathematical Society
Otto F (1977) IL 9: Pneus in nature and techniques. Wittenborn Art Books, San Francisco, pp 118–120
Otto F (1979) IL 19: Growing and dividing pneus. Institute for Lightweight Structures, University of Stuttgart, Stuttgart, pp 22–97
Perutz MF (1946) A description of the iceberg aircraft carrier and the bearing of the mechanical properties of frozen wood pulp upon some problems of glacier flow. Meeting of the Society, Cambridge
Petrenko VF (1999) Physics of ice. Dartmouth College, New Hampshire
Petrovic JJ (2003) Mechanical properties of ice and snow. J Mater Sci 38:1–6
Prerost I (1990) Heinz Isler: Seine Kuppeln tönen wie Glocken. Zeitschrift für Architektur und Design, 3:52–55

Pronk ADC, Borgart A (2014) The calculation and construction of a 30 meter span ice dome. In: Proceedings of the IASS-SLTE 2014 Symposium "Shells, Membranes and Spatial Structures: Footprints"
Pronk ADC, Dominicus M (2012) 84 ways to manipulate a membrane. J Int Assoc Shell Spat Struct 53(4):257–270
Pronk ADC, Klepper JCF (2006) Process for the production of an object from ice, prepared in this manner objects and heat exchanger. NL Patent 1031712A
Pronk ADC, Klepper JCF (2007) Method and apparatus for applying to a surface of heat or from a surface dissipating heat. NL Patent 1030399A
Pronk ADC, Osinga DR (2005) Making Igloos in September. In: Conference Paper Structural Membranes, Stuttgart, p 8
Pronk ADC, Haas T de, Cox M (2007) Heat transmitting membrane. In: International Conference on Textile Composites and Inflatable Structures, Barcelona, 2007
Pronk ADC, Maffei R, Martin H (2009) Water as a sound insulation in a membrane structure. In: International Conference on Textile Composites and Inflatable Structures, Stuttgart
Pronk ADC, Verberne THP, Kern J, Belis J (2015) The calculation and construction of the highest ice dome: the Sagrada Familia in Ice. In: Future Visions (International Association for Shell and Spatial Structures) Proceedings, Department of Structural Engineering, pp 1–13
Pronk ADC, Arntz MHFP, Hermens L (2016a) Da Vinci's Bridge in ice and other ice structures with an Inflatable mould. M Ohsaki, K Kawaguchi & T Takeuchi (eds), Proceedings of the IASS Annual Symposium 2016 Spatial Structures in the 21st Century International Association for Shell and Spatial Structures
Pronk ADC, Belis JLIF, Martens K, Lancker B van (2016b) Structural experiments with ice (composite) shells. Alphonse Zingoni (Ed.), Insights and Innovations in Structural Engineering, Mechanics and Computation: Proceedings of the Sixth International Conference on Structural Engineering, Mechanics and Computation, Cape Town, South Africa, 5–7 September 2016. CRC Press Taylor & Francis Group
Pronk ADC, Blok R, Brunschot MCAJ, Lier AJ van, Mortel FFH van de, Williams K, Arntz MHFP, Koekkoek R, Nieuwenhof T van den (2016c) Design of Da Vinci's bridge in ice. Paulo J.S. Cruz (Ed), Structures and Architecture. CRC Press Taylor & Francis Group, pp 363–375–363–375
Pronk ADC, Vasiliev N, Belis JLIF (2016d) Historical development of structural ice. In Paulo J.S. Cruz (ed), Structures and Architecture. CRC Press Taylor & Francis Group
Ramm E (2011) Heinz Isler shells—the priority of form. J Int Assoc Shell Spat Struct 52(3): 143–154
Schramek ER, Recknagel H (2007) Taschenbuch für Heizung + Klimatechnik 07/08, vol 73. Oldenburg Industrieverlag, Oldenburg
Schroedel J (2002) Ice engineering: design, construction and operation maintenance. Department of the Army, US Army Corps of Engineers, Washington
Schulson EM (1999) The structure and mechanical behavior of ice. Thayer School of Engineering, Dartmouth College, Hanover, NH
Schwartz J (ed) (1980) Standardizing testing methods for measuring mechanical properties of ice. J Hydraul Res 18(2):153–165
South DB (1999) Dome type building and method of making same. US Patent 5918438A
Stanley RG, Glockner PG (1975) The use of reinforced ice in constructing temporary enclosures. Marine Sci Commun 1(6):447–462
Stanley RG, Glockner PG (1976) Reinforced ice: its properties and use in constructing temporary enclosures. Department of Civil Engineering, The University of Calgary, Calgary
Stein D (ed) (2005) How to make monstrous, huge, unbelievably big bubbles. Scholastic, New York
Steurer-Systems. Der Igluformer. http://www.steurer-systems.com/de/produkte/schrank-systeme/igluformer/der-igluformer. Accessed 24 Jan 2017
Taylor JE (1976) The structure of singularities in soap-bubble-like and soap-film-like minimal surfaces. Ann Math 103(3):489–539
van Hove R (2007) Method for manufacturing a building construction of curable material, building construction and kit of parts therefor. WO Patent 2007061299:A1

van Hove R (2015) A process for the production of a structure of hardenable building material, inflatable body, apparatus and structure. DE Patent 102014010568:A1
Vasiliev NK, Gladkov MG (2003) Ice composites: mechanical properties and methods of creation. In: Proceedings of the 17th International conference on port and ocean engineering under arctic conditions, Trondheim
Vasiliev NK, Pronk ADC, Shatalina IN, Janssen FHME, Houben RWG (2015) A review on the development of reinforced ice for use as a building material in cold regions. Cold Reg Sci Technol 115:56–63
Veritas Bureau (2010) Ice characteristics and ice structures interactions. Marine Division, Bureau Veritas, France, Neuilly-sur-Seine

Chapter 9
Glass

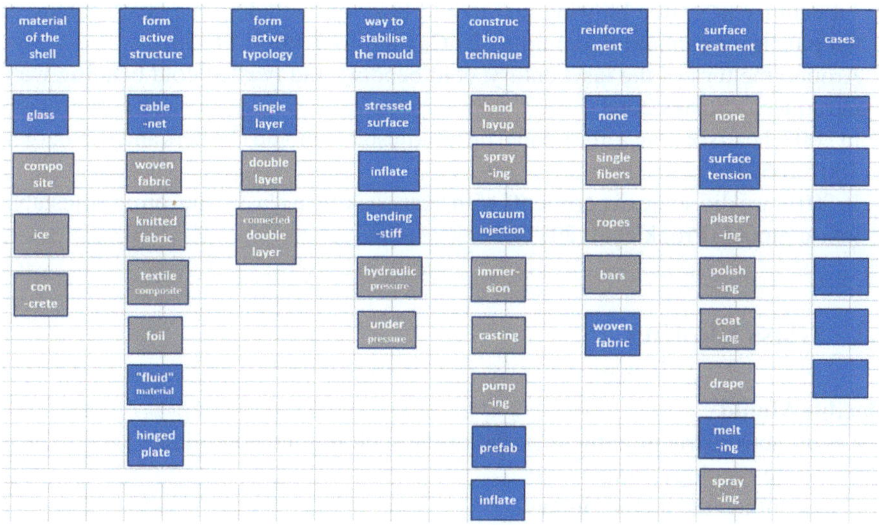

Fig. 9.1 Diagram for glass

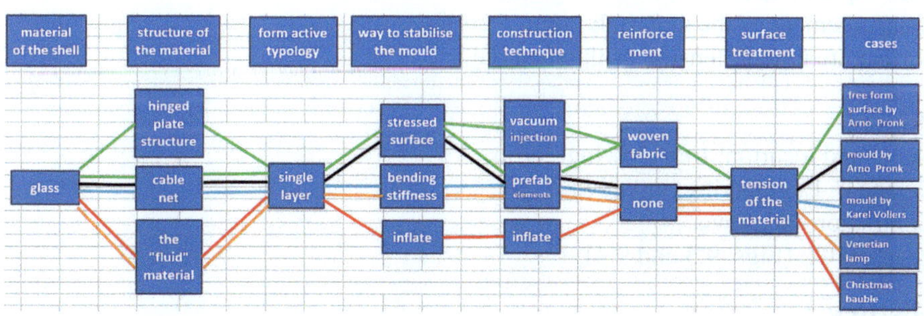

Fig. 9.2 Case studies for glass and the techniques to make fluid architecture by rigidizing form active structures

A. Pronk, *Flexible Forming for Fluid Architecture*,
https://doi.org/10.1007/978-3-030-71551-9_9

9.1 Introduction

Glass has been used for centuries and is still in use in applications like drinking vessels and window panes. This solid material is (non-crystalline) amorphous, brittle and transparent. In this research, glass is observed in a transition from an elastic form-active to a solid material with different applications for doubly curved insulated shell structures of glass (Figs. 9.1 and 9.2).

The moulding techniques to make curved surfaces treated here are:

- 1st century BCE, free blowing and mould blowing including the heat-soaking of flat panels on a mould (Cummings 2002; Lightfoot 1987: 11–18).
- Around the turn of the last century, cold bending and twisted cold bending were introduced (Staaks 2003).
- In 2007, heat-soaking a glass panel on an adjustable mould (Vollers and Rietbergen 2007:105–107).
- In 2010, we did some experiments with heat-soaking a glass panel on an adjustable fabric mould (Pronk et al. 2011).

9.2 Glass Blowing

Flexible Mould
The form-active structure: the material of the shell in fluid condition.
Form-active typology: single layer.
Way to stabilize the mould: inflate.

Composite Material
Technique to handle the rigidizing material: inflation.
Reinforcement of the material: none.
Surface treatment: tension of the material.

Project
Name and location: free blowing, Syria.
Design & Engineering: unknown.
Contractor: unknown.
Year: 50 CE (Figs. 9.3 and 9.4).

History

This paragraph will give a brief insight into free-blowing and mould-blowing of glass. The oldest pieces of glass found are from around 3,500 BCE in the region of Mesopotamia. The use of glass has been known since 50 BCE. The oldest remains are found in the temples of Old Jerusalem. Glass can be made in many different ways. For mass production, mechanical glass is made in factories, where fluid glass

Fig. 9.3 Glass blowing

Fig. 9.4 Venetian glass

is poured into standard moulds. For small scale production or precious objects, glass blowers often use manual glass blowing techniques. There are two different manual glass blowing techniques, free blowing and mould blowing. In free blowing only the handling by the blower is used for forming without any mould. It is still unknown how Mesopotamian people were able to make glass. The earliest remains of glass blowing are from Syria around 50 CE. It took several decades to evolve the glass blowing techniques that made it possible to shape the fluid mass of glass into any desired shape.

Material

Soda-lime is the most familiar type of glass, it consists of about 75% silica (SiO_2) plus Na_2O, CaO, and several minor additives. The transformation of the raw materials into glass takes place around 1,315 °C. Sodium carbonate (Na_2CO_3) is used to lower the glass transition temperature, so that the glass can be shaped at lower temperatures. When using sodium carbonate, the glass gather becomes water soluble. Usually, this is undesirable and therefore calcium oxide (CaO) is used instead. This material was gathered from mining limestone in quarries and gave the total mixture a better chemical balance. The secret behind Venetian glass (GlassWay, glassway.org) was the use of powdered quartzose pebbles from the Ticino river. These pebbles are comprised of incredibly pure silica (SiO_2). This powder was used as the vitrifying agent that causes the viscose liquid to become glass. An advantage of glass is its recyclability. Simply by reheating it, it can be reshaped into another form. When glass is completely molten down, it can be mixed with glass from other sources, allowing the glass workers to use the excess from earlier productions for new ones. Glass doesn't react with acids, bases or other chemicals, nor does it bond easily with other materials (Figs. 9.5 and 9.6).

Fig. 9.5 Silicon dioxide

Fig. 9.6 Multi-piece mould

Technique

Free Blowing

Free blowing is known as the first applied glass blowing technique. Molten glass is placed on the end of the blowpipe. The portion of glass is called a "gather". The bulb in semiliquid condition is affected by gravity and must therefore be rotated constantly to ensure an even distribution of the material. With short puffs of air, the inner cavity of the glass becomes inflated and the exterior skin is formed. The glass is cooled down by removal of the heat from the furnace, and now the glass gather can be formed into the desired shape by the glass blower. Free blowing takes a lot of time and therefore it has been used only for artistic purposes since the mass industrialization of glass production.

Mould Blowing

After the invention of free blowing, people started developing tools to make the production of glass products faster and easier by using moulds of wood and later metal. With mould blowing the final shape and texture of the glass product is determined by the design of the interior of the mould. This allowed for lesser skilled glass blowers and gave new and more possibilities for production. To develop more complicated designs, the roman glass blowers introduced the multi-piece mould. Multi-panelled segments were used and held together with clamps. Bronze moulds were mostly used for delicate work because with wood and plaster moulds such precise detail was nearly impossible.

The invention and development of mould blowing allowed the Romans to increase the production of glass products to large quantities, and made it possible to export to northern Europe. Venice lost its monopoly of glass production in the 16th century and had to respond by making more original, exclusive objects and decorations. The Venetians developed opaque white glass ("lattimo") and in 1527 Filippo Catani developed the "zanfirico" filigree with milky canes in spiral twisted crystal (Stern and Schlick-Nolte 1994; Gable 2007: 44; Cummings 2002; Mariacher 1988; Zerwick 1990; Encyclopaedia Britannica, www.britannica.com/EBchecked/topic/199967/facon/de/venise; Roman Glassmakers, www.romanglassmakers.co.uk/articles.htm; Stajda www.oldandsold.com/articles/article211.shtml).

9.3 Venetian Lamp

Flexible Mould
The form-active structure: the material of the shell in fluid condition.
Form-active typology: single layer.
Way to stabilize the mould: inflate (Figs. 9.7 and 9.8).

Composite Material
Technique to handle the rigidizing material: bending stiffness
Reinforcement of the material: none.
Surface treatment: tension of the material.

Project
Name and location: Venetian lamp.
Design & Engineering: -
Contractor: Filippo Catani.
Year: 1527.

Introduction

Venetian Lamp
With Venetian lamps, the metal frame provides only limited support for the glass; and a blower needs to get the glass to conform to the basic shape and size of the metal frame. The viscosity of the molten glass and the vitrifying process are therefore

Fig. 9.7 Venetian lamp

more critical for Venetian lamps than for free-blown or mould-blown glass. In this research, the technique is called open mould blowing.

Material

Because glass does not adhere easily to other materials and is strongly cohesive it can be blown in a mould without sticking to the inside of the mould. In combination with the viscosity of the material, the glass can adapt its form to the metal structure of the lamp, without completely enveloping it. As a result, the metal frame stays on the outside of the surface.

Technique

The technique for open mould blowing is a combination of free blowing and mould blowing. The mould provides a framework defining a basic size and shape while the craftsman has to choose how much the glass gather will be expanded beyond the framework. When the glass bulb has a certain size and viscosity it can be inflated inside the open metal structure and will become rigid by cooling. The shape and the texture of the bulb of glass is determined mainly by the design of the open metal structure. Large openings in the structure will give bigger bulges.

Fig. 9.8 Venetian lamp

Form

There is a comparison to be made between molten glass and soap bubbles. Soap bubbles form a minimal surface structure. Without the effects of gravity, a single bubble will form a perfect sphere. When multiple bubbles are combined, the adjoining surfaces will form a plane surface. Gravity and other forces will deform those surfaces. In principle, glass will behave in the same way but then there is also the changing viscosity of glass by heating it. By blowing into the pipe, rotating the glass bulb or by heating it, the tension in the glass is changed to form a new equilibrium. Both molten glass and soap bubbles form flexible and form-active structures. When the glass cools down it becomes a rigid surface-active structure. In the Venetian lamp example, the shape of the steel-checked bulges will behave like a pushed linear load: the outward pressure caused by the inflation process, balanced by the inward push of the metal framework on the bulges. The surface-active behaviour conforms to the rules formulated in Chapter 2. Soap bubbles will act purely as an isotropic material with only tensile stress. The glass in the Venetian lamp will behave slightly differently: because of the low viscosity of the molten glass it will be able to take a little bending force. Depending on the viscosity of the glass it may be anticlastic at the bulges (Charles 2001).

9.4 Doubly Curved Glass Panels

Flexible Mould
The form-active structure: cable net.
Form-active typology: single layer.
Way to stabilize the mould: prestress (Figs. 9.9, 9.10 and 9.11).

Composite Material
Technique to handle the rigidizing material: prefab elements.
Reinforcement of the material: none.
Surface treatment: tension of the material.

Project
Name and location: doubly curved glass, the Netherlands.

Fig. 9.9 Spherical mesh

Fig. 9.10 Deformation of mesh by shifting and twisting

Fig. 9.11 Adaptive anticlastic mould

Design & Engineering: Arno Pronk, Jan Belis.
Contractor: Plavercon.
Year: 2011.

Introduction

Since Lucio Blandini (Blandini 2005) developed a doubly curved synclastic shell with adhesively bonded glass components, the concept of building a self-supporting glass-only shell has come within reach. In the current contribution, a small-scale experimental concept is presented, of a self-supporting anticlastic all-glass shell scale model, created by means of an adaptable mould. First, different manufacturing parameters of relatively small shells are investigated, such as mould type, glass supporting system and dimensions, oven temperature and shell curvature. In an experiment, an adjustable mould for bending glass is developed, built and tested. With this mould it is possible to make glass panels synclastic and anticlastic in a great variety of forms. With this new moulding technique, it is possible to create different prototypes. These form the basis for an intended larger shell, composed of smaller segments. The objective is to join the latter by connecting glass panels with composites, which results in completely transparent monolithic shells. Therefore, additional experiments have been performed to explore vacuum fusion of polymer composites with doubly curved glass. Research on self-supporting all-glass structures is mainly driven by architectural aspirations, in which high transparency and/or reflections are key objectives (Bagger 2010: 54–58). Following the developments of numerical tools, 3D CAD software and a trend towards free-form design structures, glass experts face an increasing demand for doubly curved glass creations (Knaack 2008: 571–578). However only a very limited number of manufacturers are experienced in producing the latter and project costs are elevated. Consequently, continued research efforts are necessary to develop easier and more affordable curved glass technologies.

Existing techniques to curve glass include those for cold bent glass (Vákár and Gaal 2004: 95–97; Dodd and Thieme 2007: 83–86; Belis 2007: 123–146; Eekhout 2007: 213–220; Vollers and Veer 2003: 173–175), examples of which can be seen in railway station canopies such as the canopy at's Hertogenbosch (NL); cast glass as applied, e.g., in the Netherlands Institute for Sound and Vision (NL); and hot bent glass, of which the Alpenzoo Station canopy of the Nordkettenbahn in Innsbruck (AT) is a striking example. Probably the most straightforward production method for the latter technique makes use of custom-made moulds (e.g. formed by cutting and milling) with a fixed geometry, several examples of which are reported in the literature (Hobbelman 2002; Horstmeyer 2010: 2063–2074). However, usually related to construction materials other than glass, several authors have reported on different types of adjustable moulds, e.g. based on a bed of adjustable pins (Munro and Walczyk 2007: 551–565); pressurized adjustable textiles (Pronk 2009); or a bed of adjustable stainless steel rods (Raun 2013; Vollers and Rietbergen 2007: 105–107; Eigenraam 2013). The latter moulding technique has been successfully tested for doubly curved glass applications. In the research below, the main objectives are (1) to experimentally investigate the influence of several production parameters, such as glass type, thermoforming temperature in relation to glass thickness and curvature, type of thermoforming emplacement, etc., and (2) to experimentally explore geometrical concepts and technologies to join several curved glass panels to form a larger unit on scale models.

Technique

The oven cycle for (soda-lime) glass consists of heating to the deformation point (580–600 °C) and top temperature (variable), heat-soaking time at the top temperature, cooling to the annealing point (530–550 °C) and strain point (500–520 °C) and lower. The whole process has to facilitate the thermoforming of plane glass to a doubly curved surface, but has to avoid making an imprint on the surface of the glass.

Inspired by the application of the elastic properties of polymer foils in flexible moulding, several experiments with the flexibility of woven wire nets were executed. Six woven wire nets were tested, all with different wire thicknesses and mesh sizes. The ratio between mesh/wire thicknesses determines the flexibility. The tested wire nettings varied from 6 to 1.5 mm. The net with wire 0.25 mm and mesh of 1.5 mm had a high degree of freedom and could deform freely in all directions. The net was tested for heat resistance up to 800 °C with no sign of failure (Fig. 9.11).

The adaptive possibilities of this wire net (WB IO5) have been researched for application in flexible moulding for the thermoforming of glass. Extensive testing showed an inaccurate bending process when the mould was used with only one steel net. A degree of improvement was possible by adjusting the temperature and the heat-soaking time but this was limited by the need to avoid imprints on the surface of the glass.

After measuring the concave and convex sides of the upper planes in each experiment, we concluded that the top temperature is more important than the heat-soaking time to induce a deformation. Extending the heat-soaking time by 40 min meant a decrease of 20 °C in top temperature. Extending the heat-soaking time at the seam temperature produced more imprints. The glass would almost fully touch the mould at 660 °C and the result was preferred to the result at higher temperatures although it had some visual imprints.

We concluded that the sagging of the glass by gravitational forces only did not provide sufficient control. To allow a more precise deformation of the glass plates, a double steel net was proposed. By this method, the glass plate is pushed in shape by the net on top and not only by the gravitational force. The glass with the sandwich was tested with a top temperature of 640 °C. At this temperature, the 3 mm plane glass plate was deformed and then cooled straightaway. The lower top temperature and short heat-soaking time produced a smooth surface and very minor imprinting. The double steel netting was tensioned with the four corner clamps (Figs. 9.11, 9.20 and 9.21). The clamps are fully adjustable and dismountable. This makes it easy to remove the whole net and replace the glass plates. The adjustability in height makes it possible to bend several (up to 5) glass plates at once.

9.5 Structural Bonding of Doubly Curved Glass Panels

Flexible Mould

The form-active structure: cable net.
Form-active typology: single layer.
Way to stabilize the mould: prestress (Figs. 9.12 and 9.13).

Fig. 9.12 Double layered glass dome

Fig. 9.13 Anticlastic glass panel

Composite Material

Technique to handle the rigidizing material: prefab elements.
Reinforcement of the material: none.
Surface treatment: tension of the material.

Project
Name and location: structural bonding of doubly curved glass.
Design & Engineering: Arno Pronk, Teun van Dooren.
Contractor: Brakel Atmos; AGC Westland; Plavercon.
Year: 2014.

Abstract:
In this research, the connecting of doubly curved double glass elements with a translucent polymer is described together with a first prototype of a double glass structure with translucent connections without any other additional supporting structure. With this technique it is possible to make translucent shell structures from double glass elements. In this chapter we describe the connections and the testing of these connections under tensile stress.

Keywords: Free-form glass structure, lamination, polymer-glass connection, thin-shell structure, tiling (Pronk et al. 2014).

Adhesive Bonding

In our desk research we noticed the success of structural glazing in façades and the research performed by Wurm and Hobbelman and Blandini (2008: 278–282). We assumed that the structural bonding of glass panels with translucent adhesives is the obvious choice of technique (Chesnovok and Chesnovok 2011: 252–254; Fröling 2011; Goss 2002: 405–408) (Fig. 9.14).

It is a major concern and a challenge to select a suitable adhesive that satisfies all the pertinent structural, constructional and esthetical requirements, such as gap-filling properties, suitable thickness to compensate for tolerances, sufficient strength, resistance to moisture and UV, etc. Four experiments were done testing the vacuum infusion of glassfibre polymer composites to connect 4 mm thick double-layer panels of float glass: We researched (1) the connection of two plates, (2) the connection of 6 plates for a tetrahedron (Fig. 9.19), (3) the connection of 32 plates for a folded structure of 4 by 4 cm (Figs. 9.15 and 9.16) and the connection of four double-layer plates for a dome structure (Fig. 9.14).

Although it is hard to avoid leakage of the foil and the forming of little air bubbles in the composite, the technique worked out well, but the sensitive and expensive technique had to be improved to be feasible for production on site.

The experiments above led to a search for alternative possibilities for the connection of double glass plates by polymer composites. This was done in co-operation with Brakel Atmos. There are seven main ways of triggering the polymer in the curing process: (1) using two components, (2) anaerobic, (3) heat, (4) radiation, (5) moisture, (6) cyanoacrylates and (7) pressure. Of these seven main ways, irradiation with (UV) light is clearly the most favourable: the resin cures 'on command' in just a couple of minutes, only one component is needed, there is low viscosity which makes it easy to impregnate, only lighting equipment is needed on site, the high

Fig. 9.14 Connection of four double-layer glass panels for dome

Fig. 9.15 Double-layer folded glass structure

transparency is preserved and it is non-toxic. There are also some drawbacks: the resin shrinks during the curing process and little is known about the degradation of this product due to sunlight or other weather conditions (Fig. 9.17).

The connection we developed was a double glass join with a 6 mm spacer and 4 mm and 6 mm annealed soda-lime glass panes with a recessed spacer (Fig. 9.16). The

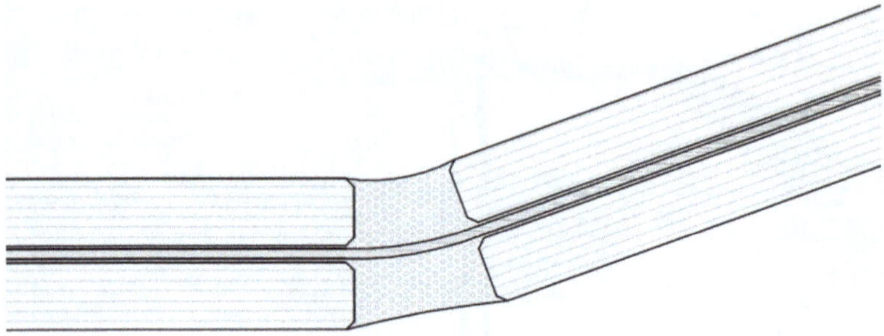

Fig. 9.16 Connection of double-layer glass panels for folded structure

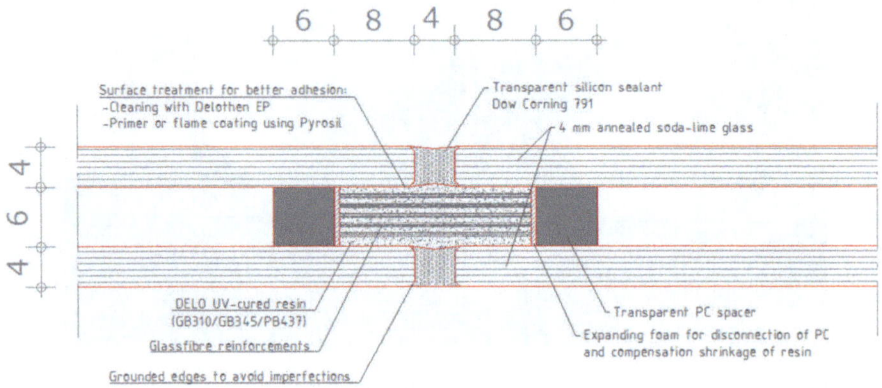

Fig. 9.17 Polymer double glass connection with recessed spacer

glass panels where produced by AGC Glass. The join had to be filled with a polymer. UV and light-curing and two-component (2C) curing adhesives are produced and sold by Delo, distributed in the Netherlands by Silo BV in Hengelo. An experiment with a 2C adhesive by 3M™ was also performed. The adhesives were selected for their properties. We compared the following parameters:

- colour.
- viscosity.
- compression shear strength glass-glass.
- tension strength.
- elongation to tear.
- Young's modulus.
- glass transition temperature.
- shrinkage.
- coefficient to linear expansion.

To give the join a greater ability to transfer tensile loads, a glass fibre reinforcement was added to the connection. For the structural experiment, a standard aluminium

Used Glass in all experiments: AGC Thermobel 6-6-4							
Used Lamp with the exception of experiment 6, in all other Omnicure© S2000 with a 315-500 nm							
	experiment 1	**experiment 2**	**experiment 3**	**experiment 4**	**experiment 5**	**experiment 6**	**experiment 7**
Used tape: Stokvis non-transparent	yes	yes	yes	yes	yes		
Nitto transparent						yes	yes
Adhesive: Delo Photobond© nr:	4468	4302	4302	4468	4468		4468
3M© nr:						DP610(2C)	
Distance lamp in mm	100	100	50	100	100	X	100
Reinforcement layers of glass fiber	1	1	1	2		1	0
glass strip in [mm]					5x20		
Time (per spot) in seconds	40	55	80	40	40	X	40
Sides	2	2	2	1	1	X	1
Clamping pressure [N/mm²]	50	30	30	30	30	30	30
Adhesion area [mm]	7,5 x 170	7 x 200	8,5 x 200	8 x 200	10,5 x 200	11,5 x 200	6,5 x 100
Adhesion area [mm²]	1275	1400	1700	1600	2100	2300	650
Highest measured stress [kN]	7,68	10,30	16,05	16,75	12,94	23,99	11,18
Elastic tensile stress peak [kN]	6,69	9,50	16,05	16,75	10,58	23,99	11,18
Elastic tensile stress per [N/mm²]	5,25	6,79	9,44	10,48	5,04	10,43	**17,20**
Elastic tensile stress per [m¹]	39,35	47,5	80,25	83,8	52,9	119,93	111,79
Total stress time until yield point [min.]	1,33	2,25	2,50	2,86	4,12	4,22	3,03
Max. deformation on yield point [mm]	0,04	0,09	0,02	0,02	0,23	0,32	**0,12**
E in connection [N/mm²]	570		704		271	116	124
E in product specification [N/mm²]	250	900	900	250	250	250	250

Fig. 9.18 Test results

spacer was used, although we are aware that more transparent spacers have been developed (for example the Swisspacer by Saint-Gobain). We made seven different connections and tested them for their structural behaviour (Fig. 9.18)

The Young's modulus (E) of the adhesives given by the specifications are 250 N/mm^2 for the DP 4468 and 900 N/mm^2 for the DP GB368. In Fig. 8 all Young's moduli are calculated and compared with the E in the product specifications. For tests 2 and 4 it is not possible to calculate a representative Young's modulus due to inaccurate measurement or mistakes. The influence of the glass fibre reinforcements is noticeable when the E of sample 1 (with glass fibre) is compared with the E of samples 5, 6 and 7 without glass fibre. The reinforcement stiffens the connection, making it more brittle. This effect cannot be detected in sample 3 where the E in the connection is lower than in the product specifications.

In the product specifications of the DP 4468, the elongation to tear, is determined by DIN EN ISO 527 at 200%. The adhesives did not stretch up to 200% before failing by tensile force. The test specimens in DIN EN ISO 527 were 1–4 mm thick and the connections tested had a thickness of 6 or 15 mm. We assume that the unusual thickness of the adhesive influenced the lower elongation to tear.

The force deformations of DP GB368 and DP 4468 have a lower tensile strength than other adhesives, but the elongation to tear is higher. The tensile strength is more than sufficient. Therefore, the toughness is preferred to the higher tensile strength.

The experiments have shown that the connection fails on shear. The tensile stresses are converted to stress per mm^2. Due to inaccuracy and tolerances during glass production, not every sample is the same and this leads to different adhesion areas. The difference between the elastic tensile stress per mm^2 of the 1st and 2nd samples and the others are due to the fact that quality of the connection was much better (cleaner).

The measured values are only rough indications, but these values are sufficiently high to conclude that from a technical point structural use is possible. The fact that samples 4 and 5 are much more brittle does not mean that these connections are not interesting for other applications. With different requirements this might even be an advantage. For this application the adhesive connection without reinforcement has the best properties.

The Delo Photobond® 4468 adhesive without reinforcement, as tested in sample 7, is favourable in all respects. It is the only adhesive that can also be cured in combination with a 15 mm spacer and has a proper transparency. The use of reinforcements proved to be superfluous during the experiments; the adhesive contains sufficient cohesive strength. Tests by J. Vervloed and W. Sobek show that the mechanical properties of DP 4468 are not heavily influenced after being subjected to façade cleaning products for 3 weeks at 45 °C. The tensile strength of DP 4468 might degrade by UV light exposure over a longer period. This is something to monitor. Tests executed by Delo show that DP 4468 will not discolour when exposed to sunlight. Furthermore, DP 4468 contains no solvents.

We found an application for an angled join with an LED-light and made a pyramid with this LED-light join (Fig. 9.19).

Warm Bending of Glass

To make synclastic shapes, three woven wire nets have been tested, all with different wire thicknesses and mesh sizes.

(1) Mesh WNIO5 1.50×0.50 B1000 (opening*diameter).
(2) Mesh WNIO5 1.40×0.40 B1000 (opening*diameter).
(3) Mesh WNIO5 1.40×0.30 B1000 (opening*diameter).

In the past we tested the heat resistance of net structures up to 800 °C with no sign of failure. Therefore, this was not tested again. Mesh number three is the most flexible one because of the diameter of the steel and the fineness. This mesh was easily deformed by the weight of the glass and created a bigger sagging than the other meshes. Mesh number one was too rigid to create the required shapes. By the

Fig. 9.19 Double glass connection with polymer and LED light for pyramid

results of the multiple tests, mesh number two was the best mesh to use for making synclastic surfaces.

In the making of anticlastic glass surfaces in the past, we concluded that the top temperature is more important than the heat-soaking time to induce a deformation. A temperature that is too high will give a visual imprint on the glass surface. The sagging of the glass only by gravitational force was not enough to control the shape. To have a more precise deformation of the glass plates, a double steel net was proposed. In this way the glass plate is pushed in shape by the net on top and not only by the gravitational force. The glass within the sandwich was tested with a top temperature of 640 °C. At this temperature the 3 mm plain glass plate was deformed and cooled forthwith. Due to the lower top temperature and short heat-soaking time there was a smooth surface and very low amount of imprint. The double steel netting was tensioned with the four corner clamps (Fig. 9.20). The clamps are fully adjustable and dismountable. This makes it easy to remove the whole net and replace the glass plates. The adjustability in height makes it possible to bend several (up to 5) glass plates at once (Figs. 9.21, and 9.22).

To test the developed polymer join we wanted to make a small dome of four equal elements with doubly curved double layered glass panels. After the dome we also made a fluid shell in glass. So far, only anticlastic glass panels have been made with

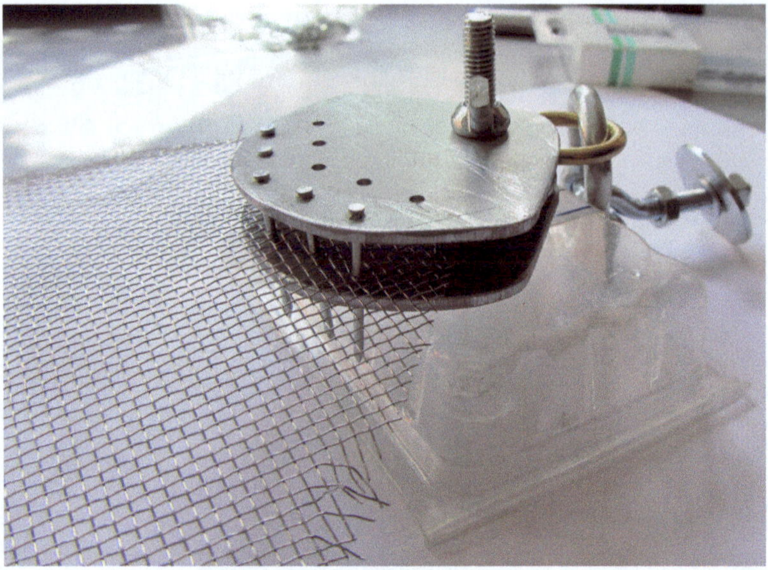

Fig. 9.20 Flexible clamp for steel mesh (single layer)

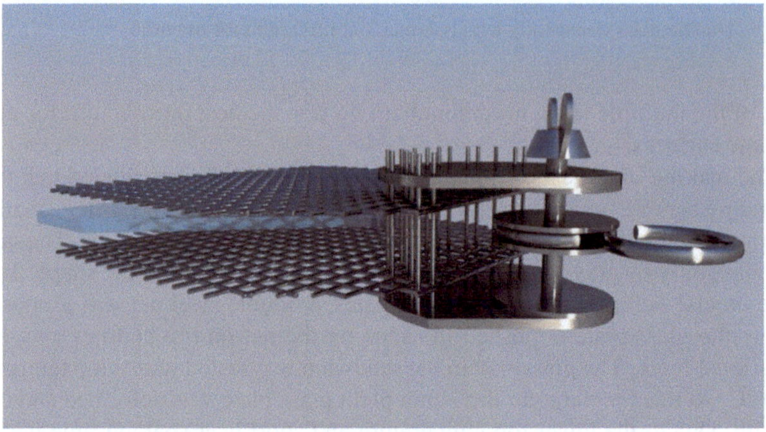

Fig. 9.21 Flexible clamp for steel mesh (double layer)

the flexible mould. Making a synclastic surface from a plane glass plate involves much more plastic deformation than making an anticlastic surface. To research this problem, we did several tests with a flexible mould. We also tried deformation by the same method of forcing that we used with the anticlastic sandwich and with pushing a top load during the heat-soaking time. But this did not give the desired deformations. After several experiments with 400 × 400 × 4 mm and 350 × 350 × 4 mm glass plates the best results were achieved with the following settings:

Fig. 9.22 Flexible mould with springs steel mesh and hinging adjusters

- Mesh WNIO5 1.40*0.40 B1000 (opening*diameter).
- five adjustment points per side.
- 20 min heat-soaking time.
- 630–650 °C heat-soaking temperature.
- Fibreglass with talcum powder underneath the plates.
- Fibreglass with talcum powder between the glass plates.
- Creating a smooth mesh by feeling the tensions in the mesh.

More research should be done to force the glass plate into the mould to reach the desired synclastic shapes. Grasshopper® was used to make a model and receive the data from the model for the mould. This is not sufficient to model the plastic deformation of the glass and a program like Marc/Mentat should be used for that.

Construction of the Dome and Fluid Shell

The dome consists of four equal panels; the shell of 6 different panels. They were placed in position on a wooden support and prepared for vacuum bag infusion. The first step was to create a dome with double-layer glass panels. The glass panels need to be airtight to prevent the occurrence of air bubbles during the gluing process. The flexible spacers are cut to the right length and the correct angle. These spacers are then placed on a glass plate. To improve airtightness, butyl is applied between the joins of the flexible spacers. After this, plasters are used on the corners of the spacers to seal the joins of the spacers. The final step is to place the second glass plate on top of the flexible spacers. For this experiment a new support structure was designed and made. This support structure was made out of wood and has the same curvature as the dome. The glass panels were placed on this support (Fig. 9.24).

Vacuum Bag Moulding of Glass

The joins in the tests above were done with the casting of the adhesive DP 4468. Curved joins are not so easy to cast. To avoid air bubbles, it is better to use vacuum bag moulding for the joins. In vacuum bag moulding, a flexible film is used to seal the curing part from the air outside. The inside air will be removed with a vacuum pump while the atmospheric pressure outside will compress the part to be cured inside. In this case the film will be connected on the rigid glass elements and sealed airtight. The void between the panels will be vacuumed and filled with a resin. During the curing the resin will have an elevated temperature and has to be held in under-pressure. In the past we have done some tests with vacuum bag moulding. This proved to be working but also turned out to be a sensitive technique. The main problem we faced was to prevent air leakage during the curing process. Another point of attention is to be totally sure the double glass panels are sealed air tight. If not, the double glass panel might collapse as a result of the underpressure that might occur in the space between the glass plates.

Conclusion

The final result has some deviations in the form of the surface and accuracy of the join. In theory all techniques were applied successfully and this opens the way to realizing fully transparent shell structures. More research on the accuracy of flexible moulding of glass and the joining of glass with polymers needs to be done before this can be achieved.

After the dome, we also made a fluid shell of 20 panels with the same technique (Figs. 9.21, 9.22, 9.23, 9.24, 9.25, 9.26, 9.27 and 9.28). The next goal for us is to realize a larger shell and to focus on the safety of these kind of structures (Pronk et al. 2015).

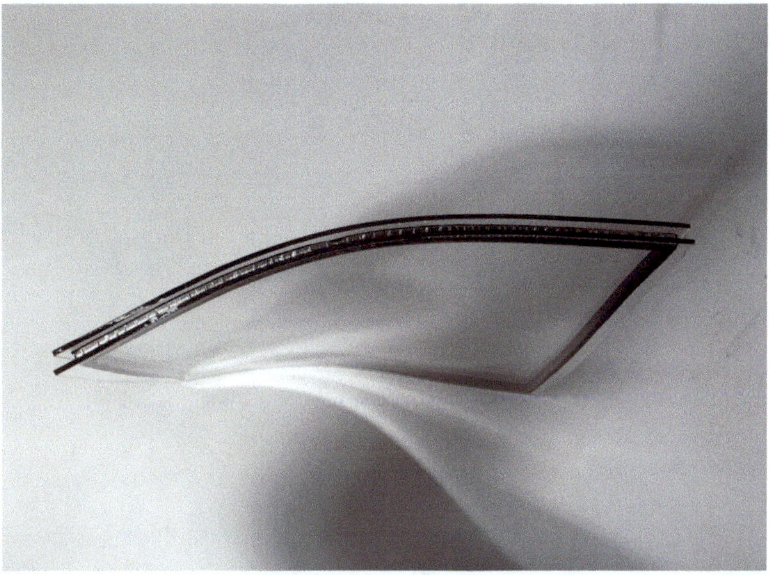

Fig. 9.23 Curved double-layer glass panel with recessed adjusters

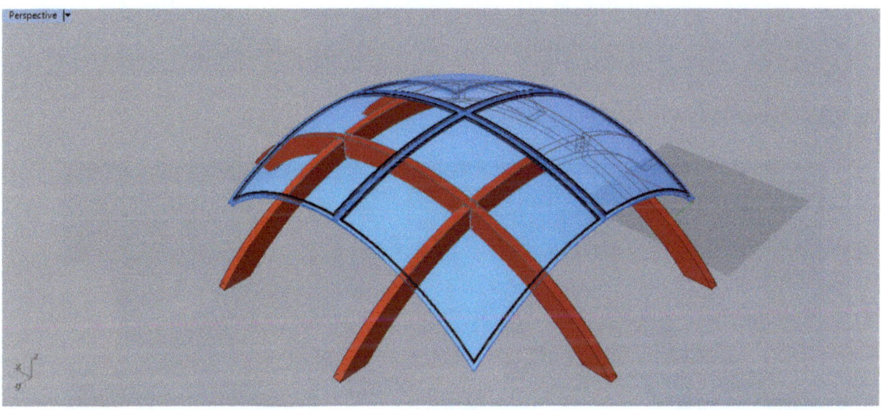

Fig. 9.24 Design and temporary support structure of the glass dome

Fig. 9.25 *Construction of fluid glass shell*

Fig. 9.26 Finished construction of fluid glass shell

Fig. 9.27 Design of fluid glass shell

Fig. 9.28 Final model of fluid glass shell with polymer connections

9.6 Summary

In this chapter on glass the research on curvature of glass is described. For this curvature a flexible mould of woven spring steel has been developed. By heating flat pieces of glass to the soaking temperature it was possible to deform the glass on the flexible mould. By gluing the parts together, it is possible to make fully laminated doubly curved glass surfaces without metal connections.

Bibliography

Bagger A (2010) Plate shell structures of glass: Studies leading to guidelines for structural design. Saarbrücken, Lap Lambert Academic Publishing

Belis J, Inghelbrecht B, Impe R van, Callewaert D (2007) Cold bending of laminated glass panels. Heron 52(1–2):123–146

Blandini L (2005) Structural use of adhesives in glass shells. Verlag Grauer, Beuren

Blandini L (2008) Prototype of a frameless structural glass shell. Struct Eng Int, Issue 3(2008):278–282

Charles B (2001) Dictionary of glass: materials and techniques. University of Pennsylvania Press, Philadelphia
Chesnovok AG, Chesnovok SA (2011) Investigation of strength properties of laminated glass with different bonding materials. Glass Performance Days, pp 252–254
Cummings K (2002) A history of glassforming. University of Pennsylvania Press, Philadelphia
Dodd G, Thieme S (2007) Comparison of curved glass and cold bent panels. Proceedings of Glass Performance Days, pp 83–86
Eekhout M, Lockefeer W, Staaks D (2007) Application of cold twisted tempered glass panels in double curved architectural designs. Proceedings of Glass Performance Days, pp 213–220
Eigenraam P (2013) Flexible mould for production of double-curved concrete elements. Delft University of Technology, Delft
Encyclopaedia Britannica (1998) Façon de Venise. http://www.britannica.com/EBchecked/topic/199967/facon-de-venise. Accessed 7 February 2017
Fröling M (2011) Strength design methods for laminated glass. Lund University, G. James, Lund
Gable CL (2007) Murano magic: complete guide to venetian glass, its history and artists. Atglen PA, Schiffer, p 44
GlassWay. glassway.org. Accessed 7 February 2017
Goss B (2002) Bonding glass and other substrates with UV curing adhesives. Int J Adhes Adhes 22:405–408
Hobbelman G, Timm B, Veer F, Swieten P van (2002) The design of a full glass dome, using a new glass/polymer composite material. LSCE Conference, Warsaw
Horstmeyer G, Adriaenssens S, Scherer G (2010) Glass hypar shells: a preliminary structural and constructional feasibility study. Proceedings of IASS Symposium 2010, Shanghai, pp 2063–2074
Knaack U (2008) Driving the future of glass. Proceedings of challenging Glass Conference, pp 571–578
Lightfoot CS (1987) A group of early Roman Mould-Blown flasks from the West. J Glass Stud 29:11–18
Mariacher G (1988) Glass: from antiquity to the renaissance. Cassell & Co, London
Munro C, Walczyk D (2007) Reconfigurable pin-type tooling: a survey of prior art and reduction to practice. J Manuf Sci Eng 129(3):551–565
Pronk ADC, ElGhazi H, Seffinga A, Schuijers NH (2015) Flexible mould by the use of spring steel mesh. Conference proceedings IASS Symposium Amsterdam
Pronk ADC, Belis J, Schuurmans W, Blancke T (2011) All-glass shell scale models made with an adjustable mould. Proceedings IABSE-IASS 2011, London
Pronk ADC, Dooren TA van, Hermans JW (2014) Transparent shell structure. Proceedings of the IASS-SLTE 2014 Symposium "Shells, Membranes and Spatial Structures: Footprints", Brasilia, Brazil
Pronk ADC, Rooy I van, Schinkel P (2009) Double-curved surfaces using a membrane mould. Proceedings IASS Symposium 2009, Valencia
Raun C (2013) Device for use in a mould for thermoforming, the moulds of the type providing a curved double-curved, dynamically reconfigurable surface. WO Patent 104369(A1)
Roman Glassmakers. Articles. http://www.romanglassmakers.co.uk/articles.htm. Accessed 7 February 2017
Staaks AP (2003) Koud torderen van glaspanelen in blobs. University of Technology Eindhoven, Eindhoven
Stajda S. Murano Glass, via Old & Sold: http://www.oldandsold.com/articles/article211.shtml. Accessed 7 February 2017
Stern EM, Schlick-Nolte B (1994) Early glass of the ancient world 1600 BC—AD 50. In: The Ernesto Wolf collection, Ostfildern. Verlag Gerd Hatje, Berlin

Vákár L, Gaal M (2004) Cold Bendable, Laminated Glass - New Possibilities in Design. Structural Engineering International 2/2004: 95–97

Vollers K, Rietbergen D (2007) Adjustable mould for annealing freely curved glass panes. Proceedings of Glass Performance Days, pp 105–107

Vollers K, Veer F (2003) Usage of cold bent glass panes as an approximation for Doublecurved surfaces. Proceedings of Glass Processing Days, pp 173–175

Zerwick C (1990) A short history of glass. Corning Museum of Glass, Corning NY

Chapter 10
All the Possibilities Researched for Composites

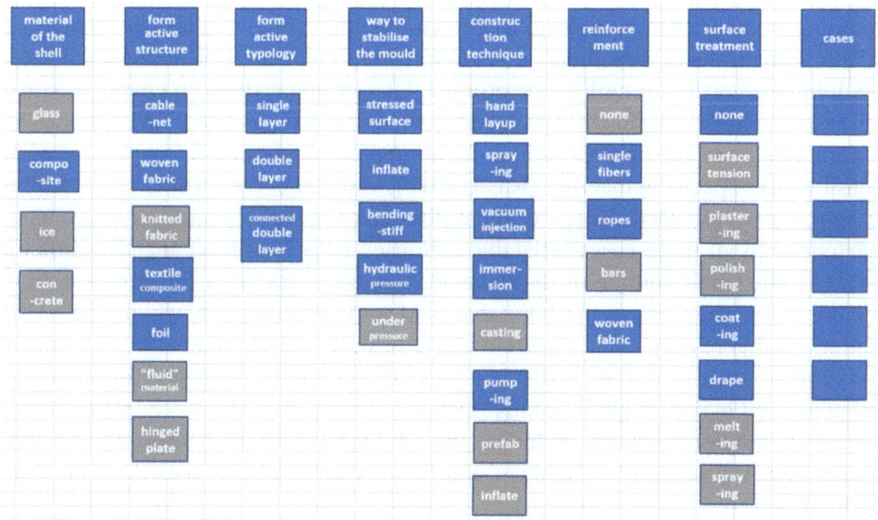

Fig. 10.1 Diagram for composites

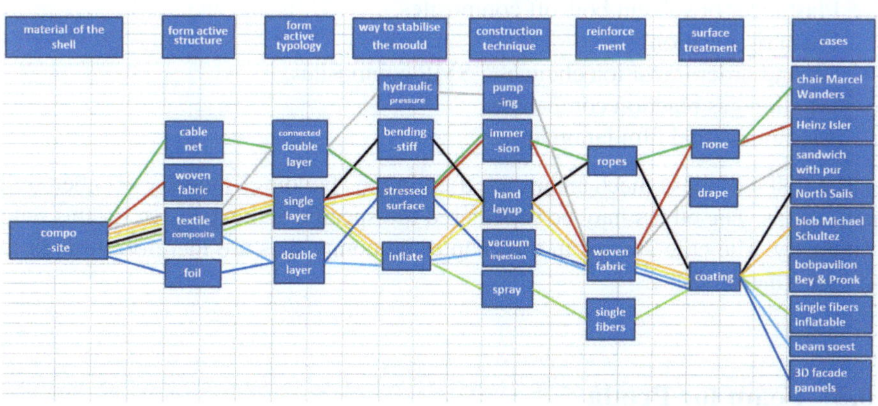

Fig. 10.2 Case studies for composites and the techniques to make fluid architecture by flexible moulding

A. Pronk, *Flexible Forming for Fluid Architecture*,
https://doi.org/10.1007/978-3-030-71551-9_10

10.1 Introduction

Composites are made by the combination of two or more materials with different properties. Cement-based composites (concrete) and ice-based composites will be treated in separate chapters. In this chapter, polymer-based resins in combination with a substrate will be dealt with. The substrate is in most cases a fibre or foam but can be of any material. The most common resins are polyester, vinyl ester, epoxy phenolic, polyamide, polypropylene and polyether ether ketone (PEEK) (Figs 10.1 and 10.2). The usual production methods for composites are hand layup, vacuum bag moulding, pressure bag moulding, autoclave moulding and resin transfer moulding (RTM). The production methods are focused on the reduction of the resin content, which influences the strength of the product. Hand layup is the worst method with a usual content of 60% resin whereas other methods can have a 40% resin content. The combination of resin composites and membrane moulding is used only on special occasions like space engineering (Rasse 2013), vascular surgery (Soc. For Vascular Surgery, www.vascularweb.org), or rehabilitation of sewers (Insituform www.insitu form.com). In all these cases inflatables are used to get at places that are difficult to reach. The small volume of the deflated package is essential for transportation. In these cases, the form of the inflatable is of minor importance, which somewhat contradicts this research. The diagram above has a broad scope to allow a comparison of all kinds of materials and techniques. The curing mechanisms for resins are not mentioned in this diagram because they are too specific. The mechanisms developed over the past 50 years, including the advanced mechanisms for use in space engineering, is impressive. Cadogan and Scarborough (2001). lists the following mechanisms:

1. Thermally cured thermoset composites,
2. Ultra-violet cured composites,
3. Plasticizer or solvent boil-off composites,
4. Foam rigidization,
5. Second order transition change & SMP composites,
6. Inflation gas reaction composites,
7. Aluminium and film laminates.

The first 4 of these mechanisms are used in the "building" industry or other civil applications. The others may be considered too specific and expensive for this field of industry. The case studies in the chapter on composites are therefore limited to the first 4 mechanisms.

10.2 Foam for Prada

Flexible Mould
The form-active structure: foil.
Form-active typology: single layer.
Way to stabilize the mould: inflate.

Composite Material
Technique to handle the rigidizing material: casting.
Reinforcement of the material: none.
Surface treatment: none.

Project
Name and location: Prada Sponge, USA, Los Angeles.
Design & Engineering: Rem Koolhaas, Ole Scheeren, Chris van Duijn, Eric Chang, Werkplaats de Rijk/Parthesius, Panelite.
Contractor: RAM Contract.
Year: 2004.

Introduction

In order to give the inside of the new Prada store in Beverly Hills a new look, the Office for Metropolitan Architecture (OMA) designed a new material OMA (OMA, oma.eu/projects/2004/prada-sponge) This new material has a sponge-like structure and can be made from a range of materials to incorporate different material properties. Ranging from translucent cast resin for shelves to bubble-textured silicone mats for floors. According to OMA, foam in its multiple and ambiguous deviations offers a new definition of functional and visual properties between artificial and natural, irregular and regular, transparent, translucent or solid, flexible or rigid (Figs. 10.3 and 10.4).

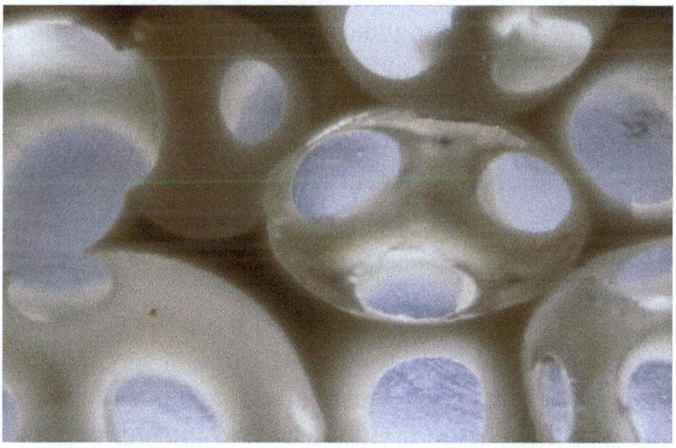

Fig. 10.3 Detail of Prada Foam wall (*Source* V. de Rijk)

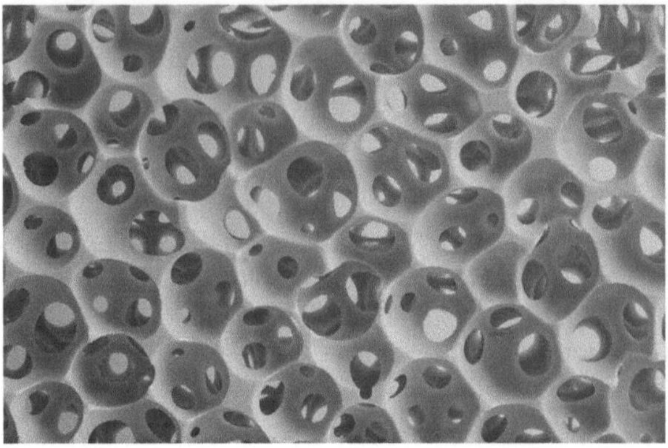

Fig. 10.4 Prada Foam wall (*Source* V. de Rijk)

At the Beverly Hills Prada store the foam is used as decoration as well as a way to showcase clothing and support shelves. The open structure is used to create a diffuse lighting effect through the walls.

Material

What defines foam is the structure, not the material itself. Foam can be cast with different plastics in gradations from hard to soft and from transparent to opaque. For the Prada store a polyurethane composite was used to ensure a sturdy material that is soft to the touch, yet rigid enough to support the weight of clothing and shelves (Vincent de Rijk, Vincentderijk.nl/products/prada-foam).

Besides the polymers that can be used in the casts, the structure of foam can also be obtained by milling and drilling into a whole range of materials such as wood or metal. Using a material like wood may give the customer an unexpected experience. The structure with large holes makes the material look flexible while the wood gives the material the ability to hold weight, which may leave someone who sees the foam uncertain about how it feels (Fig. 10.5).

Form

Foam is a cast of an aggregate condition between solid and void. It is simultaneously both an irregular and a regular structure of sponge-like consistency. The development started with an architectural model using a regular cleaning sponge. As the visual effect of this backlit texture was very intriguing, an extensive research was initiated to recreate this material in 1:1 scale. Many hundreds of tests and prototypes were hand made in order to test hole sizes, percentages of openness, translucencies, depths, colours, etc. It forms a substance out of which objects can be built, but from which entire spaces can also be carved out, which is in itself an interpretation of solid and void.

Fig. 10.5 Foam plates (*Source* V. de Rijk)

The casts are 100 × 500 × 1,500 mm plates that can be glued or melted together to create structures with a suggestion of volume and density.

Technique

At the same time, mass production and 3D computer modelling techniques were investigated that could help to translate the properties of the hand-crafted prototypes and all technical requirements into the final product. Methods from stereo lithography to CNC controlled milling processes and specific casting techniques were developed and employed. Along the way, an entirely new polyurethane composite was generated to comply with fire and building codes.

Initial designs were created with 3D computer modelling techniques and realized with 3D printers. When the desired attributes such as hole sizes, openness and depth were found, CNC milling was used to create a mould. First, the mould was created out of wood to experiment with the process and to see how the different plates would fit together. When the design was finalized, it was milled out of aluminium. This mould was used for the mass production process of injection moulding with (different) polymers.

After the mould was created, balls were placed onto it, and polyurethane composite was poured on top of those to create the distinctive foam pattern. The balls can be seen as synclastic inflatable membranes pushed against other synclastic inflatable membranes as described in Chap. 2. The curvature of the membranes decreases. In the case of the balls it might become zeroclastic and even synclastic in the opposite direction. Because the membranes are pushed against each other, there won't be any space left for the composite to flow, which creates the typical holes in the casting (Figs. 10.6, 10.7 and 10.8).

Fig. 10.6 Placing of spheres in mould (*Source* V. de Rijk)

Fig. 10.7 Casting polymer (*Source* V. de Rijk)

Fig. 10.8 The result after casting (*Source* V. de Rijk)

10.3 Heinz Isler

Flexible Mould
The form-active structure: woven fabric, originally sand, the glassfibre reinforced polymer. (GFRP) cast was later used as a serial production mould.
Form-active typology: single layer.
Way to stabilize the mould: stressed surface.

Composite Material
Technique to handle the rigidizing material: submersion.
Reinforcement of the material: glass woven fabric.
Surface treatment: none.

Projects
Name and location: Fibre reinforced polymer doubly curved shells and plate roofing.
Design & Engineering: Heinz Isler.
Year: Various.

Introduction

Heinz Isler (1926–2009) is known to be one of the pioneers of free-form shell structures (Chilton 2000: 168). As his love for nature grew in his youth, these natural shell shapes are Isler's inspiration as he interpreted the shapes as stiffened shells, doubly curved shells, rotation shells, flawlessly formed shells in countless variations and wafer-thin shells (Kotnik and Schwartz 2011: 185–190). Natural forms are always the optimum shape according to Isler (Fig. 10.9).

Fig. 10.9 Drawings of shells by Heinz Isler (*Source* J. Chilton)

Isler's innovative experimental ways and physical methods of form-finding, specifically shell forms, are what he became a well-known engineer for. In 1959, Isler presented his paper '*New shapes for Shells*' at the IASS, International Association for Shell Structures in Madrid. His paper briefly describes three innovative methods for free-form surfaces; the freely shaped hill, the membrane under pressure and the hanging cloth reversed. After the description with a brief text and a few illustrations of these three methods in the paper, he concluded the paper with a page of 39 hand drawn sketches of '*Natural hills on different edge lines*' with an "etc." at the end, implicating the infinite variety and possibilities of shell shapes (Fig. 10.9).

As mentioned, Isler's methods are based on physical experiments and physical modelling. For his work regarding shell forms, he first realized the potential for form-finding with a hanging cloth after noticing the shape of jute fabric draped over steel reinforcing mesh. His early experiments were with the hanging-cloth method, as he froze the wet draped sheets into ice structures in the cold Swiss winters. These ice structures when turned over represent the ideal shell forms according to Isler. In 1954, Isler presented a pneumatically shaped thin concrete shell concept at an international congress in Amsterdam. The following years Isler's concrete shells were applied in buildings all over Europe. More than 1400 of Isler's reinforced concrete shells were realized, mostly on industrial buildings.

Isler's experiments later developed into using unsaturated polyester resin light domes for his shells as his customers asked for a natural light enabling roof. He was also one of the pioneers for his design work with composites of fibre reinforced polymers (FRP). Light domes of FRP were applied for the concrete shells as large windows in the shell roofs as the required measurements were too large to use Plexiglas. Later in 1956 Isler received a large contract for GFRP housing, in which he used doubly curved shells, within galleries to protect pharmaceuticals from humidity.

Other than Isler's famous shells, also most of his great designs are made with FRP plates. The reason for using plates instead of shells is because with increasing shell span, the thickness also increases, which affects the desired translucency of the material that enables natural lighting. Another advantage of GFRP is its insulation valucs.

Material

The first GFRP applications of Heinz Isler's design were light domes for his concrete shells. These domes were layered as a sandwich structure. The outer layer consisted of glass woven fabrics as the core consists of glass mats. The structure was layered onto a GFRP mould based on a gypsum model. These sandwich structured domes had a total thickness of 3.5 mm with a diameter of 5 m.

For the doubly curved shells Isler designed for the GFRP housings in galleries, only GFRP was used for casting. He made the mould for the shells out of sand. The casting material consisted of GFRP and was only refined and polished afterwards. This prototype cast then was used as a mould for serial production of the GFRP elements made of unsaturated polyester resin. These elements had a thickness of 1.5 mm and a span of 6–9 meters.

Isler used a different layering process for flat roof projects than for curved shells, but still layered as a sandwich structure. A precast unsaturated polyester resin membrane was used as a mould. The second layer consisted of unsaturated polyester resin. Laid-up glass mats were used as a bottom layer. To remove air pocket voids, resin was impregnated into the glass mats. Finally, prefabricated coffers were placed and glued on. The gaps between the elements were filled with a spackling compound. The material is still as translucent as the light domes and doubly curved shells after this layering process, because the customers requested a natural light enabling roof. The roof measured 319 m^2. At the time, in 1960, this project was the largest GFRP plate made.

Form

The forms Heinz Isler is best known for are his thin shells which could span great distances. His form-finding methods are quite similar to Frei Otto's work, with the three most practiced ones being the *freely shaped hill*, *pressurized membrane* and *reversed hanging cloth* (Chilton 2010: 64–71; Ramm and Schunk 1993: 112; Chilton 2009: 12, 2012). Aside from these methods, Isler also experimented with soap film, however, this method is not suitable to directly create full scale moulds. According to Isler one can create an infinite number of shapes with these methods. Of the three main form finding methods only the sand hills and hanging cloth methodologies have produced a composite shape (Figs. 10.10, 10.11 and 10.12).

Firstly, the hills of sand or earth were used to serve as an inverted swimming pool mould (Fig. 10.10), as well as the base for a repetitive mould for a 9 m (1.5 mm thick) spanning *arch*, created to cover the inside of a WWII tunnel. The latter was non-structural, but served only as a protection against moisture from the mountain. The swimming pool surface qualifies as a combination of non-manipulated *monoclastic* and *synclastic* surfaces, with multiple monoclastic curves (the steps) combined with a synclastic surface which originates on both sides. These monoclastic shapes are not manipulated.

Fig. 10.10 Swimming pool by Isler (*Source* J. Chilton)

Fig. 10.11 Sand mould for thin shell (*Source* J. Chilton)

Fig. 10.12 Composite shell (*Source* J. Chilton)

Secondly, the hanging cloth method can be used to create inverted domes or shell shapes, which are often non-manipulated *synclastic* shapes with Heinz Isler, used to form full scale gypsum model as a moulding base. A simple example are the light domes to supplement his concrete structures (Figs. 10.13 and 10.14). These domes did not exceed 8 meters since a larger span requires more material leading to less translucency.

The hanging cloth form-finding method also demonstrates a great benefit of composite construction over concrete, often used by Isler. In contrast with concrete, when producing a shell with a hanging membrane, no mould has to be formed, saving a lot of material. These hanging membranes were occasionally manipulated locally with stitches and pinches.

Thirdly, around 1960 Heinz Isler created a flat surface (*zeroclastic*) with composites. These large surfaces consist of multiple composite (glass fibre reinforced plastic) cubes together (Figs. 10.15 and 10.16). They were used as a roof structure for a gas station (Fig. 10.17) and pitched roofing for a German school.

Fig. 10.13 Dome shell (*Source* J. Chilton)

Fig. 10.14 Dome shell on concrete shell (*Source* J. Chilton)

Fig. 10.15 Test of composite plate (*Source* J. Chilton)

Fig. 10.16 Construction of flat composite (*Source* J. Chilton)

Fig. 10.17 Flat composite roof (*Source* J. Chilton)

Technique

Shell Fabrication

Isler used the natural draping of woven fabrics as a tool for the definition of the morphology of eight hundred of his built concrete shells. By hardening a hanging fabric and inverting it, an optimized shape is found for a shell. This method can also be applied to FRP membrane manufacturing, the difference is that, finally, the hardened membrane will be the base for the structure itself instead of serving as a model.

Before the manufacturing comes the design phase. Simulation of the manufacturing process plays a significant role for modelling the draping of the hanging membrane under gravity. When the desired shape is created and simulated, the membrane is tailored according to the design, different techniques such as stitches, cuts and pinches can be used to generate various forms as shown in Fig. 10.9.

The actual process of manufacturing for producing a shell with membrane consists of 3 basic steps: hanging, curing and inverting. The materials used in the process are pre-pregnated or wetted-out woven reinforcement fabrics.

After the last step, the tailored membrane is firstly hung on anchoring devices (Fig. 10.10). This step should be accurate and controllable, because it will determine the transfer of stresses over the membrane into the ground, the next step is curing and it has two variants: dry layup and wet layup.

Dry Layup

In the dry layup method, an oven is required. The hanging membrane from the hanging step is moved into a pre-heated oven for curing. After several hours' curing, the object will become light-weight and self-supporting. The material used in the process could be three layers of carbon 0/90 pre-preg in satin weave. The object was produced in 8 h, by two persons, with a curing time of 16 h in an oven (Blonder 2008: 8). In this method, the preliminary joining of segments could be avoided by taking advantage of the material's natural tack at certain temperatures. However, the consolidation in this method is not perfect as there is still a small amount of air between the layers. The reason is that the object is exposed to the air; therefore, no vacuum was applied to the hanging piece while curing. Because extreme material performance is not required, this method can still be considered satisfactory for the application.

Wet Layup

Wet layup is another method for composite manufacturing. Liquid resin can be applied to the hanging membrane and fibre reinforcement is placed manually on top. A metal laminating roller is used to impregnate the fibre with resin and remove any trapped air. Several steps are repeated until a suitable thickness is reached (Fig. 10.18).

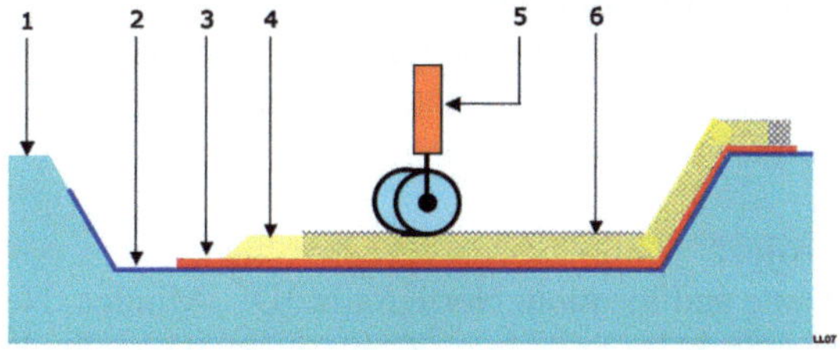

Fig. 10.18 Layup method (1 mould, 2 wax, 3 top coating, 4 fibres, 5 roller, 6 resin)

Dry layup and wet layup both have advantages and disadvantages. In wet layup, more manual labour is required and some additional technical difficulties for handling a large impregnated membrane may occur. On the other hand, more manual labour can also offer simple application and minimum tooling. The dry layup has the most rapidity and simplicity of installation; however, the oven increases the cost of the manufacturing and also leads to a possible limitation of size.

Plate Manufacturing

Isler also surprisingly created a flat surface with composite materials. His successful solution is a plate which has a surface similar to a honeycomb (Meier 2011: 6–8; 2012: 12). The translucent coffer roof of a gas station is a great example showing the manufacturing techniques.

In GFRP plate manufacturing, different layers are either installed or applied to build the final plate product. At the beginning, a pre-cast membrane of unsaturated polyester resin is placed on a clean and even floor. On top of this layer of membrane, a layer of unsaturated polyester resin is applied with laid-up glass mats as bottom, which are rolled carefully to improve the impregnation with the resin and to avoid unwanted air staying in this layer. On top of the wet surface, prefabricated coffers are distributed. After the distribution, another layer of coffers is glued to the bottom coffers and the gaps between the two are closed with spackling compound. After the gluing, a flat surface is formed and finally resin impregnated glass mats are applied as the top layer. With this method of manufacturing, a translucent GFRP plate can be made to be used as a building roof.

10.4 Strong Tapes for High Performance Sails

Flexible Mould

The form-active structure: coated fabric.
Form-active typology: single layer.
Way to stabilize the mould: stressed surface.

Composite Material
Technique to handle the rigidizing material: hand layup.
Reinforcement of the material: single fibres.
Surface treatment: coating.

Project
Name and location: North Sails 3DL, Minden, Nevada.
Design & Engineering: - Contractor: 3Di technology
Year: 2007.

Introduction

Fibres are used in a broad spectrum of applications nowadays. Traditionally, fibres are most often applied in twisted form, such as in yarns and ropes, are woven, as in fabrics and cloth, or are used in cut form, e.g. in non-woven fillings and in composites. Still, there are also opportunities to make more out of these fibres and use them in a different way (Figs. 10.19, 10.20 and 10.21). Then we call them high-performance fibres (Figs. 10.22 and 10.23).

In order to employ high-performance fibres more efficiently and to better exploit their ultimate potential, laminates of multifilament strands and spread multifilaments have been developed. Striking examples in the arena of high-performance sails are products made with the three-dimensional (3DL) technology by North Sails Group (Lefevre 2008). 3DL technology is one of the latest techniques in the field of the production of sails. It all starts with a three-dimensional CAD/CAE design file created

Fig. 10.19 Sails reinforced with different tape layers (photos: North Sails 3Di technologies) (*Source* North Sails)

Fig. 10.20 Machine to reinforce sails with different tape layers (photos: North Sails 3Di technologies) (*Source* North Sails)

Fig. 10.21 Sails reinforced with different tape layers (photos: North Sails 3Di technologies) (*Source* North Sails)

Fig. 10.22 Jacks for adjustable mould (*Source* North Sails)

Fig. 10.23 Production of sail on adjustable curved mould (*Source* North Sails)

by a North Sails sail designer. North's proprietary design software creates a custom "mould" file for each individual sail. A computer program reads the design file, then instructs an articulating mould to form the designed shape. After a base layer of Mylar film is draped over the mould and tensioned, a 6-axis fibre head suspended from a computer controlled overhead gantry then applies structural yarn onto the surface of the base film. Once the yarns are laid, a second film is positioned on top of the base film and yarn, tensioned, and then covered with a large vacuum bag that compresses the laminate. Finally, a heat technique cures the pressurized laminate tightly to the mould. With this technique the following result can be reached:

- it combines different kinds of fibres;
- it makes tapes out of these fibres; and
- it is combined with an adaptive mould to realize optimal forms.

Material

The use of polymers in demanding mechanical applications has surged since the development—starting in the late 1960s—of high-performance fibres, which by now have found a broad spectrum of applications, including bullet proof vests and panels, cut-resistance gloves, reinforced tires, ropes, composites in aviation and aerospace, etc.

Combining Different Polymers

Before the 3DL technique was discovered and applied, the sails mainly consisted of single fibres only. The properties of these single fibres were responsible for too many disadvantages. The disadvantages led to the invention of a new technology to improve the sails. Sails made up of tapes turned out to be a better solution. Polymers of carbon, Dyneema and aramid are the main materials used in these tapes (Fig. 10.24).

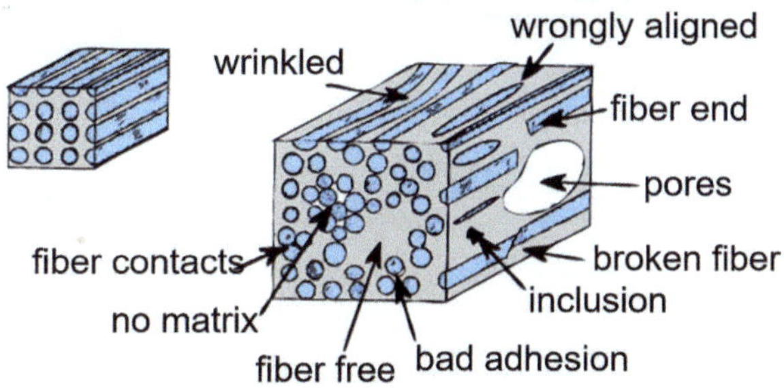

Fig. 10.24 Disadvantages of the use of single fibres (*Source* North Sails)

Material Properties

Carbon

Carbon fibres are produced from a variety of polymer precursor filaments, e.g. polyacrylonitrile (PAN), or directly melted and spun from petroleum or coal-tar pitch. The precursor fibres first are oxidized at elevated temperatures (200–400 °C) and in a second step at higher temperatures (2,700–3,000 °C) to a certain extent graphitized. The thickness of this fibre is between 5–10 micrometres and consists of elongated carbon crystals that lie parallel to the fibre axis. The fibre is characterized by very low stretch and has the property that it does not oxidize under the influence of water and oxygen, it also has a low density, great tensile strength and low expansion coefficient.

Dyneema

Dyneema fibre, also known as Ultra-high-molecular-weight polyethylene (UHMWPE) has extremely long chains, with a molecular mass usually between 2 and 6 million daltons (u). Long chains transfer the load more effectively to the polymer matrices making them very tough and preserving a low friction coefficient. Dyneema is odourless and tasteless, and is nontoxic. However, it is not resistant to oxidizing acids. It is self-lubricating and highly resistant to abrasion.

Aramid

Aramid fibres are a type of nylon of which the molecular structure is composed of linked benzene rings and amide bonds. They are heat-resistant and the chain molecules are oriented along the finer axis. It has no melting point, low flammability, good fabric integrity at elevated temperatures. Para-aramid fibres, which have a slightly different molecular structure, also provide outstanding strength-to-weight properties, high tenacity and high modulus. Some major aramid fibre uses are: flame-resistant clothing, protective vests and helmets, composites, asbestos replacement, hot air filtration fabrics, tire and mechanical rubber goods reinforcement, ropes and cables, sail cloth, sporting goods. The best-known aramid is Kevlar. For making

	material	chemical structure	E		σ		ε	ρ	flex-l
			GPa	N/tex	GPa	N/tex	%	g/cm³	%
	Carbon Fiber Tenax UMS3536	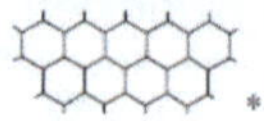	430	238	4.5	2.5	1.1	1.81	100
	Ultra-high molecular weight polyethylene UHMWPE Dyneema® SK76	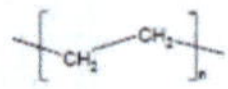	116	120	3.6	3.7	3.8	0.97	0
	Poly(*p*-phenylene terephthalate) PPTA Kevlar® 49	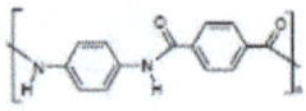	113	78	3.0	2.1	2.5	1.44	28

Fig. 10.25 In-axis properties of Carbon/Dyneema and Aramid polymers (*Source* North Sails)

optimal sails it was important to combine the specific requirements of the above-mentioned composites (Fig. 10.25).

Form

Several Sail Parts

The company has chosen to produce several smaller sail parts that are easier to handle when moved to the mould. They also get better utilization of the equipment by having several smaller areas rather than few large ones. The layout software makes it easy to seamlessly connect the different layers of tape on the mould. The tapes are sticky enough to stay together when the sail is being handled, but not so sticky that they cannot move when it is shaped on the mould.

Figure 10.26 shows an image of the sail design software. A big advantage of this program is the possibility to determine the places which need more stiffness, the sails are then made with some extra layers on these spots. The software is the most important tool with regard to the form of the sail. In the end, the different pieces only need to be put in the right places.

Fibre Forms

In common composite structures fibres are not (all) loaded in the axial direction. It is well known and readily understood that this leads to a more or less severe reduction of the mechanical properties of the final objects into which they are incorporated. Therefore, tapes with a unidirectional fibre direction were developed to ensure that the performance will be better (Fig. 10.19).

Technique

Process

The process begins with untwisted yarns of carbon, aramid and Dyneema. They are run through a machine, dubbed the "Pregger", which spreads the individual fibres. Another machine makes three bundles of these fibres. This machine has different paths and technologies for different materials. Then the fibres are coated with a thin

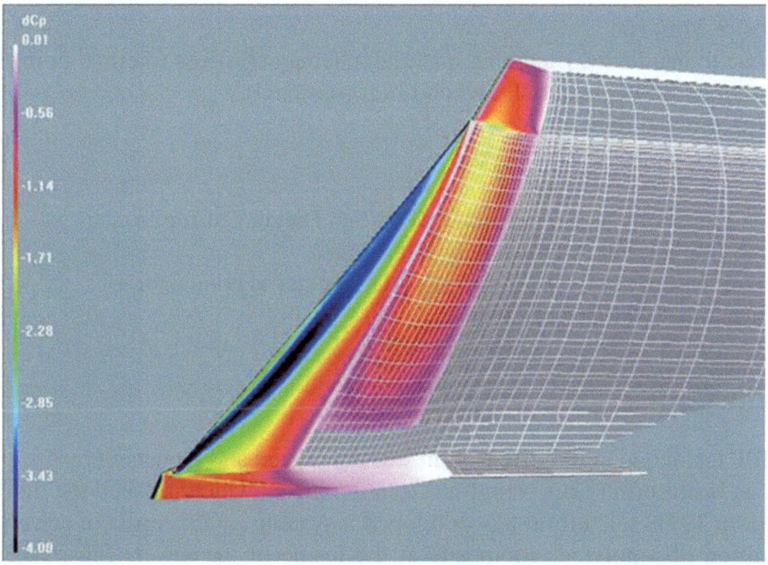

Fig. 10.26 Sail design software by North Sails

coat of thermoset adhesive to hold them together, both as a tape to build the sail and to consolidate all the parts forming the final sail. Then the fibres are placed on a backing paper, dried, rolled up and cut into narrower width rolls. Depending on the mix of fibres, the tapes can be run two or three times through the machine. The next step is to weigh the material. As sails are built up of many layers of tape, the tolerances are very small in order to keep the overall weight down. Then the tape-laying head goes back and forth over at flat floor to lay out all the tapes according to the design. It collects the backing paper and cuts the tapes. The pieces are then transported to the moulds. The actual shape is achieved when the tapes are stretched on the mould and heat is applied. The whole sail is then sealed in a vacuum bag and all layers are consolidated with heat into an almost isotropic structure (resistant to force in all directions). After the sails have consolidated and hardened, they're finished on the floor. As batten pockets and reinforcements are already included in the sail itself, this requires less work (SailSelect, www.Youtube.com/watch?v=co2gu-WRn_w; The Ocean Race, www.youtube.com/watch?v=kyqIlm5EPaI; Gustafsson 2010).

10.5 Insituform Cured in Place Pipe Rehabilitation Technology

Flexible Mould

The form-active structure: resin-impregnated plastics.
Form-active typology: double layer.
Way to stabilize the mould: inflate, pressurize.

Composite Material
Technique to handle the rigidizing material: chemical process (thermal or drying).
Reinforcement of the material: carbon fibres (optional).
Surface treatment: none.

Project
Name and location: Airport terminal CIPP, San Diego, California.
Design & Engineering: Charles King, Insituform.
Contractor: Insituform (subcontractor), AOC (material composition).
Year: invention 1971, project 2010.

Introduction

The CIPP technique in its current form was developed and pioneered by a company known as Insituform (Latin term for "form in place") which initiated the first CIPP program in 1971. Insituform has developed a system for the repair of sewers with the addition of a polymer composite at the inside of the sewer (Figs. 10.27 and 10.28). After cleaning and repair of the sewer, the membrane hose is brought in place and inflated. Then the polymer composite is cured and after rigidization, the hose is deflated. The system invented in 1971 is still in use and has been successfully adopted in many countries. Currently Insituform holds a dominant position in the global CIPP market, although other companies have also appeared on the horizon since the CIPP process became public domain (Insituform, www.insituform.com; Wood 1972).

CIPP stands for Cured in Place Pipe Rehabilitation Technology, a relatively new technique in the civil engineering field that allows for repairing or strengthening damaged or deteriorated pipes. CIPP differs from traditional pipe repair methods

Fig. 10.27 CIPP composite (*Source* Insituform)

Fig. 10.28 Insituform sewer renovation system (*Source* Insituform)

in that CIPP is entirely trenchless as repairs through an entire pipe segment can be done via an existing opening or manhole. This eliminates the need to dig trenches to expose whole pipe segments, which can be a very labour intensive, costly and logistically problematic undertaking.

CIPP applicability is favoured over traditional repairs in situations where time constraints or limited space are a key factor in the rehabilitation. An average CIPP project can take a matter of days or weeks as opposed to months of work in traditional pipe repairs. The CIPP technique can be compared to the medical procedure of placing a stent, a mesh "tube" inserted into a natural body like a blood vessel.

The CIPP rehabilitation process consists of the following steps:

- Pre-insertion cleaning of existing pipelines, by means of high-pressure water or by using a camera and remote robot.
- Insertion of the liner and resin material into the damaged pipeline.
- Pressurization to adhere the liner to the existing pipeline throughout the whole pipe segment.
- Chemical curing of the resin liner.
- Clean-up work and material removal at lateral openings and the insertion point, using a camera and remote robot.

Form

Shaping a CIPP form into an existing pipeline is done by means of insertion of the CIPP components into the pipe innards by means of one or more openings, either a manhole or pipe end (Fig. 10.29). CIPP can be applied to pipes ranging from several

Fig. 10.29 CIPP application within pipelines (*Source* Insituform)

centimetres to several meters, the application methods depend on the pipe-section size and complexity.

The linings can travel through, align and fit through the pipe innards by means of inflation and pressurization by pumping air or water through the pipes. After the linings are properly fitted and aligned within the pipes the final stage, the curing process, can be initiated, the curing process will turn the resin into a rigid adhesive material that will bind to the existing pipes. The curing process can be done by means of a catalyst or hardener. This can be anything that evokes a chemical reaction in the resin, either with thermal curing by means of hot water or hot steam, photosynthesis by UV or by ambient curing. The curing process takes a few hours depending on the methods. Afterwards the linings are fully strengthened and rigid.

Application of CIPPs is common among sewage pipe maintenance, the process allows for repairing/strengthening deteriorated pipes of any size, ranging from small pipes to larger tubes.

Lining must be applied inside pipes with a tight fit to ensure maximum strength enhancements.

Loose liners are preferably avoided, in cases where these must occur, they must be stronger by a certain degree than tightly fit liners that adhere to the existing pipe surface.

Material

CIPP consists of two main components: The resin consists of an epoxy or polyester\styrene material, this will cure by means of a chemical process (Fig. 10.30). The lining sleeve surrounding the resin, consisting of a soft plastic material like polyester/vinyl. A strengthening material, consisting of woven carbon fibres, this may not be a prerequisite, depending on strength requirements. The lining material, which forms the basis of the CIPP process, consists of a pliable, stretchable plastic material. The material must easily fit through a pipe segment with a certain degree of complexity. Another important factor is leak proofing, as infiltration or exfiltration can easily ruin the whole CIPP process. This lining is evenly impregnated with a resin. It is important that the resin is evenly distributed as pockets of air can ruin the whole

Fig. 10.30 CIPP bag impregnated with a resin before construction and curing on site (*Source* Insituform)

CIPP process. The resin-impregnated lining material is usually pre-applied in an industrialized process. Reinforced liners offer much improved strength as compared to non-reinforced materials, however reinforced liners do not offer the high flexibility of non-reinforced counterparts, which tend to have an additional thickness as well. The relation between thickness and strength varies, it ultimately depends on the application of the strengthening material (Jaganathan et al. 2007).

Technique

CIPP liner can be applied to pipes by means of inversion or pull-in methods: the inversion method involves pushing the CIPP liner through the pipes inside-out, meaning that the end is connected to the start point of the pipe segment and liner is pushed into the pipes by pumping water through the segment.

The pull-in method involves pulling the liner through the pipes, then sealing it off, by either closing the ends or creating a closed circuit, then using air pressure to inflate the liner and fitting it into the existing pipeline.

Both methods require some preparation work beforehand (Fig. 10.31), by clearing the pipes of uneven surfaces, debris or dirt, and post-installation work by creating the proper openings for junctions and ends and/or removing excess foils. This work is usually done in smaller pipes by driving a wheeled mechanical robot arm through the pipes (Fig. 10.32).

Lining strength depends on the pipe condition and pipe strength, the exact lining must be calculated based on pipe diameter/dimensions and strength requirements of the pipe segment.

Fitting of the CIPP material must be carefully executed and the procedure is not without risks. Stretching the material too far results in weakened strength due to compromised integrity when the liner and resin densities become less than optimal. Underpressurizing may result in lack of adhesion, also resulting in weakened strength.

An effect of weakened pipeline strength can be loose liners, pockets and wrinkles, posing a threat to the structural integrity as well as the flow. Generally, these anomalies occur at stress points due to underpressurizing, inadequate pipe clean-up before

Fig. 10.31 Inversion push-through method (*Source* Insituform)

Fig. 10.32 Wheeled robot junction opening (*Source* Insituform)

applying the CIPP material, or due to the improper lining material fitted inside a pipe.

Examples of ensuring a proper lining application with the stress areas of the existing pipes can be taken from Chap. 2, where general practices of membrane geometry manipulation are explained and classified. The form effects are visibly evident in the CIPP liners around stress points and corners (Fig. 10.33).

Proper application of CIPP material results in long term durable pipe repair, the claim is that in sewage pipes CIPP applications can last 25–50 years with little or no maintenance, only after that period will the CIPP material deteriorate noticeably (Doherty 2005).

Fig. 10.33 Proper application CIPP (*Source* Insituform)

For an expansion project at San Diego airport, the existing sewage pipelines that ran for more than 500 meters underneath the area needed to be strengthened to allow for the weight load of the newly constructed terminal. Because of the difficulty of this task; the small size of the pipeline and the inability to dig trenches, which would disrupt airport traffic, traditional pipeline replacement methods were inapplicable.

For the San Diego airport pipeline repairs, not only was the challenge to cover a 500 m long segment with few existing manholes to spread the coverage, but the CIPP material needed to be strong yet thin enough not to disrupt the water flow capacity of the current sewage pipelines, which meant the CIPP liner had to have a thickness of 50 mm at most. This challenge sparked a unique solution with fibre woven resins that could cover the whole 500-meter segment in two parts. 250 meters from one side and 250 meters from the other (Rush 2010; The Int. Soc. for Trenchless Technology, www.istt.com).

10.6 Knotted Chair by Marcel Wanders

Flexible Mould
The form-active structure: cable net.
Form-active typology: connected double layer.
Way to stabilize the mould: stressed surface.

Composite Material
Technique to handle the rigidizing material: submersion with epoxy resin.
Reinforcement of the material: ropes; A thread constructed of aramid and carbon fibres.
Surface treatment: none.

Fig. 10.34 Knotted Chair

Project
Name and location: Knotted chair, the Netherlands.
Design & Engineering: Marcel Wanders in cooperation with Droog Design/Delft University of Technology.
Contractor: Cappellini (Manufacturer).
Year: Design: 1995, the Netherlands. Manufactured: 2005, Italy.

Introduction

Marcel Wanders's Knotted Chair, made of rope wound around a carbon core, hardened with epoxy, was widely publicized as soon as it was made, and rapidly came to define a certain way of designing in the 1990s (Figs. 10.34 and 10.35). It had the appearance of hand-knotted macramé but also incredible strength and durability. The chair grew out of a project called Dry Tech, a collaboration which aimed to combine the rough-and-ready low-tech aesthetics of Droog Design with the high-tech materials of the Aviation and Space Laboratory of Delft University of Technology. Once the fibres had been hand-knotted and soaked with epoxy, the form was draped upside-down over a mould to get the shape of the seat and back. The designer produced the first versions of the chair himself, before production was taken over by Cappellini in Italy (V&A, collections.vam.ac.uk/item/O115748/knotted-chair-chair-wanders-marcel).

Fig. 10.35 Detail Knotted Chair

Epoxy

Epoxy is a mixture of liquid resin and hardener which cures to a high strength plastic solid. It has high moisture resistance and bonds with wood fibre, fibre glass, reinforcing fabrics and certain metals. Resin is the base ingredient of epoxy compounds. It is a clear, pale yellow, low-viscosity liquid. Hardener, depending on the temperature of the working environment, can affect the cure speed. Fillers are used to thicken the mixture. These produce certain handling characteristics or cured physical properties. Additives are used instead when specific coating properties are desired.

Form

At the studio of Marcel Wanders there was an intern whose idea it was to make furniture with knotted ropes. Wanders took this idea to Delft University of Technology and there the old and new techniques were combined. Wanders was inspired by the architect Gaudi and his design for the Sagrada Familia that uses gravity to determine the shape of the design. He took the wet chair and hung it upside-down in a mould. Then he let it dry for 24 h and that is how the shape came about (Fig. 10.36).

Technique

Marcel Wanders' acclaimed Knotted Chair is manufactured from aramid fibres. Wanders used a macramé technique, knotting several hundred metres of aramid braid around a carbon-fibre core.

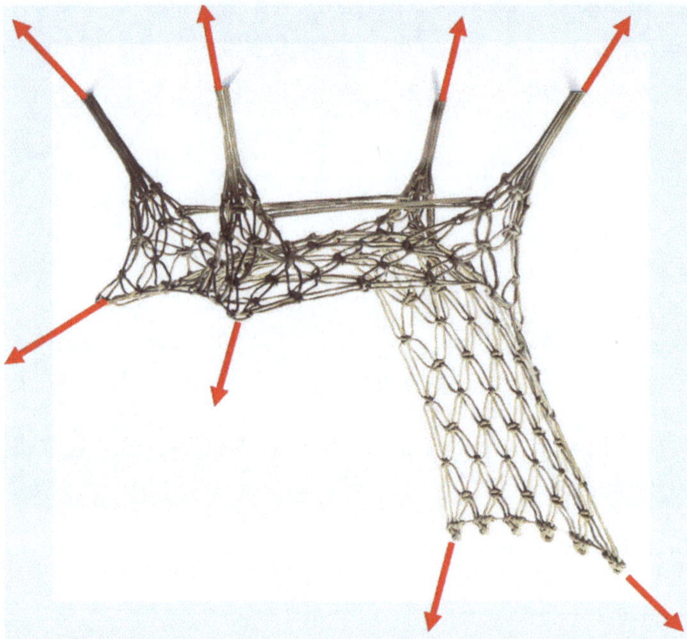

Fig. 10.36 The way the chair was hung to dry

Macramé

Macramé is a technique to make a kind of net-structures by knotting (no weaving or knitting). There are various ways to make the knots but the most used knot is the square knot. Sailors used it in the past to decorate anything from knives to bottles to parts of their ship. In the 1970s it became very popular for decoration in homes with wall hangings, window coverings and lamps. The cords used in macramé can be of any flexible material but the most obvious materials are cotton twine, linen, hemp, jute, leather or yarn.

Marcel Wanders combined the well-known macramé with new materials and techniques by curing the resin-saturated prestressed rope structure (Fig. 10.37) (Beukers and Hinte 2005: 149).

10.7 Pack Chair by François Azambourg

Flexible Mould
The form-active structure: woven fabric.
Form-active typology: connected double layer.
Way to stabilize the mould: inflation with foam.

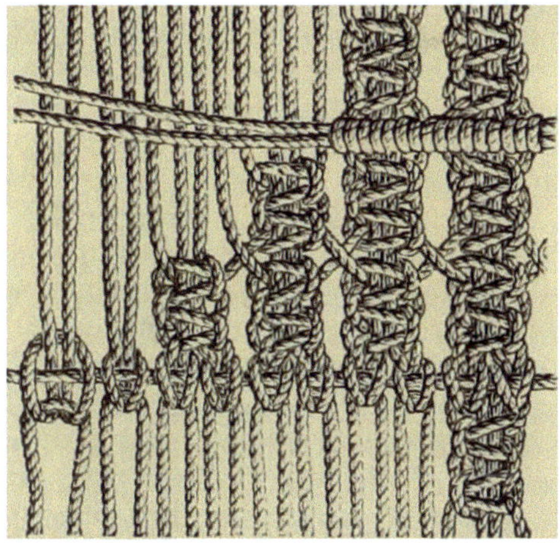

Fig. 10.37 Knotting the threads in to Macrame

Composite Material
Technique to handle the rigidizing material: chemical reaction of the foam.
Reinforcement of the material: inflation.
Surface treatment: none.

Project
Name and location: pack chair, France.
Design & Engineering: Francois Azambourg.
Contractor: Francois Azambourg.
Year: 1999.

Inventor and designer, Azambourg combines beauty with the technique of craftsmanship. He is a true adept of applied arts. The Pack Chair is made from 3D textiles (Figs. 10.38 and 10.39). In its compact state, this chair is as small as a soda canister. The 'pack chair' gets its final shape when the distance fabric is inflated with a liquid polyurethane foam. The foam rigidizes when it comes into contact with an emulsion of two chemical products on the surface of the membrane. On contact, the foam rigidizes within a few seconds. Francois Azambourg claimed this design as an invention and obtained a patent in 1999 (Azambourg 2002). Distance fabric has been known since the development of an inflatable airplane by Goodyear (Goodyear Aerospace Corp 1961) in 1956. It is hard to imagine that in the period 1956–1999 no one investigated the inflation of distance fabric with foam, since the inflation with foam was researched for space applications by NASA (Cadogan and Scarborough 2001). On-site foam inflation can have a wide-range of applications in design and industrial processes because it allows large savings to be made in transportation costs and maintenance of the inflatable.

The Pack chair is made of 185 g/m^2 polyester sailing canvas, structured by polyester strings. They are hand sewn and filled with rigid two-components

Fig. 10.38 Pack chair

polyurethane foam. A foam-filled distance fabric structurally behaves like a sandwich. Sandwich structures behave like a beam, plate or shell consisting of at least three layers: two face-sheets and one core (Plantema 1966; Zenkert 1995). They are efficient lightweight structures. The structure consists of a core with a low to moderate stiffness, with at the surfaces a material with a higher stiffness and in general better material properties. A sandwich has a higher shear stiffness, bending stiffness and tensile strength to weight ratio compared to a homogeneous beam.

10.8 Adjustable Mould for Composite Façade Elements

Flexible Mould
The form-active structure: foil.
Form-active typology: single layer.
Way to stabilize the mould: stressed surface.

Composite Material
Technique to handle the rigidizing material: vacuum infusion.
Reinforcement of the material: woven fabric.
Surface treatment: coating.

CONCOURS

François Azambourg récompensé

■ Sa chaise autogonflable a remporté le premier prix du concours Tops Plastiques.

PAR GAËLLE VACHER

Pour la deuxième année consécutive, un jury de professionnels s'est réuni chez Colette pour y décerner les Tops Plastiques, qui récompensent des designers travaillant le plastique.

Le premier prix a été décerné à François Azambourg, designer et enseignant à l'école Boulle, pour sa chaise extensible autogonflable baptisée Pack. C'est une chaise réalisée en tissu polyester tridimensionnel et qui comprend une bombe de mousse polyuréthanne expansive. Sa construction est un jeu d'enfant. Il suffit d'appuyer sur un bouton pour mettre en contact les deux composants, secouer la chaise pour bien les mélanger, et attendre… La chaise gonfle d'elle-même, et, quelques minutes plus tard, le temps que la mousse expansée durcisse, elle est prête à l'emploi. Vendue repliée sous la forme d'un pack de 3 litres, elle peut être distribuée en grande surface ou en VPC. Deux autres projets ont été distingués : la lampe Punch Light, de Jean-Marc Gady, objet en silicone qui s'allume et s'éteint quand on tape dessus, et la Lampe Baladeuse de Xavier Moulin, une lampe molle réalisée avec le gel utilisé pour les sièges de VTT.

Le concours Tops Plastiques est organisé par Profession Plastique, qui réunit les acteurs français de la filière. Sa mission est de promouvoir l'utilisation de cette matière dans l'univers du design. Mais aussi de donner la possibilité à de jeunes designers de se faire connaître et éditer. Ce fut le cas, l'année dernière, d'Elsa Francès et de Jean-Michel Policarle, concepteurs d'un Dé Lumineux, aujourd'hui édité par Ligne Roset, et récompensé lors du Salon du Meuble. ■

La chaise Pack de François Azambourg.

Fig. 10.39 Prize winning design (*Source* F. Azambourg)

Project
Name and location: Doubly curved façade element, the Netherlands.
Design & Engineering: Arno Pronk, Ivo van Rooy and Pieter Schinkel.
Contractor: Buitink Technology.
Year: 2009.

Introduction

This chapter describes the development of an adjustable mould. This moulding technique is based on the manipulation of mechanically prestressed elastic foil. This chapter describes the way the elastic foil is manipulated and gives the possibilities for making three-dimensional curvatures. The technique developed was tested by the creation of a doubly curved façade of 12 panels.

Moulding techniques for the construction of curved panels can be divided into three different categories (Helvoirt 2003).

- The static mould. The most common mould of this type is the EPS/PS (polystyrene) mould, which is 3D formed using a CNC miller device. This technique has a high accuracy but produces a lot of waste material.
- The reusable mould, for example a mould made of clay, wax or sand (Oesterle 2012: 52). There is no waste material and therefore the method is a lot more environmentally friendly. It is labour intensive and cannot be used in all circumstances.
- The flexible mould. This type of mould can be used for a lot of differently shaped elements because the mould will form itself to the elements that will be produced. Examples of this kind of mould are pinbed moulding (Munro and Walczyk 2007) like the FlexiMould by Boers, the adjustable mould by Rietbergen & Vollers and later by Roel Schippers, and the flexible mould by Adapa adaptive moulds (Raun et al.) (Hawkins et al. 2016: 911–935).

Pinbed moulds contain a field of height-adjustable pins. In most cases this is done by a computer- automated machine. The pins are covered with a polymer to smoothen the surface. It is important to avoid local distortion of the product surface by the pins. The downside of these flexible moulds is the high investment. The expenses can only be recovered when a large volume of elements will be produced using this particular moulding technique.

Material

To avoid the high investments required for a complex bed of pins and to avoid the problems with local distortion of the surface, the authors developed an adjustable mould without pins on the surface of the mould. Instead of the pins, it is possible to use the capability of doubly curved mechanically prestressed membranes to make a stable surface. This surface can be manipulated in 18 different ways. Every manipulation has its own characteristic form. Combining the different manipulations leads to a sheer endless number of possibilities. In this way, the benefits of a flexible mould (adjustable to a lot of different forms) are combined with the benefits of a reusable mould (no waste materials and lower investment costs). The first step of this research was to make one panel on a prestressed membrane fixed at the edges and manipulated on the surface with an inflatable. To produce one element, an EPDM rubber foil was used for the mould and a glass fibre reinforced epoxy resin composite was cured by vacuum infusion. Apart from the usual benefits of vacuum infusion it is possible to push an inflatable against the surface from both sides without disturbing the infusion and curing process. The infusion process takes one hour. After that, the panel needs to harden over time (up to three hours, depending on the temperature). Finally, the composite panel is sprayed with a primer and shiny coating (Figs. 10.40 and 10.41).

After the success of this experiment, the next step was to develop an adjustable mould based on the flexible and elastic properties of a stretched foil over an edge. The deviation in curvature and measurement perpendicular to the surface of the panel can be quite big in fluid architecture before it will be perceived. The perceived deviation at the edges and the voids between the panels is more critical. This depends on many factors, but is mainly dependent on the width of the void. Therefore, the boundary of the mould is made fully adjustable. It consists of several height-adjustable bars that support a boundary that is formed like a spline (de Boor 2001: 87). The spline will flow gently along the height-adjustable bars. To guarantee a smooth transition between two panels, the foil is clamped between two adjustable edges of EPDM rubber that can be placed into a desired angle. The upper and lower edges are pushed against the stretched foil to force it gently into the correct angle (Fig. 10.42). The

Fig. 10.40 Curved composite façade panel made with a membrane mould

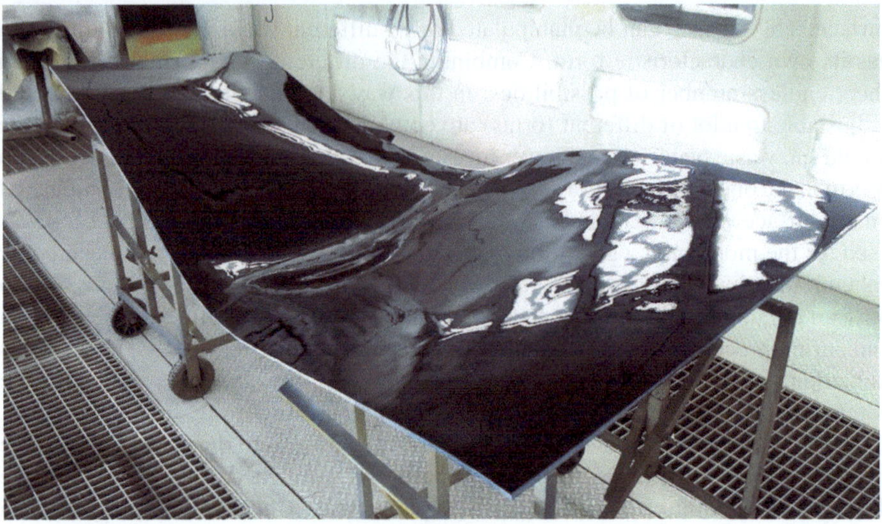

Fig. 10.41 Curved composite façade panel, final result

Fig. 10.42 Curved composite façade panels made with an adjustable mould

edges are tapered, thinner and flexible to avoid the edges pressing into the foil. The result is a fluent transition to the correct angle (Fig. 10.43).

To hold the bars in place, a steel structure has been made. This steel structure does not only support the bars but is also used to prestress the membrane. Figure 10.44 shows the design of the steel structure. The bars are represented in blue and the rubber edges in black. The green beams are used to fix and stretch the membrane.

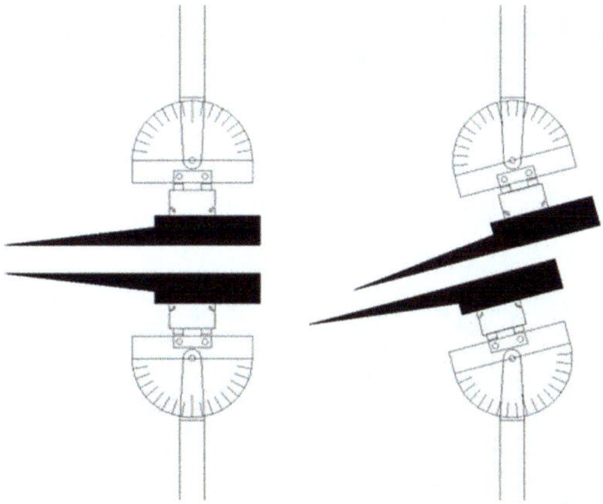

Fig. 10.43 Two adjustable edges of rubber

Fig. 10.44 Design of adjustable steel frame

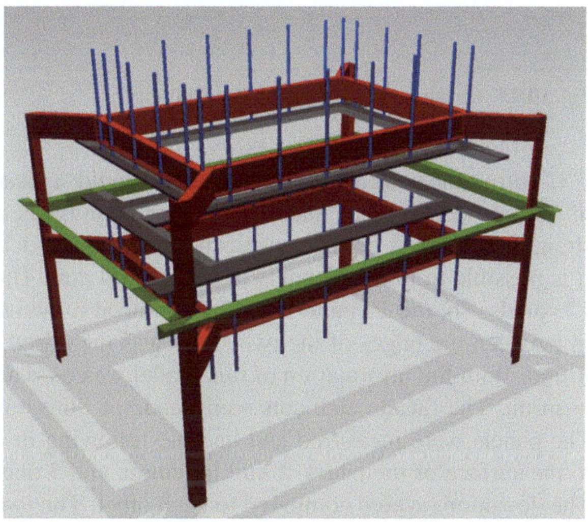

They can also be placed on the corners at different heights. The difference in length of the green and black beams is bridged by the overlap at one side. The maximum surface that can be created at one time is 1,824 × 1,216 mm.

It is important to stretch the membrane (PU-foil) over a continuous length. The team used a system developed by Buitink Technology. An inflatable tube is placed on four sides of the foil. By increasing the pressure in the tube, it deforms from a lens shape to a circular cross-section. In this way the membrane is stressed by the tubes (Fig. 10.45).

Once the membrane has been tensioned between the adjustable boundaries, the shape can be manipulated by placing inflatables or other objects below or on top of the membrane.

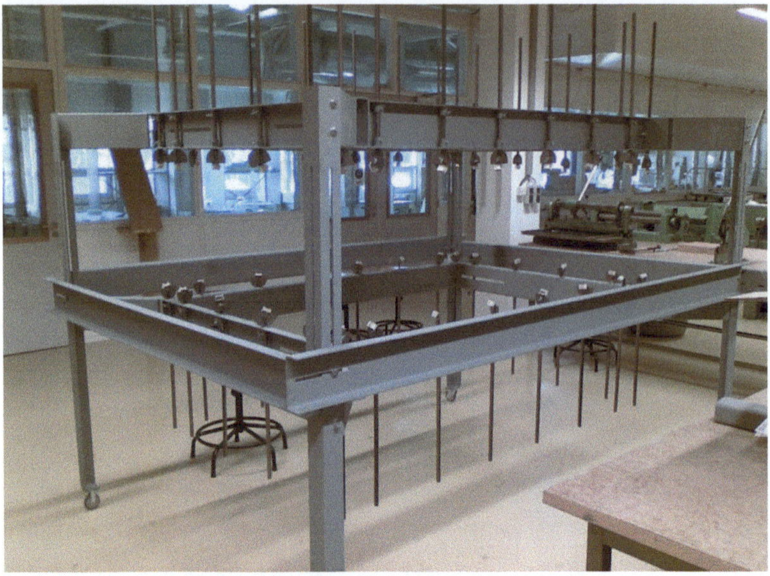

Fig. 10.45 Construction of adjustable steel frame

To prove the working of the adjustable mould, a façade of 12 doubly curved panels of 600 × 600 mm was made. The outer boundaries of the 12 elements were kept straight, making it possible to fit the mock-up model into an existing, flat façade. It is possible to make 6 panels of 600 × 600 mm. The mock-up façade consists of 12 panels. So, the 12 panels will be produced in two shifts. With this mock-up, the fit between the edges of the two shifts was measured. The façade was modelled in Rhino. With the information of this model we could adjust the mould in the desired position. The façade elements were made of fibre-reinforced polyester composite. The panels were measured and had the following deviations; 3 mm perpendicular to the surface of the panel, 2 mm in length, and 5 degrees in the angle of the edge. The deviations where compared to each other. The two big panels were cut into the desired dimensions of 600 × 600 mm. They can be cut using a saw or even a water or laser cutter. In this case it was done by a hand controlled circular saw. The void between the panels is 8 mm. After sanding, the panels can be coated or painted. As in Fig. 10.42, the mock-up was coated with a metal coating which gives the panels a rusted steel look (Fig. 10.46).

Conclusion

In the mock-up, the deviations of the panels were not perceived. The cutting of bigger panels into smaller elements was almost without deviation. With this moulding system it is possible to make a large continuous mould. The products made will be cut into panels afterwards. If the elements produced for the mock-up had been made in a conventional way, the price would have doubled. The main advantages

Fig. 10.46 Construction of the composite panels

of the technique are its low cost and the simplicity of the mould. It is also possible to optimize the process by introducing computer controlled bars to speed up the adjustment of the mould and 3D monitoring to allow better control of the 3D surface.

10.9 Free Forming or Bubble Thermoforming of Thermoplastics, Bubble Chair by Eero Aarnio

Flexible Mould
The form-active structure: air.
Form-active typology: single layer.
Way to stabilize the mould: air pressure (no mould present).

Composite Material
Technique to handle the rigidizing material: air cooling.
Reinforcement of the material: none.
Surface treatment: none.

Project
Name and location: Bubble Chair by Eero Aarnio, Finland.
Design & Engineering: Eero Aarnio (b. 1932).
Contractor: Adelta.
Year: 1968.

Introduction

Eero Aarnio designed the ball chair made of glass fibre reinforced polyester in 1963. He noticed the curved polymer skylights in buildings. He approached a manufacturer to realize a transparent ball by heating and blowing up acrylic to the size of a ball chair. This transparent ball chair is named the Bubble Chair and is still in production by the firm Adelta. To realize this chair, thermoforming techniques are used. Thermoforming is a collective term for deforming polymers and comprises techniques like billow forming, cavity forming, drape forming, plug-assist vacuum forming, billow drape forming or snap-back forming (Schwarzmann 2001). Bubble forming, or free forming, is the deforming of a thick-gauge plastic sheet without the use of a mould using heat and air pressure. This technique is used when special requirements are set as for the surface and its transparency (Fig. 10.47). This technique is mainly used for the production of skylights. Although free-form thermoforming is limited to creating "bubble"-form parts, there is a wide variety of forms that can be made with (manipulated) bubble forms. A closer look at the polymers used and subsequently the form, techniques and processes, is given below.

Material

Three types of plastics are appropriate for this free forming technique. These polymers are polymethyl methacrylate (PMMA), cellulose acetate (CA) and polycarbonate (PC) (Gruenwald 1998). All three of them are amorphous polymers and therefore have good characteristics in terms of tenacity and thermos-elastic behaviour. Moreover, these polymers have useful optical properties. In combination with the free-form thermoforming technique, a very transparent surface with minimal optical distortion can be achieved. The maximum of deformation of these materials is reached in the thermos-elastic range. The optimal strain is reached below the melt temperature. When the surface temperature is too high, damage to the surface will occur. Because these three polymers have a broad elastic range, the change on damage is small and it is easier to control the process. Because the material is transparent and not subsequently treated, any damage on the surface is immediately visible. Moreover, amorphous materials have a low thermal expansion. Consequently, the heating process has minimal influence on the shape of the sheet. These plates are produced by extrusion and supplied in discrete pieces. These plates may be already textured or decorated (Throne and Kutz 2011: 333–358).

Fig. 10.47 Bubble chair by Eero Aarnio

Form

In comparison with other thermoforming processes, the starting materials are thick-gauge sheets (thickness >3 mm) and not bulk materials. The use of this semi-finished product makes the free-form thermoforming process more expensive. Thick-gauge sheets are used because of the structural requirements for skylights. These products are non-uniform in thickness due to the blowing process. Because thermoforming is a differential stretching process, the material will always thin out during the process. This is inherent to thermoplastic materials. The thickness of the sheet is adjusted to the local areas where the material must meet a minimum thickness. This results in an inefficient use of the material.

The height of the shape can be controlled by the amount of air that is blown under the sheet, or is sucked out of the plenum chamber. A large variety of "bubble" shaped objects can be created by the manipulation of the surface but most of the time only spherical dome shapes are made. Because of the reduction of thickness of the material at apex during the shaping process, the height of the spherical shape is limited. For a height of the sphere of 0.2 × width, the material thickness is 80% of the original thickness of the sheet. This is a non-linear process because with a height of the sphere of 0.5 × width, the material thickness at the apex is reduced to 33% of the thickness of the original sheet.

The shape of the perimeter can be varied as well as the height. Flat and three-dimensional openings can be realized. It is not possible to control the air pressure on specific areas in order to bulge a part of the sheet. It possible to make irregular formed surfaces by changing the thickness of the sheet, by adding material, by changing the form of the boundaries, by locally heating the surface, etc.

Technique

Free-form thermoforming can be realized by blowing air into a vacuum box or by creating a "vacuum" (underpressure) under the sheet. Most of the time, the sheet is uniformly heated to avoid irregular bulges. Bubble-formed parts are made with a vacuum box. The heated sheet will not touch the surface of the vacuum box and therefore the surface has minimal distortion and a high transparency. The first step in the process is to heat the plastic and bring the sheet in a thermoelastic state. There is no clear dividing line between the thermoplastic and thermoelastic states of the material. For most plastics the thermoelastic range is between 80 and 160 °C. The optimal temperature range is called the thermoforming window (Cantor et al. 2011: 195–203). To avoid damage to the surface of the material, it is important to heat the sheet homogeneously. Convection ovens are preferred for free-form thermoforming to heat the sheets. They are preferred because these ovens distribute the heat uniformly (Rosen 2002). With these ovens it is possible to increase the temperature gradually in order to keep the temperature gradient in the material low (Michaeli and Kopp 2001: 3239–3247). Convection ovens are often used for thicker sheets, in particular, in order to avoid large inner material strains. There are two types of machines for thick-gauge sheets: the shuttle presses, which can form large sheets and are versatile but economically inefficient; the rotary presses, which are energy efficient but expensive. The use of a second oven is recommended when the polymer PMMA is formed in order to increase the productivity of the machine. The sheet dimension, depth of draw and nature of the motive power for moving the sheets and clamping the sheets are all aspects that need to be considered when choosing a particular type of machine.

The shape can be formed when the sheet is heated. This is done in a vacuum box. The plastic sheet is clamped to a mould and inflated. The sheet is clamped pneumatically on four sides. The height of the bubble can be measured with a photoelectric eye. Then the shape must be rigidified. This is done by cooling the sheet with fans. To avoid surface damage, it is important to cool the sheet at a very low rate.

Conclusion

Free-form thermoforming is a basic technique that has a small range of applications compared to the other thermoforming techniques like cavity forming, plug-assist forming and billow drape forming. Free-form thermoforming is mostly used for the production of skylights, because of the clear optical properties and minimal surface distortion. The reasons for this are some characteristic limitations of this technique like the limited form possibilities and the slow production speed. Compared to other thermoforming techniques, the use of thick-gauge sheets is expensive, given that the free-form thermoforming technique uses only the bubble forming method. Now, if a technique can be developed that makes use not only of heat and air pressure but also of yield stress and edge distortion, then the plastic plates can be used in more applications. This new moulding method can be economically interesting if used to produce free form façades and could compete with free form concrete and glass façades.

10.10 Free-Form Thermoforming of Anticlastic Thermoplastics

Flexible Mould
The form-active structure: the material of the shell in viscous elastic state.
Form-active typology: single layer.
Way to stabilize the mould: inflate or stressed surface.

Composite Material
Technique to handle the rigidizing material: cooling.
Reinforcement of the material: none.
Surface treatment: none.

Project
Name and location: Eindhoven.
Design & Engineering: A. Pronk, E. Rijbroek, M. Verboord and F. Gard.
Contractor: TU/e.
Year: 2013–2014.

Introduction

The research goal for this project is to develop free-form anticlastic surfaces in thermoplastics. The elements produced will be cheap and durable and this will be a fast way to create transparent and doubly curved surfaces. The technique can be used for façade elements or for moulds for the production of (concrete) façade elements (Figs. 10.48 and 10.49).

Material

A thermoplastic is a polymer that becomes flexible above a specific temperature, and returns to a solid state when it cools down. Most thermoplastics have a relatively high molecular weight due to their large molecular chains, which interconnect

Fig. 10.48 Free forming by tensioning of a thermoplastic

Fig. 10.49 Free-form anticlastic surface

through intermolecular forces (Throne 2008). It is this property which makes that thermoplastics can be remoulded or reformed, since the intermolecular connections are restored when the material cools down. Between the glass transition temperature

(T_g) and the melting point (Tm), the physical properties of the material, such as stiffness and strength, change before the phase change occurs. Within this temperature range, most thermoplastics are rubbery or leathery and are able to change shape.

Thermoplastics exist in different chemically related groups whose general material properties can be described as follows (Boser 2013).

Acrylics

Acrylic, a polymer called *polymethyl methacrylate (PMMA)*, is also known as Lucite, Perspex and Plexiglas. They are well suited for thermoforming because of the high hot strength and a wide temperature range for processing, from 143 to 177 °C. Because of the excellent properties such as optical clarity, outdoor stability and rigidity, they are a good substitute for glass. Acrylics often are the preferred materials for thermoforming but are also expensive thermoplastics. They are applied in for instance: aircraft windows, aquariums, motorcycle helmet visors, lenses of exterior lights of automobiles, viewing ports of submersibles, and to replace eye lenses (Fox 1962: 111–128).

Cellulosics

Cellulosics are thermoplastics that are based on biomass materials. All the cellulosics are chemical reaction products of cellulose, which can be obtained from wood or cotton linters. Because the cellulose macromolecules need to be broken down for the production and must be combined with many different other materials, their properties can vary greatly. The material most used for thermoforming is cellulose-acetate-butyrate (CAB). The cellulosics can be transparent but are also easy to colour or decorate (Gooch 2011: 128; Victorius 1984).

Styrene Polymers

Polystyrene is one of the oldest known plastics but, due to its fragility, it became less popular when other plastics with better impact properties became available. Polystyrene has a very low specific heat and it quickly regains rigidity when cooled after forming. This makes very short and cheap processing possible. Some improved styrene-containing plastics maintain the benefits of polystyrene but also have better heat, impact and weather resistance, and transparency properties. These are used, e.g. for plastic models of cars and the casings of all kinds of electronic devices but also for packaging materials and drinking cups (Morawetz 2002).

Vinyl Resins

Polyvinyl chloride (PVC) is characterized by excellent mechanical toughness, weather and chemical resistance and it is flame retardant. PVC is quite difficult to process because when it is heated to the thermoforming temperature, the material becomes very soft and some of the hot strength will be lost, especially when it is slightly overheated. In particular, it is difficult to achieve the desirable form when vacuum, billow or bubble forming is used, because it shrinks or stretches back when the pressure is released. PVC is used, e.g. for drainpipes, gutters, roofing sheets, hoses, tubing, electrical insulation, coats, water beds and pool toys.

Polycarbonate

Polycarbonate (PC) has a very high impact strength over a wide temperature range, compared to the other thermoplastics, good clarity and a higher glass transition temperature. It is used for objects that need to withstand a lot of physical impact such as industrial guards, amusement arcade game housings and bullet proof glass. It is also used for eyeglass lenses.

Other Materials

There are many other materials that can be thermoformed. To list them all would be unnecessary and impossible in the context of this research. Such materials may be specially made to meet specific and sometimes extreme requirements. The materials most commonly used for thermoforming are listed above. Among these, polycarbonate turns out to have superior properties for anticlastic free-form forming, because the viscous elastic behaviour of polycarbonate is more predictable and it is easier to tool.

Form

Generally speaking, thermoforming thermoplastics is limitless in form possibilities, although what forms can be derived depends on the technique that is used. There are four main categories into which the form techniques can be divided; *vacuum thermoforming, pressure thermoforming, mechanical thermoforming* and *free-form thermoforming*. The first three forming techniques use a mould to form the plastics. A thermoplastic can be made into every form for which a mould can be made (Figs. 10.50 and 10.51).

For free-form thermoforming it is not as simple. This technique is about forming without a mould. The most common way is to work with a *plenum chamber* (vacuum or pressure box) on which a thermoplastic plate can be clamped. This is described, e.g., in the paragraph Free-Form or bubble thermoforming of thermoplastics, Bubble Chair by Eero Aarnio.

An alternative way to free form the heated sheet is to handle it like a textile, clamp certain points and let it form by hanging, or stretching it into the desired form. The problem in this case is that when the sheet is removed from the oven, the stabilizing should be done within seconds. Tried solutions are to stabilize the form in the hot medium (air, water, oil, etc.) or simply to stabilize it very quickly. Both ways present multiple difficulties. For this project, it was decided to stabilize the thermoplastic very quickly by clamping the sheet and pushing it over the edges of the desired form.

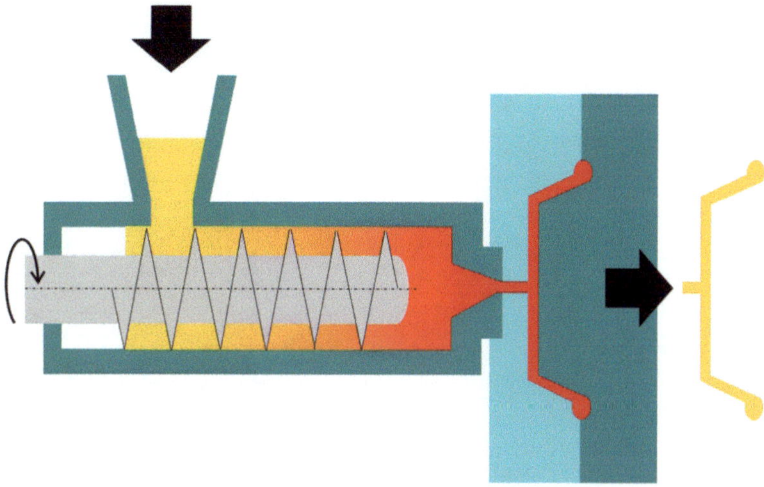

Fig. 10.50 Injection moulding

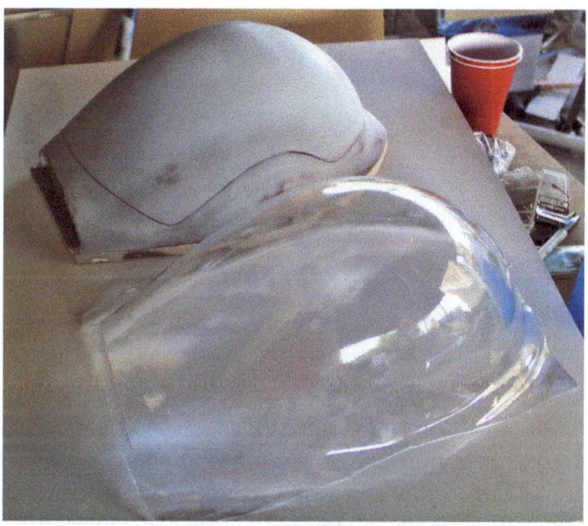

Fig. 10.51 Vacuum thermoforming

Technique

There are four main categories into which the forming techniques can be divided: *vacuum thermoforming, pressure thermoforming, mechanical thermoforming* and *free thermoforming*. Only the free thermoforming of anticlastic surfaces is discussed in this paragraph (Figs. 10.52 and 10.53).

Thermoplastics thin out non-uniformly; the centre of the shape will be thinnest and near the clamp the sheet will be thickest.

Fig. 10.52 Design of free anticlastic mould

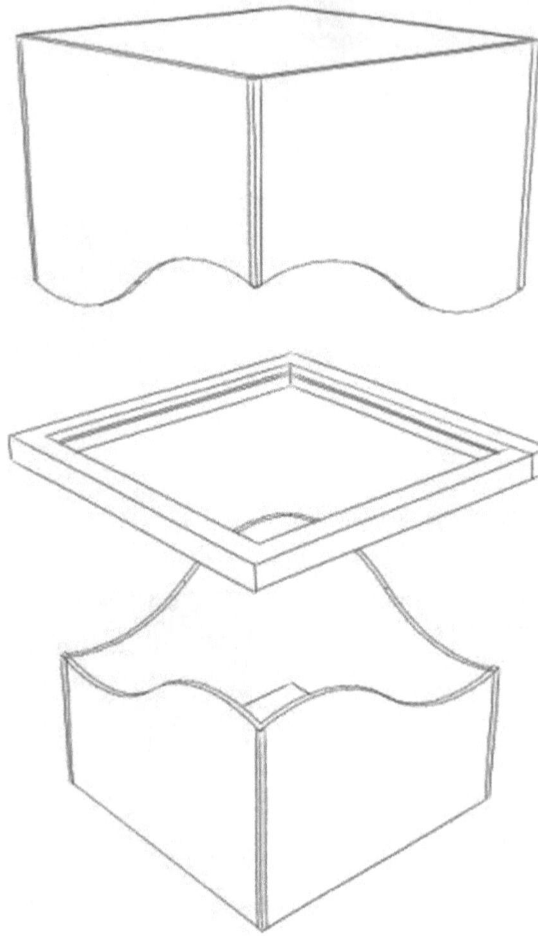

Because small temperature differences effect the elasticity of a material, the sheets need to be uniformly heated to avoid odd shapes. In fact, many thermoplastics are not suited to this technique. Only a few are suitable because of their coherence or tensile strength and thermo-elastic behaviour; *polymethyl methacrylate (PMMA), cellulosics, oriented polystyrene (OPS), polyvinyl chloride (PVC) and polycarbonate (PC)* are the most frequently used transparent materials for this technique (Gruenwald 1998). Heavy, free-form shapes are used in skylights and aircraft windscreens, light free form shapes have been used to produce blister packaging.

The technique that is used for quick free thermoforming is to push a rigid element with the desired form into a zeroclastic surface. The thermoplastic acts similarly to an elastic membrane when its temperature is between the glass transition temperature (Tg) and the melting point (Tm). By pushing a rigid element (mould) into a heated and prestressed thermoplastic, the plastic will adapt to the surface of the rigid element.

Fig. 10.53 Free anticlastic mould

In this case the rigid element (mould) only consists of the edges of the desired form. By pushing the mould into the heated and prestressed thermoplastic, it will adapt to the edges and will form a new surface in force equilibrium in between the edges. The production of a façade element with this technique in combination with a mould with reconfigurable edges is discussed below.

The free thermoforming surface is clamped between the edges of a reconfigurable mould. The edges require the best accuracy for tiling purposes since deviations will be visible at the connections of the elements and not so much in the middle. This set-up allows the surface to be "freely" formed in between the edges (Figs 10.54, 10.55, and 10.56). The edge settings of the panel are transferred to the fluid mould device. After clamping a thermoplastic sheet between the edges of the mould device, the thermoplastic is heated above its glass temperature. Once the heat has been equally distributed in the thermoplastic, the sheet is visco-elastic (rubbery). Then the sheet can be formed by adjusting the pins in the reconfigurable edge to the proper height and angle. The desired shape/curvature cannot always be obtained simply by adjusting the edge. Sometimes a pretension or other manipulations are needed to obtain the required shape (Fig. 10.57). The required amount of pretension can be calculated by simulation in the FEM-software Marc/Mentat. When the desired shape is obtained, the mould can be cooled so the sheet becomes rigid again.

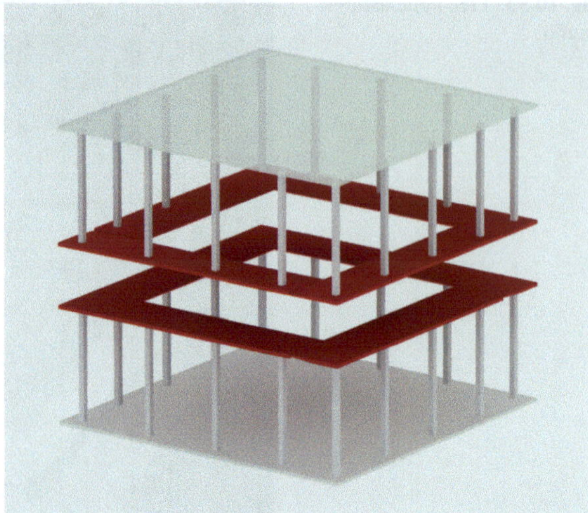

Fig. 10.54 Adjustable moulding system

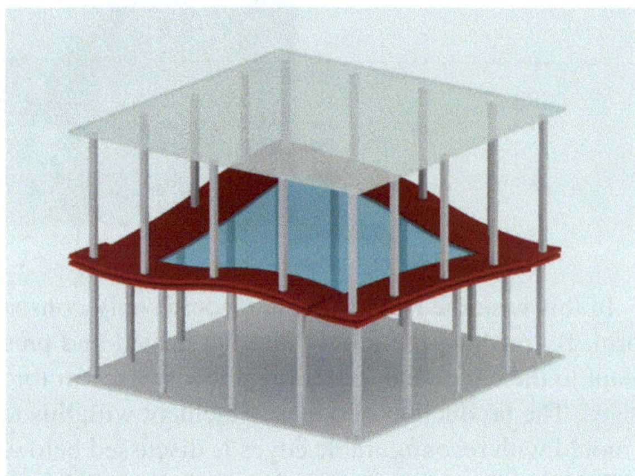

Fig. 10.55 Result

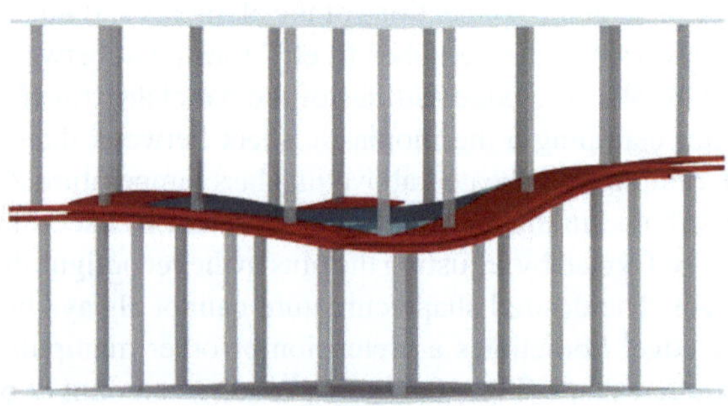

Fig. 10.56 Clamping of the sheet

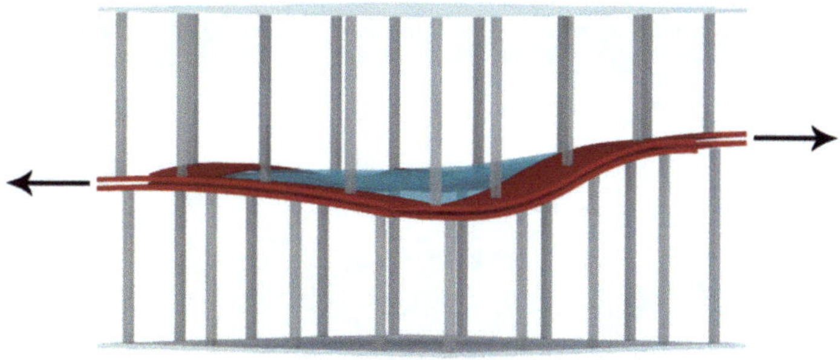

Fig. 10.57 Tensioning of the sheet

10.11 Single Fibre Inflatables: Cocoon Mothballing

Flexible Mould
Form-active structure: Coated Fabric.
Form-active typology: Single layer.
Way to stabilize the mould: Inflate/Stressed surface.

Composite Material
Technique to handle the rigidizing material: Spraying.
Reinforcement of the material: Single fibre.
Surface treatment: Coating.

Project
Name and location: Chair in the Netherlands, Healy Guest House, USA.
Design & Engineering: Jurgen Bey, Paul Rudolph.
Contractor: - Studio Makkink & Bey, Cocoon.
Year: 1999, 1951.

Introduction

Cocooning was invented by Frederick J. Myers and Jack D. Robinson in 1943 (patent US 2400054 A) (Myers and Robinson 1946). Richard M. Hollingshead (1932) explored the technique, named cocoon mothballing, for the packaging and storage of machines. After World War II, this technique was used especially by the army and the navy in de US. Aircraft and ships were 'wrapped' to protect them against environmental influences. By wrapping the aircraft and ships with a sprayable vinyl plastic coating, the cocoon, they were kept dry and protected against sand, dust and pollution. A certain range of temperature and air-humidity inside the cocoon was capped to prevent corrosion of the equipment. When the aircraft and ships had to be used, the plastic coating would be removed and the conveyances could resume their functions (Figs. 10.58, 10.59, 10.60 and 10.61).

Fig. 10.58 Prada transformer made with cocoon on a steel frame (*Source* Cocoon)

Fig. 10.59 The use of cocoon mothballing in pieces of art, creating hybrid furniture covered with cocoon by Studio Makkink and Bey (*Source* Makking and Bey)

Fig. 10.60 The Healy Guest House with concave roof made of insulation with a sprayed cocoon (*Source* Cocoon)

Fig. 10.61 The Healy Guest House sprayed with cocoon (*Source* Cocoon)

Cocoon mothballing was used in architecture by the architect Paul Rudolph in 1950. He had seen the cocoon technique in the US Navy for the mothballing of ships. In 1950, he designed a one-story guest house for in-laws. A year later, during the construction of this house, he applied the *cocoon technique* on the concave roof. The *Healy Guest House*, by its original name, is better known as the *Cocoon House* (Gator Preservationist, gatorpreservationist.wordpress.com/2011/01/25/cocoon-house).

The Cocoon House is an experiment in structure and technology. It uses steel straps tensioned between side walls. In this way a curved catenary roof structure is shaped. The steel straps are fastened to flexible insulation boards and sprayed with Cocoon. The experimental roof has been leaking from the beginning and has had to be renewed. The concave roof was the first insulated polymer membrane roof. In 1958, Eero Saarinen also designed a tensioned concave roof. This time in concrete, for Dulles International Airport in Washington (Figs. 3.42 and 3.43).

The technique of cocoon mothballing was rediscovered in design by Jurgen Bey, co-founder of *Studio Makkink & Bey* (Studio Makkink & Bey, www.studiomakkinkbey.nl/info/about), a studio which works in various domains of applied art. The design strategy of the studio is to re-appropriate what is already present in the context of an assignment, with a strong emphasis on the process. In the past, Bey also used the cocoon technique in collaborations with *Droog*® (Droog®, www.droog.com/store/studio-work/kokon-furniture—double-chair), a Dutch design foundation.

Bey uses the cocoon mothballing technique for his art-pieces. In line with the design strategy of *Studio Makkink & Bey*, he used old furniture for his creations. For example, in 2001 he sawed a couple of old chairs and tables into pieces and joined them together. After creating a piece of *hybrid furniture*, he covered it with cocoon as shown in Fig. 10.58.

The application of cocoon mothballing is increasing slightly, but is still mostly used for its primary goal: protecting items against environmental influences. Nowadays, not only are aircraft and ships wrapped, but also cars, weapons systems and even (parts of) buildings, see Figs. 10.58 and 10.62.

Material

Cocoon-types

The cocoon mothballing technique can be done in two ways: with *fabric membranes* and *sprayable coatings*, both will be discussed below.

Fabric membranes consist of expanded polytetrafluoroethylene (*ePTFE*). They are produced as a three-layer fabric that combines several benefits into a single material. Figure 10.58 shows an application of a fabric membrane. These types of cocoons are like sleeves which you pull over an object. Sprayable coatings are *polymers*, e.g. polyvinyl or polyvinyl-chloride solutions. The coatings are sprayed directly onto the object. In some cases, the cocoon is not sprayed directly onto the object. The object will be wrapped with foil parts which are custom made. After that the cocoon will be sprayed.

Fig. 10.62 Rooftop restaurant in Amsterdam sprayed with cocoon (*Source* Cocoon)

Properties of cocoon

Cocoons protect an object against environmental influences. By describing the properties of cocoons, it will become clear why they are used as (long term) protection:

Moisture: dependent on the size and the preservation duration, the cocoon is provided with a *desiccant* or is connected to automatically controlled air dryers. In this way, a relative humidity below 45% can be guaranteed so corrosion will not occur.

Temperature extremes: cocoons can stand temperatures from –42 to +85 °C.

Chemicals & Pollution: cocoon is proof against salt, sand, dust, pollution, alcohol, oils, gasolines, fats, bacteria and most industrial acids. It protects the object against parts of up to 0.3 micron.

UV-rays: cocoon is proof against UV-rays. Because of this, the colourfastness of the coating is guaranteed. The colour can be white, transparent and generally any colour of the rainbow.

Seamless finish: Because you can spray the material, there will be no seams. Fungi and moulds will not grow on the inside of the cocoon in combination with the control of the relative humidity.

Self-extinguishing: cocoon does not burn, but extinguishes by itself.

Tough material: cocoon can stretch up to 2.5 times its original length. The tensile strength is 7.45 MN/m^2 given a thickness of 0.5 millimetres. It has a lifespan of several decades.

Adhesion: cocoon can be sprayed onto almost any (building) material due to the fact that it has a high adhesion coefficient.

Drying time: the drying time of a single layer of coating can be as little as 5 min.

Nowadays there are different kinds of cocoons on the market. The values of the properties discussed above can be taken as averages.

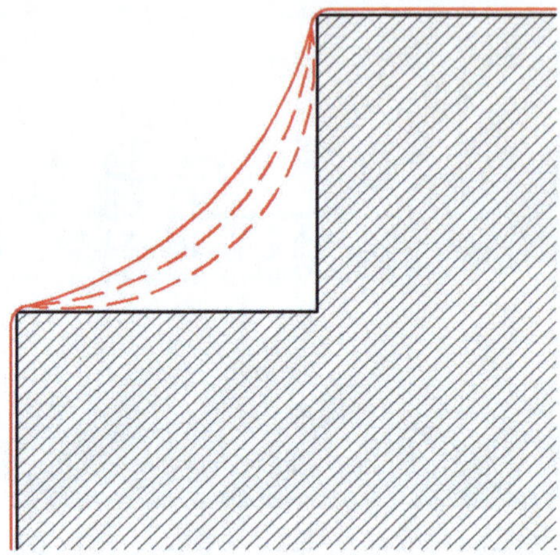

Fig. 10.63 The ability of cocoon to bridge gaps and voids with curved surfaces

Form

When using cocoon to make fluid architecture, it is important to discuss the possibilities with regard to form. By studying various cases where cocoon is applied, the following statement can be made: *generally speaking, any form can be created with cocoon, unless it has (sharp) inside corners.* The case of inside corners will be discussed later on.

The combination of spraying with a high adhesion coefficient makes it possible to follow every form. Cocoon has the ability to bridge gaps and voids. Because of the technique that is used while applying the material and the properties of the cocoon, there can be curved surfaces at protrusions and gaps. Cocoon and the cocoon mothballing technique can be used in fluid architecture to create smooth curved surfaces and for mouldings.

The form of a cocoon

The free-floating cocoon surface is a mechanically prestressed membrane. According to the rules in Chap. 2, the free-floating surface must be *plane* or *anticlastic*. In any other cases, it is adapted to the form of the supported element.

Curvature in Inside Corners and Gaps, Adhesion in Outside Corners

By studying several applications of cocoon, it is noticeable that a curvature arises where there are inside corners (Fig. 10.63) and gaps. In outside corners and straight surfaces, the cocoon is stressed tight against the surface.

Because of the curvature at inside corners and gaps and the adhesion at outside corners and straight surfaces, it can be assumed that the cocoon shrinks while drying. In most cases this will lead to the anticlastic forms as shown in applications of cocoon.

Technique

In application of direct spraying of the material this will result in a seamless wrap. If gaps and voids are too wide to bridge, it is necessary to prepare a grid on which the cocoon can be sprayed. Mostly, this grid is created with paper masking tape because this is easy to remove. Other kinds of masking tape are also suitable. The ability of cocoon to bridge gaps is due to its webbing structure. While applying sprayable cocoon, the gaps and voids will close slightly. The application of cocoon has to be done by skilled labourers. If not, then the quality and lifespan of the material can be poor (Andek, www.andek.com/hans/cocoon/cocoon2009b.html; Cocoon Holland, www.cocoon.eu).

10.12 "Blob" Pavilion: Doubly Curved Surface on Membrane

Flexible Mould

The form-active structure: coated fabric.
Form-active typology: single layer.
Way to stabilize the mould: stressed surface, inflate.

Composite Material

Technique to handle the rigidizing material: hand layup.
Reinforcement of the material: woven fabric.
Surface treatment: coating.

Project

Name and location: The BLOB Art Pavilion, Eindhoven, the Netherlands.
Design and Engineering: Jurgen Bey (Studio Makkink & Bey, www.studiomakkinkbey.nl), and A.D.C. Pronk et al.
Contractor: Buitink Technology and Polycel (Buitink Technology, www.buitink-technology.com; Polycel, www.polycel.nl).
Year: 2003.

Introduction

The Blob pavilion in Eindhoven was designed as a moving art exhibition pavilion on the university campus (Pronk and Houtman 2003: 1–6; 2005: 305–322). The pavilion has two positions, horizontal and vertical. In the horizontal position, the pavilion has a projection screen on the flat side. In the vertical position, the flat side serves as the base (Figs. 10.64, 10.65, 10.66, and 10.67).

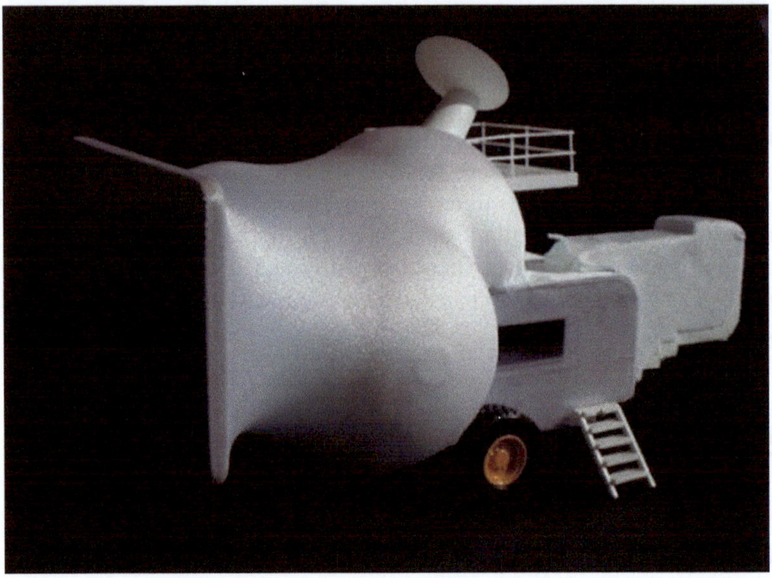

Fig. 10.64 Blob pavilion: the scale model

Fig. 10.65 Process of the Blob pavilion manufacturing

Fig. 10.66 Blob pavilion: the end result

The pavilion consists of a caravan and a segment of fluid architecture, which are combined as one volume. Fluid architecture is known for its smoothness, irregularity and doubly curved skin. The construction can be simply compared with the human body where the membrane is the skin stretched over the body of bones and muscles. In this case the "body" is formed by the frame and inflatables that support a mechanically prestressed membrane. This membrane is a smooth, irregular and doubly curved mould for the composite surface for the pavilion. The reason for the realisation of this pavilion is the testing of this building method. The designing artist (Jurgen Bey) was made familiar with form-active modelling and had to combine an inflated inner membrane with a tensioned outer membrane in a physical model (Fig. 10.64). From the model that is shown above a 3D scan was made. It was represented as an IGES file, which made it possible to generate cross-sections. This digital model with cross-sections was used to analyse the form in general and the inner and outer membrane. Below, you will find a description of the form, the materials and the technique that were used to realize this pavilion.

Form

The first form analyses were done in the program Rhino. Here, the size and position of the inflated elements were derived from the 3D scan. The inflated elements were imported into the program EASY of Technet GMBH. To be able to form-find a fabric over the inflated elements, a new module for the program was developed. This program is called Conformer Alpha and is based upon a sliding support system and barycentric coordinates. This makes it possible to investigate the interaction between the inner inflated elements and the outer tensioned surface. Both inner and outer

Fig. 10.67 Vertical position of the blob pavilion

surface will need to be translated into cutting patterns for production. Figure 10.68 shows the cutting patterns of the inflatable. Figure 10.69 shows the cutting patterns of the outer membrane. Figure 10.70 shows a drawing of the connection between the tensioned outer skin and the inflatable.

The free-form surface of the pavilion consists of synclastic and anticlastic surfaces. The synclastic surface is realised by the inflatable and the anticlastic surface is realised by a free floating mechanically prestressed membrane (Fig. 10.71). The natural shape of an inflatable is spherical and can be manipulated in different ways as postulated in Chap. 2.

The inflatable is formed by one surface that is a combination of different spheres into one object. At the connection of the spheres there is an internal membrane that pulls on the inflatable surface. The advantage of this single inflatable is its stability and the need to have only one air compressor. The inflatable and mechanically stressed membrane are made by a PVC-coated polyester membrane (Fig. 10.72). The patterns were high frequency welded.

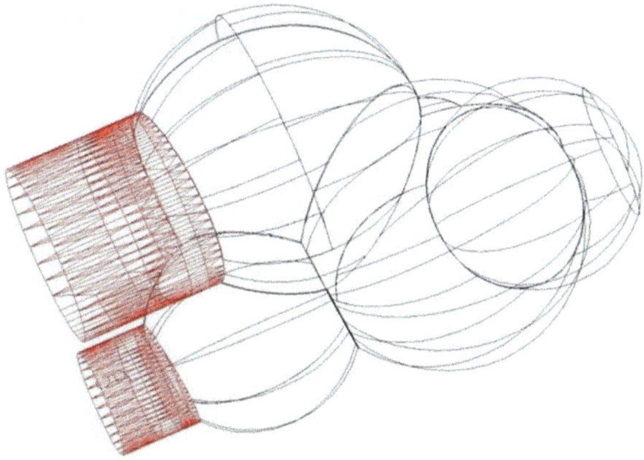

Fig. 10.68 Cutting patterns of the inflatable

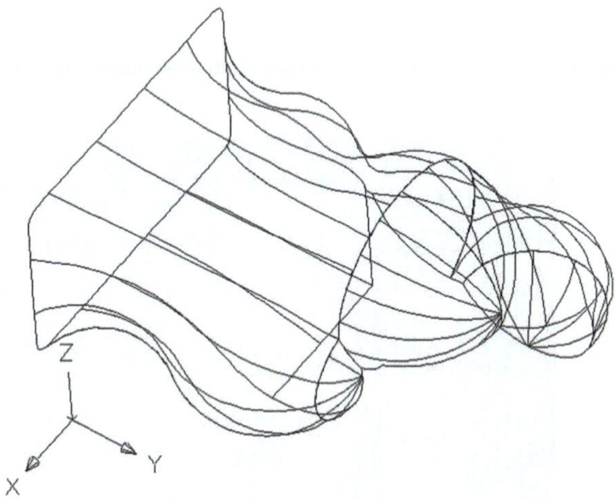

Fig. 10.69 Cutting patterns of the mechanically prestressed membrane

Technique

The 3D scan made from the model (Fig. 10.64) was also used for the structural analysis of the surface. This was done with the finite element program Marc/Mentat. Marc/Mentat is able to import the 3D scan in STL format. The STL file consists of triangular elements. In Marc/Mentat it is possible to calculate with triangular elements but in order to calculate with a thick shell element, quadrangular elements were needed.

We did not succeed in transforming the triangular elements into quadrangular elements automatically. Therefore, we had to redraw the mesh of the model by hand in AutoCAD. The benefit of the program Marc/Mentat is that if the model works, it is quite easy to calculate different types of materialization for the skin of the pavilion. Another benefit is that the calculation time of the model is short in comparison with

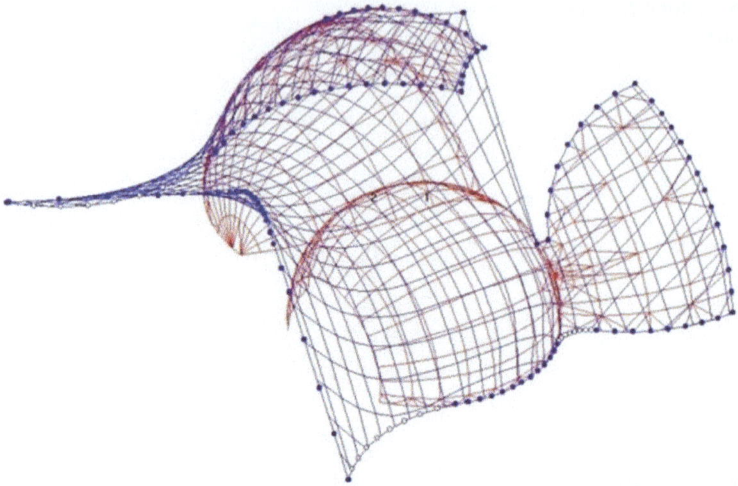

Fig. 10.70 Combination of inflatable with tensioned membrane

Fig. 10.71 The combination of the inflatable

other finite element programs. The results are shown in the pictures in Figs. 10.73 and 10.74.

Fig. 10.72 Several spheres combined in one with the tensioned membrane large inflatable

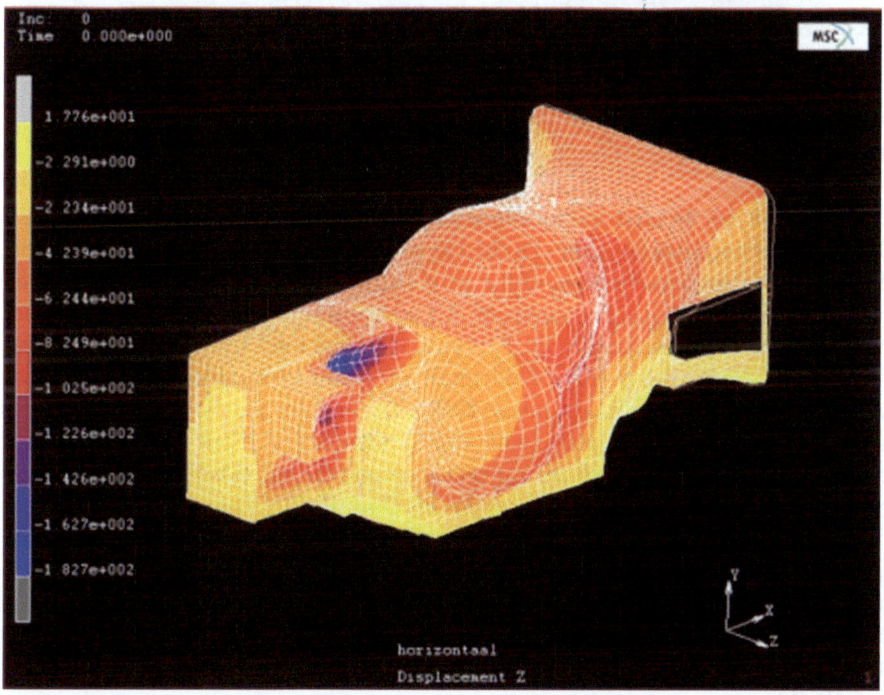

Fig. 10.73 FEM analyses of shell in Marc/Mentat: horizontal displacement

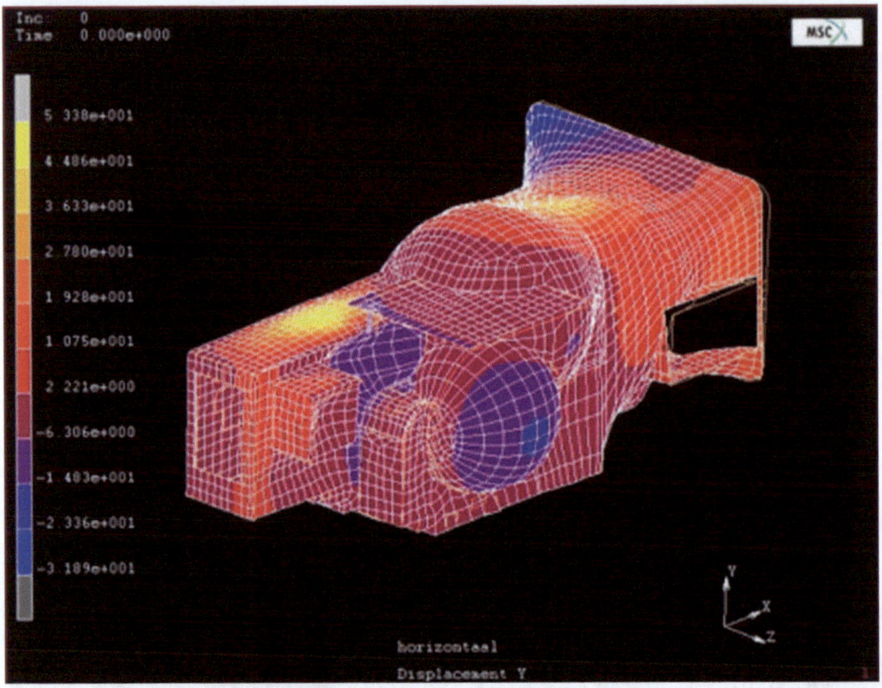

Fig. 10.74 FEM analyses of shell in Marc/Mentat: vertical displacement

Material

In the model, we divided surfaces into translucent and non-translucent parts. Two factors determined this outcome. The first was the structural behaviour of the pavilion. The second factor was the input of the artist (Jurgen Bey). The translucent parts are made of 5 mm single layer composite; the other parts are divided in 60 mm thick non-translucent sandwiches for the bottom and 30 mm thick translucent sandwiches for the rest. Both with a surface thickness of 3 mm composite. The core of the sandwich is made of polyisocyanurate (PIR). To curve the plates, they are sawn, for 2/3 of the thickness of the plate, with a quadrangular grid. To avoid delamination, both surfaces of the sandwich are connected around the individual polyisocyanurate plates with glass fibre-reinforced composites (Fig. 10.77). The surface of the glass fibre reinforced polyester composite is applied by hand layup. The outdated hand layup technique was chosen for the experimental setup to allow maximum freedom in the composition of layers, colours and surface treatment. All kinds of materialization experiments were done, such as an experiment to the structural behaviour of ropes and the visual effect of lace and translucency in a polyester composite (Figs. 10.75 and 10.76).

Fig. 10.75 Ropes for structural stiffness of the blob pavilion (*Source* Polycell)

Fig. 10.76 Net structure in polyester surface

Fig. 10.77 Sandwich structure with PIR foam (*Source* Polycell)

The polypropylene ropes have a much lower Young's modulus (1.5–2 GPa) compared to glass fibre reinforced polyester (17 GPa). Therefore, the ropes only function as a spacer to form structural ribs with a higher moment of inertia with glass fibre reinforced polyester.

Conclusion

With the combination of form, technique and material it is possible to create fluid architecture of the highest level. This case proves that the combination of inflatable moulds with mechanically prestressed membranes in combination with any other structural typology is able to form a doubly curved irregular and smooth mould for the realisation of fluid architecture. Although the possibilities and results are good, it must be said that the construction by hand layup is a time and labour consuming technique and provokes a high emission of styrene. It shows that the combination of this moulding technique with other techniques for the construction of a rigid surface is worth researching (Figs. 10.77, 10.78 and 10.79).

Fig. 10.78 Finished interior of the Blob pavilion

Fig. 10.79 Design of tensile structure with a floating beam (*Source* H. Werkman)

10.13 Vacuum Infusion of an Inflatable Beam

Flexible Mould
The form-active structure: textile composite.
Form-active typology: double layer.
Way to stabilize the mould: inflate.

Composite Material
Technique to handle the rigidizing material: vacuum infusion.
Reinforcement of the material: woven fabric.
Surface treatment: coating.

Project
Name and location: Tensile stage covering open-air theatre, Soest, the Netherlands.
Design & Engineering: Master thesis of H. Werkman; Tutors: A.D.C. Pronk, R. Houtman and A. Borgart (TU Delft) (Werkman et al. 2003: 244–249).
Contractor: Buitink Technology.
Year: 2002.

Introduction

For the open-air theatre in Soest (NL), a curved concentric beam (Fig. 10.80) was realized as part of a semi-permanent membrane structure that covers the stage of an open-air theatre. An inflatable mould was used to realize this beam. The mould was rigidized with a polymer composite by vacuum infusion. A suspended PVC-polyester membrane structure hangs on two highpoints which are connected by cables to the top of two columns at either side of the stage (Fig. 10.79). The membrane is connected to the foundation by 6 cables. The columns and arch transmit compressive loads. The tensile loads and the stabilization of the whole structure are effected by the

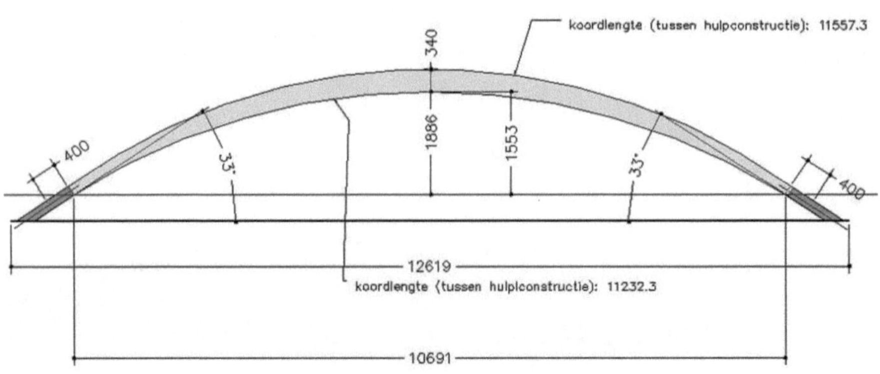

Fig. 10.80 Design of the beam with an inflatable mould (*Source* H. Werkman)

prestressed membrane and cable structure. The membrane and beam look as if they float above the stage; in this way, the visual relation with the natural environment is preserved. The form of the membrane is influenced by the curvature and length of the beam in such a way that the covered area of the stage is increased. Because of the position and function of the beam it had to be strongly optimized and light weight. Optimization was achieved by:

- a combination of different materials within the sections;
- a compensated curvature of the beam before comprehensive loading; and
- variable moments of inertia (Iy) at different sections.

Material

Conventional construction materials such as steel and aluminium and accompanying production technologies are not suitable for making conical light-weight arches. The material properties and production methods of polymer composite match the requirements for the arch. The advantages of polymer composite are, amongst others: rigid and light-weight construction possibilities, fatigue resistance, chemical and corrosion resistance, freedom in design and form and the possibility to integrate parts. Disadvantages are the relatively high cost prices of material, mould, production (labour) and engineering. In the case of complex shapes, for example a conical arch, approximately 50% of the production costs consist of moulding. By using an inflatable mould the moulding costs could be reduced considerably.

Combination of Materials

The arch consists of a synthetic matrix of fabrics cured by injecting a thermoset resin. By varying the thickness, direction, structure and material of the fibres a range of E-moduli can be obtained. A change in layer composition was introduced in order to be able to produce the arch with the determined stiffness. Eventually the E-module of 210 GPa (210.000 N/mm^2, comparable to steel) was used, but was not sufficient. By adjusting the E-module from 210 to 60 GPa a new stiffness was found (E $\times$ Iy $= 5.14 \times 10^{12}$ N/mm^2). This led to the use of carbon fibres at the top and the bottom of the beam section and glass fibres in the other parts (Fig. 10.81).

Form

Variable Moments of Inertia

For the structural engineering of the membrane structure the software package EASY by Technet was used (Figs. 10.82, 10.83, and 10.84). To examine the shape of the structure and its reaction to external forces, the structure was first modelled with an arch set up as a spatial truss with a defined stiffness. This model was used for the structural analysis of the comprehensive structure (membrane, beam, supporting cables and columns). Deformations caused by extreme loading (wind and snow loads)

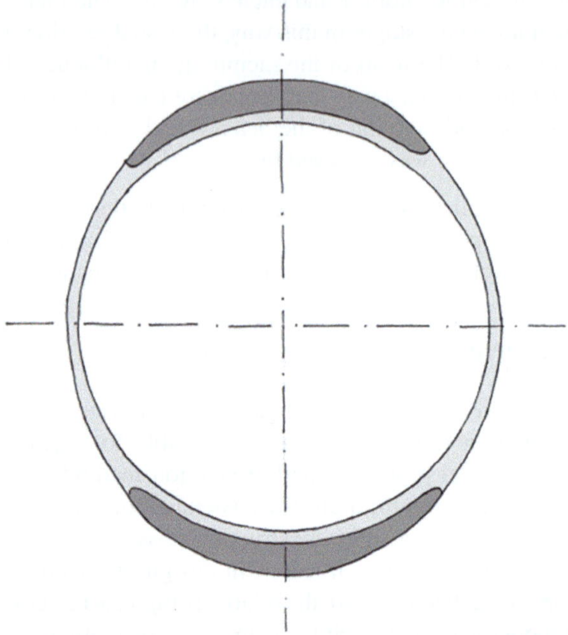

Fig. 10.81 Section of the composite beam with an inflatable mould (*Source* H. Werkman)

Fig. 10.82 The stage covering for the Soest open-air theatre in the Netherlands (*Source* Buitink Technology)

were examined. The form of the beam can be seen as part of a curved tube (anticlastic on the inside, synclastic on the outside). The beam was optimized in such a way that the diameter varies between 150 mm at both ends to 360 mm in the middle. The

Fig. 10.83 An arched beam ensures an increase of the protective area and the curvature of the membrane structure (*Source* Buitink Technology)

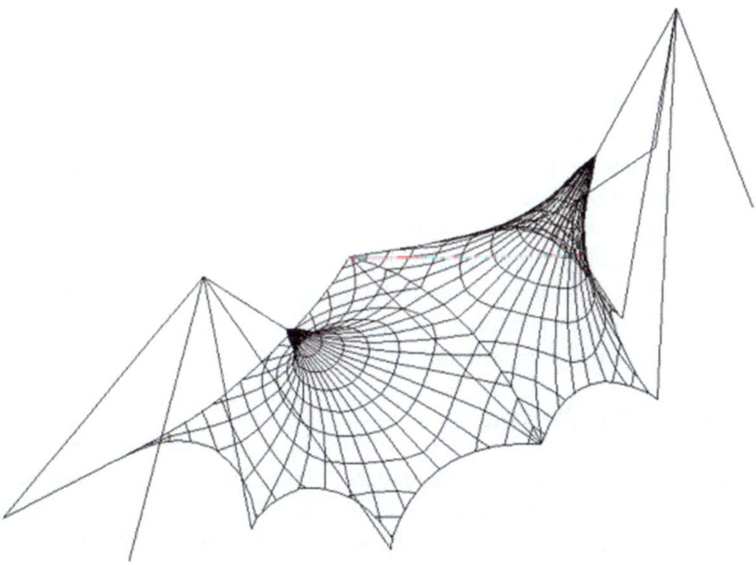

Fig. 10.84 Computer model of the tensile structure (*Source* Tentech)

moment of inertia (I_y) varies between 8170 × 10 mm^4 at the ends and 16,170 × 10 mm^4 in the middle section. The same computer applications (EASY, FEM-based

Fig. 10.85 Pneumatic mould supported by the auxiliary structure (Tentech /Buitink Zeilmakerij) (*Source* Buitink Technology)

software) and production technologies were used in the production of the mould as were used for the development of the membrane structure. After modelling and form finding in EASY, cutting patterns were generated and used for the production of the mould. The internal overpressure ensures the rigidity of the inflatable mould. The general dimensions, such as the distance between the supports, are controlled by an auxiliary structure (Figs. 10.79 and 10.85). The surface of the beam was slightly manipulated by the dead load of the polymer before rigidizing and the pattern of the inflatable membrane. The effect on the overall form is negligible and within the tolerances of tensile building structures.

Sag Compensated Curvature

The prestress in the membrane and boundary cables causes an axial compression in the arch. Hence the curve of the arch will increase by sagging. In order to find the desired shape, the released arch was made with less curvature to compensate the sag of the arch in loaded condition.

Technique

Vacuum Infusion

A rigidizable inflatable structure can be described as a structure that is flexible when inflated and becomes rigid after exposure to an external influence. After rigidizing it is not necessary to maintain the overpressure. There are several ways to rigidize and new techniques are being researched. They can be divided into three categories: thermosetting composite systems, thermoplastic composite systems and aluminium/polymer laminate system. Advanced rigidizing systems used for space applications are designed for specific structures and can be very expensive. In civil engineering, vacuum injection, which is a thermosetting composite system, is a feasible way for rigidizing membranes (Figs. 10.86, 10.87 and 10.88). Alternate layers of fibres and resin are placed around the inflatable mould. To create a closed system, the whole package is wrapped with an air-tight and protective foil. In the closed system a pressure differential is applied that impregnates the fabric with resin. The pressure differential of the technique is obtained by means of a vacuum. The

Fig. 10.86 Production of the conical arch, by using the vacuum injection method. Both glass and carbon fibres are applied (photos: Rep-air Composites) (*Source* Rep-air)

Fig. 10.87 Production of the conical arch, by using the vacuum injection method. Both glass and carbon fibres are applied (photos: Rep-air Composites) (*Source* Rep-air)

injection has to take place within the curing time of a resin. The following formula represents the filling time (tfill) as a function of the porosity (φ) and permeability (κ) of the reinforcement, viscosity (η) of the resin, flow distance (l), and applied pressure (ΔP).

$$t_{\text{fill}} = \frac{\varphi \cdot \eta \cdot l^2}{2\text{k}\Delta\text{p}}$$

The objective is to design a channel layout that ensures full wetting of the fabric at each location. Three distinctive injection strategies for a three-dimensional object can be followed, viz. edge injection: downward, upward and sideways. Downward injection is sometimes disadvantageous because bubbles will be trapped more easily and there is an increased risk of dry spots due to race tracking by the runner channels. The choice between the other two injection strategies depends on the geometrical shape of the product. Factors that are of influence are the number of inlet ports and the total injection time that, when minimal, are both at an optimum. In this case upward injection was used.

Fig. 10.88 Vacuum injection of conical arch (photos: Rep-air Composites) (*Source* Rep-air)

10.14 Vacuum Infusion of Inflatable Truss Beams

Flexible Mould
The form-active structure: textile composite.
Form-active typology: single layer.
Way to stabilize the mould: inflate.

Composite Material
Technique to handle the rigidizing material: hand layup.
Reinforcement of the material: woven fabric.
Surface treatment: none.

Project
Name and location: Vacuum Infusion of Inflatable Truss Beams, Eindhoven, the Netherlands.

Design & Engineering: Master thesis of S.H.M. van Dijck and W.J. van de Koppel
Tutors: A.D.C. Pronk, J. Lichtenberg and M. Dominicus.
Contractor: TU/e.
Year: 2013.

Introduction

In this part the best combination of construction, material and typology for truss structures with fabric formwork is researched (Pronk et al. 2013a). The results are inflated three-dimensional (open cell) structures rigidized and structurally optimized. The structure is about 40% lighter than conventional beams of the same material and was realized and tested in prototypes and full-scale models. The form-finding process initiated with the topological optimization of the four section-active systems was defined in a theoretical framework by Engel (Engel 1997). Its objective was to derive a case, on which the proposed production method is based, by identifying the general morphological features of the optimized section-active structures. Thorough empirical case studies were performed using SolidThinking Inspire 9.0 (SolidThinking, www.solidthinking.com) to determine which optimized structure reflected the most of the general morphological features. Subsequently, shape and size optimization were performed using the ParaGen method, Altai OptiStruct and Hypermesh (Pronk et al. 2013b; SolidThinking, www.solidthinking.com; Palmer 2011). The goal was to combine two principles: inflatables and open cell structures. In line with Tensairity and Tensegrity, we want to combine the strength of an inflatable and the strength of an optimal structure. In this research we studied the processes of producing optimal structures without additional falsework through the application of rigidizing inflatables.

For validation purposes, topological optimization on the section-active structures was also performed using Topostruct (Sawapan, sawapan.eu), which uses the homogenization method as a solver. The optimized three-dimensional beam was used as a basis for the development of the parametric model, which in turn forms the basis for the ParaGen method. The model consisted of 46 nodes, of which 42 were parametric. The nodes were bound to the surface of a hypothetical tube, limiting the number of free variables to 2: the elevation in the x-direction (x) and the angle (α).

The next phase was to calculate the size optimization. The ParaGen method (von Buelow 2012) used entails a cyclic process, which starts with the breeding of parents. At the start of the cycle, there are no parents to select, therefore the 'no parent' breeding method is used. This implies that all of the values are randomly generated and chosen to fit within the range required for each variable used in the parametric modeller. The ParaGen-method is an iterative process which, in combination with StaadPro, gives the optimum thickness of the plates (Figs. 10.89, 10.90, 10.91 and 10.92).

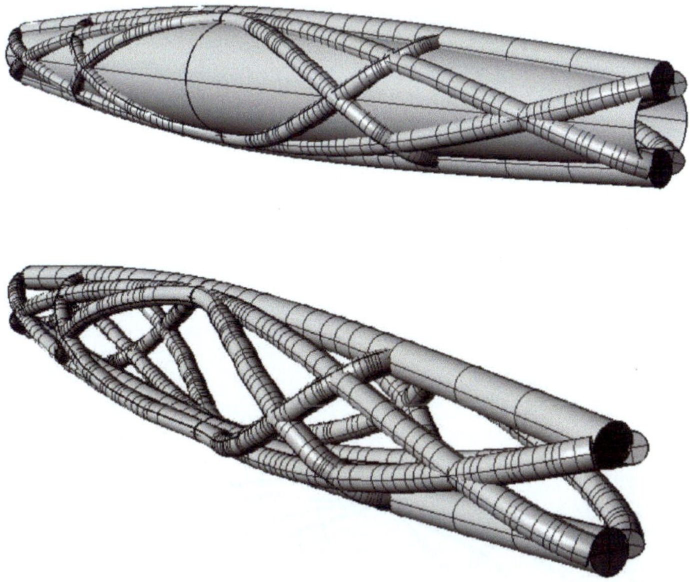

Fig. 10.89 Composite truss beam with an inflatable mould

Fig. 10.90 ParaGen method by Von Buelow

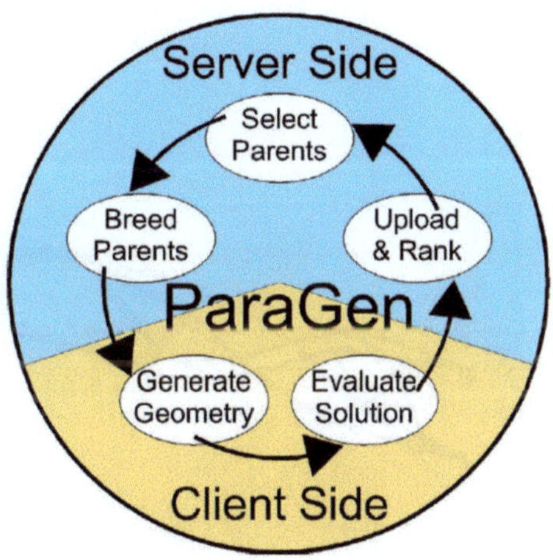

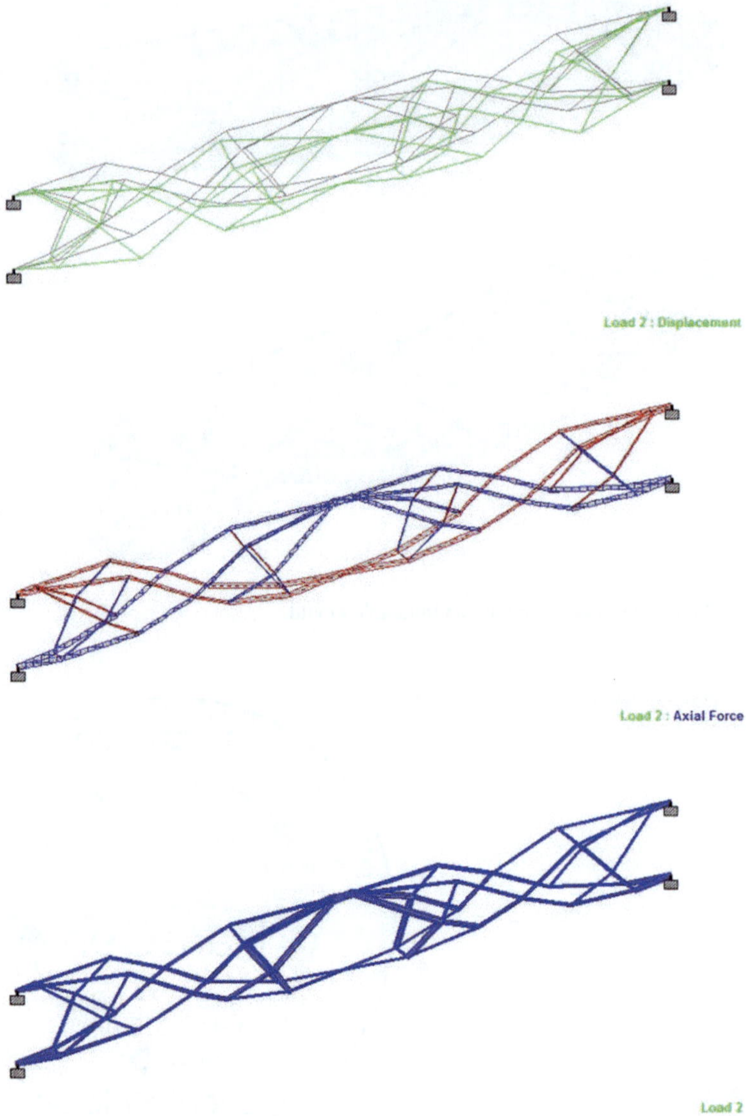

Fig. 10.91 Breeding of parent for a composite truss beam (*Source* P. Von Buelow)

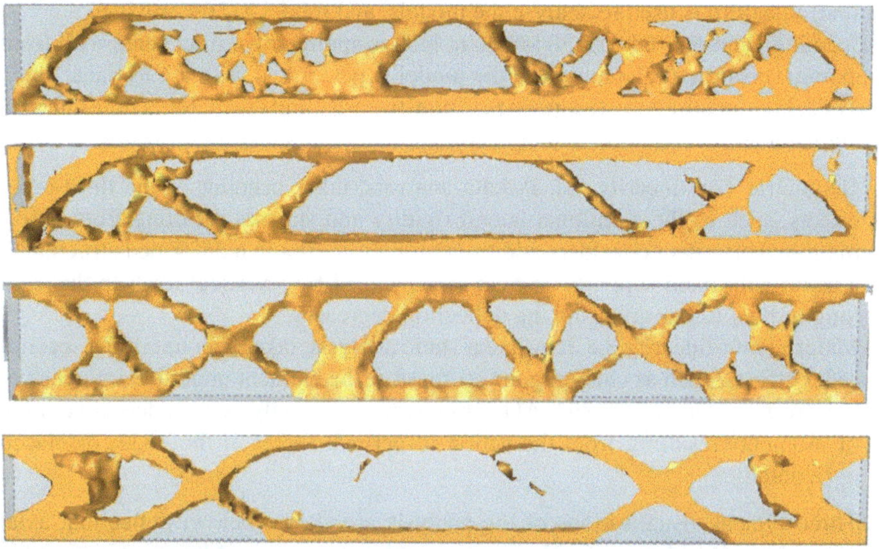

Fig. 10.92 Optimization of beams with the program SolidThinking Inspire

In a 2002 project, an inflatable mould was used to realize a beam. The mould was rigidized with a polymer composite by vacuum infusion (Werkman et al. 2003: 244–249) (see paragraph 10.13). The material properties and production methods of polymer composite match the structural requirements of this (curved) beam. In the case of complex shapes, for example a conical arch, approximately 50% of the production costs are for moulding. By using an inflatable mould, the moulding costs can be reduced considerably.

A rigidizable inflatable structure can be described as a structure that is flexible when inflated and becomes rigid after exposure to an external influence. After rigidizing it is not necessary to maintain the overpressure. To have the curing of the fibres at any desired moment is an interesting property for several applications when transportation and fast erection are an issue, like in space engineering, military and emergency relief applications. The company Concrete Canvas realized a cement-based shelter that can be inflated and rigidized on site (Concrete Canvas, concretecanvas.co.uk). In the following experiments we focused our research on form optimization (Pronk et al. 2013a). Budget wise we realized the models with cement-bound composites. The material of the trusses we are developing can be optimized for different applications.

Form

This method is performed with a parametric model with linear members that connect the nodes. This is a simplification of the final model. Therefore, a structural analysis of a linear member model as well as a curved member model need to be performed.

At first, StaadPro is used to analyse the models, and model #1269 provided the best performing geometry when a 50 kN point load is applied on a steel beam with fixed supports. The structural analysis of the model is performed with a non-linear static analysis. For our structure, the geometric non-linearity and the material non-linearity are applicable.

- Geometric non-linearity, i.e. P-delta analysis, takes account of the third-order effects, such as the additional lateral rigidity and stresses resulting from deformation. This effect considers additional forces arising in a deformed structure at a beam with fixed supports on both ends, loaded by a vertical load. In this case longitudinal forces arise and the deflection decreases.
- Material non-linearity, i.e. non-linear static analysis, takes account of the second-order effects, such as changing the stiffness of the element under the influence of the stress state in the element. At the same time, this analysis considers generation of moments resulting from the action of vertical forces at the nodes displaced horizontally.

Three typical visualisations of the analysis are shown below; deflection, axial forces and combined stresses. These visualisations give a good interpretation of the behaviour of the beam under load.

The deflection is visualised in Fig. 10.93, and shows the highest deflection at the point load in the upper part of the beam. The maximum deflection at this point is 13.22 mm. It is remarkable that the lower centre part has a small deflection, possibly caused by the compressing of the beam to a more oval shape. The axial forces are shown in Fig. 10.94, where the red and orange suggest tension forces, and the blue and green suggest compressive forces. If this image is compared with the moment diagram

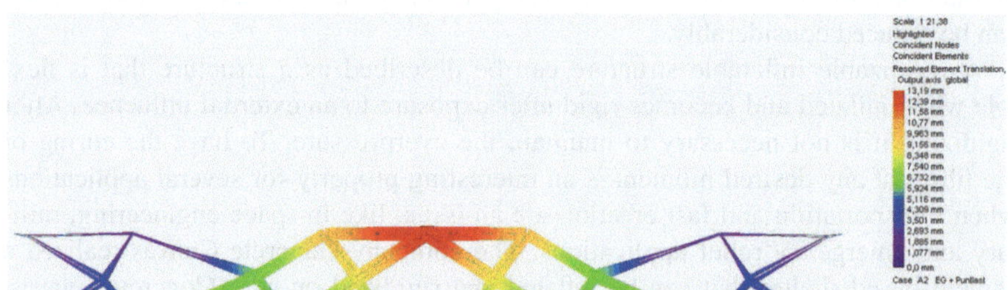

Fig. 10.93 Deflection of beam structure

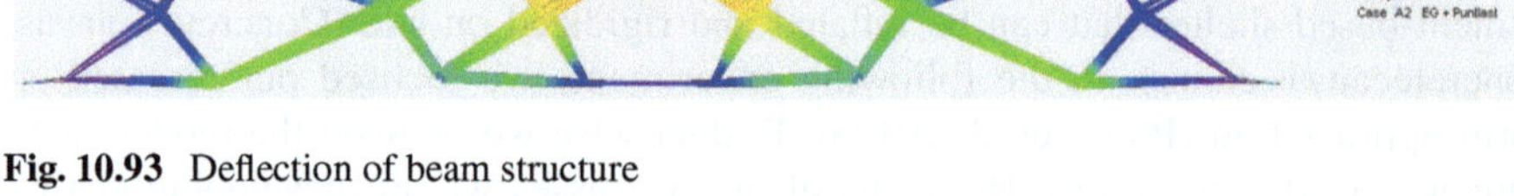

Fig. 10.94 Axial forces in beam structure

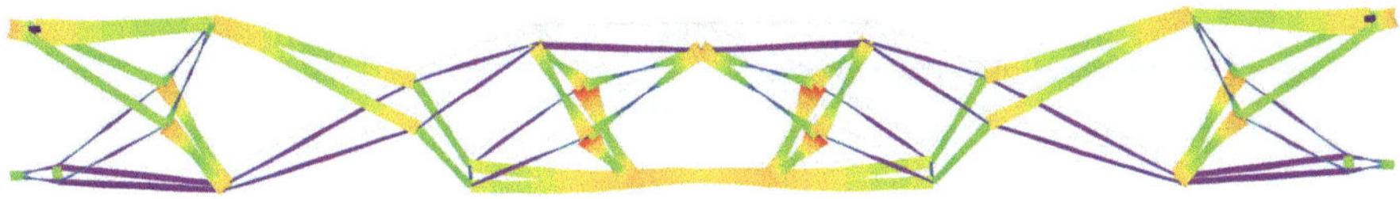

Fig. 10.95 Combined forces in beam structure

of the 2D beam in chapter 2.5 at the point where the moment is zero, the tension and compressive forces change sides in this model. The tension and compression lines are clearly visible and show that the tension starts at the top and is at the bottom in the centre with the point load. Figure 10.95 shows the combined forces in the model. So, bending, shear and axial forces are combined in this image, and the largest moment occurs at the intersections close to the centre of the beam. Also, the tension and compressive lines are clearly visible in this figure. The maximum force that occurs is 303.4 N/mm^2. By comparison, steel qualities are in between 235 and 355 N/mm^2.

Form

According to Rozvany it is possible to optimize a structure by three types of optimization: (i) size, (ii) shape and (iii) topology (Rozvany 2001: 90–108). They address different aspects of a structural design problem, e.g. the goal may be to find the optimum thickness distribution of a linearly elastic plate or the optimum member areas in a truss structure. The aim of shape optimization is to find the optimum shape of its domain. Topology optimization aims at finding the optimum layout of a structure within a specific domain. Structural optimization seeks to achieve the best performance for a structure while satisfying various constraints such as a given amount of material (Bendsoe and Sigmund 2003).

To perform a structural optimization, three variables are required:

I. Objective function; a function used to classify designs, which for every possible design, returns a number that indicates the 'appropriateness' of that design (Christensen and Klarbring 2009).
II. Design variable: a function or vector that describes the design and which can be changed during optimization.
III. State variable: represents the response of the structure.

Optimum structures require non-orthogonal geometries. Fabric formwork should be seen as the technology that offers a new paradigm with a clearly emerging parametric requirement for form, process, precision and complexity of optimized structures (Pedreschi 2011). To find a suitable geometry for the beam, a method was chosen which combines parametric form generation with multi-objective shape optimization. The method used was ParaGen, a genetic algorithm (GA) based program developed at the University of Michigan. ParaGen uses commercial software on a cluster of PCs connected through a web interface to a server that maintains a solutions database from which new solutions are bred (Figs. 10.96, 10.97 and 10.98).

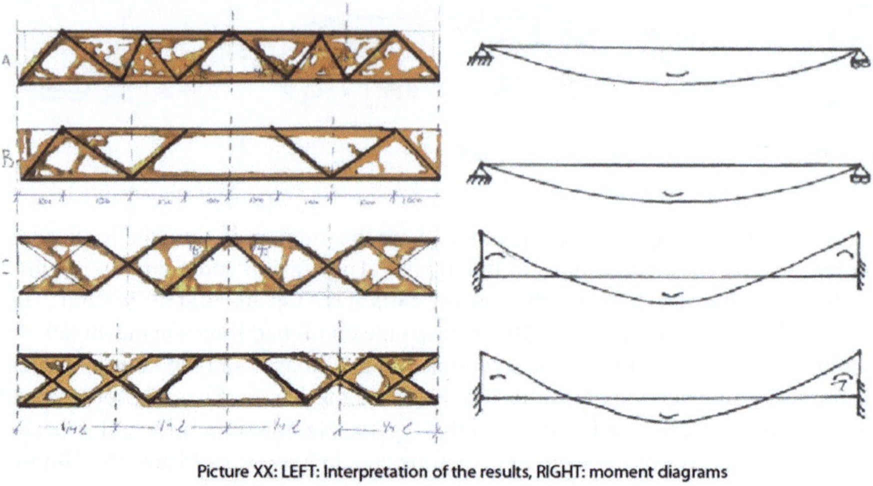

Fig. 10.96 Interpretation of results (left) and moment diagrams (right)

Fig. 10.97 The structural optimization process (*Source* Rozvany)

$$(\mathrm{so}) = \begin{cases} \textit{minimize } f(x, y) \textit{ with respect to } x \textit{ and } y \\ \textit{subject to} \begin{cases} \textit{behavioral constraints on } y \\ \textit{design constraints on } x \\ \textit{equilibrium constraint.} \end{cases} \end{cases}$$

An optimized three-dimensional beam was used as a basis for the development of the parametric model, which in turn has formed the basis for the ParaGen method. The model consisted of 46 nodes, of which 42 were parametric. The nodes were bound to the surface of a hypothetical inflatable tube, limiting the number of free variables to 2; the elevation in the x-direction (x) and the angle (α). Compared with a model using 3 free variables, this led to better output with respect to the proposed production method, and it also reduced the necessary computational capacity. In addition, due to computational limitations, the model was schematized using linear instead of curved members. The FEA model used a centre point load of 50 kN, fixed supports, continuous members with fixed moment connections and the material properties of ASTM A-36 steel. Pipe profiles were used with selected diameter and wall thicknesses resulting from the structural analysis.

A total population of 1276 individual solutions were created algorithmically with this set up and stored in the SQL database. ParaGen optimized for two different fitness functions, i.e. minimal weight and highest modal frequency (stiffness). A scatter diagram was created with the plot function in order to find the most promising solutions that performed well for both objectives (Fig. 10.98).

The three most promising solutions were analysed according to their morphology and the proposed production method. Solution #1269 was assessed to be the best

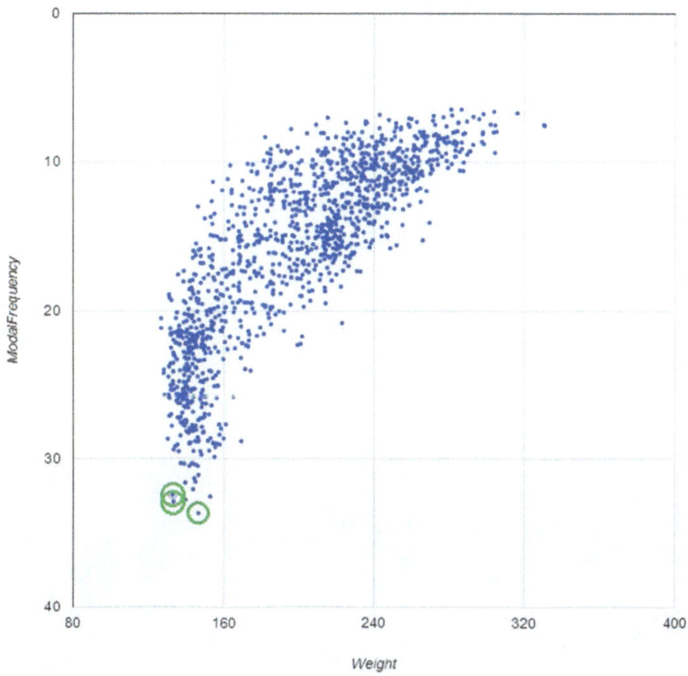

Fig. 10.98 Scatter diagram for weight vs. stiffness (*Source* P. Von Buelow)

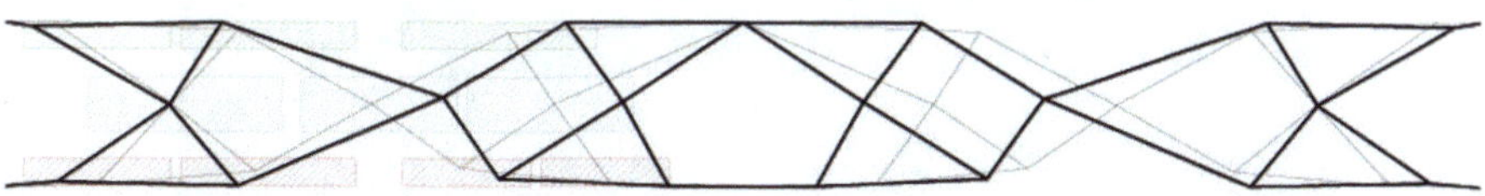

Fig. 10.99 Solution #1269 with the original parametric model and domain in the background (*Source* P. Von Buelow)

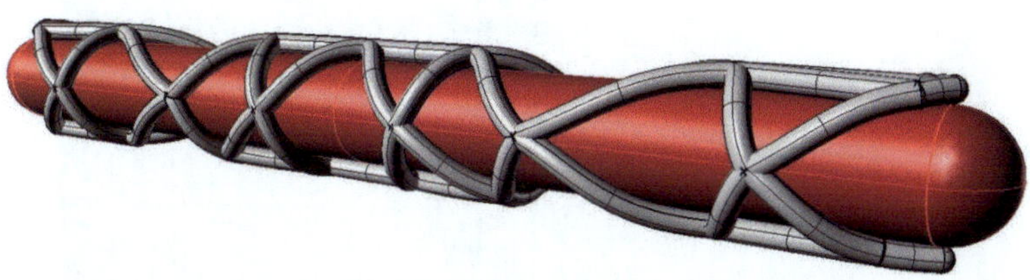

Fig. 10.100 3D image of the prototype of solution #1269

solution, since it deviated the least from its curved counterpart (Figs. 10.99, 10.100, 10.101, 10.102, 10.103 and 10.104).

Fig. 10.101 Table with concrete composition for 10 litres

Material	Amount (kg)
CEM I 52.5 R	9.5
Limestone flour	3
Glenium 27 con.20%	0.45
Water	2.66
Polypropylene fibre (12 mm)	0.114
Sand 0.125-0.25	2.144
Sand 0.25-0.5	3.369
Sand 0.5-1	0.612

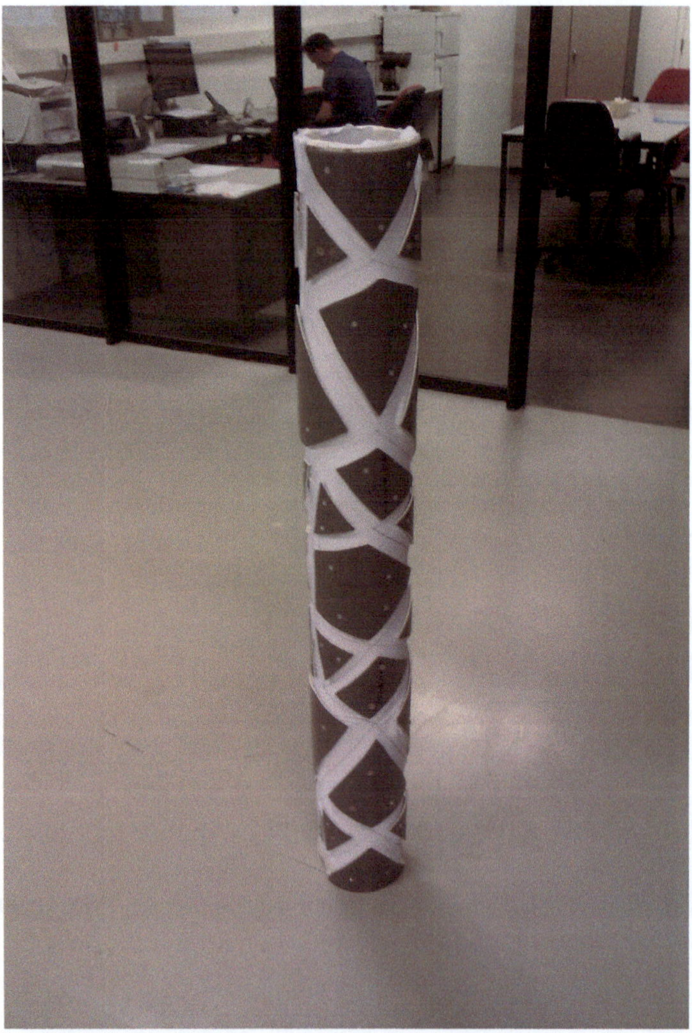

Fig. 10.102 Mould and falsework

Fig. 10.103 Concrete scale models

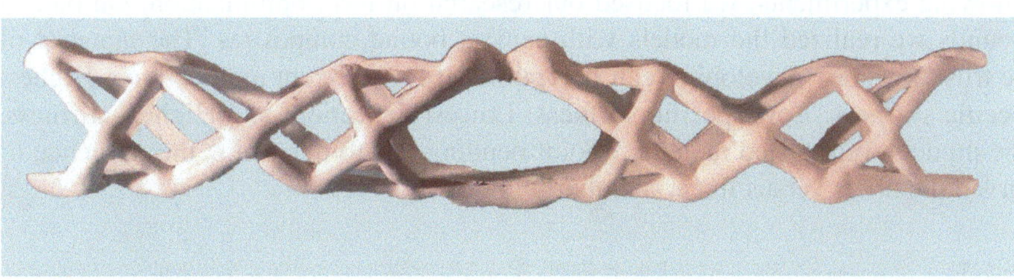

Fig. 10.104 Concrete scale model #1269

Solution #1269 was also realized with 3D software to assess the final result of the prototype.

Rigidizing Inflatables

In the final model, the inflatable core was rigidized with polymer composites in strings at the outer surface. The development of rigidization mechanisms with polymer composites over the last 50 years, including advanced mechanisms for use in space engineering, has been impressive.

Material

The advantages of polymer composites are, amongst others: rigid and light-weight construction possibilities, fatigue resistance, chemical and corrosion resistance, freedom in design and form and the possibility to integrate parts. Disadvantages are the relatively high cost prices of material, mould, production (labour) and engineering. In the case of complex shapes, for example a conical arch, approximately 50% of the production costs consist of moulding. By using an inflatable mould the moulding costs can be reduced considerably.

There are several ways to rigidize inflatables and new techniques are being researched. They can be divided into three categories: thermosetting composite systems, thermoplastic composite systems and aluminium/polymer laminate system. Advanced rigidizing systems used for space applications are designed for specific structures and may be very expensive. Cardogan comes to the following mechanisms: (i) Thermally cured thermoset composites, (ii) Ultra-violet cured composites, (iii) Plasticizer or solvent boil-off composites, (iv) Foam rigidization, (v) Second order transition change and SMP composites, (vi) Inflation gas reaction composites, (vii) Aluminium and film laminates (Cadogan and Scarborough 2001). The first 4 of these mechanisms are used in the "building" industry or other civil applications, the others might be considered too specific and expensive for civil purposes. The Insituform company (see paragraph 10.5), specialized in the rehabilitation of pipeline systems, is the biggest application for thermoset composites in civil applications. It is an interesting property to have the curing of the fibres at any desirable moment for several applications when transportation and fast erection is an issue, like in space engineering, military and emergency relief applications. The Concrete Canvas company realized a cement-based shelter that can be inflated and rigidized on site. In the following experiments, we focused our research on form optimization. On budget grounds we realized the models with cement bound composites. The material of the trusses we are developing can be optimized for different applications: Highest specific stiffness; Low specific stiffness; Lowest deflection; and Highest stiffness. The models were tested using a 5-point bending test to determine the influence of curved members versus linear members.

Manufacturing Method

The manufacturing method was developed with a view to controlling the different parameters as much as possible. In this method, the only difference between the models is the geometry. Every geometry was manufactured three times to give significant results. The models were cast in 2 weeks, casting one model per day, and were tested using a 5-point bending test in sets of 3 after 13, 14 or 15 days. Also, standardized mortar bars were made with every cast to test inconsistencies in the concrete composition (NEN, www.nen.nl/nen-3835-1991-nl-878). The low viscosity concrete mixture was as follows:

The production technique used for the fabrication of the scale models was based on a method described in the paragraph on open cell structures in concrete. The moulds for the concrete beams were constructed of multiple layers of PVC piping. Initially, clamps which represent the inverse of the beam were sawn out of standard PVC pipes with a diameter of 160 mm. Two layers of these clamps were mounted onto a PVC pipe of 1.4 meters long. Then, two layers of high tenacity Lycra fabric were pulled over the PVC, followed by a third and final layer of clamps. The mould was then mounted in a steel bracket keeping it in place during the casting and allowing the model to rotate about its axis to prevent sagging of the uncured concrete. Threaded rod was mounted into the top and bottom ends of the model, allowing every model to be fixed in identical ways in the bending test. Since the model was pinned at the top and bottom of each side, the two form a force couple.

Results

All the scale models were tested in a 100 kN pressure bench in the van Musschenbroek laboratory at the TU/e campus. The weight of the beams varied between 23 and 33 kg due to member size fluctuations in the moulds. However, the distribution of the material within the model influenced the outcome more than the weight. The results of the mortar bars showed no significant variations with respect to the material density and yield strength. However, the flexural strength of the concrete mixture for #1249 was 27% less than the mixture for the other geometries due to the use of a different plasticizer (Glenium 51 con. 35%). #1269 failed in all the beams in the tension zone which runs helically around the beam. This is natural given the properties of concrete, and means that there are no clear weak parts in the geometry. In every geometry, the members always fail near the nodes. The structural analysis performed in GSA and StaadPro showed that the stresses are highest near the nodes due to the fixed moment connections, which explains the crack pattern of the scale models. #1259, and to a lesser extent #1249, showed a concentrated crack pattern near the cross members at the supports. The morphology of the members prevents the formation of a stable cross locally, causing the node to fail due to buckling and not because the yield limit of the material is reached. Therefore, large angles between members should be avoided to prevent the nodes from local buckling, and the member length in a transverse direction should be minimized.

Fig. 10.105 Average linear elastic regions of the tested geometries

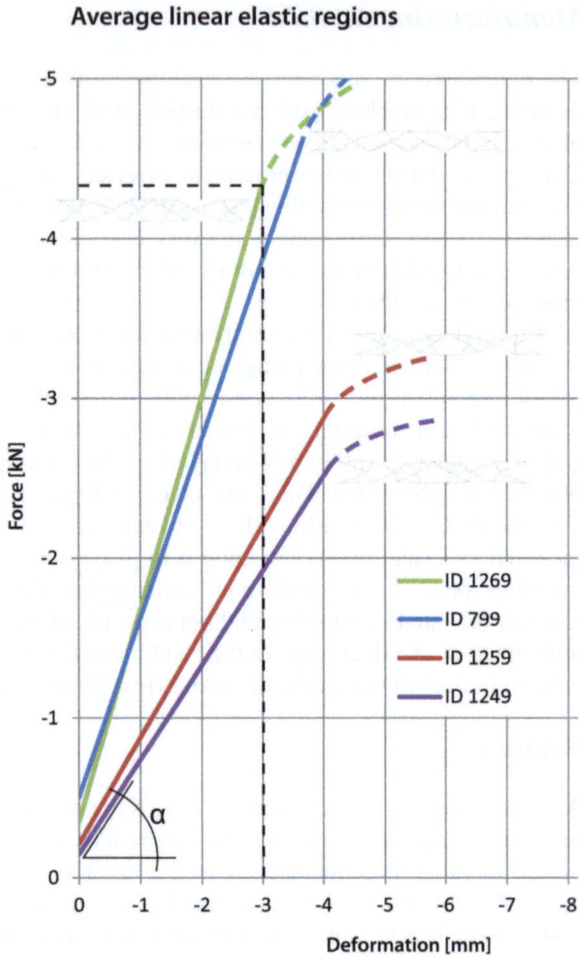

To determine the stiffness of the four different geometries, the linear elastic region of every beam was determined using trendlines. In this region, according to Hooke's law, the deformation ε is proportional to the force *F*. All the beams showed this linear behaviour until a first crack occurred, which caused the deformation-force diagram to behave unpredictably due to the use of polypropylene fibres.

Figure 10.105 shows the average linear elastic region of the four geometries, where the angle α is used as a measure for the stiffness. This confirms that #1259 indeed has a disadvantageous geometry with respect to curved members, since it was initially optimized for straight linear members. Now, #1269, which is also the case for the remainder of the research, has the highest stiffness. It has to be noted that #799, which had the lowest theoretical stiffness, is the second stiffest geometry according to these results. This might be due to the theoretical optimization with steel and linear members, while the scale models have curved members and are produced with concrete.

Conclusion

The aim of this research was to develop and refine new production techniques for producing optimum structures. Optimum structures can be described as structures with a minimal material use and maximal strength and characterized as an open cell and complex structure. In this research, we use different techniques to generate optimum structures, SolidThinking Inspire was used for topological optimization and ParaGen was used for size and shape optimization. A gap was noticed in this research between the production and generation of optimized structures. Traditional framework cannot be used to produce this kind of structures, because of the high amount of material (formwork) and the necessary labour hours. Fabric formwork can provide an alternative. Current traditional fabric formwork is combined with additional (wooden) falsework. In this research we described a process of producing optimum structures without additional falsework by applying rigidizing inflatables. The scale models we produced with fabric formwork and concrete showed us a smooth surface and a high accuracy to the digital models. The scale models are tested with structural analysis programs (StaadPro and GSA) and physical tests (pressure bench). The digital geometry with the highest stiffness from the structural analysis program produced similar results in the physical tests. The combination of structural optimization and production method with rigidized inflatables is promising. Applications of rigidized inflatables are found in contemporary architecture such as emergency relief and military purposes. More research to validate and optimize the production technique is necessary. Later a full-scale model in composite with an inflatable mould was realized (see Figs. 10.106, 10.107, 10.108 and 10.109)

Fig. 10.106 Construction of composite model

Fig. 10.107 Photo shop of application

Fig. 10.108 Final result of composite truss

10.15 Summary

In this chapter on composites the following moulding systems have been discussed:

I. an adjustable mould based on the pretension of membranes in combination with inflatable membranes for façade elements;
I. the fluid forming of tensioned anticlastic thermoplastics;
II. the construction of the "blob" pavilion on a pre-tensioned membrane mould in combination with an inflatable;

Fig. 10.109 Final result of composite truss

III. the vacuum infusion of an inflatable beam for an outdoor stage covering; and
IV. the structural optimization of a vacuum-infused truss made with inflatable tubes.

Acknowledgements The author likes to thank Peter von Buelow for his kind cooperation in this research.

Bibliography

AEGION. Insituform. www.insituform.com. (redirects to https://www.aegion.com/about/our-brands/insituform). Accessed 23 Feb 2017

Andek. Cocoon (2009) http://www.andek.com/hans/cocoon/Cocoon2009b.html. Accessed 29 June 2013

Azambourg F (2002) Manufacture of armchairs, etc. includes injecting expandable plastic foam into sealed envelope with interconnected cavities that take up form of chair when pressurized with expanded plastic. FR Patent 2813550A1

Bendsoe MP, Sigmund O (2003) Topology optimization. Springer Engineering.

Beukers A, Hinte E (2005) Lightness: the inevitable renaissance of minimum energy structure. 010 Publishers, Rotterdam, p 149.

Blonder A (2008) ell FABRICation: an alternative process for the manufacturing and design of free form composite shells. CompositesUK - annual conference papers 2008:8

de Boor C (2001) A practical guide to splines. Springer, New York, p 87

Boser LM (2013) Thermoforming manual and trouble-shooting guide, http://www.plastimach.com/pmblog/wp-content/uploads/2013/04/VACUUM-FORMING-MANUAL.pdf. Accessed 23 Feb 2017
von Buelow P (2012) ParaGen: performative exploration of generative systems. University of Michigan, Ann Arbor
Buitink Technology. https://www.buitink-technology.com/nl/. Accessed 23 Feb 2017
Cadogan DP, Scarborough SE (2001) Rigidizing materials for use in Gossamer space inflatable structures. AIAA gossamer spacecraft form 2001.
Cantor KM, Watts P, Kutz M (2011) Plastic processing. Applied plastics engineering handbook: Processing and materials. Elsevier, Amsterdam, p 195–203
Chilton JC (2000) Heinz Isler. Thomas Telford Publishing, London, p 168
Chilton JC (2009) 39 etc…: Heinz Isler's infinite spectrum of new shapes for shells. Proceedings of the International Association for Shell and Spatial Structures (IASS) Symposium 2009, p 12.
Chilton JC (2010) Heinz Isler's infinite spectrum: form-finding in design. Architectural Des 80(4):64–71
Chilton JC (2012) Form-finding and fabric forming in the work of Heinz Isler. University of Nottingham, Department of Architecture and Built Environment, Nottingham
Christensen PW, Klarbring A (2009) An introduction to structural optimization. Springer
Cocoon Holland. Coatings I Conservation I Mothballing. http://www.cocoon.eu/. Accessed 23 Feb 2017.
Concrete Canvas. Concrete Canvas: Erosion Control & Containment. http://concretecanvas.co.uk/. Accessed 21 June 2013
Doherty I (2005) Trenchless design engineering: quality controls are key to successful CIPP lateral lining applications
Droog®. Studio Work: Kokon furniture-double-chair. http://www.droog.com/store/studio-work/kokon-furniture—double-chair/. Accessed 29 June 2013
Engel H (1997) Structure systems. Hatje Cantz, Berlin
Fox TG (1962) Properties of dilute polymer solutions III: intrinsic viscosity/temperature relationships for conventional polymethyl methacrylate. Polymer 3:111–128
Gator Preservationist. Cocoon House. https://gatorpreservationist.wordpress.com/2011/01/25/cocoon-house/. Accessed 29 June 2013
Gooch JW (2011) Cellulose acetate butyrate. In: encyclopaedic dictionary of polymers. Springer, New York, p 128
Goodyear Aerospace Corp (1961) Airmat materials investigation of one-place inflatoplane GA-468. Goodyear Aerospace Corp, Defense Technical Information Center, Akron
Gruenwald G (1998) Thermoforming: a plastics processing guide, 2nd edn. Techonomic Publishing Company inc., Lancaster
Gustafsson P (2010) North Sails 3Di, http://www.blur.se/2010/04/26/north-sails-3di-english-version/. Accessed 23 Feb 2017
Hawkins W, Herrmann M, Ibell T, Kromoser B, Michaelski A, Orr J, Pedreschi R, Pronk ADC, Schipper R, Shepherd P, Veenendaal D, Wansdronk R, West M (2016) Flexible formwork technologies: a state of the art review. Structural Concrete 17(6):911–935
Helvoirt J (2003) Een 3D blob huid. Master Thesis 3370. TU Eindhoven, Eindhoven
Hollingshead Jr. RM (1932) Drive-in Movie Theatre. US Patent 1909537 A
Jaganathan A, Allouche E, Baumert M (2007) Experimental and numerical evaluation of the impact of folds on the pressure rating of CIPP liners
Kotnik T, Schwartz J (2011) The architecture of Heinz Isler. Journal of the international association for shell and spatial structures: Heinz Isler—50 years of "new shapes for shells", 52: 185–190
Lefevre JP (2008) Ultra-high-performance polymer foils
Meier U (2011) Heinz Isler: world famous for his concrete shells - however little-known for his pioneering work in FRP. FRP International 8(2):6–8

Meier U (2012) Examples of the long-term reliability of fibre reinforced polymers (FRP) in civil construction. Conference of Civil Engineering Infrastructure Based on Polymer Composites, p 12.

Michaeli EW, Kopp J (2001) Forming procedures in polymer component manufacturing. Encyclopaedia of Materials: Science and Technology, 2nd edn. Elsevier, Amsterdam, p 3239–3247.

Morawetz H (2002) Polymers: the origins and growth of a science. Courier Corporation, North Chelmsford MA

Munro C, Walczyk D (2007) Reconfigurable pin-type tooling: a survey of prior art and reduction to practice. J Manuf Sci Eng 129(551)

Myers FJ, Robinson JD (1946) Polydispersions containing curable synthetic rubber. US Patent 2,400,054

NEN (1991) NEN 3835,1991: Mortels voor metselwerk van stenen, blokken of elementen van baksteen, kalkzandsteen, beton en gasbeton. https://www.nen.nl/nen-3835-1991-nl-878. Accessed 23 Feb 2017

Oesterle S (2012) Zero waste free-form formwork. Proceedings of the Second International Conference on Flexible Formwork, ICFF 30. CICM and University of Bath, Dept. of Architecture and Civil Engineering, Bath, p 52

OMA. Prada Sponge. http://oma.eu/projects/2004/prada-sponge/. Accessed 23 Feb 2017

Palmer P (2011) Optimizing next-generation automotive structures using Altair OptiStruct. Plastics in lightweight and electric vehicles

Pedreschi R (2011) The use of fabrics as formwork for concrete structures and elements. International conference on textile composites and inflatable structures, Structural Membranes, Barcelona

Plantema FJ (1966) Sandwich construction: the bending and buckling of sandwich beams, plates, and shells. Jon Wiley and Sons, New York

Polycel. Polycel Home. www.polycel.nl/. Accessed 23 Feb 2017

Pronk ADC, Dominicus MMT, Buelow P von, Dijck SHM van, Koppel WJ van de (2013a) Rigidized inflatable structures a production method for optimized structures. Proceedings of the International Association for Shell and Spatial Structures (IASS) Symposium: "BEYOND THE LIMITS OF MAN"

Pronk ADC, Dominicus MMT, Buelow P von, Dijck SHM van, Koppel WJ van de (2013b) Producing optimized structures with inflatables structural membranes 2013. E. Onate, K.U. Bletzinger & B. Kroplin (Eds.), VI International Conference on Textile Composites and Inflatable Structures STRUCTURAL MEMBRANES 2013, October 9–11, 2013, Munich

Pronk ADC, Houtman R (2003) Blob architecture, art and expo pavilion at the Technical University Eindhoven. In: Grundig L (ed), Proceedings textile roofs, p 1–6

Pronk ADC, Houtman R (2005) making blobs with a textile mould. In: Eugenio O, Koplin, B (eds), Computational methods in applied sciences, vol 3, Textile Composites and Inflatable Structures. Springer—Ecomas, p 305–322

Ramm E, Schunk E (1993) Heinz Isler: Schalen: Katalog zur Ausstellung. Zurich, ETH Hongger-berg, p 112

Rasse B (2013) TRP innovative curing system. ESTEC/Contract N°4000102125/11/NL/RA

Rosen SR (2002) Thermoforming: improving process performance. Society of Manufacturing Engineering, Dearborn

Rozvany GIN (2001) Aims, scope, methods, history and unified terminology of computer aided topology optimization in structural mechanics. Struct Multidisc Optim 21:90–108

Rush J (2010) Trenchless technology, San Diego Regional Airport Authority Fibre-Reinforced CIPP Project

SailSelect (2012) SailSelect string sails. http://www.youtube.com/watch?v=co2gu-WRn_w. Accessed 23 Feb 2017

Sawapan. http://sawapan.eu/. Accessed 9 Sept 2013

Schwarzmann P (2001) Thermoforming: a practical guide. Hanser, München

Society for Vascular Surgery | SVS. www.vascularweb.org. Accessed 23 Feb 2017

Solid Thinking. http://www.solidthinking.com. Accessed 25 June 2013

Studio Makkink & Bey. About. http://www.studiomakkinkbey.nl/info/about. Accessed 29 June 2013

The International Society for Trenchless Technology. The International Society for Trenchless Technology (ISTT). https://www.istt.com/. Accessed 23 Feb 2017

The Ocean Race (2013) The Ocean Race Episode 11: Sailmaking | Volvo Ocean Race 2014–15. https://www.youtube.com/watch?v=kyqIlm5EPaI. Accessed 23 Feb 2017

Throne JL (2008) Understanding thermoforming, 2nd edn. Carl Hanser Verlag, Munich

Throne JL, Kutz M (2011) Thermoforming. Applied plastics engineering handbook: Processing and Materials, Elsevier, Amsterdam, p 333–358

Victorius C (1984) High solids color coat containing alcohol soluble cellulose acetate butyrate. US Patent patent 4,451,597

Vincent de Rijk. Prada Foam. https://vincentderijk.nl/products/prada-foam. Accessed 23 Feb 2017

V&A Search the Collections. Knotted chair. http://collections.vam.ac.uk/item/O115748/knotted-chair-chair-wanders-marcel/. Accessed 23 Feb 2017

Werkman H, Bogart A, Pronk ADC, Langeveld E, Houtman R (2003) Making geometrical complex structural elements—a lightweight stage covering for an open-air theatre. In: Kroplin B, Onate E (eds), Proceedings, CIMNE Congress Structural Membranes, Barcelona, p 244–249

Wood E (1972) Method of lining a pipe. US Patent 4,009,063

Zenkert D (1995) An introduction to sandwich construction. Engineering Materials Advisory Services Ltd, Worcestershire

Chapter 11
Concrete

The diagram (Fig. 11.1) gives an overview of all the technical possibilities for concrete. The next diagram (Fig. 11.2) gives all the recent techniques researched in relation to concrete. In the last column, case studies are mentioned. Many older examples were already mentioned in the historical overview in Chap. 7.

11.1 'Curvy' Concrete

Flexible Mould
The form-active structure: concrete (Figs. 11.3 and 11.4).
Form-active typology: double layer.
Way to stabilize the mould: hydraulic pressure.

Concrete Material
Technique to handle the rigidizing material: casting.
Reinforcement of the material: bars.
Surface treatment: none.

Project
Name and location: Quilt-point method and Frame method
Design and Engineering: Kenzo Unno.
Contractor: Kenzo Unno.
Year: various.

Introduction

Concrete is not usually seen as a curved or bent material. Big flat surfaces made in plywood casts and reinforced with steel are what concrete is best known for. The way Kenzo Unno uses the concrete gives an elegant and beautiful finish to it.

A. Pronk, *Flexible Forming for Fluid Architecture*,
https://doi.org/10.1007/978-3-030-71551-9_11

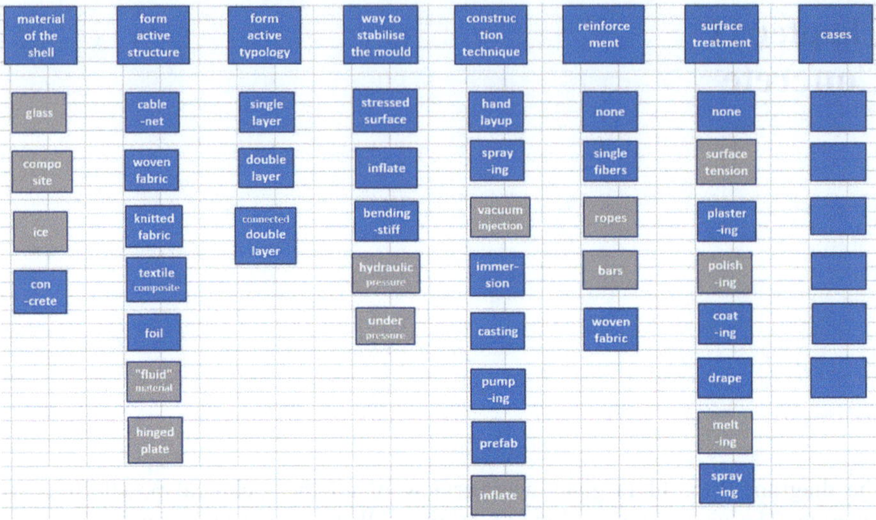

Fig. 11.1 Diagram of materials and techniques to make fluid architecture by rigidizing form-active structures

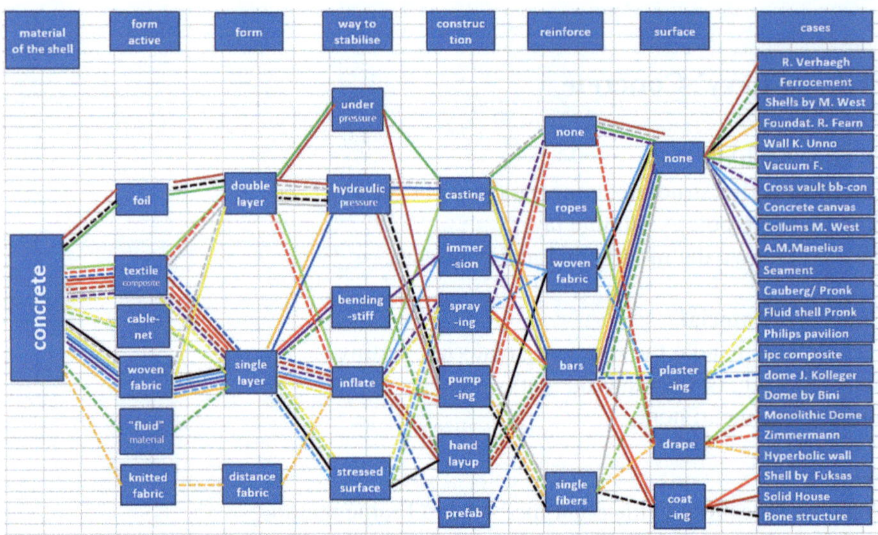

Fig. 11.2 Diagram for the material concrete and the techniques to make fluid architecture by flexible forming

Fig. 11.3 Appearance of a fabric formwork finished wall by Kenzo Unno (*Source* K. Unno and M. West)

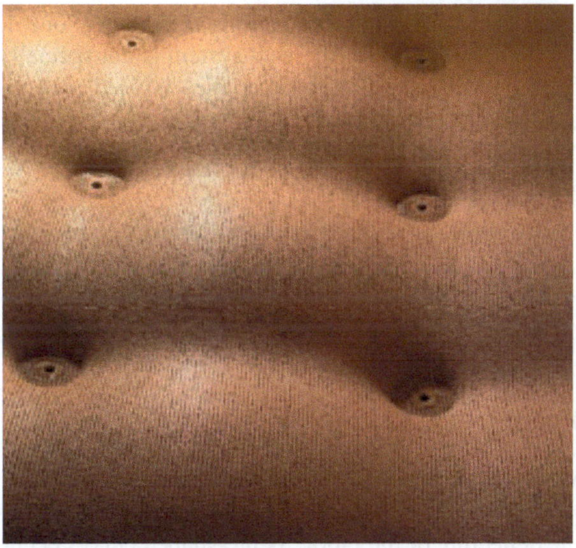

Fig. 11.4 Close-up of a fabric formwork finished wall by Kenzo Unno (*Source* K. Unno and M. West)

'Unno Reinforced Concrete' is the name of a number of methods for making concrete fabric-formed walls that was invented and developed by Kenzo Unno. He is the principal architect of the Tokyo based architectural practice, Umi Architectural Atelier. In his mostly single-family house projects, he acts not just as an architect but also as a contractor. Besides, he also advises other architects and contractors about his fabric-form methods. In Unno's wall system less material is wasted and it is even called a 'zero'-waste formwork system. The focus in this text will be on two of his methods for the cast-in-place fabric-formed concrete walls; the frame method and the quilt-point method.

Material

Concrete can withstand tensile stress of about 10% of its compressive strength. By adding reinforcement, the concrete is also able to withstand higher tensile stresses. Because this reinforcement is added in the Kenzo Unno wall, the finished wall will obtain a strength comparable to that of a 'standard' plywood cast wall.

The finished fabric-formed wall does not consist of reinforced concrete only. It is a system that, in the end, forms a complete insulated (loadbearing) concrete wall.

Combined Material

The other materials in Unno's wall system form the boundaries for the concrete. The other materials used in the wall system are; furring or studs made out of metal or wood, a rigid polystyrene insulation, steel form-ties, steel reinforcement, and of course the netting membrane. Depending on the method, vertical elements or washers will be added to the system. These give boundaries to the fabric and therefore to the concrete as well, in order to ensure that not just one deflection occurs.

The fabric used for the wall system is not specially produced for Unno's fabric-formed walls. The fabric he applies is used on construction sites to enclose the scaffolding. Instead of the netting fabric, geotextiles are a possibility as well. The original thought of Kenzo Unno was to allow people to build their own houses with these easy building methods.

> The netting fabric allows a percentage of the cement paste to escape outwards through the weave of the fabric. This presents a problem for cleaning up afterwards. Some types of geotextiles, such as those that Mark West uses, will only allow excess water and air bubbles to escape through the fabric and these fabrics can be reused in other wall systems or put to their original use as geotextiles. (Unno 1998: 124–127; 1999: 1–9; Yokota et al. 1989)

Form

When filling the fabric formwork with concrete, the fabric will deflect under the pressure that the concrete exerts on it. Because both systems use elements to restrain the deflecting of the fabric, bulge-shaped curves occur.

Because of the added rigid insulation and the studs on the back of the wall system and reinforcement, the deflecting will only occur on the finished side with these efficient tension curves. The reverse side with rigid insulation will have the same function as the plywood in standard plywood casings except in this case, where it will not be removed after construction but will function as the building's insulation. By adding different kinds of elements to keep the fabric and hence the concrete in place, different shapes can be made. Kenzo Unno commonly uses steel scaffold pipes, wooden studs and washers in his method.

The aforementioned quilt-point method by Kenzo Unno, in which the washers shape the boundaries of the membrane, relate to the 'pushed in element' and the 'point load' method of manipulating the membrane. In the same source a 'pushed linear load' is mentioned. This membrane manipulation can be recognized in Unno's Frame method.

> A zeroclastic mechanically prestressed membrane and an anticlastic mechanically prestressed membrane will develop an anticlastic surface at the boundaries of any rigid element pushed in or out as long as the boundaries of the object are not parallel to the boundaries of the membrane. (Chapter 2)

In Unno's wall system where only vertical linear elements are added, the curves will only occur in a horizontal direction. In the Quilt-point method, the singular elements allow the netting to deflect in a vertical direction as well, generating a combination of anticlastic and synclastic surfaces. The anticlastic shape occurs around the washers, and the synclastic shape occurs in between these 'pushed in elements', because of the pressure that the concrete generates on the netting fabric.

The scaffold netting Unno uses in his methods is produced in rolls of 2 meters wide, while geotextiles can be produced in rolls of around 5 meters wide. To resolve this difference, Kenzo Unno gives the sheets an overlap. The overlapping parts of the fabric are pushed firmly together when the concrete is poured in, leaving only a small line on the finished concrete. This line occurs because the two fabrics have more capacity to restrain the concrete than the part that does not overlap. This overlapping part will generate a deliberate imperfection in the synclastic and anticlastic surfaces.

Other elements can be added to the fabric to give it a different appearance, like pre-cast concrete elements or tied objects. For example, by adding a wooden beam, a standard separation wall between rooms can easily be added to the finished wall. Also, stones or other objects can be added to the wall. For this, a hole is made in the netting fabric and the stone is put in. The netting fabric will be tied around the stone and the stone will be part of the finished wall. These objects tied into the fabric have little impact on the formation of the curves in the fabric formed concrete wall as these are not pushing the fabric in or out but creating a small imperfection in the fabric and small deviations will occur only around the edges of the object. The size of the deviation depends on the object and how tight the element is tied in the netting fabric.

By changing the 'pushed in elements', the additional elements, the direction and the placement of these elements, the wall system of Unno can get many variations and appearances.

Technique

The wall systems used for fabric-formed walls can be prefabricated and placed in between existing elements, like beams, walls or other building objects. The fabric will not be fixed to these existing elements. Because of the flexibility of the membrane, the concrete will just adapt slightly to these elements. When pouring the first few centimetres of concrete in the fabric cast, the fabric should be held in place, to ensure that the concrete will not escape. After this beginning, the fabric is under pressure and the concrete ensures the deflection of the fabric. The vertical edges also will be formed by the pressure of the concrete on the fabric. The top of the fabric-formed wall system is supplemented with a beam to enclose the top part and form an extra boundary (Figs. 11.5 and 11.6).

Quilt-point and Frame method

The differences between both methods are on the finished side. Both consist of the furring or studs made out of metal or wood, a rigid polystyrene insulation, steel form-ties, steel reinforcement, and of course the netting membrane. The differences occur in the restraining of the fabric. As both have these form-tie elements, different elements can be added. As for the quilt-point method, only washers are added. In the frame method on these form-ties, wooden or metal studs are added. Variations on both techniques can be made.

As in standard plywood casings, the concrete needs to be vibrated to get the air bubbles out and thereby give the concrete its strength. In Unno's system this has to be done as well, but in this case, the fabric allows air bubbles to get out easily through

Fig. 11.5 Fabric formwork wall (*Source* K. Unno)

Fig. 11.6 Insulated formwork wall system. Fabric formwork wall (*Source* K. Unno)

the meshes of the netting. But to ensure that this happens, the fabric will be poked with a small beam to make the bubbles move to the fabric side of the wall and escape there (Chandler 2008).

General Technical Elements

The so-called zero-waste formwork is almost finished after the pouring of the concrete. When it is hardened, the netting, washers and studs will be removed. No surface treatment is necessary.

Using a geotextile has some more advantages. A geotextile can be washed and reused for another wall or for its original use as a geotextile. The scaffold netting, on the other hand, is not always reusable for other projects. Because it can be torn apart more easily while being removed and because it is more difficult to wash after use if it allowed a percentage of cement paste to escape through the fabric. Another reason why Kenzo Unno uses the netting instead of the Polyolefin is the availability in Japan, since the geotextile is not sufficiently available in the Japanese market and also more expensive.

Whereas standard plywood casings require the use of oil to allow the plywood to come off easily, in Unno's method no oils are needed for the fabric. The netting fabric, like the geotextile, does leave a weave pattern as a finish. This in combination with the curved surface gives the concrete a beautiful and elegant appearance and makes it look like fabric (Univ. of Manitoba, www.arch.umanitoba.ca/cast/2.kenzounno/; West, fabwiki.fabricformedconcrete/lib/exe/fetch.php?media=chile:thin_shell_panels.pdf).

11.2 The Ambiguous Chair

Flexible Mould
The form-active structure: Woven fabric (textile) (Figs. 11.7 and 11.8).
Form-active typology: Double layer connected.
Way to stabilize the mould: Hydraulic pressure.

Fig. 11.7 The Ambiguous Chair (*Source* A. Manelius)

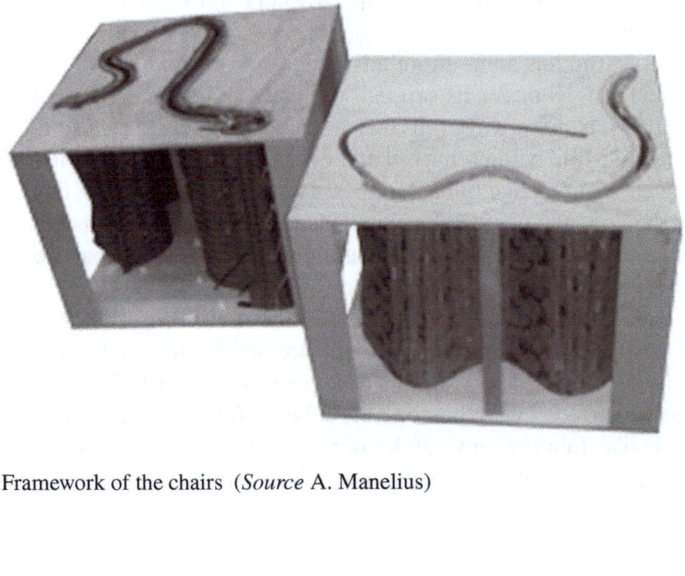

Fig. 11.8 Framework of the chairs (*Source* A. Manelius)

Composite Material
Technique to handle the rigidizing material: Pouring concrete into casting.
Reinforcement of the material: Reinforced with PVA fibres and with 6-mm rebar.
Surface treatment: textile patterns 'printed' on the concrete surface.

Project
Name and location: The Ambiguous Chair, Copenhagen, Denmark.
Design and Engineering: Anne-Mette Manelius (Manelius 2012).
Contractor: Centre for industrialized Architecture (CINARK), Institute of Technology at the Royal Danish Academy of Fine Arts, School of Architecture.
Year: 2009.

Introduction

The Ambiguous Chair is essentially a curved wall turned on its side. The chair was made for the exhibition CINARK Trade fair. The formwork principle was based on Kenzo Unno's (2007) quilt-point method, and findings from student workshops at the University of Edinburgh. The shape of the chair is created using two plywood sheets, the plywood sheets are cut to the desired shape, the plates are then placed one above the other. Between these multiplex plates, the fabric is stretched, which forms the formwork. By using the quilt point method by Kenzo Unno it was possible to reduce the amount of material and give an artistic shape to the chair.

Materials

Fabric: The formwork textiles were not made specifically for concrete casting. There were two requirements for the chosen fabric: First, the textile had to provide good permeability for the transportation of water and air so the concrete would slide in easily. If the material is non-porous then it is important, first, that the concrete mix is moulded in without air bubbles or gravel accumulation, And, second, the pattern of the textile would have to be transferred clearly onto the concrete surface as an aesthetic feature in accordance with the attempted upholstered look.

Concrete (reinforcement): ECC 2 (refers to Eurocode 2: Design of concrete structures, equal to C90/105) concrete was used for its high strength level. Because of the unusual shape of the formwork, the concrete needed reinforcement. This was provided by two different techniques. First: 6-mm rebar was shaped into the form that was cut into the plywood sheet and second, the concrete was mixed witch PVA (Poly Vinyl Alcohol) Fibre.

Framework: The framework was made from plywood. These plywood sheets were placed one above the other. Between the sheets, columns of wood were placed.

Fig. 11.9 The pattern of the textile 'printed' onto the concrete by Remy and Veenhuizen (*Source* Remy and Veenhuizen)

Form

The shape of the Ambiguous chair is based on the quilt-point method of Kenzo Unno and experiences of studio workshops at the University of Edinburgh. A plane surface is manipulated by a hydraulic pressure and stabilized by fixing points. The quilt-point method of Kenzo Unno gives the Ambiguous Chair the classic look of a Chesterfield chair (Fig. 11.10). The concrete surface is combined with the surprising effect of patterns and shades of grey; this effect is realised because of the different patterns in the textile (Figs. 11.9 and 11.11).

Through the eyes of observers

After completion of the two seats, several people were asked to give their opinion. The comments and opinions of various observers were alike, particularly in terms of the experience of the chair. People expected a soft, well-fitting seat, but since the whole was executed in concrete this proved to be a misconception. The reactions of the various observers showed that the chair has a deceptive appearance, because the fabric still seems to cover the chair.

Technique

See Figs. 11.12, 11.13, 11.14, 11.15, 11.16, 11.17, 11.18, 11.19, and 11.20.

Fig. 11.10 Chesterfield chair (*Source* A. Manelius)

11.3 Columns by Mark West

Flexible Mould
The form-active structure: woven fabric (Fig. 11.21).
Form-active typology: single pressure.
Way to stabilize the mould: hydraulic pressure.

Composite Material
Technique to handle the rigidizing material: casting.
Reinforcement of the material: bars.
Surface treatment: none.

Project
Name and location: Columns, Manitoba, Canada.
Design and Engineering: Research by Mark West and Ronnie Araya.
Contractor: C.A.S.T. and University of Manitoba.
Year: 1984.

Introduction

Fabric formwork has been used since the mid-1800s to form concrete. Nowadays, there are some new developments with regard to the process of producing concrete

Fig. 11.11 On the concrete surface (*Source* A. Manelius)

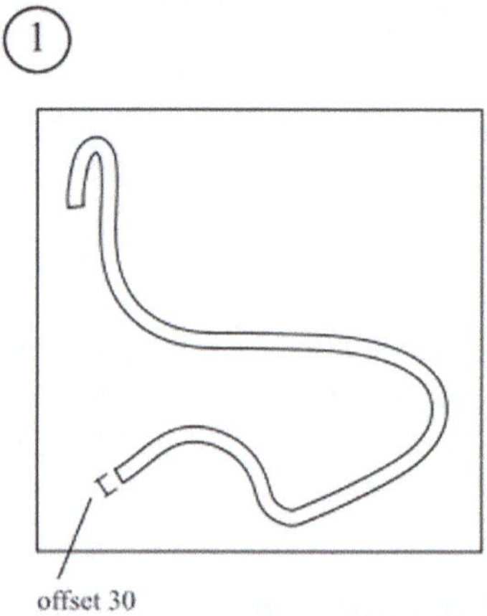

Fig. 11.12 The desired shape is cut into the plywood sheets, the measurement between the two lines is set to 30 mm (*Source* A. Manelius)

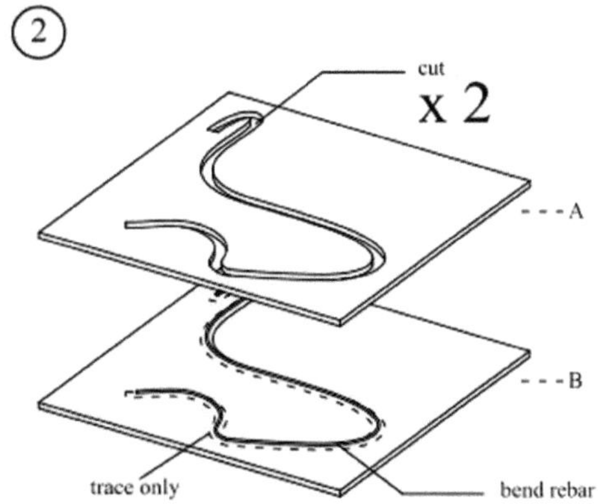

Fig. 11.13 Make a copy of the first plywood sheet and bend the rebar in the same shape (*Source* A. Manelius)

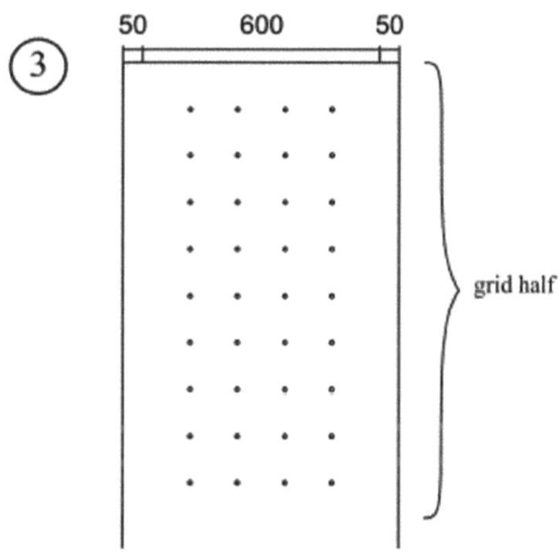

Fig. 11.14 The height of the textile becomes the width of the chair. Set marks on the textile for the required grid (*Source* A. Manelius)

structures. The steel and wooden moulds are replaced by flexible moulds. Mark West (Fig. 11.21) did a lot of research and invented a series of techniques for constructing fabric-formed walls, beams, trusses, columns, slabs, panels, and thin-shell vaults (West 1995; 2004: 42–45; 2007). All these methods use flat untailored fabric sheets.

Mark West is internationally regarded as one of the inventors of different methods of building concrete sculptures by using flexible fabric moulds. His practical and architectural education gave him the experience and know how to start research and

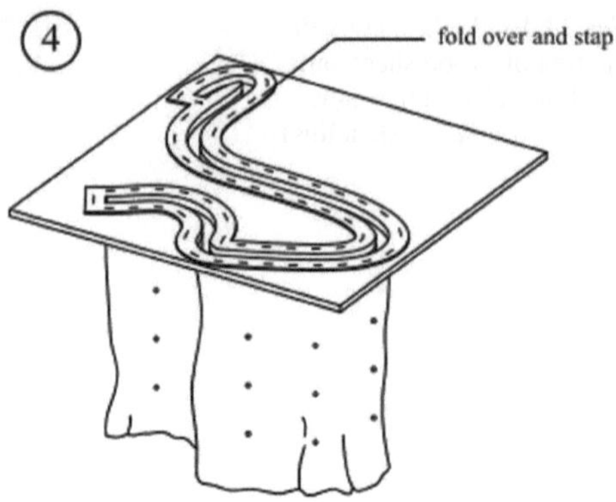

Fig. 11.15 Staple the fabric on to the plywood sheets. Make sure the same offset measurement is used (*Source* A. Manelius)

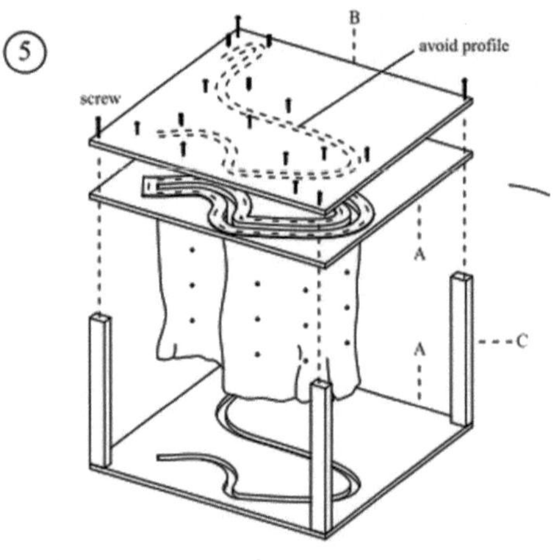

Fig. 11.16 Put the framework together and close the bottom with a plywood sheet (*Source* A. Manelius)

experiments on fabric-formed constructions. Since 1984, Mark West's drawings and sculptures have been exhibited internationally, including public sculptures in New York, NY, Bar Harbor, ME, Tallahassee, FL, Ottawa, Canada, and a one-person show at the Storefront for Art and Architecture in New York. His most recent exhibitions include shows of photographs and fabric forms at the Royal Institute of British Architects (RIBA) Gallery, London, England (2002), drawings and photographs at

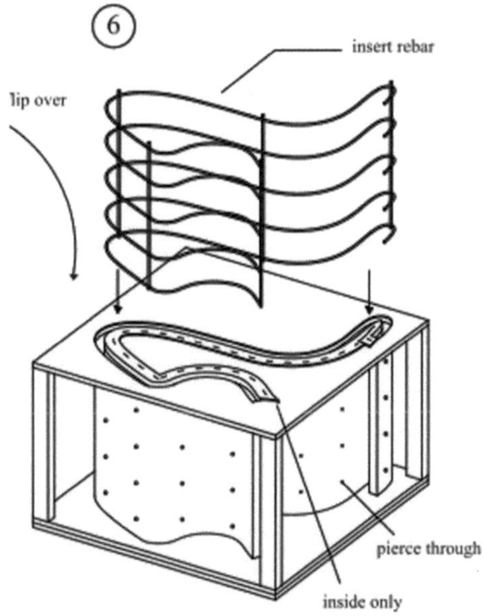

Fig. 11.17 Flip the framework over and put the rebar in. Pierce trough the fabric on the grid points (*Source* A. Manelius)

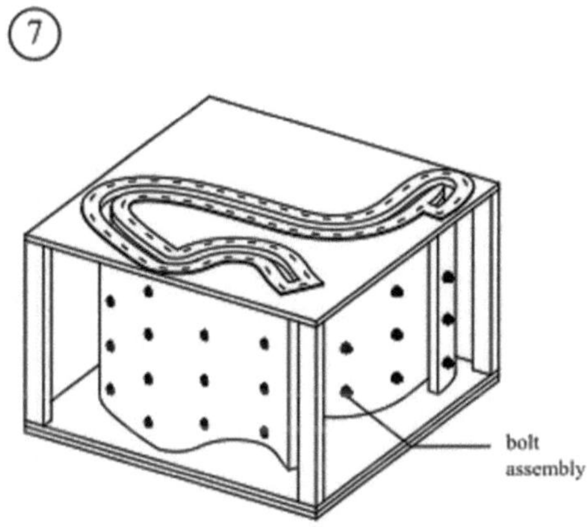

Fig. 11.18 Bolt through the pierced holes. And leave a desired space between the two sides of the fabric (*Source* A. Manelius)

the Eric Arthur Gallery, in Toronto, Ontario (2002), and fabric-formed beams at the National Building Museum in Washington, DC (2004–2005).

Material

Concrete is one of the most used materials in the building industry nowadays. It consists of a mixture of water, cement, sand and gravel. The amount of water that is used to bind the other elements has a large influence on the quality of the mixture. A lower water-cement ratio usually makes a stronger concrete. Also, the design constituents, the mixing, the placement and the curing methods have their influence

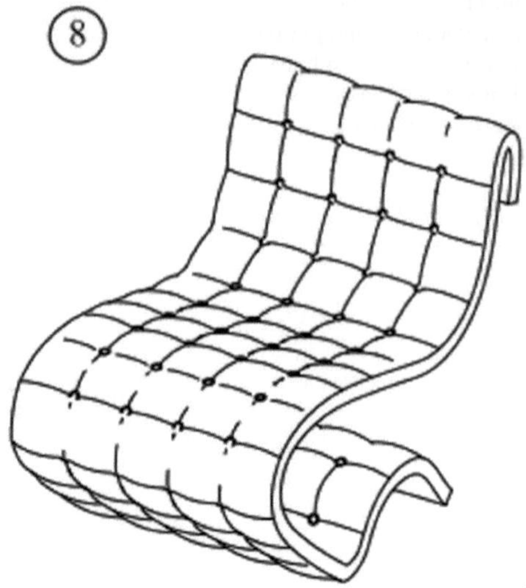

Fig. 11.19 Pour concrete into the casting, wait a few days after die-casting and this is the result (*Source* A. Manelius)

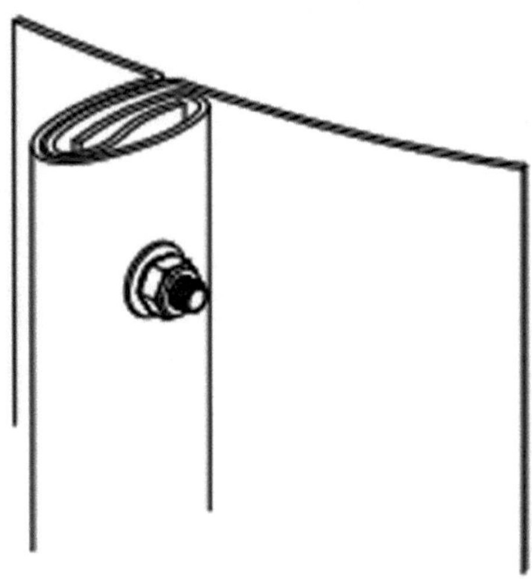

Fig. 11.20 A close-up of the closing detail of the textile (*Source* A. Manelius)

on the quality of the concrete. Adding some additives to the concrete can affect and improve the properties of the concrete. These components define the different properties such as strength, water requirement, shrinkage, abrasion and density of the concrete.

Fig. 11.21 Mark West (*Source* M. West)

Concrete objects are mostly designed in such a way that most of the forces that occur within the material are compressive forces. Floors and pavements, however, are subjected to tensile forces and have to be designed with extra attention to the reinforcement of the mixture. Because the material is liquid when used, it's relatively easy to create special forms, but it is also ideal for creating large and simple forms (Figs. 11.22, 11.23, and 11.24).

Fabric formworks for reinforced concrete construction and architecture is an emerging technology with the capacity to transform concrete architecture and reinforced concrete structures. Different fabrics will have different properties and qualities with regard to form and surface finishing.

The new formwork used is light, thin and easy to form and manipulate. The natural tension geometries given by flexible fabric membranes provide light and inexpensive formworks, which use up less material than conventional formworks, and some provide less wasteful formwork systems. The flexibility of a fabric formwork makes it possible to produce a multitude of architectural and structural designs from a single, reusable mould. The use of a permeable formwork fabric can produce smooth surface finishes with high strength concrete.

A flexible fabric mould awakens concrete to its original wet, plastic nature by naturally producing concrete members with complex sensual curvatures. The freedom

Fig. 11.22 Column (*Source* M. West)

that is granted by the flexibility of the formwork offers new opportunities for architects and constructors. Research has produced different methods for forming beams, trusses, panels, vaults, slabs, and columns.

Form

Fabric formwork uses a flexible textile membrane instead of the rigid formwork materials usually used in concrete construction. The thin formwork membrane is flexible and allows the concrete to flow into places where it is needed. This means that the prestress level of the fabric and the qualities of the concrete determine the shape of the column: places with a lot of tension in the membrane will be flatter and with less tension will be more curved and will contain more concrete. This produces naturally efficient structural curves, unprecedented sculptural forms, and extraordinary surface finishes.

Fig. 11.23 Beam (*Source* M. West)

Fabric formworks can be used to form columns, walls, beams, trusses, slabs, panels, and thin-shell structures in both precast and in situ construction. The formwork itself is very light, but needs to be reinforced until the concrete has cured completely. Especially with the great columns, it's very important to keep the formwork in place, due to gravity pulling the whole formwork and the concrete down. The most important thing is to support the membrane at the top to ensure its position.

By pouring wet concrete into a flexible, permeable membrane, it is possible to produce a strong column with a smooth surface.

Due to the permeability of the membrane and the pressure of the concrete, the water and the air bubbles in the mixture are pushed outside. The result is the creation of small ridges of the cement at the surface of the membrane.

Forming of the column can be achieved by prestressing the membrane. Different stress-levels will give different degrees of resistance to the concrete, making some places easier to reach than other ones.

A problem occurring during the realization of columns is that the pressure of the concrete at the bottom is much more than the pressure in the top of the formwork. This is caused by gravity and the weight of the concrete. To prevent all the concrete descending, it is important to think about the different prestress levels when designing the mould.

The main form of the column can be affected in different ways. The membrane used is a mechanically prestressed membrane, making it an anticlastic surface. By

Fig. 11.24 Beam (*Source* M. West)

changing the prestress in a certain area, by applying an external load, by pushing another form-active surface into the membrane or by pushing or pulling a rigid object into the membrane, different forms can be achieved.

Mark West influences the forms by changing the prestress levels of the fabric, controlling the concrete by adding pressure in different places, using ropes or his hands to guide it (Figs. 11.25 and 11.26).

Technique

The basic strategy for an in situ fabric-cast column is as follows: the bottom of a fabric tube is attached to a footing, or in this case to the floor slab, while the top is stretched vertically to a scaffolding structure above. No other intermediate or lateral supports are required. The mild vertical pre-tensioning, provided by hand tightening of the formwork, is all that is required to laterally stabilize a vertical column of wet concrete. The roof-slab formwork is used as scaffolding for the column formworks. The framing of this deck was standard, except for some special attention to the location and direction of framing members directly above the column locations. The roof-slab formwork decks of beams were used as "sheaves" to pull the outer

Fig. 11.25 Closing of the fabric (*Source* M. West)

polyethylene jackets upwards. This arrangement allowed the same jacket design to be used for all columns; differences in height were simply taken up by pulling or retracting the jackets (in other circumstances the formwork can simply be connected at the appropriate height, with the excess fabric being furled out of the way at the top or bottom).

The columns all use standard steel and concrete designs. The only essential difference from standard construction is the change in formwork materials. The flexible fabric tubes thus installed entirely replace the rigid steel or plywood box forms traditionally used to cast reinforced concrete columns. Due to the great efficiency of tension membrane formworks, the formwork itself is very light. Thin-shell structures are formed from simple fabric sheet moulds. Further reductions in materials consumed in construction can be achieved by more efficiently shaping the concrete members formed in these flexible moulds. These savings provide a direct pathway towards more sustainable architectural and engineering constructions in both advanced and basic building economies (West 2016).

Fig. 11.26 Manipulation of the fabric (*Source* M. West)

11.4 Open Cell Structures

Flexible Mould
The form-active structure: foil (Figs. 11.27, 11.28, 11.29, and 11.30).
Form-active typology: double layer.
Way to stabilize the mould: hydraulic pressure/inflate.

Composite Material
Technique to handle the rigidizing material: pumping.
Reinforcement of the material: single fibre.
Surface treatment: coating/none.

Project
Name and location: Producing open cell structures, Eindhoven, the Netherlands
Design and Engineering: Master thesis of M.M.T. Dominicus; Tutors: A.D.C. Pronk, J. Beetz and J. Lichtenberg, Elective of H. Sioncke; Tutor: A.D.C. Pronk.
Contractor: TU/e, KU Leuven.
Year: 2009–2012.

Fig. 11.27 Open cell structures with flexible moulding (*Source* M. Dominicus)

Introduction

Open cell structures are (often) organic structures characterized by their low own weight and optimized load bearing capacity, therefore they can be very useful in an architectural context. Outside the architectural context they can contribute to a more sustainable build environment (waste reduction). We can describe an open cell structure as a polyhedral structure which consists of struts and plates, the mass is formed around the cells. Open cell structures are used for their optimized load-bearing capacity in combination with a minimum usage of necessary material. Open cell structures are optimal structures—in an architectural context—and can be generated

Fig. 11.28 Façade with open cell structures (*Source* M. Dominicus)

Fig. 11.29 Cylindrical open cell structures with membrane moulding (*Source* KU Leuven)

Fig. 11.30 Y-shaped open cell structures with membrane moulding (*Source* KU Leuven)

by applying an optimization process, this process is already a well-known method in the car industry, mechanical engineering and aeronautical engineering. Preconditions for the optimization process are: (i) the boundaries, (ii) load case, (iii) material properties. Using these conditions, the optimization process generates an optimal structure based on maximum strength and minimal amount of material. According to Rozvany it is possible to optimize a structure by three types of optimization: (i) size, (ii) shape and (iii) topology (Rozvany 2001: 90–108).

Structural optimization integrates structure and form in a way similar to natural or biological optimization. Thus, open cell structures can be described as *optima forma*. The research to open cell structures is based on the literature of Otto and Rozvany. The use of open cell structures in the built environment leads to a reduction in material use and a more appealing architecture but it remains restricted by the existing formwork. A flexible mould in combination with concrete can be used to produce these kinds of structures. For this research, we combined membrane moulding with different (filling) materials like concrete, gypsum and resin in different configurations.

Three main phases within this research are: modelling of a bone structure as a façade element, optimization of the structure and construction of models with gypsum. These projects are made by using a membrane mould (flexible moulding).

Simulation

In these projects we used programming and optimization programs to generate 'open cell' structures. In the first project (bone) we modelled a script inside Rhinoceros (Dominicus et al. 2012: 1–8). The script was based on the Voronoi-algorithm and is capable of generating an open cell configuration, based on several parameters, such as density, the location of the voids and the location of the points. The Voronoi algorithm is transformed, so it works like a lattice algorithm. The objective of the script is to generate a continuous surface suitable for analysis and production. Application of the script allows the user to quickly generate a surface with an open cell structure and manipulate it (Fig. 11.31).

Production Techniques

As can be seen in Fig. 11.32, wooden discs were used to define the layout. At this point our method shows an important difference to that of both Zimmermann (see below) and West (see above). They both use the primary casting material to achieve the desired form, whereas we achieved our form by modifying, not the primary casting material (EPDM rubber), but its support (wooden discs). In this way we reduced the complexity of the production process and introduced a certain level of recoverability of the mould (Figs. 11.33, 11.34, and 11.35).

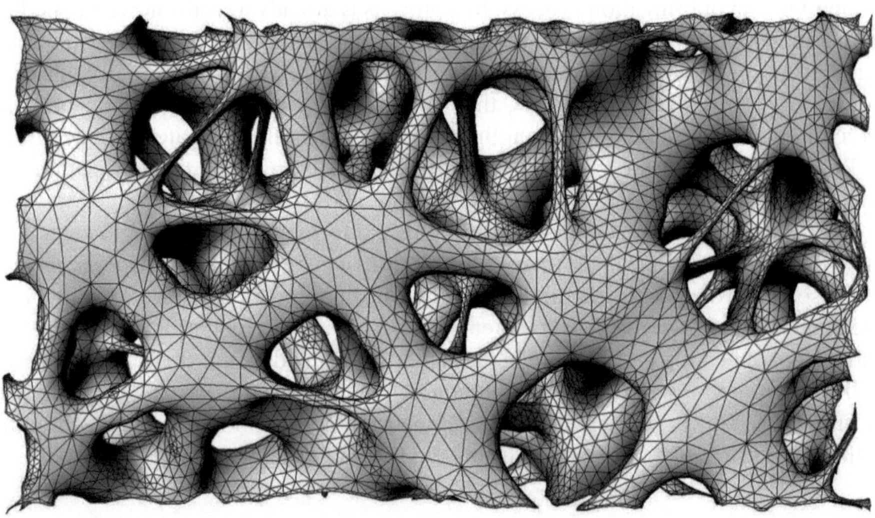

Fig. 11.31 Simulating open cell structures (*Source* M. Dominicus)

Fig. 11.32 Bone structure with EPDM membrane

As can be seen in Fig. 11.36, the façade element is divided into three separate sections. The inner layer functions as the bearing structure. The two elements have been connected through a steel connecting element. Between the elements, a cavity remains, which functions as a thermal break and is filled with a transparent sheet that is pressurized.

This division increases its functionality in relation to the building process. The first element is placed during construction of the bearing structure, while the final element is placed later in the process, when the façade is closed.

The goal of this project was to create a free-form concrete façade element with a membrane mould based on a microscopic bone structure. Initially the attempt was made to achieve this form by using stiff membranes and by welding a mould out of custom-made patterns. This proved undesirable, however, because of the irregularities in the end model due to the welds and the labour-intensive process. Also because of the rigidness and inelastic characteristics of the mould, the mock up was

Fig. 11.33 Concrete bone structure

not fluent and curved enough. It would be possible to create a more fluent end model by increasing the number of patterns but this would prove to be far too difficult.

Another disadvantage of a welded and stiff mould was the fact that the mould itself was unrecoverable. In order to remove the mould, it would have to be cut into pieces. Again, this was undesirable for the fact that for each individual element another mould would have to be constructed. After some consideration it was decided to try creating a mould with two continuous sheets of flexible material. This differs from the known processes to create a somewhat similar model cast in textile or Polystyrene moulds (Pedreschi 2011; Pronk and Houtman 2004; Cauberg et al. 2008). After comparing some flexible rubber-like materials it was decided to use EPDM (ethylene propylene diene monomer (M-class) rubber) known for its elastic characteristics. The material is used as a seal for multiple purposes but mainly as a roofing membrane. The slabs come in different thicknesses varying from 0.8 mm to about 2 mm and in 2 types: the American and European type which differ in elasticity and surface evenness.

Fig. 11.34 Casting with membrane mould

The American variety was chosen because of its smooth characteristics, leaving no imprints in the concrete. The concrete used was a polypropylene-fibre-reinforced concrete. The casting of the mould took place in an almost horizontal position due to the hydrostatic pressure of the concrete. This ensured that the deflection and expansion of the EPDM-slabs at the base would not differ from the top of the façade element.

The second project was produced with the help of flexible fabric. The method which was used for the bone was not suitable in this case. A flexible mould was modelled by using cutting patterns. Using Rhino, we were able to generate a structure and the different cutting patterns. The cutting patterns were sewn together. The cutting patterns are generated in such a way, that the seams are not an inconvenience. By using this method, the 'three-leg' shape can be produced by flexible moulding. This method is more 'shape predictable', because of the stiffness of the fabric mould. In the cutting-pattern method, seams are inevitable (Figs. 11.37, 11.38, 11.39, and 11.40).

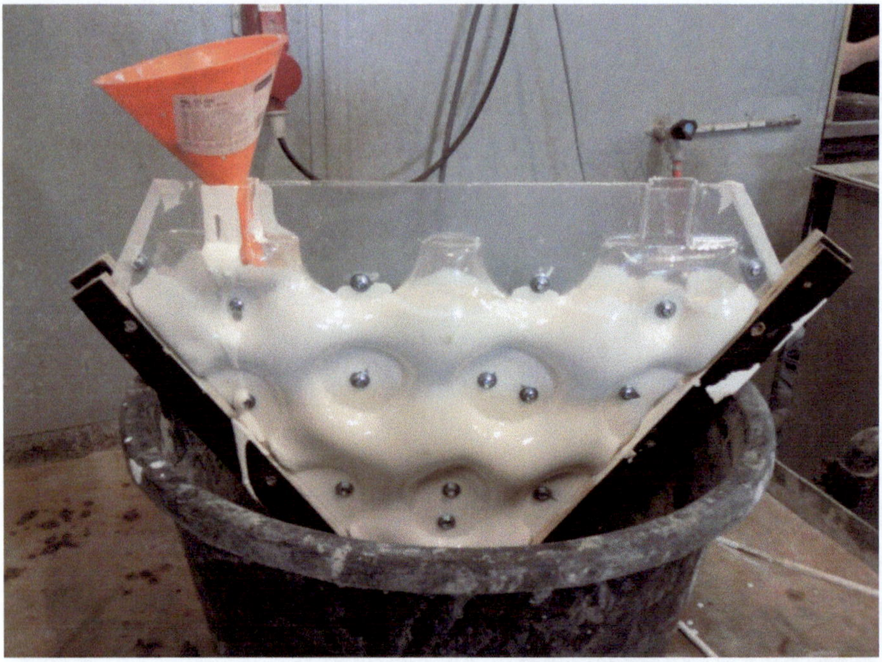

Fig. 11.35 Casting process

Fig. 11.36 Layered bone structure

Fig. 11.37 Membrane mould (*Source* KU Leuven)

Fig. 11.38 Stretching of the mould (*Source* KU Leuven)

Fig. 11.39 Casting (*Source* KU Leuven)

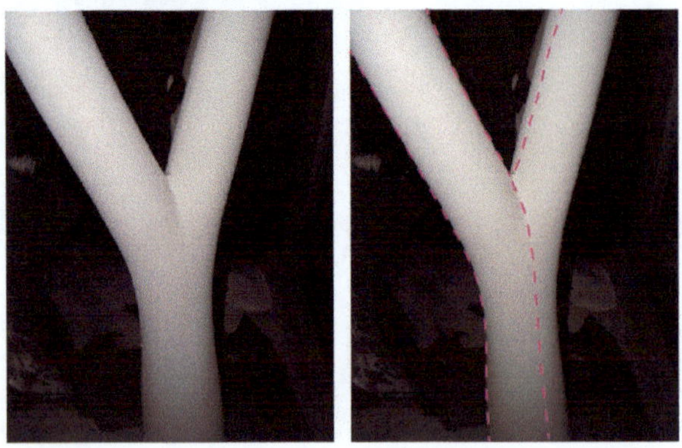

Fig. 11.40 Cutting patterns (*Source* KU Leuven)

11.5 Concrete Canvas

Flexible Mould

The form-active structure: woven fabric (Figs. 11.41 and 11.42).
Form-active typology: connected double layer.
Way to stabilize the mould: inflate.

Fig. 11.41 Concrete Canvas (*Source* Concrete Canvas)

Fig. 11.42 Distance fabric (*Source* Concrete Canvas)

Composite Material
Technique to handle the rigidizing material: submersion.
Reinforcement of the material: woven fabric.
Surface treatment: none.

Project
Name and location: Concrete Canvas Shelter.
Design and Engineering: Peter Brewin and William Crawford.

Contractor: Concrete Canvas Ltd.
Year: 2004.

Introduction

Concrete Canvas (Concrete Canvas, www.concretecanvas.com) (CC) is a geo-composite that combines the flexibility of geotextile fabrics with the durability of concrete. This flexible, double layered, cement-impregnated fabric hardens into a thin and rigidized concrete layer upon submersion (Fig. 11.43).

Because of this specific technology, Concrete Canvas is used in various civil as well as military applications. One example is the integration of this technology in a rapidly deployable building called a Concrete Canvas Shelter (CCS) for humanitarian aid (Fig. 11.44). These Concrete Canvas Shelters are designed as part of a modular system and only require air and water for construction: The mould is stabilized by inflation of air, and after the building has reached its actual size only water is required to make its shell completely solid. This solid concrete shell provides environmental protection and safety of the inner environment against theft and natural hazards. Because of the modularity and thermal properties, it is possible to create arrays of structures that satisfy both civil and military operational requirements, such as storage of food or medical capacity (Figs. 11.44 and 11.45).

Fig. 11.43 Cement-impregnated Concrete Canvas (*Source* Concrete Canvas)

Fig. 11.44 Concrete Canvas Shelter (*Source* Concrete Canvas)

Fig. 11.45 Internal view of shelter (*Source* Concrete Canvas)

Once a Concrete Canvas Shelter is constructed it has a design life of approximately ten years (Brewin and Crawford 2010).

Material

The structure of Concrete Canvas consists of a cement impregnated distance fabric. Distance fabric is a structure textile that was developed in the nineteen fifties and is characterized by two parallel skins of fabric, which are integrally connected by large

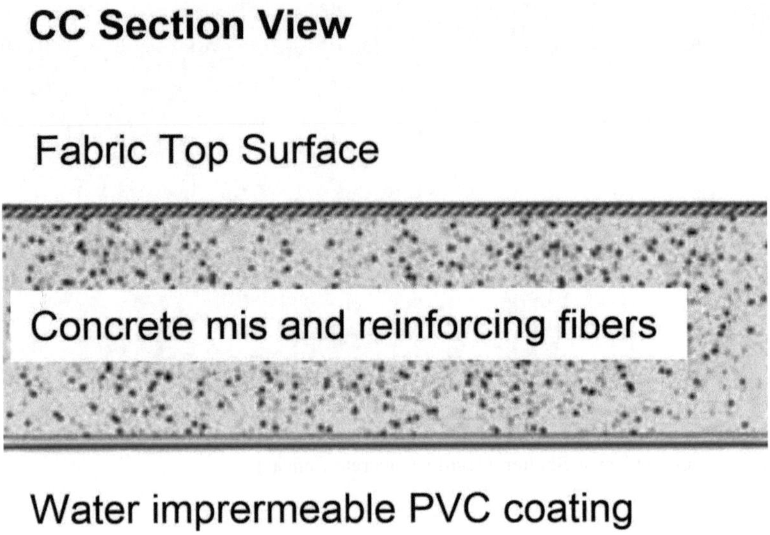

Fig. 11.46 Section view Concrete Canvas (*Source* Concrete Canvas)

numbers of threads (Yang and Li 2011: 647–650). Because of these characteristics, distance fabric is often used in inflatable boats manufactured by Zodiac. In Concrete Canvas, these two skins of fabric enclose a substance of both a concrete mix and reinforcing synthetic fibres (Fig. 11.46). This concrete mix consists of high early strength concrete with a limited alkaline reserve. The synthetic fibre reinforcement prevents cracking of the concrete and absorbs energy from impacts, while the PVC-coated fabric ensures that the material is water impermeable. Once the structure is rigidized, the Concrete Canvas structure is fire proof and does not contribute to the surface spread of flames.

Form

The key characteristic of Concrete Canvas Shelters is the use of inflation to create a specific form-active surface. Once the form-active structure has reached its force equilibrium it can also be manipulated in four different ways: by changing the (pre-)stress in a certain area, by an external load, by pushing other form-active surfaces into a membrane or by pushing or pulling a rigid object into a membrane. In their paper, an overview of the different parameters for manipulating form-active surfaces and their main geometric effects is presented in a matrix.

The monoclastic, Nissen-hut-shaped form-active structure of a Concrete Canvas Shelter is optimised for compressive loading and can also be covered with sand or earth in a process called berming. The external thermal mass reduces heat loss, maintains a steady indoor air temperature and provides extra protection against blasts or fire.

Art

Concrete Canvas is basically a material that consists of all the elements of concrete, but it is flexible enough to be turned into any shape before it is submerged in water to harden. Because of this flexibility, Concrete Canvas can also be applied in an artificially shaped element, like the stitching concrete shapes of Fig. 11.47, made by German designer Florian Schmid (Florian Schmid, www.florianschmid.com/stitching_concrete_collection.html). In his designs, the softness of cloth, together with the stability of concrete, are both represented in an artificially shaped stool: There is the illusion of cloth, stitched together with coloured thread, that would not provide enough strength to sit on. However, because of the concrete inside the fabric the stool is strong enough to use it like any other stool.

Technique

Humanitarian aid

Due to its innovative structure a Concrete Canvas Shelter can be rapidly deployed in four stages: delivery, inflation, submersion and setting. Firstly, the Concrete Canvas Shelters are supplied folded in polyethylene sacks which are water and air proof. Arrived on site, the Concrete Canvas Shelters do not require foundations or hard surfacing for deployment. However, it is recommended that the ground is relatively flat and that rocks or vegetation are removed. Maximum 5 h before inflation the concrete needs to be hydrated. Secondly, the plastic bladder needs to be inflated using a 2.6 PSI electric air pump of 110 V or 240 V to lift the structure to its designed shape

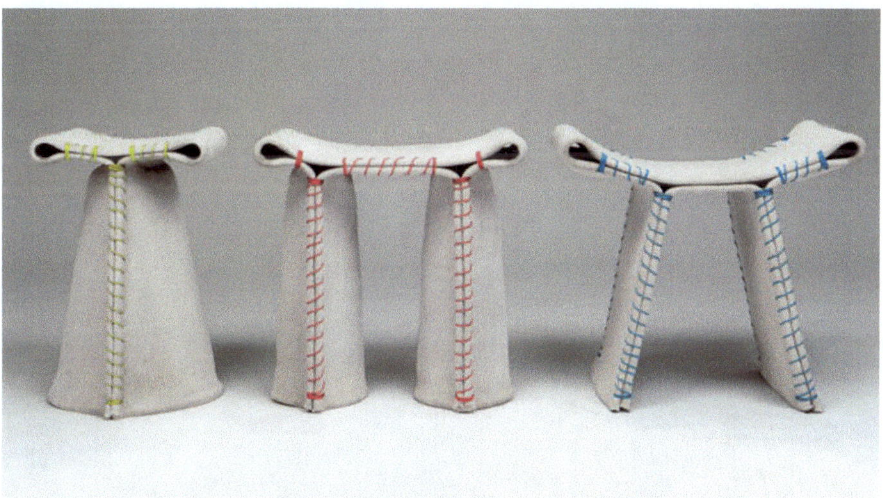

Fig. 11.47 Stitching concrete, made by Florian Schmid (*Source* Florian Schmid)

until it is self-supporting. The shelter is then secured to the ground with anchors. After inflation of the shelter, the Concrete Canvas needs to be hydrated with more water to cure. Lastly, it takes about 24 h until the shelter is ready to use. No surface treatment is needed afterwards (Figs. 11.44 and 11.45). Nevertheless, it is possible to make small openings for services by cutting or drilling holes into the Concrete Canvas shell after deployment. Once a Concrete Canvas Shelter is constructed it has a design life of approximately ten years, and the structure can be demolished using basic tools.

Civil Applications

As was said before, due to its technology, Concrete Canvas is not only used in shelters to provide humanitarian aid, but it is also applied in various other civil applications. The fabric is used notably for creating complex, irregular shapes to strengthen the soil of dikes, benches, retaining walls and dam reinforcements, and therefore provide erosion control (Fig. 11.48). Nevertheless, Concrete Canvas can also be used for the construction of canals or pipeline protection against impact damage and chemicals. Once the Concrete Canvas is hydrated it has a working time of 1–2 h and will achieve 80% strength after 24 h. After 10 days the compressive strength of the rigidized canvas is 40 MPa, and its bending strength 3.4 MPa. Depending on the application of the Concrete Canvas, there are several options to secure the fabric together, such as screwing or using hog rings. In a dry, unhydrated state an adhesive sealant can also be applied on the flap of the fabrics. In an already hydrated state mortar is used. The

Fig. 11.48 Concrete Canvas applied for slope protection with a spreader beam (*Source* Concrete Canvas)

most watertight connection can be achieved by welding the fabric thermally together by heating the PVC-coated fabric with hot air and then pressing the parts against each other. Depending on the ground conditions, Concrete Canvas can be secured to the ground using drilled bolts or nails with a washer to ensure that the head of the nail does not penetrate through the surface of the Concrete Canvas (Fig. 11.49). For ground surfacing applications such as ditch lining, slope stabilization or erosion, control pegs are used or the edges of the Concrete Canvas are buried with soil or aggregates (Milliken, www.foundationperformance.org/pastpresentations/MillikenPresSlides-Nov12.pdf).

Fig. 11.49 Synclastic curved 3D fabric (*Source* T. v Engelen)

11.6 Fabric-Formed Concrete with Distance Fabric

Flexible Mould
Form-active structure: woven fabric.
Form-active typology: connected double layer.
Way to stabilize the mould: inflate.

Composite Material
Technique to handle the rigidizing material: casting.
Reinforcement of the material: none.
Surface treatment: drape.

Project
Name and location: various projects.
Design and Engineering: Arno Pronk and Tom van Engelen.
Contractor: Eindhoven University of Technology.
Year: 2001–2005.

Introduction

A distance fabric is a dual membrane structure (Fig. 11.50). In order to inflate a dual wall structure, both walls need to be fixated. A distance fabric is a 3D fabric

Fig. 11.50 3D fabric with steel bar. Fabric with concrete (*Source* T. v Engelen)

Fig. 11.51 Testing of 3D concrete fabric beam (*Source* T. v Engelen)

that connects two surfaces with parallel threads perpendicular to the surfaces. The vertical drop threads ensure that the distance between each surface remains constant. Distance fabric combined with inflation is mainly used for plane structures with a constant thickness (Fig. 11.51).

Picture 11.52 shows a layout of a Parabeam distance fabric (Bakker 1951). The distance between each facing can be varied to meet the needs of the application. If both facings are coated, an airtight space is obtained which can be inflated. This inflatable structure was already used in the past. In 1956, Goodyear developed a prototype of an inflatable airplane (Inflatoplane model 466) using a rubberized "air mat" (Phane 1954; Brown et al. 2001). The sandwich consists of two neoprene-coated nylon membranes connected by a mesh of nylon threads. As soon as the nylon was exposed to air, it absorbed and repelled water (as it stiffened), giving the aircraft its final shape and rigidity. Structural integrity in flight was retained-by continually circulated air (pressure) pumped by the aircraft's motor (Fig. 11.53).

Figure 11.54 shows a section of an inflatable wing for the Goodyear inflatoplane. This inflatable wing shows the possibilities of variable distance threads forming a discontinuous section (Cocke 1958).

Form

In 2001 (TU Delft) and 2005 (TU/e) several studies concerning the curvature of continuous distance fabric were executed. The material used was a PVC-coated

Fig. 11.52 3D fabric Goodyear's Inflatoplane

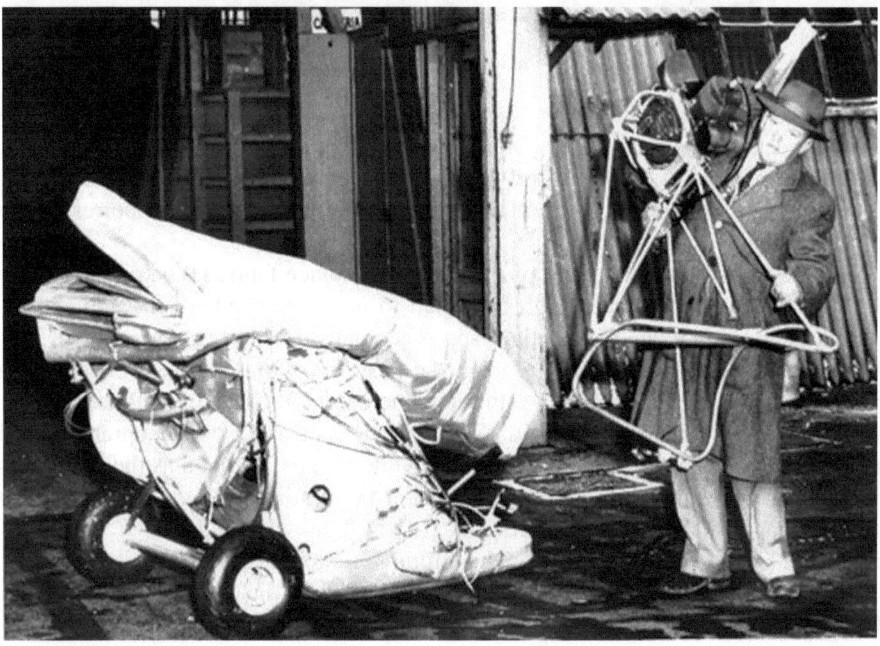

Fig. 11.53 Inflatoplane before inflation

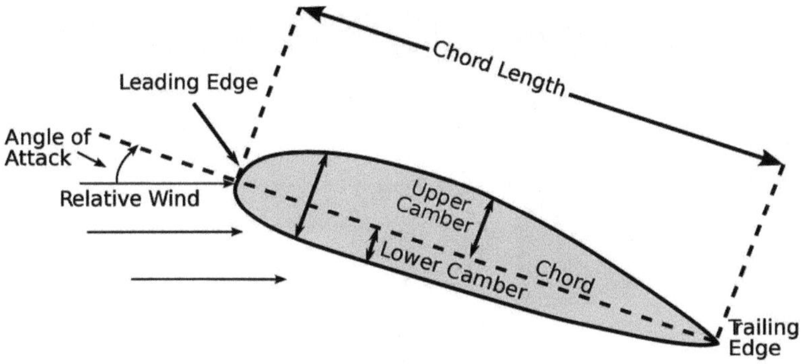

Fig. 11.54 Section of wing

polyester with a spacing (length of the threads) of 100 mm. In the first research the goal was to make curved forms by cutting out small pieces on one side of the distance fabric. After closing the gaps (by rejoining the cutting edges), one side had become smaller than the other. By removing the small parts of one surface, the surface on the other side showed a clear bubble with a circular section. This local bulging out of the fabric was already found by Goodyear and patented to combine the distance fabric flat mattress with integrated structural beams (Cross 1965).

The second experiment was extended by making a doubly curved model with the same distance fabric. The 3D model of the curved surface is generated with the software package EASY (Technet Gmbh, technet-gmbh.de). To model the inner layer, the outer layer made an offset of approximately 100 mm. Because the shape is not a sphere, the distance of each point to the centre is not the same and therefore the position of the inner layer is only approximated. The cutting patterns are determined by means of geodesic lines. To create the patterns, we used geodesic lines on one half of the surface. Afterwards they were mirrored to obtain the full cloth. Picture 15 shows the result of the 3D cutting patterns. The 3D patterns are transformed into 2D strips by using the Cut&Grow procedure of EASY. Finally, we got a set of individual strips were welded together to create the surface envelope. The strips are shifted as much as possible towards each other, which results in a pattern layout as shown in Fig. 11.55. The surplus membrane outside the edges of the cutting pattern were folded out and covered by a straight strip of fabric.

In this way we could avoid the bubbles at the outer surface. Pictures 11.56 and 11.57 shows the result of the experiment. The inside of the surface is smooth as expected. The outside is less smooth, which can be caused by the fact that scaling of the outer fabric was not completely appropriate and consequently the outer fabric was too large.

Fig. 11.55 Cutting of the distance fabric

Fig. 11.56 Inflated curved model

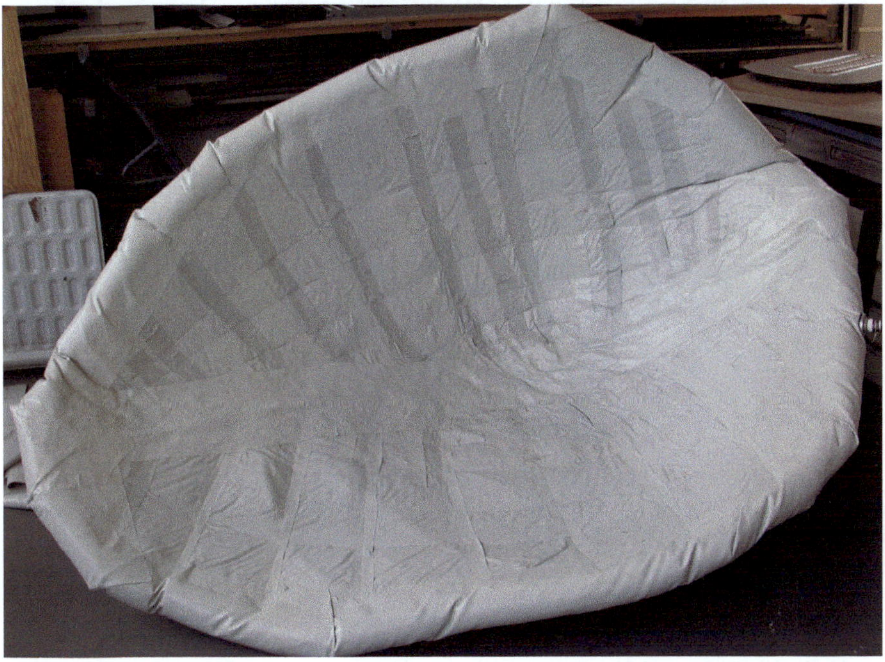

Fig. 11.57 Inflated curved model from below

The next experiments were done to investigate the problems that occur when making an irregular doubly curved surface with a distance fabric. Although it is technically possible to weave a curved distance fabric, there is no manufacturer for small quantities and irregularly curved surfaces (Fig. 11.58).

The first model was made by combining two surfaces made out of PVC polyester membrane strips covering both sides of a doubly curved polyurethane slab. After the surfaces of the distance fabric were stitched together to create a doubly curved shape, the polyurethane was removed with acetone. The inner surface was closed and inflated. The distance fabric was filled with concrete. Figure 11.59 shows the result (Fig. 11.60).

The second model is of a hyperboloid distance fabric as a mechanically stressed membrane. The two surfaces were connected by stitching in a flat position and were moved into position afterwards. The space between the membrane was filled with concrete and reinforced with straight bars. In this prototype the elasticity of the membrane allows the deformation that occurs when it is made into a doubly curved surface. The pretension in the membrane combined with a dense matrix of connecting threads provokes a smooth surface of the concrete shell (Figs. 11.61 and 11.62).

Fig. 11.58 3D PUR mould for weaving of distance fabric (*Source* T. v Engelen)

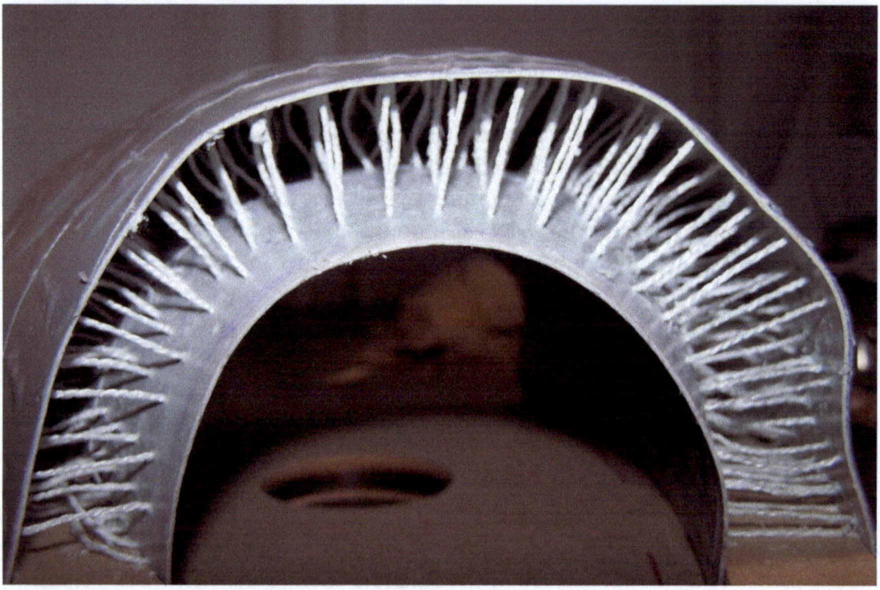

Fig. 11.59 Synclastic 3D fabric (*Source* T. v Engelen)

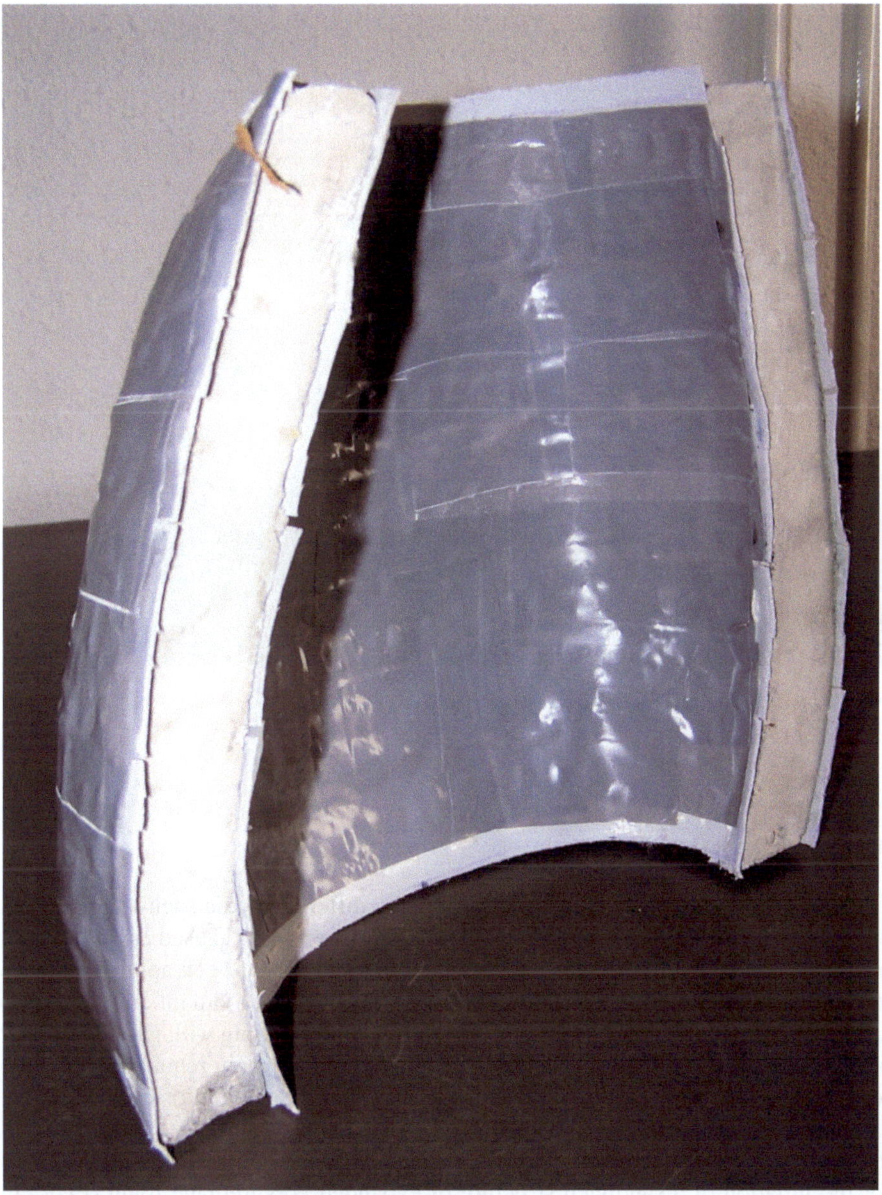

Fig. 11.60 Synclastic 3D concrete fabric (*Source* T. v Engelen)

Fig. 11.61 Curved 3D fabric (*Source* T. v Engelen)

Material

The inflation of a distance fabric can be done by different media such as gas, liquid or foam. A sandwich construction can be created when foam is used as an inflation medium. Research into the use of foam inflation methods has been applied in space industry since 1956 (Weiss 1956). In aerospace engineering, foam inflation provides a way of rigidizing the structure to prevent the need for inflating with gas over longer periods of time. A variety of foams can be applied but the main requirement for the foam is that it has a long reaction time to allow for controlled inflation of the product. The inflation strategy should be such that the entire product is filled with foam. In the mid-1970s Thomas Herzog with his students researched building applications for the inflation of a sandwich structure of expanded polyurethane foam (PUR) in combination with standard available membranes (Herzog and Bacon 1977).

The inflation with water was researched for cooling purposes (Pronk et al.2007) and sound insulation (Pronk et al. 2009). In this case the distance fabric was cast with concrete. The best-known concept of a double-layer membrane filled with concrete is the Bini shell concept. Since 1911, there have been some examples of connected double-layer membranes filled with concrete in civil engineering (Condie 1911; Chenoweth 1916; Store 1922). Other applications include the dome concept by

Fig. 11.62 Anticlastic 3D concrete fabric (*Source* T. v Engelen)

Zimmermann and Teichmann (2004), the work by West (2008) and Concrete Canvas by Brewin and Crawford (2005) (please see other case studies for more information).

For this case study we did 5 material experiments testing a distance fabric filled with concrete and different reinforcements: (1) plain concrete beam without distance fabric and reinforcement, (2) polyester fibre without additional reinforcement, (3) with glass fibres, (4) with aramid fibres and (5) with steel bars. The purpose of the test was to get an indication of the possibilities and outcome of the combinations. The membrane used is HEYtex H2586 and the concrete BEAMIX 104 (Figs. 11.63, 11.64, 11.65, 11.66, and 11.67).

Product	HEY-TEX 2586 (distance fabric)
Material distance fabric	Polyester
Length distance fabric	67 mm
Coating	PVC
Weight of fabric	780 g/m^2
Weight of total	1,900 g/m^2
Tensile strenght	± 3,700 N / 50 mm
Elasticity modulus	± 12,000 N/mm^2
Elongation at fracture	± 14%
Fracture stress	± 1,200 N/mm^2

Fig. 11.63 Properties of distance fabric (*Source* T. v Engelen)

Kind of concrete:	Cement
2250 - 2400 kg/m^3	CEM III / B 42,5 N LH HS
B 35	NEN EN 197-1 / NEN 3550
Strength: 35 N/mm^2	Fractures: NEN 5905
Classification: 1, 2, 5a	Maximum gains size ± 4 mm
Water-concrete factor: < 0,55 kg/kg	Super plasticizer NEN 3532

Fig. 11.64 Properties of concrete (*Source* T. v Engelen)

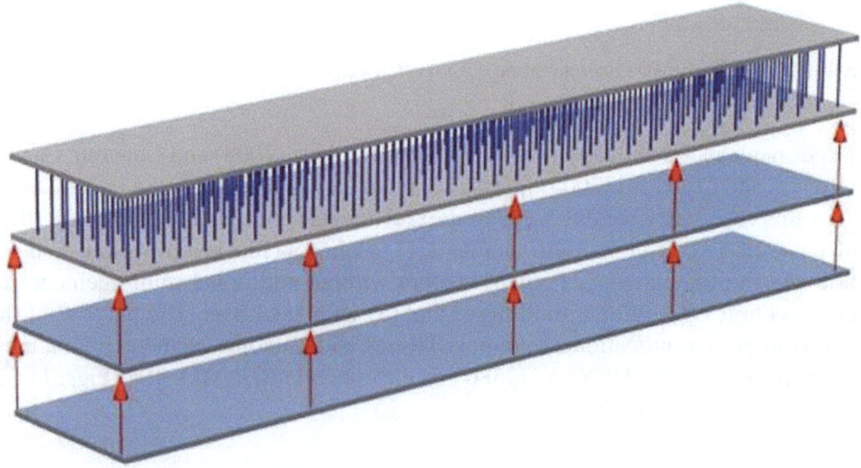

Fig. 11.65 Beam with aramid (*Source* T. v Engelen)

Fig. 11.66 Failure of beam with aramid (*Source* T. v Engelen)

Fig. 11.67 Concrete beam with distance fabric and two steel bars (*Source* T. v Engelen)

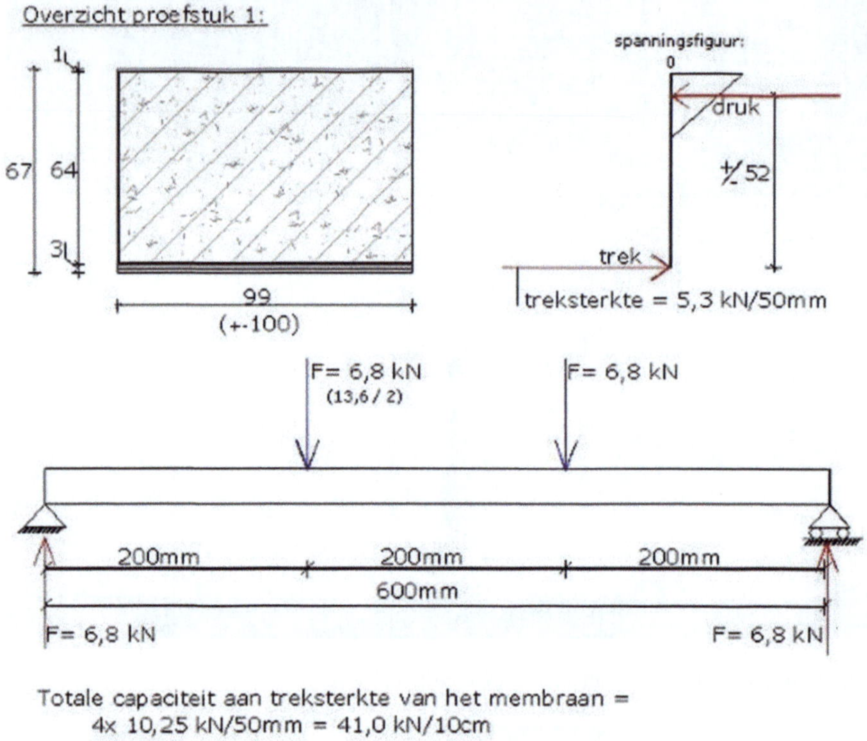

Fig. 11.68 Test of concrete beam with distance fabric and aramid fibre composite at the bottom (*Source* T. v Engelen)

A standard 4-point bending test will test the bending stress of a cross-section. Below, the results for Stress and Rupture with aramid and steel (Fig. 11.68).

The stress capacity of aramid is 41.0 kN. The stress capacity of the aramid rupture beam is down to 13.6 and 12.7 kN, because the rupture strength of concrete at 24.6 kN is about 60% of capacity of the aramid fibre (Young's modulus of aramid is 70–112 GPa). The beam with two 6 mm steel rebars ruptured at 18.4 kN. The stress capacity of steel is 24.6 kN. We found a rupture at 29.4 kN. This is about 115% of theoretical capacity of the steel bars (Young's modulus of steel is 200 GPa).

The adhesion between the concrete, distance fabric and additional reinforcing material is sufficient. The relatively low Young's modulus of aramid causes rupture of the beam for bending at 60% of the capacity of the aramid fibres. It is assumed that reinforcement with a material with a higher Young's modulus like carbon or dyneema would give a result with a higher efficiency in the use of the material. Also, a hybrid mix of steel bars and fibre reinforced polymers on the outside of the distance fabric is assumed to be an option. Both options could be used for the structure of buildings. In case of hyperbolic surfaces, the steel bars can be straight in two directions while the fabric would make a doubly curved surface.

Technique

With distance fabric there are two ways to keep the membrane straight during the rigidizing process. One way is to inflate the distance fabric; the other way is to stretch at least one side of de distance fabric by inflation or by mechanical prestressing. The other side of the membrane will be stressed by the hydraulic pressure of the concrete. In the experiments above, we chose the second way.

11.7 Flexible Casting with Vacuum Structures

Flexible Mould
The form-active structure: foil (Figs. 11.69 and 11.70).

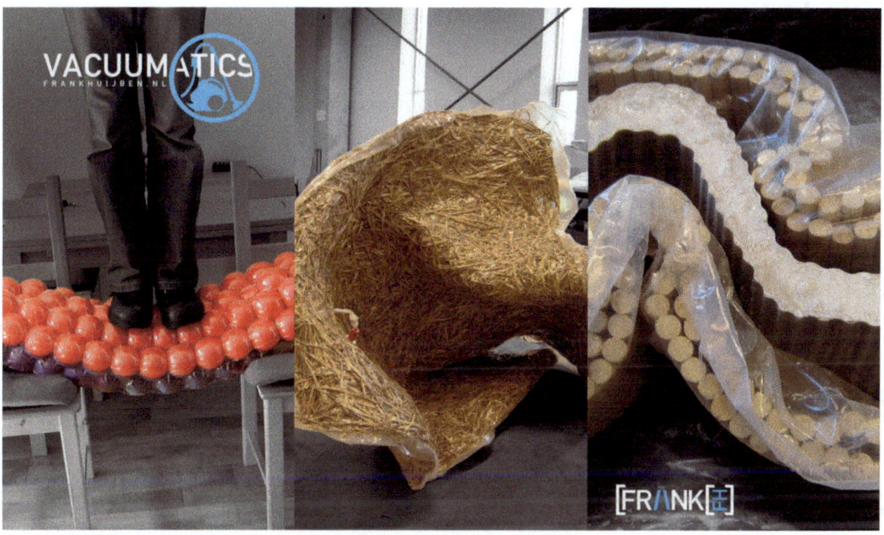

Fig. 11.69 Vacuum moulding (*Source* F. Huijben)

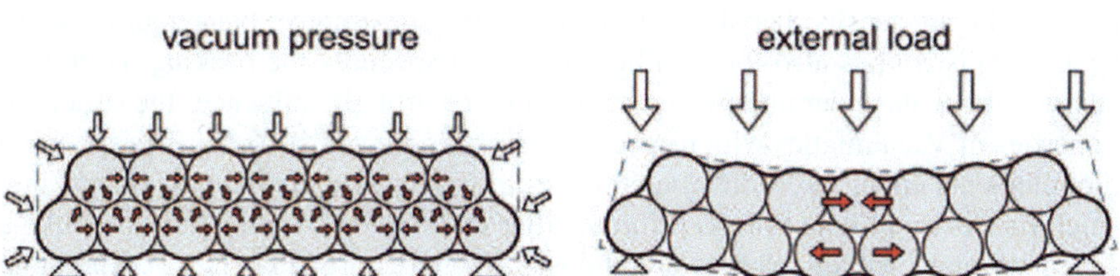

Fig. 11.70 Structural prestressing principle of Vacuumatics (*Source* F. Huijben)

Form-active typology: double layer.
Way to stabilize the mould: underpressure.

Composite Material
Technique to handle the rigidizing material: casting.
Reinforcement of the material: none.
Surface treatment: none.

Project
Name and location: PhD Research 'Vacuumatics 3D Formwork Systems'
Design and Engineering: F. Huijben (TU Eindhoven & ABT), F. van Herwijnen (TU Eindhoven & ABT) and R. Nijsse (TU Delft & ABT).
Contractor: Eindhoven University of Technology.
Year: 2014.

Introduction

As part of the PhD research 'Vacuumatics 3D Formwork Systems' carried out at Eindhoven University of Technology (TU/e), commissioned by ABT Consulting Engineers, the idea was put forward to utilize Vacuumatics as a flexible formwork system to produce customized 'free-form' concrete structures. Vacuumatics are prestressed structures that can be defined as the rigidifying of a flexible system of enclosed structure elements by extracting the air from inside the structure. The principle is similar to prestressing of concrete, with the important difference that the pressure of Vacuumatics is multi-directional and, when the underpressure is removed, the full flexibility of the original structure is restored. There are some related projects, like the underpressure bridge by the Delft University of Technology (Fig. 11.71) (Knaack et al. 2008). In this project, a material is packed into a pre-shaped bag. The ideal arch form can be found in an upside-down hanging position. After evacuation of the air, the construction has become stiff and can be turned right side up.

Material

The strength, stiffness and cohesion of the vacuum constructions will depend on the filler materials used and the enclosure. The enclosure (flexible skin) not only provides the essential air density, but also contributes to the constructive behaviour of Vacuumatics. This applies also for the filler material. Especially the packing fixation of the filler elements under multi-directional pressure strongly influences the structural properties of Vacuumatics with respect to their strength and stiffness. The spherical filler elements under multi-directional pressure will be arranged in a thick packing, which results in a high stiffness. Fibrous filler material, by contrast, behaves like a woven structure, which leads to a composite-like structure (Fig. 11.72). The large variety of configurations and combinations of the several filler materials is applicable to a wide range of architectural applications.

Fig. 11.71 Underpressure bridge in Delft (*Source* A. Borgart)

Fig. 11.72 Straw structure (*Source* F. Huijben)

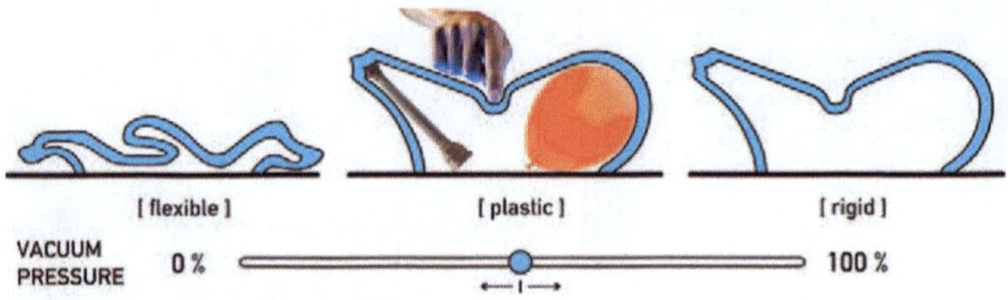

Fig. 11.73 Flexibility control (*Source* F. Huijben)

Form

The flexibility and adaptability of the shape are the main advantages of Vacuumatics. This makes it possible to create from one single system a wide range of three-dimensional forms. The adaptability of the form will be determined by means of something like "flexibility control" (Fig. 11.73). In the initial situation, without vacuum prestressing, the filler elements could move freely in the flexible enclosure. Then pumping the air from the enclosure (creating a near vacuum) improves the cohesion of filler elements and the construction shows plastic behaviour. In this phase, the vacuum construction could be easily formed, while it still has enough stiffness to maintain this new shape. A further increasing of the difference in air pressure (maximum possible atmospheric pressure) leads to a completely rigid construction, depending in part on the material properties of the filler elements and enclosure used. A unique property of Vacuumatics is the adaptive stiffening. The original flexibility of the construction returns when the underpressure in the construction decreases, so that the vacuum construction can be remodelled again and again.

In order to mould Vacuumatics literally into a special form, it requires a morphological treatment, that forces the filler elements into the desired position. Especially the possibility to locally adapt the geometry of the vacuum constructions to the new demands increases the potential of Vacuumatics as a custom-made construction element. The fact that Vacuumatics presents very clearly defined outlines of the filler material used and is able to adopt the geometry of a contra form (form fitting), results in a sensorial experience of the vacuum poured concrete sides. The creation of the different forms is also reflected in Chap. 2. This type of manipulation is classified as number 13 (Inflatable—Synclastic surface), number 16 (Pushed out element—All kinds of boundaries) and number 17 (Pushed in element—positive boundary curve) (Fig. 11.74).

Fig. 11.74 Form-fitting (*Source* F. Huijben)

Technique

The constructive operation of Vacuumatics is based on the principle that filler elements in a closed flexible enclosure will be multi-directionally prestressed. This form of prestressing arises when air is extracted from inside these flexible enclosures, which results in a uniform underpressure. The atmospheric pressure exerts a multi-directional force on the enclosure, whereby this skin is tightly wrapped around the surface of the filler elements. The filler elements will be compressed and fixed in their position, which leads to a rigid element with a certain strength, stiffness and cohesion. The prestressing so created causes the filler elements at the tensile zone of the construction to be pressed against each other when the construction is externally

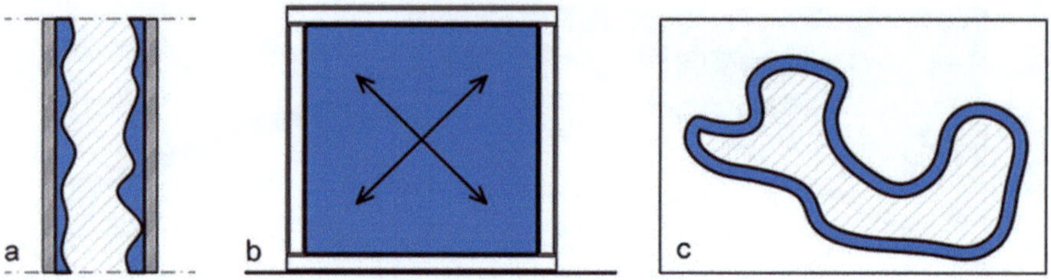

Fig. 11.75 Formwork techniques (*Source* F. Huijben)

loaded. This principle of a material with a low tensile strength and shear strength that will be prestressed to improve the bearing capacity is not new. The technique of the prestressing of Vacuumatics is in a certain way comparable with prestressed concrete, with the difference that de prestressing of Vacuumatics is multi-directional.

There are three types of formwork techniques of Vacuumatics, namely (Fig. 11.75):

Addition to standard formwork system

The vacuum construction is added to a standard formwork system and leaves a mark behind in the concrete surface. when the concrete becomes hard it's easy to demould by letting the air into the system. In contrast with the use of rubber profile mats, the Vacuumatics texture so created can be easily adapted by (arranging) the filler elements.

In-filled-frame formwork system

In the in-filled-frame formwork there will be two separate vacuum constructions attached to an external framework into which the concrete is poured. Because the edges of the frame are fixed, the connection of the concrete elements to the other construction elements is simplified, although the two flexible vacuum constructions could be produced individually or changed into new shapes. The vacuum constructions are easy to disconnect and reuse for other formwork systems.

Self-supporting formwork system

The self-supporting formwork system functions as a single closed vacuum construction as a self-dependent formwork into which concrete is cast. Besides the opportunity to change the complete geometry of the vacuum construction, it's also possible to edit the surface texture in line with the new requirements. Figure 11.76 is an example of what you could do with this type of formwork system (Huijben 2009: 78; 2014; Gilbert et al. 1970).

Fig. 11.76 Self-supporting formwork system (*Source* F. Huijben)

11.8 The Fabrication of Free-Form Concrete Segments Using Fabric Formwork

Flexible Mould

The form-active structure: foil (Fig. 11.77).
Form-active typology: double layer.
Way to stabilize the mould: underpressure/hydraulic pressure.

Concrete Material

Technique to handle the rigidizing material: pumping/casting.
Reinforcement of the material: none.
Surface treatment: none.

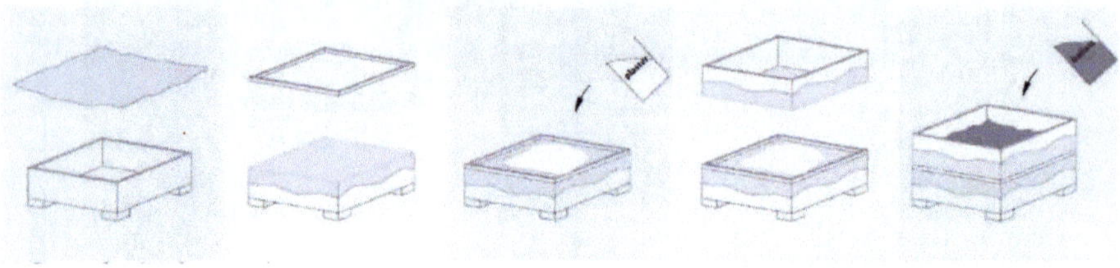

Fig. 11.77 Process of casting concrete segments (*Source* R.W.A. Verhaegh)

Project
Name and location: Free forms in concrete, Eindhoven University of Technology, the Netherlands.
Design and Engineering: Master thesis of R.W.A. Verhaegh.
Contractor: Mastertrack Structural Design: Mastertrack Building Technology TU/e.
Year: 2010.

Introduction

The goal of this research was to test if a formwork made out of prefabricated segments produced with fabric formwork can perform better than existing formwork systems in building free forms in concrete (Verhaegh 2010). The case-study project was the 40 m Spencer Dock Bridge in Dublin (Garcia 2010: 106–113) (SDB) (Figs. 11.78), this case study was the motive to start researching free-form concrete segments using fabric formwork.

Fig. 11.78 The Spencer Dock Bridge—Dublin, Ireland (*Source* W. Murphy)

Material

The case study bridge is made out of concrete. The purpose of this study was to make a form that is curved like the bridge but made with a different technique. Because no external construction is possible, the segments had to be able to resist the load of the pouring concrete. Reinforcing the concrete segments in the traditional way with a net of steel bars would introduce difficult problems in casting the segments. Such a net would have a complex form, which would make the steel fixer's job very difficult. After the structural concrete is poured and cured, it is capable of taking all of the loads bearing on the bridge. The strength and stiffness of the segment depend on four factors: form, thickness, material and reinforcement. The form of the segment is determined by its architectural design.

UHPC is a suitable material for casting segments for a number of reasons:

- The high compressive (and tensile) strength of UHPC offers opportunities of casting very light and thin structures. It allows omitting the traditional steel reinforcement or replacing it with steel or synthetic fibres. Also, a complex reinforcement net is not necessary so the segments can be thinner.
- The density of UHPC (Walraven 2004: 853–864) is much higher than that of ordinary concrete. This makes UHPC very durable as it has much fewer pores for aggressive substances to penetrate. It's also less easily polluted and is easier to clean.
- The surface quality of UHPC is stunning; it does perfectly mirror the formwork's surface.
- UHPC is a self-consolidating concrete, this means that the use of a concrete vibrator is no longer necessary. The use of a concrete vibrator is also not possible on a vulnerable fabric surface.
- UHPC has a low viscosity which means that it's possible to push the concrete from lower points to higher points in the mould (Fig. 11.79).

Beside these advantages of UHPC, there is also a downside; the price is about 20–30 times higher than the price of regular concrete anno 2010.

With the use of this UHPC, it is possible to create a concrete form with a high-quality smooth surface and a low thickness which is still strong enough to bear the weight of the construction.

Form

The design for the segment is a form with four equally curved edges, each edge is 800 mm long and has a symmetrical arc with a rise of 80 mm.

The form of the symmetrical arc was created in AutoCAD using the 'edge surf' command, this command creates a smooth surface based on four edges. The surface resulting from this command to produce four arcs is an exact cut-out of a sphere with a 1,080 mm radius. The radius of the curves is the same on all edges (because it's a segment of a sphere). Of course, this method/form has some restrictions; firstly, it's very hard to create a watertight mould, secondly, it's very hard to maintain an even thickness through the element.

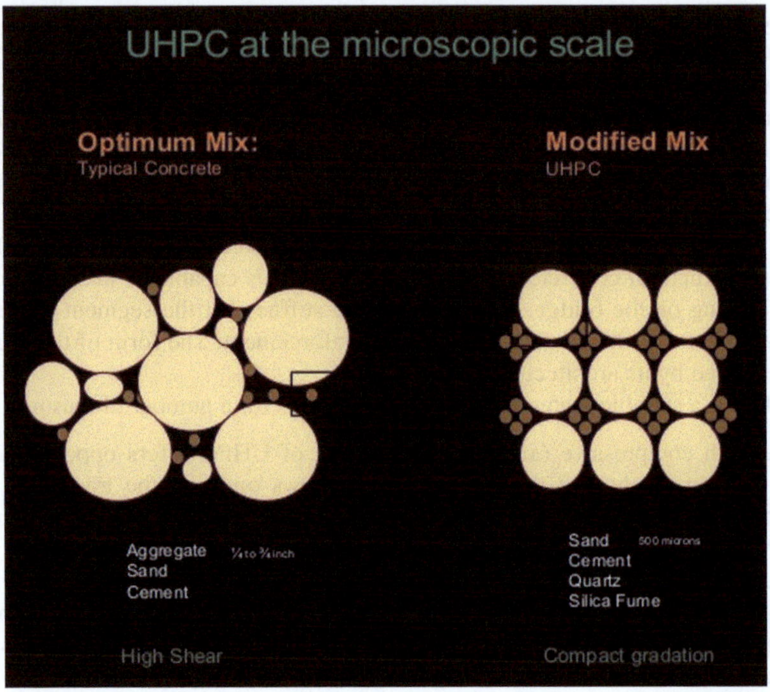

Fig. 11.79 The difference between UHPC and normal concrete (*Source* R.W.A. Verhaegh)

As can be seen in the concrete/membrane matrix this method is performed with a foil which is pumped under pressure to create the spherical form. This is a typical way to create this kind of segments which is quite unique. In Chap. 2 it is classified as a zeroclastic surface with a mechanically prestressed membrane and a pushed-in element with a parallel boundary curve (Figs. 11.80 and 11.81).

Technique

The process to make a thin curved form out of concrete is quite difficult, the challenges are to make a smooth curved mould on the bottom but also on top of the poured concrete. Also, the demoulding is difficult because of the curved shape.

The mould consists of a bottom mould, a spacer and a top mould. The bottom mould has to provide counter-pressure; therefore, it will be filled with water only (air has to be removed). This means that an air- and watertight connection between the mould and the foil has to be made.

The spacer is placed onto the bottom mould and determines the thickness of the edges of the segment, it also determines the distance between the bottom and top mould. The form of the spacer needs to be the same as the bottom mould to make a perfect fit.

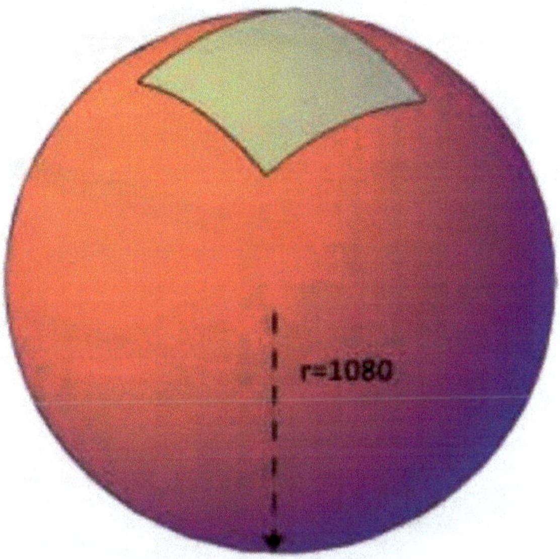

Fig. 11.80 Segment is cut out of a sphere (*Source* R.W.A. Verhaegh)

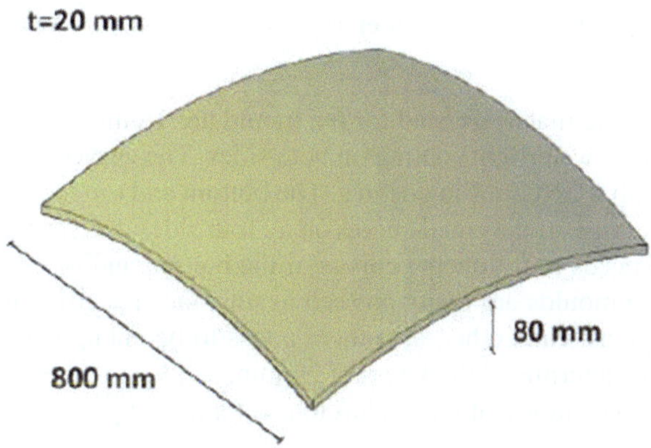

Fig. 11.81 Designed segment (*Source* R.W.A. Verhaegh)

The top mould fits perfectly on the spacer and has a layer of foil fixed into it, in this mould sand has to be poured to put pressure on the concrete and bottom foil (Figs. 11.82 and 11.83).

When creating spherical forms like the one in this case study, the biggest challenge is to make an air- and watertight mould which would also disassemble easily. To make this happen a rubber cord is placed between the bottom mould and the spacer, between this the foil is clamped (Fig. 11.86). To make sure that the rubber cord and the foils are pressed together a steel frame is designed which makes it possible to press the top mould and the bottom mould together on the spacer. When pressed together an air- and watertight connection is made possible by the rubber cord and the foils.

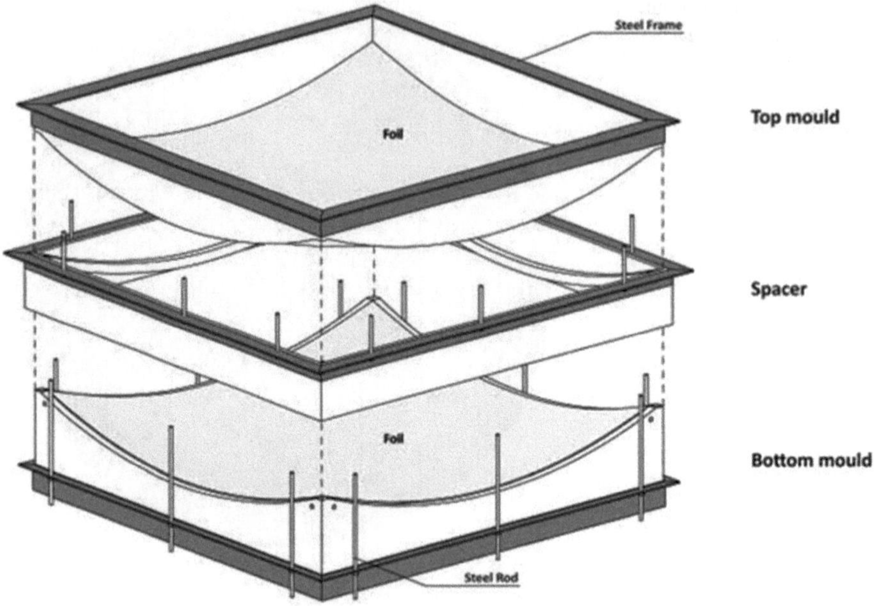

Fig. 11.82 Exploded view of the mould (*Source* R.W.A. Verhaegh)

The materials used for the mould are 18 mm concrete casting boards; this board had a watertight coating on both sides. The curved forms are milled from the boards with a CNC-milling device. The bottom and top moulds are relatively simple to mill because of the equal curves on all four sides, the spacer is more complicated because it needs to follow the curves of the bottom and top mould. Because the bottom and top moulds are being pressed against the spacer it's hard to let the edges not shear against each other, also there needs to be enough space to make the segment with the determined thickness of 20 mm.

The fabric of choice has to meet a few requirements:

- The fabric has to have a high elasticity, so that any wrinkles will be stretched out of it.
- It should be strong enough to handle the stresses; however, the counter-pressure of the water decreases the stresses in the fabrics.
- The fabric should be available in dimensions above 800 × 800 mm.
- The fabric needs to be watertight.

For this mould, ETFE-foil was chosen, turning the 'fabric formwork' into a 'foil formwork'. This transparent foil is very smooth so the concrete out of the mould will also be very smooth. At first the foil is taped to the edges, after adding the spacer the foil is attached with a clamped connection.

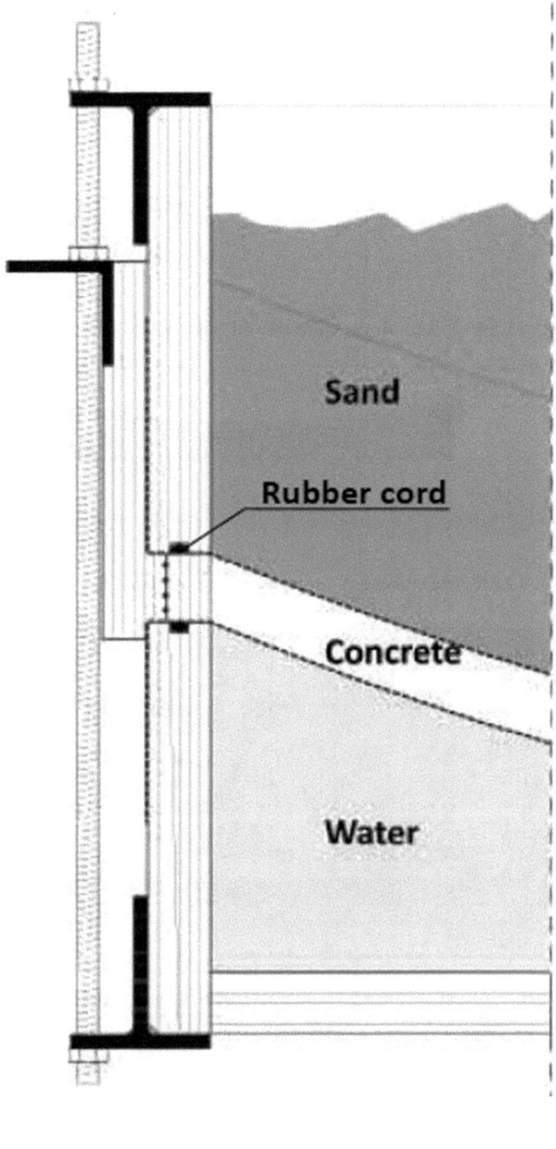

Fig. 11.83 Detailed view of the mould (*Source* R.W.A. Verhaegh)

The bottom mould is filled with water to create a good support for the concrete, when adding the water from a tub with a hose four tubes let out the air under the foil. This creates a full water bulb without any air bubbles (see Figs. 11.84 and 11.85).

After this process the concrete has to be poured into the mould, the thickness should be around 20 mm. Now the top mould has to be clamped on the spacer, and the top mould has to be filled with sand. Now the top of the concrete will also follow the curved form of the bottom foil. After about 20 h the segment can be demoulded (Fig. 11.86).

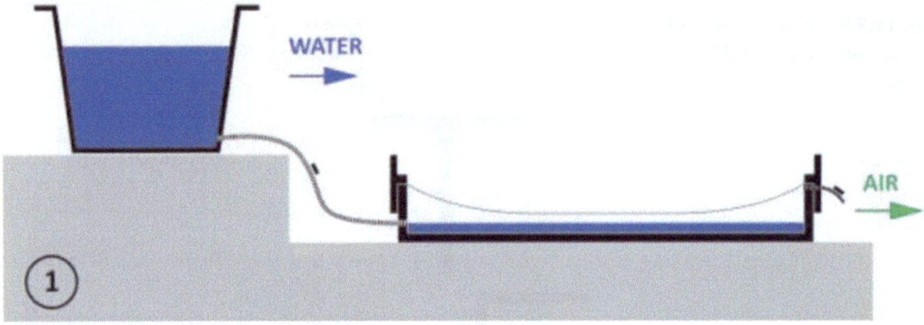

Fig. 11.84 Filling the bottom mould with water (*Source* R.W.A. Verhaegh)

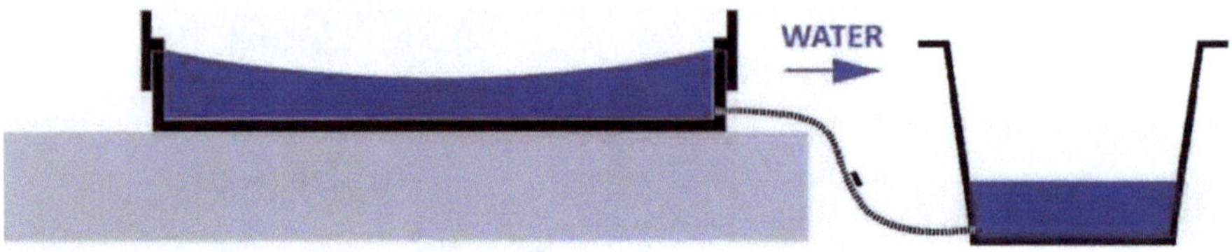

Fig. 11.85 Membrane in under-pressure (*Source* R.W.A. Verhaegh)

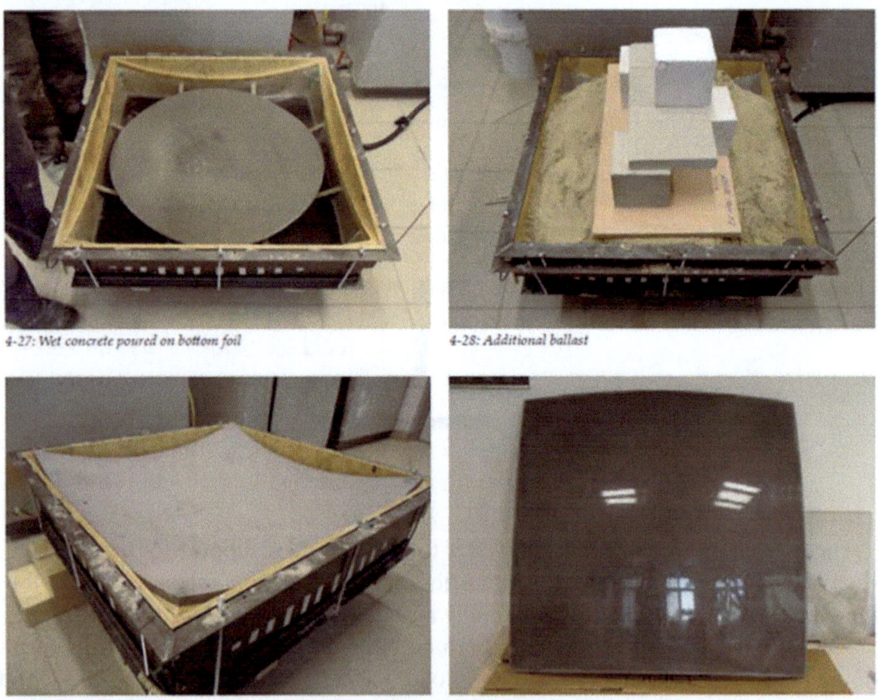

Fig. 11.86 Casting and demoulding process (*Source* R.W.A. Verhaegh)

This is a very difficult technique to perform with an optimum quality, however it is the best way to create a smooth spherical form in concrete with an even thickness. The downside of this technique is that it costs a lot of time to cast an element, so mass production is not really an option. Also, it is almost impossible to create segments with uneven edges. The downsides notwithstanding, it is a great and unique way to create perfectly smooth curved concrete surfaces.

11.9 Zero Waste Free-Form Formwork

Flexible Mould

The form-active structure: wax formwork: adjustable rubber foil/concrete: wax formwork.
Form-active typology: wax formwork: single layer/concrete: double layer.
Way to stabilize the mould: wax formwork: bending stiffness/concrete: wooden structure.

Materials

Technique to handle the rigidizing material: wax formwork: casting/concrete: casting.
Reinforcement of the material: wax formwork: none/concrete: iron rebar (possible).
Surface treatment: none.

Project

Name and location: Eidgenossische Technische Hochschule Zürich.
Design and Engineering: S. Oesterle, A. Vansteenkiste and A. Mirjan.
Contractor: Architecture and Digital Fabrication (ETH Zürich).
Year: 2012.

Introduction

To form a free-form concrete structure, a formwork is required. Through the different forms/shapes the formwork is usually for one-time use only and ends up as waste. Clearly, this new construction method saves construction material, which compensates for the additional formwork and waste. Or if a free-form structure is chosen for the shape, the additional formwork and cost can still be acceptable. To handle the weight of the liquid concrete, a lot of formwork is required. Besides the amount of waste that is created (the environmental impact) it is also a large part of the construction costs. These costs consist of the cost for the new materials, the cost for the manufacturing of the formwork and the cost for the discharge of the waste. To minimize the amount of waste created by the formwork and to minimize the cost, it would be ideal to use a reusable zero waste free-form structure material (Oesterle et al. 2012: 52).

Zero Waste Free-Form Structure

A zero-waste free-form structure uses a formwork which can be fully re-used or recycled to prevent waste and additional costs (new materials and the discharge of waste). A material with which it's possible to achieve this is wax. The material properties of the wax will be considered below. What are the possibilities that exist for the different types of shapes and curvatures? There are more techniques to fabricate the formwork, one possibility is shown in Fig. 11.87. In this figure, the technique is used with a mechanical adjustable mould to create the shape for the formwork. Other techniques for creating the wax formwork are also possible and these will also be explained. The main goal of a zero-waste formwork is to prevent waste and thereby reduce the cost of constructing the free-form structure.

Material

With regard to the materials, we shall look at the types of wax, the possibilities with the wax, the supporting structure for the wax elements and the type of formwork that is made.

Wax

For this type of zero waste free-form formwork wax is used. If a wax with a high melting point is used, there is a large chance of cracks in the formwork. If a wax with a low melting point is used, there can be problems during the use of the formwork.

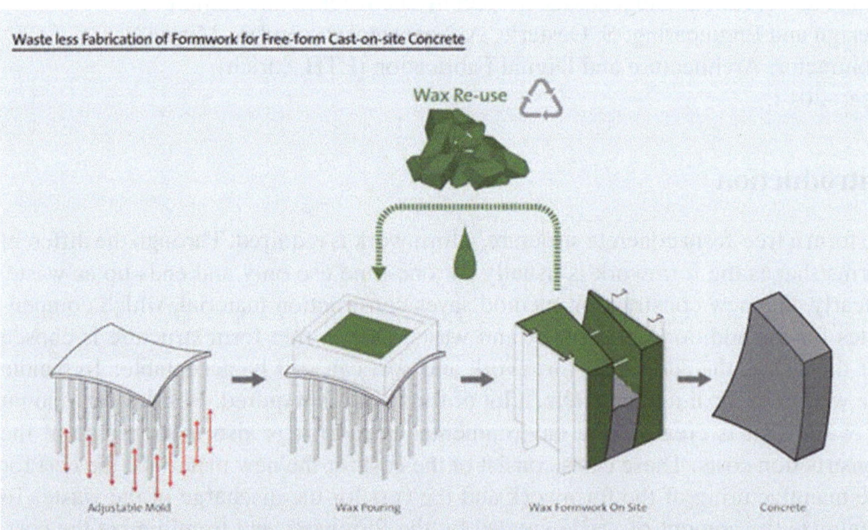

Fig. 11.87 Re-usable formwork without waste, for an on-site free-form structure production (*Source* S. Oesterle)

Such problems can be that the wax formwork melts or gets too weak through the heat of the sun. It's also possible that through the heat of the chemical reaction of the hardening concrete, the melting point of the wax is reached. To realize the best result, it's crucial that the formwork doesn't develop cracks, or deforms because of the wax softening. When the right type of wax has been chosen then the formwork can be made, see the chapter on Technique for further information. After the use of the formwork it's possible to melt the wax again and form it into the next shape. With this material it's possible to reuse the wax to prevent additional cost for the disposal of the used formwork and new material.

Wax Formwork

With the wax, the formwork is made to form and hold the liquid cast concrete. Because the concrete exerts great forces on the formwork it's necessary to strengthen the wax with a support structure. It's possible to use wood as a support structure. The structure has to be capable of withstanding the forces and weight of the concrete. When a free-form concrete structure is made, it's usually cast between two layers. The wooden support structure has to be made on the building site and onto this structure the wax elements can be mounted. After the production of the formwork, it's possible to put the rebar between the wax formwork elements and to cast the concrete to form the free-form structure.

Form

The wax can be cast into any form desired. When a larger free-form structure is required, it's possible to form multiple wax formwork elements and "melt" them together on the building site. The joins can be melted (sealed) together by using a heat source and some additional wax. If the joins are closed this will prevent seams in the concrete structure after casting. With his production method it's possible to create large free-form structures without the seams remaining visible.

Technique

There are two major techniques to manufacture the wax formwork. The first technique is with an adjustable mechanical machine which forms a rubber foil into the requested form. When the rubber foil is in the right shape the wax is cast on it, this technique is shown in Fig. 11.88. It's also possible to adjust the pins under the rubber foil manually, for this technique it's not absolutely necessary to have a high-tech machine. The second technique is to form the formwork for the wax with sand. This sand will be placed, shaped and compressed by a robot, with this technique it's necessary to have a high-tech robot to manipulate and form the sand.

After the production of the wax formwork it's transported to the construction site. On the construction site a wooden support structure has to be made to support the wax elements and the cast liquid concrete. The wax element can be mounted on the wooden structure. After the mounting of the wax elements, the joins can be

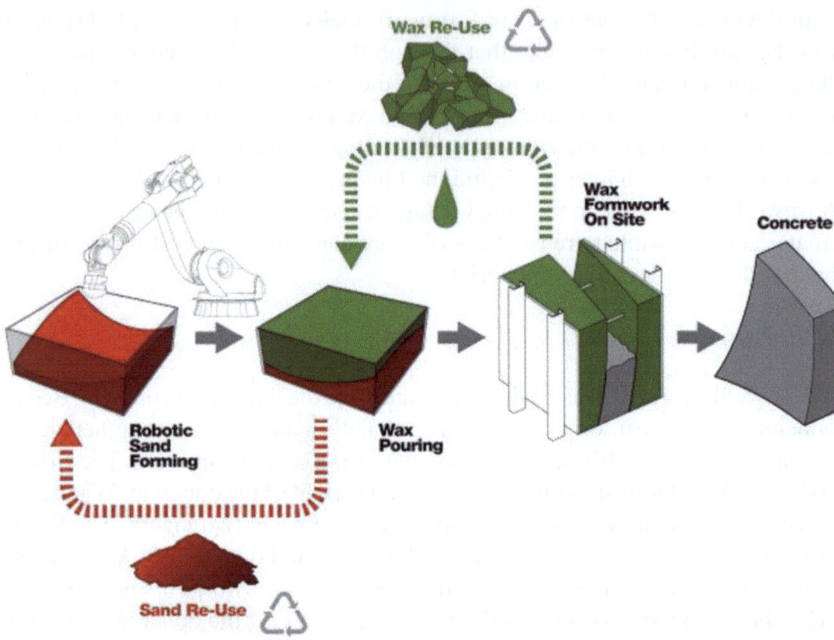

Fig. 11.88 Reusable formwork without waste, for an on-site free-form structure production (*Source* S. Oesterle)

sealed to prevent seams in the free-form concrete structure. The concrete will be cast between two layers of wax formwork. In between these two layers of wax elements it's possible to put iron rebar for reinforcing the free-form construction. After the rebar has been placed the construction can be closed, by placing the second side. When the formwork is completed the concrete can be cast. After the hardening of the concrete the formwork can be removed. The wooden support structure can be disassembled and reused straight away. The wax can be collected and returned to a factory for recycling (melting) and reuse in a new free-form formwork. With his technique waste will be prevented during the entire construction process.

11.10 Eco Dome

Flexible Mould
The form-active structure: coated fabric (Fig. 11.89).
Form-active typology: single pressure.
Way to stabilize the mould: inflate.

Fig. 11.89 Eco Domes in Sri Lanka (*Source* Solid House Foundation)

Composite Material
Technique to handle the rigidizing material: hand layup.
Reinforcement of the material: bars.
Surface treatment: coating.

Project
Name and location: Pottuvil, Sri Lanka.
Design and Engineering: Neff, Monolithic constructor.
Contractor: Solidhouse foundation.
Year: 1942–2009.

Introduction

Wallace Neff (1895–1982) patented (Neff 1942) the first concrete dome with an inflatable formwork. He developed a low cost airform house made by reinforced concrete cast on site over an inflatable balloon. Openings such as doors and windows were made by frames placed against the inflatable form. In 1942, he made a 4.5-meter diameter dome with a height of 3.35 meters as a result of experiments between 1934 and 1941. The system was used in the 1940s and 1950s. He built over 1,200 domes mostly in West Africa, Egypt and Brazil. He reported improvements in his construction method in a 1945 patent (Neff 1945) and in a 1953 patent (Neff 1953).

The SolidHouse Organization based their design on that of the EcoShell I, patented by David B. South of the Monolithic organization in 1977 (South 1979; Monolithic, www.monolithic.com). They renamed it SolidHouses, to impress the client with the safety factor they receive when choosing domes rather than regular shelters. Compared to the construction principle of Monolithic Domes, EcoShells are, in general, not insulated and therefore—when used as a residence—not suitable in colder climates. The difference between an EcoShell I and an EcoShell II is in the fact that in case of an EcoShell I the shotcrete is applied on the exterior of the inflatable and in case of the EcoShell II the Shotcrete is applied on the inside of the inflatable. In order to explain the construction principle of SolidHouses, an EcoShell project in Sri Lanka has been chosen. This extensive facility, located in Pottuvil Sri Lanka includes a complex of monolithic EcoShell I domes as long-term rentals. The EcoShells are constructed with an inflatable mould. In the following illustrations the construction principle is given (Figs. 11.90, 11.91, 11.92, 11.93, and 11.94).

Fig. 11.90 Foundation (*Source* Solid House Foundation)

Fig. 11.91 Ground floor (*Source* Solid House Foundation)

Fig. 11.92 Inflatable mould with reinforcement (*Source* Solid House Foundation)

Fig. 11.93 Window and door (*Source* Solid House Foundation)

Fig. 11.94 Finished structure (*Source* Solid House Foundation)

Material

Except for the more common building materials used in the above-mentioned method, the flexible moulding material is characteristic of this building method. An EcoShell I is a concrete thin shell dome whose construction process includes a relatively new technique called inflatable moulding.

The inflatable used in the construction of an EcoShell I is removable and reusable. With proper care, it can be used 100 or more times, and that feature significantly reduces construction costs for projects involving more than one structure.

For an EcoShell I, concrete is layered onto the exterior of the inflated mould. Usually, the shell is not insulated, but can be by including polystyrene beads, vermiculite or perlite into some of the concrete as it's applied. This insulation and the thermal mass of the concrete does provide some thermal protection.

The EcoShell I is a super-strong structure that can withstand hurricanes, tornadoes, earthquakes, fire, termites and rot.

Form

This technique uses an inflatable that's made of a high-strength, high-tech fabric. When inflated, the inflatable mould, looking like a semi-rigid balloon, creates the EcoShell's dome shape. It's made of PVC-coated nylon or polyester fabric, available in several weights. The material of the shell consists of concrete, whereas the form-active structure consists of coated fabric. The mould is regulated with single pressure. The aforementioned pressure is induced by inflating the mould. The addition of the concrete is done by hand layup. The concrete is reinforced by placing bars on the inflatable and finally the concrete shell is coated on the outside.

Technique

When constructing an EcoShell, proper inflation of the inflatable mould and continuous regulation of air pressure are as important as clean foam and Shotcrete application. The measurement of air pressure referred to here will be in inches of water column. This is the amount of air pressure differential needed to raise water in a column.

Since the inflatable is most vulnerable when partially inflated, inflating should not be done in strong winds—not more than 10–15 mph. Less than 5 mph is certainly desirable. If the mould is inflated during mild windy conditions, it is extremely important that all exposed surfaces on equipment inside the inflatable be padded.

For small EcoShells, inflation usually takes less than 30 min. During inflation, continual inspection of the inflatable is absolutely essential.

In a nutshell, here is how to properly inflate the inflatable mould:

- Inflator fans are started and the inflatable mould is being inflated. Air pressure must be regulated. As the mould inflates, the easiest way to adjust the pressure is by opening the airlock doors. When the inflatable first becomes tight, the inside air pressure should be held at a minimum pressure of about one to two-tenths of an inch of water column.

- The inflatable mould should then be checked for weak spots, holes, etc. Inflatable mould tie-down should then be completely checked.
- At this point, a check should be made on the pressure gauge. The pressure gauge (manometer) can be as simple as a clear drinking glass, half-filled with coloured water, and a small (1/8″) clear tube running from the water, through a hole in the inflatable, to the outside. The additional air pressure inside the inflatable will force water up through the tube to a level higher than the water in the glass. The difference in vertical distance is the actual air pressure reading in inches of water column.
- Air pressure can then be gradually increased until it measures two inches of water column. While this is being done, the inflatable and tie-downs should be checked for any problems (An EcoShell is usually built to withstand 6″ of pressure). Whenever possible, let the inflatable stand at least 12 h before continuing construction. This gives it time to stretch.

11.11 Monolithic Domes

Flexible Mould
The form-active structure: coated fabric (Figs. 11.95 and 11.96).
Form-active typology: single layer.
Way to stabilize the mould: inflate.

Composite Material
Technique to handle the rigidizing material: spraying.

Fig. 11.95 Sketchup model of a monolithic dome house (*Source* J.C. Eneroth)

Fig. 11.96 Pensacola Beach house, by Julia Christina Eneroth. Florida (*Source* J.C. Eneroth)

Reinforcement of the material: bars.
Surface treatment: drape.

Project
Name and location: Monolithic Dome Institute, Italy.
Design and Engineering: David B. South and his brothers Barry and Randy.
Contractor: Monolithic Dome Institute.
Year: 1976.

Introduction

The Monolithic Dome technique is a method to realize dome shaped buildings. The Monolithic Dome can be attributed to David B. South (1939–) and his brothers Barry and Randy. When David B. South was a teenager, he heard architect-inventor Buckminster Fuller speak about the innovative geodesic dome that he had developed. South began experimenting and built the first Monolithic Dome in Shelley, Idaho in 1975. This dome was used as a potato storage facility and is 32 m in diameter and 10.7 m high. The largest Monolithic Dome in the world at this moment is the home of Faith Chapel Christian Center in Birmingham Alabama. This dome has a diameter of 85 m and is 22 m tall. The floor area of this dome is 6,920 m^2 in two levels.

Today there are more dome shaped buildings in existence, for personal and public use, which have been built with this technique. Monolithic Domes are constructed according to a method that requires a tough, inflatable airform that is connected to the foundation. The inflated air form is the basis for the dome. The Monolithic Dome is built on the inside of the airform. Polyurethane foam is sprayed against the formwork. The reinforcing rebar is then attached to the foam by rebar hangers. Finally, shotcrete is sprayed against de foam. The airform remains on the outside and can be finished with an exterior coating. The Monolithic Dome gets its strength from the natural synclastic shape.

Material

The inflatable airform is the base material to realize a Monolithic Dome. The airform is made of PVC-coated nylon or polyester fabric, available in several weights. Fabrics for air forms must be evaluated to meet the requirements of strength, elasticity, fabrication, ruggedness, durability and desired surface characteristics. Strength requirements are dictated by the air pressure, shape and size of the structure. Durability is important because the air form must be able to survive the construction process and must be able to withstand degradation by ultraviolet rays and weather exposure. The capacity of the fabric to resist the tension, pressure and seal strength has been tested in the factory. Fabric tension in a pipe or cylinder is calculated using the formula Tension = Radius (measured in inches) × Pressure (measured in PSI). For homes, 18 oz. of fabric per square yard is used, for larger domes 28 oz. of fabric per square yard to 40 oz. per square yard is used.

The inside of the air form is covered with polyurethane foam. For the polyurethane, two-pound density foam is used. If a lighter density foam is used, it will not be strong enough to hold the rebar and provide structural integrity with the building's shape. If a higher density foam is used it will give a stronger outside surface but a lower insulation value. It is also possible to strengthen the dome with shotcrete on the outside of the air form. The main construction materials of a Monolithic Dome are concrete and steel. Concrete has compressive strength; steel has tensile strength. Concrete can resist compression or squeezing, steel can withstand the tensile forces. The rebar is suspended from the rebar hangers which are anchored in de foam. In general, #2 (2/8″ or 1/4″) rebar with Grade 60 (or 60,000 PSI) is applied in the construction of Monolithic Domes. This rebar can be freely formed as the dome itself has a perfect shape for strength.

The concrete with a 4,000 PSI for the dome shell can hold 4,000 lb per square inch before the weight causes it to lose structural integrity. Because the concrete is sprayed against the foam, the aggregate has to have a diameter of three eighths of an inch or smaller. For the dome it is important to use the correct water/cement ratio. Different additives are needed for the shotcrete. The first additive is an accelerator. This will make the concrete set and rigidize faster. The second additive is a liquefier. The liquefier makes the concrete mixture more liquid and manageable without requiring more water. The third additive is fibres. These are needed to reduce the effect of thermal and shrinkage micro cracks in the concrete.

Fig. 11.97 Caterpillar domes

Form

The surface of the Monolithic Dome is curved equally in all directions. This shape, a synclastic surface, is made by inflating a membrane. The inflation process determines the shape of the concrete. This structure can be combined with a monoclastic surface. It is also possible to combine it with a tensile structure (anticlastic surface). By combining these 3 fundamental shapes it is possible to produce an infinite number of combinations and building shapes (Fig. 11.97).

The final shape of the building can be altered with minor adjustments. These alterations can be done to give the building its own unique design. For normal window and door frames and other openings a special frame is connected to the air form. This frame will move into its desired position when the air form is inflated. The frame is not covered with concrete.

This will leave an opening in the dome when the air form is removed. A window frame can then be used to make normal windows in the dome. With this technique it's possible to make bay windows or dormers in the Monolithic Dome. When larger shapes are needed, they are connected to the air form. Many different shapes and sizes of domes are possible because of the many possibilities of combination of building techniques (Zimmerman 1988: 41–58; Domebuilder, www.domebuilder.com). Extra shapes and forms can be given to the structure by putting elements in or on the air form.

Technique

A concrete ring foundation reinforced with steel rebar is the foundation of a Monolithic Dome. The reinforcement that extends outward from the foundation is connected with the Monolithic Dome. The custom factory-made airform is unfolded and attached to the foundation. Inflation of the airform is done by the means of electrical fans. This process will take up to 30 min for a small dome. When the airform is inflated, the builders go inside through airlocks. During and after inflation continual inspection of the airform is very important. It is also important that during the application of the foam and the reinforcement of the concrete the air pressure remains constant at 2″ of water column pressure. The two-pound density foam is sprayed against the inside of the air form in a three-inch thick layer. The rebar hangers are pinned into this foam layer. The mesh can be hung on these rebar hangers. The two-pound density of the polyurethane foam ensures that the connection is strong enough. The rebars are tied together with steel wire. It is recommended to let the bars overlap when they are tied together, but to leave at least two bars of free space in between so that the concrete can get in between. The shotcrete is a special high-strength spray mix of concrete. The steel rebar is embedded in the concrete and when about three inches of shotcrete is applied, the Monolithic Dome is finished. For structural integrity, the dome is thicker at the bottom. When the concrete has set, the electrical fans can be removed. A coating on the outside of the airform provides for protection. The inside of the dome can be coated optionally. The water-tightness of the connections is a determining factor for the insertion of windows and doors.

Conclusion

The synclastic shapes give the Monolithic Domes a strong and minimal construction. The reinforced concrete and the foam are applied on the inside of the inflated airform. This method creates a strong, weatherproof and durable building which is capable of withstanding any climate. The inflated airform gives the dome a synclastic surface. The air form is specifically customized at the factory and by combining different elements and buildings, the possibilities will be infinite. It is difficult to build within exact specifications with this technique.

11.12 Binishell

Flexible Mould
The form-active structure: textile composite (Fig. 11.98).
Form-active typology: double layer.
Way to stabilize the mould: inflate.

Composite Material
Technique to handle the rigidizing material: casting.
Reinforcement of the material: steel bars.
Surface treatment: drape + polishing.

Fig. 11.98 The Margaret Preston Hall, or Bini Shell, as pictured at Ku-ring-gai High School in 2019 (*Source* Wikimedia)

Project
Name and location: Dome by Bini, Binishells.
Design and Engineering: Dr. D. Bini, Universita degli Studi di Firenze (Italy).
Contractor: Binishells S.p.A.
Year: 1967 First construction, Patent 3462521 12-DEC-1966.

Introduction

The dome by Bini is a patented (Bini 1969) concrete forming technique dating back to 1966. The technique is developed by Dr. Dante N. Bini from the Universita degli Studi di Firenze in Italy (Bini 1967, 1974). The dome by Bini is a product called the Binishell. The technique is based on the erection of a dome constructed with reinforced concrete. The building production process involved is a low-cost technique with minimal material, equipment and labour needed. The inflatable structure will always follow the shape of the air pressure. Binishell domes (Fig. 11.99) require specific reinforcement to obtain their final shape. A major benefit of this technique is the construction time. A dome can be erected within four days, depending on its size and shape. The concrete is poured on a membrane lying on a prefabricated foundation. The dome can be built using one membrane layer. It depends on the specifications and the location of the project if a second membrane is required. In some climates or due to product sizes, the outer membrane is needed to keep the concrete in place while it hardens. After the concrete has set, the inner membrane is removed. The optional second membrane can be kept in place to provide for a roofing cover.

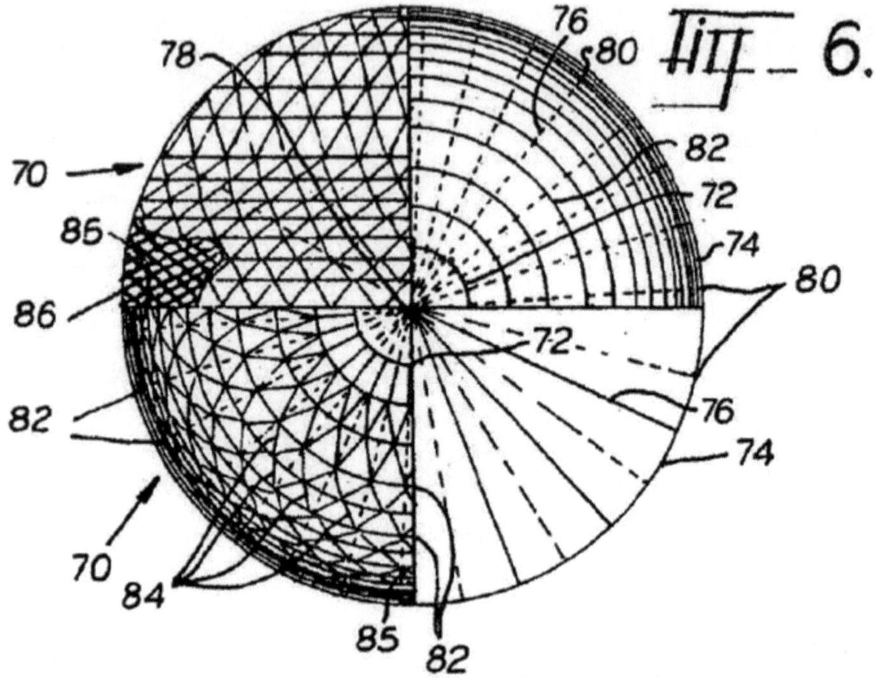

Fig. 11.99 Reinforcement mesh design, US-Patent 3,462,521- p. 2

Material

The Binishell dome consists of construction concrete with steel reinforcement. The concrete mixture contains a higher percentage of small particles than regular concrete. The aggregates can be up to 0.5 inch in size. The aggregates can be acquired locally for the specific project or they can be artificial. Preferably light weight aggregates are used to optimize the structure. It has a high slump rate of up to 6 inches and it is particularly fluid. Setting of the concrete starts within 1 h after the concrete has been poured on the membrane. The concrete provides for both a mouldable and a low-cost building material. Another advantage of the application of concrete is that the shell structure is weatherproof by itself. Depending on the climate it can be required to insulate the dome. An insulated Binishell can be made using the double layer technique. The insulation is not necessary structurally, however. When reinforcement is applied, it consists of a steel mesh. The mesh is specifically designed and documented in the 1966 Patent by Bini. The reinforcement has a specific design (Figs. 11.99) because it is important that the steel bars must be able to follow the shape of the dome before, during and after the inflation of the membrane. For all domes, double reinforcement steel bars are required in the boundary outline. This boundary must be reinforced to maintain the circular (or other) desired shape of the dome while inflating. These double bars are also used around openings cut in the dome. With the use of the steel reinforced concrete, domes with a diameter over 37 meters (120 feet) have been made.

Form

The thin concrete shell of the Bini domes illustrates a natural structural shape. The natural shape is created by the combination of gravity and air pressure while the dome is inflated. The shell is a perfect roofing solution due to the shell being completely self-supporting without any interfering columns or bearing walls. One of the advantages of the Bini dome is that no manual shaping of the concrete is required. The shape is contained by the interaction of the reinforcement and the membranes. The air pressure forces the dome to its final shape. A disadvantage is that when the membrane is inflated, the concrete cannot be controlled anymore. The shape has to be defined prior to the moulding process. Deformations of the final form will occur only after the concrete has been poured and the shape is inflated. Gravity will make the fluid concrete sink to the boundary lines of the shell. The shell will therefore become wider around the boundary and thinner at the top. This effect causes the shell to vary in thickness across its surface. The double layer membrane reduces this effect but does not counter it completely. The top membrane only makes sure that the concrete cannot lose its roughly intended shape. The exact shell thickness can therefore not be controlled by the membranes. Deformations in the concrete surface can also be due to the setting of the concrete. Some parts of the concrete may be less or more fluid than other parts. This will cause parts to shift while the concrete sets. The concrete can be sprayed to reduce the effect, but again the effect cannot be countered completely. Traditionally the Binishell dome was designed to create circular dome shells. The patented technique describes primarily the circular Binishell shape. The design of the steel reinforcement bars allowed for only the circular shape. By modifying this design, it was also possible to construct square, trapezoid or hexagonal shapes. The hexagonal shape has been further developed into the Binix system. The Binix system allows for domes to span from 40 up to 64 meters. The essence of the Binix design differs from the Binishell. The Binishell is a Monolithic Dome structure. The Binix system is constructed by a concrete wireframe. This wireframe consists of equal triangles.

Technique

Binishell domes are basically concrete shells. The first step for the construction of a Bini dome is excavation of the foundation. The outline of the shape must be weighted and needs to be finalized before the membrane is placed on top of it. The area inside the outline will become the floor. The floor will generally be a normal concrete floor. The concrete floor is poured together with the foundation beam. An air tube is placed underneath the foundation and floor. This tube will be used later on to inflate the membrane. The diameter of this tube can vary depending on the air pressure source. As is described in the patent, only a minimum of air pressure is required to inflate the dome. Air pressure in the order of 0.03 atmospheres is enough for inflation to occur. When the foundation concrete has hardened together with the floor, the first and inner membrane layer is spread out over the shape. It is imperative that it is completely smooth without any folds or wrinkles because it will stay visible in the final shell. The outlining is then reinforced with double steel bars. Double steel bars are required because the outlining will define the overall shape of the dome. The rest of the surface is covered with predesigned reinforcement steel. This wire frame of steel bars is able to move radially. This means that when the membrane is inflated, movement of the radial steel bars allows the shape to be erected and guides it to its dome shape.

The concrete can be poured when all of the reinforcement steel is in position. When pouring the concrete, it must be spread evenly over the shape within the boundary lines. Another provision of this building technique is that it requires but a minimum of surface on the building site for both the construction and the foundation. Desired openings in the shell can be cut out of the reinforced concrete afterwards. It is also possible to do that when the concrete has set or when the dome is complete. When larger openings in the shell are required, it is mandatory to further reinforce them with double steel bars (Figs. 11.100, 11.101, 11.102, and 11.103).

Cutting of the openings can be done with a circular diamond blade saw. A disadvantage of the organic shape is that it is hard to position the openings exactly.

The quality of the concrete shell should be the same over the entire surface. If there are any minor deformations, they will be visible in the finalized dome. If the design demands it, or the external climate, a second and outer membrane can be used to weatherproof the construction. The Binishell dome is now ready to be inflated. Air pressure can be generated by a simple blower with just 5 hp. After 24 h the outer membrane can be removed. Now the surface quality and overall thickness of the shell

Fig. 11.100 Preparation close-up

Fig. 11.101 Preparation overview

Fig. 11.102 Inflation of the mould

Fig. 11.103 Cutting of the opening

must be checked. Any deformations can lead to structural integrity flaws and need to be addressed before the dome is put to use. By spraying water on the concrete, the aesthetic surface quality can be increased. After another 24 h the concrete will have set enough to be self-bearing. The inner membrane can now be deflated and removed. The membrane system must be cleaned before the concrete has completely hardened. When the membrane is clean it can be used again. Openings can then be cut into the dome. Windows, doors and other openings can be cut out of the shell with a circular saw. The nature of the dome shape has a disadvantage in that the position of openings in the dome cannot be specified. Due to deformations in the curvature of the dome, the openings have to be made to fit after the dome has been erected.

The surface of the shell must be cured while the concrete sets. To do this, a polishing tool is used. This device is connected to the top of the dome with a cable. The vibration of the polishing head makes sure that the device travels along the surface of the dome by itself. This is primarily used to close the pores by compacting the concrete. Only when the concrete has completely set can the device be turned off and the cable removed from the shell. This technique can only be used on domes. After 3 days, the inner membrane is deflated and the dome is now ready for further processing.

Conclusion

The Binishell domes are a low-cost technique to build and construct concrete domes. The technique makes efficient use of the implemented materials and available surface area. The process of construction only takes 4 days and a minimal number of workers are needed. Construction time of a Binishell can fit within 24 h. Preparations and prefabrication take up the majority of the total construction time. The created dome can be built for all climates and weather conditions. The simplest form is the single layer concrete shell. Where this is needed, there is the double layer and insulated concrete shell with only a minimum increase of labour and materials. With this technique it is difficult to build to exact specifications. This is also true for the thickness and quality of the concrete shell. Windows, doors and other openings have to be cut out of the shell by hand.

11.13 Membrane Concrete Grid Shells

Flexible Mould
The form-active structure: coated fabric (Fig. 11.104).
Form-active typology: double layer.
Way to stabilize the mould: inflate.

Composite Material
Technique to handle the rigidizing material: pumping.
Reinforcement of the material: single fibres.
Surface treatment: drape.

Project
Name and location: University of Kassel.
Design and Engineering: Ing. Gregor Zimmermann.
Contractor: department Structural Design at the FB6 Architecture, Urban- and Landscape Design—University of Kassel.
Year: 2007.

Introduction

Membrane Concrete Grid Shells (Zimmermann 2007; Kassel Institut für Konstruktiven Ingenieurbau 2007) (MBG—Membran Beton Gitterschalentragwerke) are an invention of the department for Structural Design at the FB6 Architecture, Urban and Landscape Design of the University of Kassel. In the nineteen-fifties, improvement of materials gave traditional materials new applications and created new ideas for engineering with geotextiles. Richard Fearn, Kenzo Unno and Mark West used the developments as a source of inspiration for architectural applications and this led to the fact that the technique is still used by the company Fabriform. The invention of Membrane Concrete Grid Shells combines the advantages of these traditional technologies, like membrane formwork, and new materials, like ultra-high-performance

Fig. 11.104 MBG construction assembly (*Source* G. Zimmermann)

concrete, which leads to fast buildable, economic and ecological structures. It is a redeveloped construction method based on air-halls, inflatables and grid shells of ultra-high-performance concrete (UHPC).

Ultra-High-Performance Concrete, developed in France during the 1990s, is a material with great strength. By using UHPC for the structures, the amount of concrete needed to cover a construction can be reduced. By using the membrane formwork with a grid shell, the use of concrete can be reduced up to 60%.

With a special technology, it is possible to create 'UHPC' with local resources, so the material doesn't need to be imported. This 'green concrete' with the use of local materials leads to a decrease of 30% of the cement needed, also a completely different self-hardening material can be used.

Compared to the 'traditional' Monolithic Domes, this invention makes it possible to generate new designs that weren't possible before. Figure 11.105 shows parameters for the formwork membrane. Inflatables and membrane formwork are traditionally used for building monolithic, spherical domes. This means that the form giving construction with steel or wood isn't needed anymore and costs and time are saved.

Membrane Material

The materials that are used by Gregor Zimmermann in this case are:

PVC-coated fabric, air, Ultra High-Performance Concrete (UHPC), midget-fibre reinforced.

Concrete is obtainable in many different compositions, in the case of MBG structures, concrete with integrated reinforcement should be used. The idea behind the application of the extremely expensive Ultra High-Performance Concrete is the reduction of weight in order to minimise the concrete load during the assembly of the MBG inflated system. Another aspect in favour of the application of UHPC is to minimize the amount and thickness of concrete for structural and architectural

Fig. 11.105 Detail of grid shell with membrane concrete (*Source* G. Zimmermann)

design aspects. This type of concrete, the UHPC, is developed especially for the MBG structures by a research cooperation of the University of Kassel. An analysis of the material shows that the Ultra High-Performance Concrete is self-compacting, has a high Young's modulus (about 55,000 N/mm^2) and a high permissible load, also due to the integrated steel fibres. The compressive strength of UHPC could be between 200 N/mm^2 and 400 N/mm^2, a splitting tensile strength of up to 17 N/mm^2 and a bending stiffness of 39 N/mm^2. The concrete should be executed in the following ratios, w/b = 0.2 (water/binder ratio) and w/z = 0.28 (water/cement ratio). The material could be pumped with a piston pump and in that way allows simple handling.

There are different materials available that can be used to make air-halls or inflatables. Possibilities are PTFE Foils (Polytetrafluoroethylene, or Teflon), PTFE-coated fibre fabrics, PVC-coated Polyester fabrics (Polyvinylchloride) or ETFE Foils (Ethylene tetrafluoroethylene). In the case of large inflatable structures with high loads, it is most common to use PVC-coated polyester fabrics as membrane. Five different classes of membranes can be distinguished with regard to permissible stress, from TYPE I with low permissible stress (3,000 N/5 cm) up to TYPE V with high permissible stress (9,800 N/5 cm). It is typical for MBG structures that the internal pressure, the concrete load and the resulting tensile stress are very low. That's why it is possible to use the membranes of TYPE II or TYPE III depending on the size of the structure. In contrast to the extremely expensive concrete, the connections and membrane details are very simple and easy to manufacture, which makes it quite cheap. High frequency welding or steel clamping plates are very simple and cost-effective. It is even possible to manufacture the membranes in a removable way by using hook and loop fasteners or other zip connections (Fig. 11.106).

Form

The operation of making and perfecting a structure that meets all the necessary preconditions and object functions, as well as architecture and design parameters is quite complex. The outcome of the designing process results in just a couple of

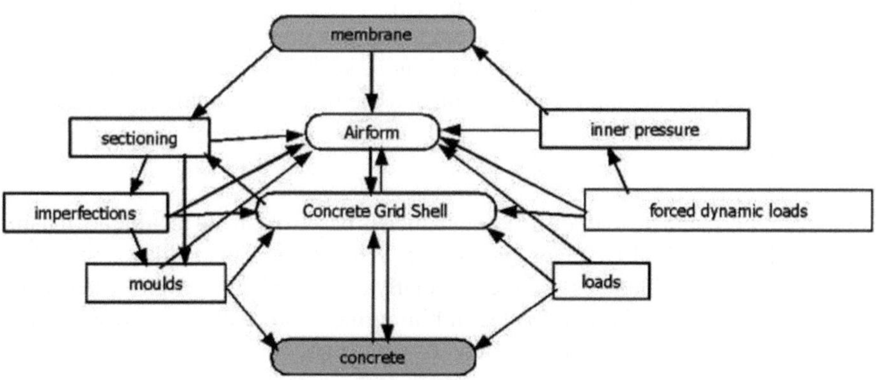

Fig. 11.106 MBG construction interdependencies (*Source* G. Zimmermann)

solutions and the process itself is very time consuming. By using 'Generative Design' during the structure design process, time is saved and more (different) solutions are shown.

The main shape of the MBG depends on the dimensions of the two membranes and also on the way the shape is fixed to the floor. This can be specified into an inflatable with a synclastic surface. In this case there is a double-layered form-active surface or membrane, the MBG. By attaching the outer membrane in a fixed grid to the inner membrane there will arise a pattern on the membrane. The space in between the weldings will act as concrete chambers. This pattern follows the main form because it forms a uniformly distributed load on the membrane. The pattern of the MBG depends on the connection between the two membranes, made by high frequency welding or clamp plates. The mesh could vary depending on the surface and in that way generate different patterns. For example, the thickness of the MBG ribs depends on the degree of overdimensioning of the second membrane. It is possible to manipulate the geometry of the membranes.

Technique

Zimmerman's technique consists of combining the fibre reinforced Ultra High-performance Concrete (UHPC) and air supported membrane chambers respective membrane moulds. The double-layered membrane is spread out on the ground with the edges fixed at the circumferential foundation. The membrane chambers are created by two PVC-coated fabrics which are layered and directly connected through welding. Other possibilities of connecting the two membranes are applying hook and loop fasteners or other zip connections, which makes it possible to re-use the membranes. The first membrane is being used as an air-hall and airform with chambers which are being defined by the second layer. This makes it possible to generate a continuous chamber system and a contemporary geometry for the concrete grid shell. The system will be erected by inflating the air form. When the inner maximum pressure is achieved, the second membrane system could be inflated. After creating the membrane chambers, the chambers can be filled with UHPC, by pumping it from bottom up into the chambers, the system will be filled in sections. These sections will be defined and filled with UHPC from the foundations to the top. Almost every kind of concrete grid mesh on an airform-shaped surface is possible because of the simple construction, the high compression and the flexural strength of the UHPC.

Conclusion

Gregor Zimmermann developed a system which is based on an old technique of using an inflatable covered with a layer of concrete. What Zimmerman did was to make a combination of the inflatables and the especially developed Ultra High-Performance Concrete with single midget fibres. Combining these two techniques resulted in a large amount of saved material, concrete as well as polymers for the inflatables. By using a grid, it wasn't necessary anymore to make a complete solid dome. Gregor Zimmermann developed a technique to use generative design in the design process.

The design process was complex and time consuming before, this technique results in many possible concept solutions and is time saving. Some improvements still need to be made, because the software is not able to implement all the different aspects and parameters. The technique of pumping UHPC into an inflatable mould isn't put into practice, due to the extremely high costs of the Ultra High-Performance Concrete. The method developed by Zimmermann could be linked to the 1965 research by GoodYear (Cross 1965). In this research a distance fabric is combined with a local grid of beams. By leaving out the connectors, the membrane surfaces will bulge out and form ribs to make the structure stiffer (Fig. 11.107).

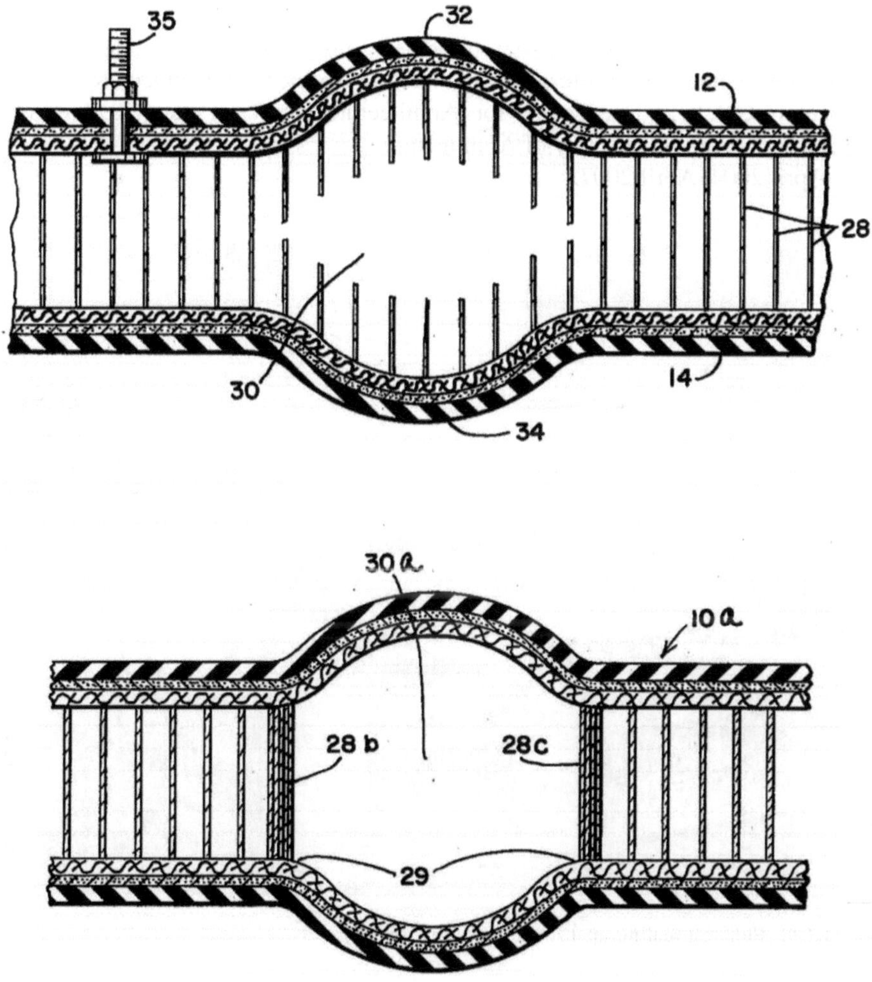

Fig. 11.107 1965 patent 3205106 by W.B. Cross

11.14 Precast Concrete Thin-Shell Vaults

Flexible Mould

The form-active structure: woven fabric (Figs. 11.108, 11.109, and 11.110).
Form-active typology: single layer.
Way to stabilize the mould: stressed surface.

Composite Material

Technique to handle the rigidizing material: spraying and hand layup.
Reinforcement of the material: woven fabric.
Surface treatment: none.

Project

Name and location: Precast thin-shell funicular vaults, Winnipeg, Canada.
Design and Engineering: Professor Mark West, department of Architecture.
Contractor: C.A.S.T. The Centre for Architectural Structures and Technology, University of Manitoba Faculty of Architecture.
Year: April 2004–April 2007.

Fig. 11.108 First test scaling-up a small (*Source* M. West)

Fig. 11.109 Double-curvature funicular shell funicular model. cast directly from a single hanging flat sheet of fabric (*Source* M. West)

Introduction

Precast thin-shell funicular vaults are made using moulds made from hanging flat sheets of fabric. These fabric sheets are allowed to deflect into naturally occurring funicular geometries, producing moulds for lightweight funicular compression vaults and stiff double curvature wall panels. These methods were developed at the Centre for Architectural Structures and Technology (C.A.S.T.) at the University of Manitoba's Faculty of Architecture (Univ. of Manitoba, umanitoba.ca/faculties/architecture/facstaff/faclist/west.html).

These methods were developed to provide lightweight wall and floor structures for the Taller de Obras project, a new architectural atelier building at the Open City in Ritoque, Chile, designed by architects Miguel Equem and David Joll.

The Taller de Obras floor structure is a cast-in-place reinforced concrete slab supported by a 3 m by 3 m grid of supporting beams (plan below). The project proposes to reduce the amount of reinforcing steel this slab requires by using pre-cast thin-shell funicular vaults (approximately 3 m by 1 m), made from fibre-reinforced concrete to serve as stay-in-place formwork pans for a cast-in-place floor slab designed to span by compression between its integrally

Fig. 11.110 Folded membrane shell (*Source* M. West)

formed beams (West, fabwiki.fabricformedconcrete.com/lib/exe/fetch.php?media=chile:thin_shell_panels.pdf) (Figs. 11.111 and 11.112).

Material

For the production of the first model, an inexpensive woven polypropylene Geotextile was used as cast. A uniform thickness of glass fibre-reinforced concrete (GFRC) is applied to the hanging fabric by spray application.

For the production of the second model plywood and a woven polyethylene fabric was used as cast. A sprayed fibre-reinforced concrete was then used to form the cross-vault. For reinforcing this shotcrete, again glass fibre is used.

Form

For the making of the fibre-reinforced, thin-shell concrete funicular vault, a mould is used made from hanging flat sheets of fabric. These fabric sheets are allowed to deflect into naturally occurring funicular geometries, producing moulds for lightweight funicular compression vaults. The way that the membrane is manipulated in this case can be described as a non-inflated monoclastic surface. The membrane forms without the help of any cables, external loads, elements or other membranes.

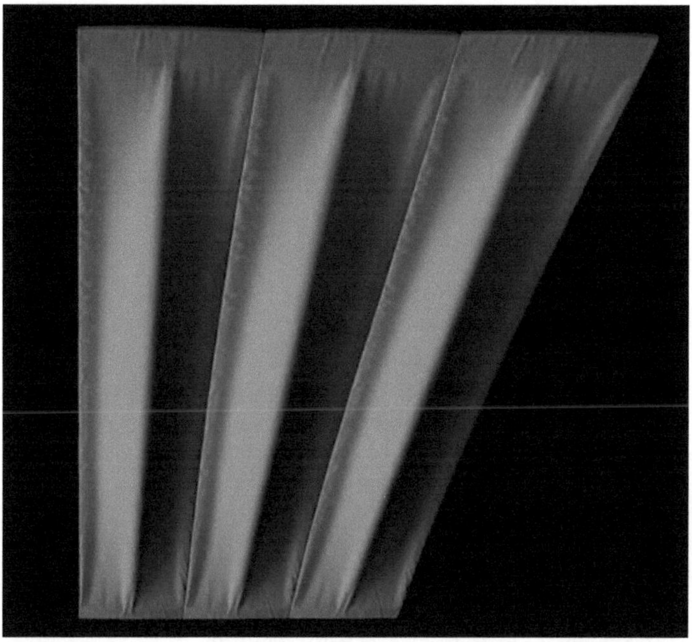

Fig. 11.111 Design of the compression vault floor slab (*Source* M. West)

For making the fibre-reinforced, thin-shell concrete funicular cross-vault, a simple frame is produced which is covered by a polyethylene fabric. The fabric is fixated to the frame and since the fabric is fixated on all four sides it will eventually form a funicular vault in both length and width of the frame, hereby forming the cross-vault shape. The manipulation of the membrane can be described as a zeroclastic surface with an equally divided load consisting of the own weight of the membrane and the shotcrete. This load forms a duoclastic surface in the rectangular frame, again without the help of any cables, external loads, elements or other membranes (West and Arraya 2009).

Variable forms

Using the same principle as the first model, an alternative vault was made in May 2009. By curving the surface on which the fabric is attached a doubly curved vault was created.

Another alternative for that model was made after that. Thereby a keel was used to control the bottom curve of the vault creating two flat surfaces originating from the curved surface. The membrane forming the vault is formed using a monoclastic frame. Within this frame the membrane is stretched and pinched together making a V-shape (Fig. 11.113).

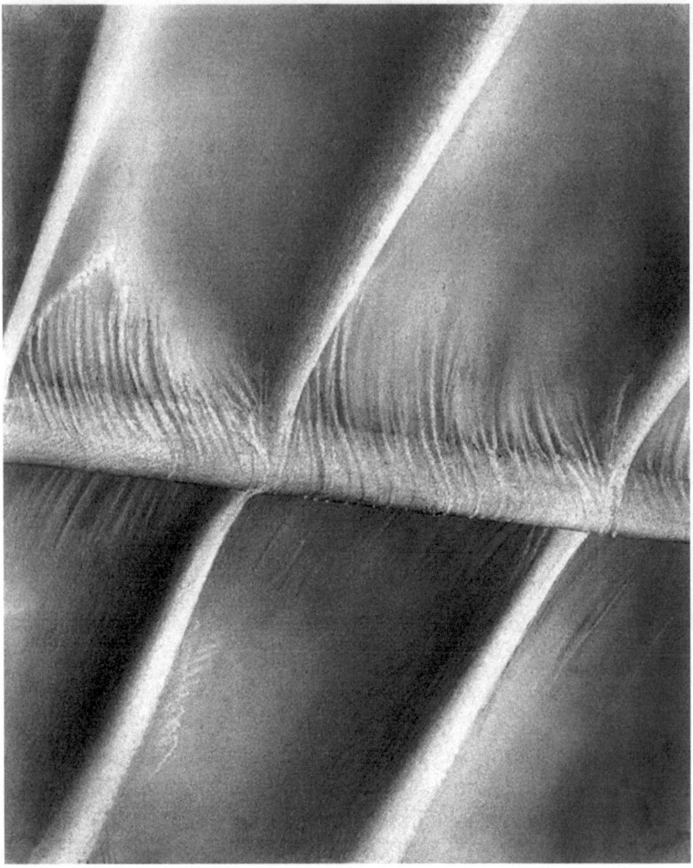

Fig. 11.112 Prototype of the compression vault floor slab (*Source* M. West)

Technique

For the making of the first upscaled model in April 2004, an inexpensive woven polypropylene Geotextile fabric was simply hung from two sides forming the basic funicular vault. A uniform thickness of Glass Fibre Reinforced Concrete (GFRC) is applied to the hanging fabric. The resulting funicular shell is inverted to make a mould for precast production. In this test, the fabric was removed to expose the concrete surface, which was then sealed and oiled to provide a release surface for the mould. A single thin-shell GFRC vault was produced from this fabric-formed mould.

In the second test in April 2007 a simple open frame was constructed by using plywood, built to the full-scale dimensions of the proposed Taller de Obras formwork pans. A coated woven polyethylene fabric is stapled to plywood strips and stretched over the opening to remove any wrinkles that might form in the sheet. The wood

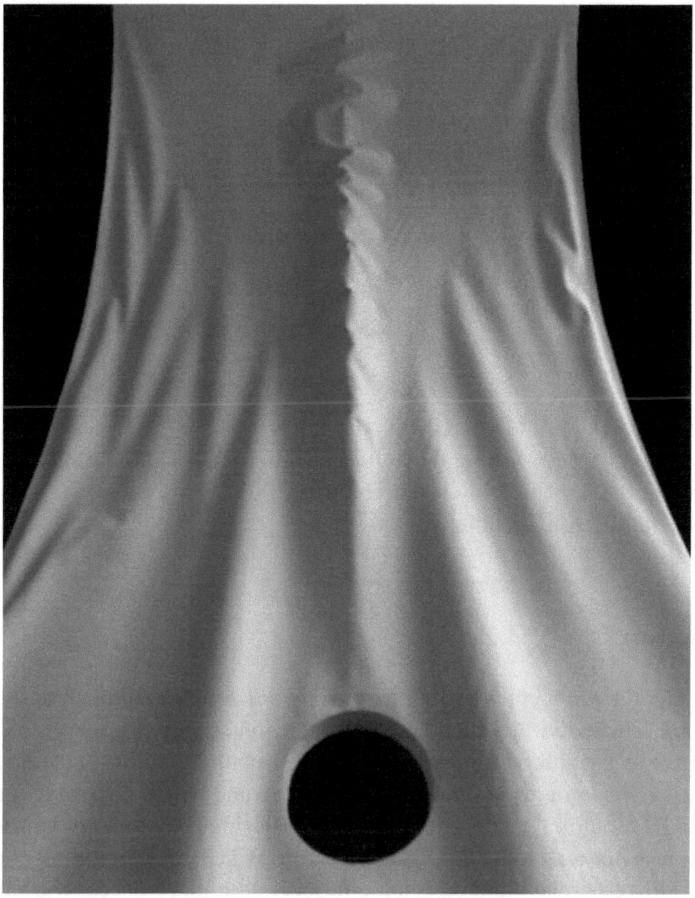

Fig. 11.113 Funicular shell (*Source* M. West)

strips are screwed to the frame, thus fixing the fabric in place. A uniform thickness of fibre-reinforced concrete is placed over the fabric. In this case a sprayed shotcrete is used, but hand application of the concrete is also possible. The edges are reinforced with steel rebar. Woven high density polyethylene or polypropylene fabrics can be manufactured with a smooth waterproof coating on one side, and a fuzzy non-woven fabric welded to the other. If the concrete is applied to the non-woven side, the fabric will permanently adhere to it, providing a smooth, permanent, plastic-coated release surface for the mould. The funicular shell thus formed is removed from the frame to be used as a plastic-coated mould for precast production of fibre-reinforced thin-shell funicular compression vaults. These vaults can be used directly as structure, or as stay-in-place formwork pans (West 2016).

11.15 Park Garage

Flexible Mould
The form-active structure: membrane (Fig. 11.114).
Form-active typology: double layer.
Way to stabilize the mould: prestress.

Composite Material
Technique to handle the rigidizing material: casting.
Reinforcement of the material: bars, fibres.
Surface treatment: none.

Project
Name and location: Park Garage the Netherlands.
Design and Engineering: Master thesis Toine Bullens and Tiddo Folmer Tutor: Arno Pronk (Pronk et al. 2007a; Bullens and Folmer 2007).
Contractor: Bam Beton Technieken.
Year: 2006.

Introduction

The "Park Garage" concept is a structure of standardized square concrete shells of 8.1 × 8.1 meters. The individual shell is supported by one concrete column in the centre. When the shells are linked together, a vaulted structure creates a continuous deck. A park can be made on top of it with grown trees and below a parking garage. We researched the moulding technique and logistics during construction. Unfortunately, the project was not realized. Niki Cauberg of the WTCB in Belgium in cooperation with the VUB researched and cast some models with more or less the same shape (Cauberg et al. 2009). This Park Garage concept has been developed for situations where there is a high density of buildings and a problem with the storage of rain water.

Technique

The mould for a doubly curved shell of 8.1 × 8.1 meters is expensive. Due to transportation limits, the prefabricated shells cannot be made in one piece. Therefore,

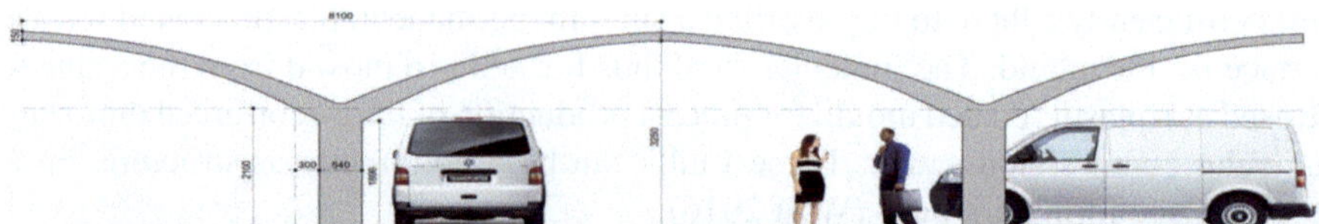

Fig. 11.114 Design of section of Park Garage (*Source* BAM)

we studied onsite production with membrane moulding as an alternative. Shotcrete in combination with a chain linked net structure as for the Philips Pavilion and Barbapapa was too expensive and too slow for the reproduction of shells for a garage. Therefore, we looked at casting between a double layered membrane. The upper layer will be covered with earth therefore the form-finding of the upper layer had no aesthetic demands. The lower layer had to carry the dead load of the concrete and had to be smooth view work. Both membranes are clamped at the borders to a steel structure. The steel structure and membranes can be assembled and mechanically prestressed on site. The shells will be produced close to the building site and placed on site by a mobile crane. The dead load of concrete on the lower membrane is too high, therefore this membrane has to be supported. This can be done by inflation with a gas or liquid, or by rigidizing the membrane with a supporting material like shotcrete or PUR foam. If the project would have been realized we would have chosen for the last-mentioned option. PUR rigidizing in combination with casting of concrete on a membrane was used and proved by the Crushedwall project by Jack Walter in 2012 (Walter 2012).

Form

The form of the shell was developed to realize an optimized shell structure with the purpose to carry a heavy load of earth and stored water. A fluent transfer from shell to column is needed to prevent local tensions in the concrete. The distance of the columns in a square grid of 8.1 meter is chosen to be able to park three cars in between the columns. The height of the shell is 1.35 m (by 8.1 × 8.1 meters) The thickness of the shell changes from 75 mm at the edge to 125 mm in the middle. The form of the membrane would have been done by welding PVC-polyester membrane cutting patterns in a diverging way from the middle. The columns have a diameter of 0.54 meters and a height of 1.9 meters. The total height inside the "Park Garage" is 3.25 meters (Figs. 11.115, 11.116, and 11.117).

Matcrial

The shell structure was calculated in the FEM program Marc/Mentat with the following properties; specific gravity 2,400 kg/m^3 Young's modulus 10,000 N/mm^2 Poisson ratio 0.2. permanent load at edge 0.0138 and in the middle 0.0432 N/mm^2 Variable load 0.0075 N/mm^2. The calculations gave a maximum deflection of a standalone shell in the corners of 37 mm and with an asymmetrical load a maximal deflection of 62 mm. If 9 shells will be structurally combined the maximum deflection will become 47 mm at the corner. Only in de edge a limited tension of 8 N/mm^2 will occur. The rest of the shell will have pressure with a maximum load of 2 N/mm^2. The tension in the edge can be resolved with traditional reinforcement bars that can be straight, the rest of the surface that is doubly curved can be resolved with fibre reinforced concrete (Figs. 11.118 and 11.119).

Fig. 11.115 Model of the Park Garage

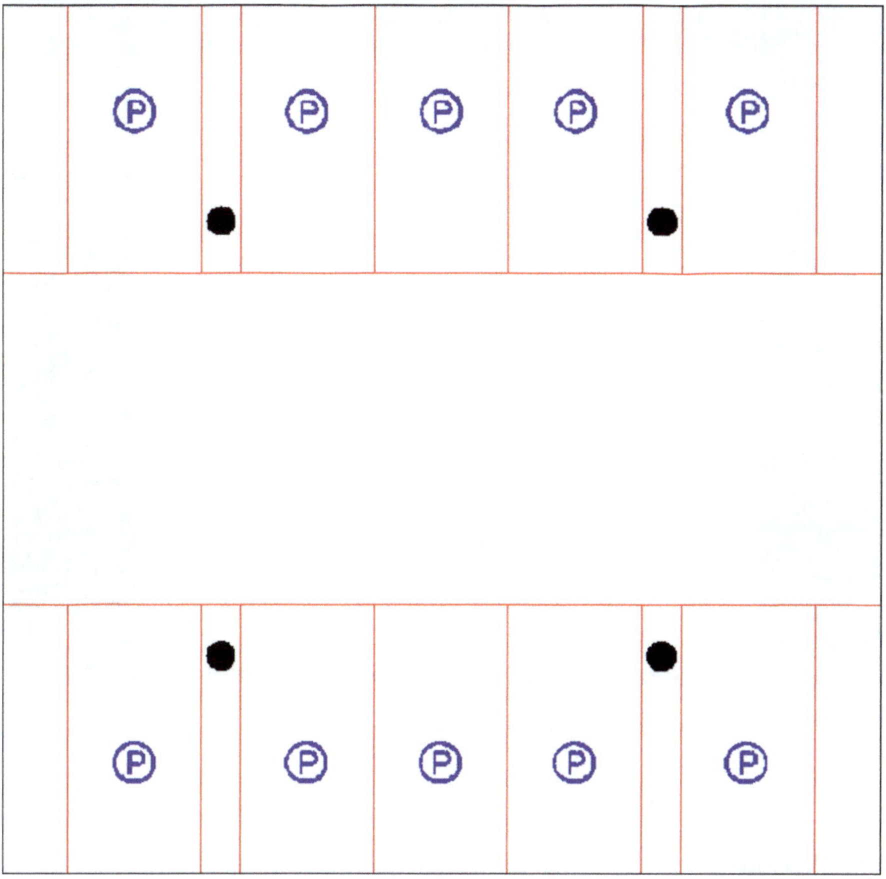

Fig. 11.116 Floor plan

Fig. 11.117 Construction method

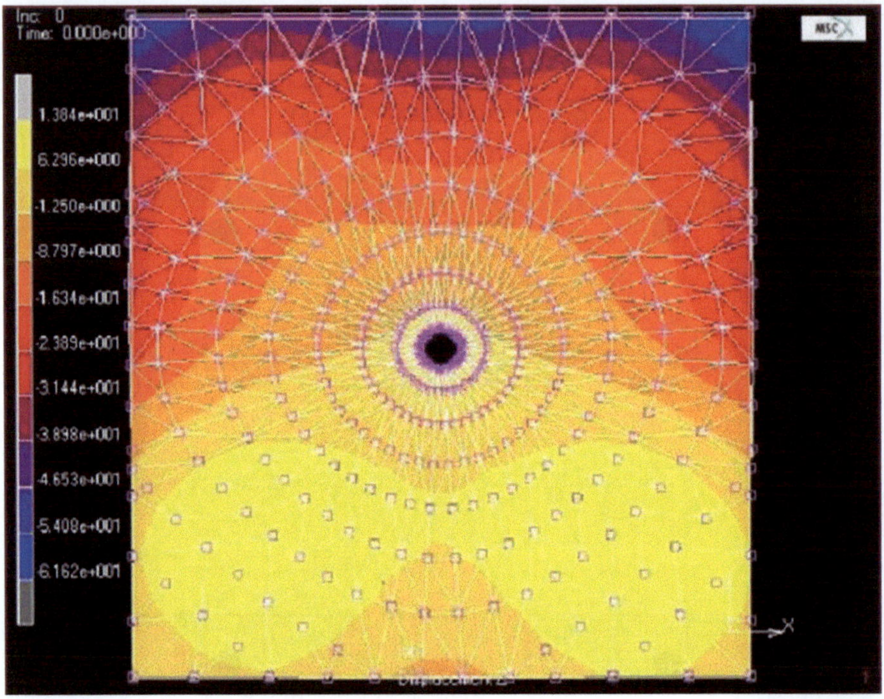

Fig. 11.118 Deflection of asymmetrically loaded shell

Conclusion

This technique seems to be economically feasible. It was disappointing that this system for the production of shells for parking garages was never realized. We calculated the environmental benefit of the production of doubly curved shell structures on site compared with prefabricated elements on bending for this “Park Garage” concept. There are two benefits. The first is less material by thinner concrete structure (59% savings on concrete) and less reinforcement (73% savings on steel). The other is less transportation because the transport of prefabricated elements would be less compact and the distance between factory of prefab elements and building site is longer in the Netherlands than for a concrete factory and building site. Bulk transport of raw material in the Netherlands is by boat which also has less environmental impact (altogether 91% savings on transport).

In 2016 we were able to realise a shell with students of the TU/e and the support of Geelen Beton. Please see the pictures of this shell below (Figs. 11.120 and 11.121).

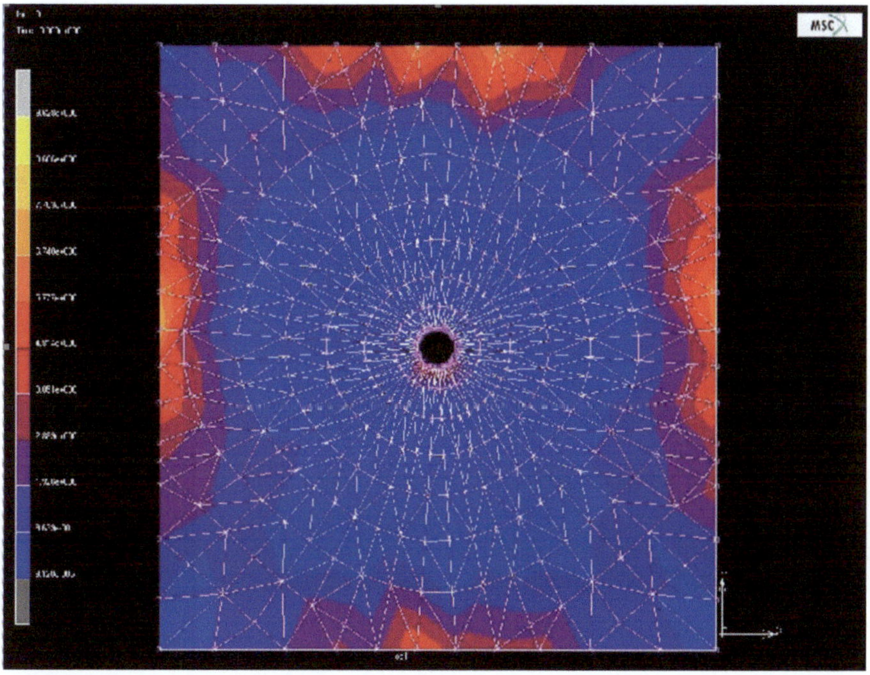

Fig. 11.119 Deflection of symmetrically loaded shell

Fig. 11.120 Membrane mould with wooden clamps

Fig. 11.121 Final result at Geelen Beton

11.16 Barbapapa Object

Flexible Mould
The form-active structure: membrane + net structure.
Form-active typology: single layer (Figs. 11.122, 11.123, and 11.124).
Way to stabilize the mould: inflate + prestress.

Composite Material
Technique to handle the rigidizing material: spraying.
Reinforcement of the material: bars.
Surface treatment: plastering.

Project
Name and location: Barbapapa object.
Design and Engineering: Arno Pronk, Toine Bullens and Tiddo Folmer.
Contractor: Bam Beton Technieken.
Year: 2006.

Introduction

Arno Pronk developed a system for the realization of fluid shell structures. In 2006, it was patented in the Netherlands and by 2008 as a European patent (Pronk 2008a).

Fig. 11.122 Barbapapa structure

Fig. 11.123 Inflatable mould for Barbapappa object

The system consists of a combination of inflatable or other objects covered with a mechanically stressed net structure. The net structure is used as a mould for making a fluid shell structure. The (standardized) objects below the net can be combined in different ways. With this method it is possible to make endless series of combinations resulting in an endless diversity of curved surfaces with the same objects

Fig. 11.124 Construction of Barbapapa object

(Fig. 11.125). To prove the system, an object was designed and made in cooperation with the students Toine Bullens and Tiddo Folmer and Bam Beton Technieken (Fig. 11.126).

Technique

The technique used is based on shotcrete and the same as for the case studies with the houseboat and reconstruction of the Philips Pavilion. The technique and shape refer

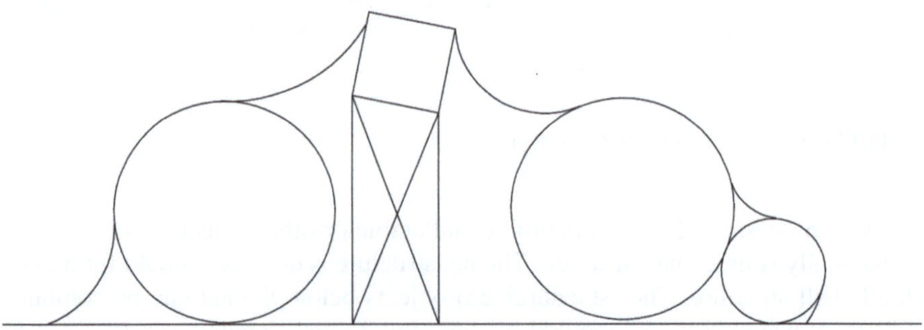

Fig. 11.125 Section one of a mechanically stressed net structure supported by (inflatable) objects

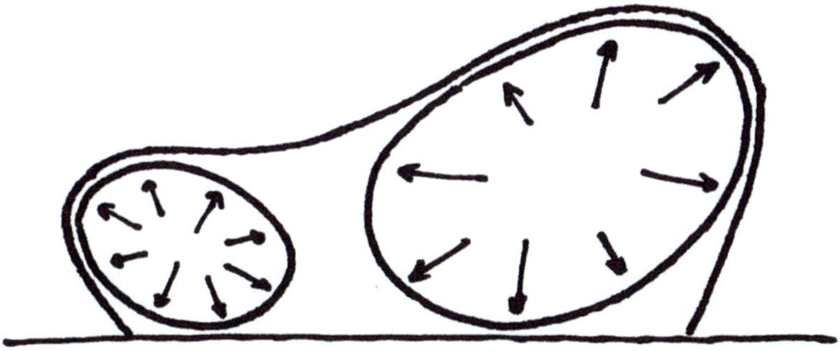

Fig. 11.126 Section of a mechanically stressed net structure supported with (inflatable) objects

to the animated movies of Barbapapa. In the movies the housing of the Barbapapa family is solved by using their adaptive body as a mould for shotcrete. Maison Bulles by Antti Lovag (also early 1970s in France) looks the same way as in the animations of Barbapapa (Fig. 11.127 and 11.128).

Form

Inflatable structures are most useful for creating shapes derived from circles or circle segments. In most cases it leads to (spherical) synclastic or cylindrical shapes. With prestressed tent structures it is possible to construct (hollow) anticlastic shapes.

Fig. 11.127 Barbapapa

Fig. 11.128 Window at Maison Bulles by Antti Lovag (*Source* G. Zetbase)

Applying a combination of inflatables and overlaying prestressed membranes it is possible to construct a variety of different shapes. After obtaining the desired shape, the membrane can be used as a mould for the shell structure. Referring to the research in Chap. 2 about the manipulation of form-active structures the contact between the inflatable spheres can be qualified as SI 13 and the contact between the net and the inflatables as SI 11. In case of pushed in or out elements in the mechanically prestressed membrane AS 16–19 or at the inflatable SI 16–19. All the manipulation methods concerning the prestress in a certain area of the net or membrane 1–5 and all the manipulation methods concerning an external load 6–10 are possible.

The object was used for children to climb on and play with. Following the NEN-EN 1176-1 the object has a maximum height of 2 metres to avoid the requirement of a fence against falling off the object and holes may not be V shaped to avoid the harming or jamming of children. The Design is based on 4 spherical inflatables combined together and covered with a continuous surface. The surface has three circular windows.

Material

The calculation of the shell was done in the program Marc/Mentat. The form-active behaviour was not calculated. The values for the calculation were: E-modulus 10,000 N/mm^2, Poisson ratio of 0.2, Weight 2,400 kg/m^3, variable load 1.5 kN/m^2. Calculating with a thickness of 30 mm at the top and 40 mm for the lower parts a displacement of 0.15 mm was found with maximum tension of 0.6 N/mm^2. In this object with limited measurements, the thickness of the structure is determined by the construction method and material and not so much by the structural optimization. The outside of the shell was made of 40 mm shotcrete on the net structure and 10 mm on the inside (Fig. 11.129).

The inflatables are made with PVC polyester. The patterns are welded. The chain linked net by Bekaert is galvanized without coating and has a mesh of 50 mm. The net is 1.75 meters wide; the inflatables were covered with 3 strips. The reinforcement was bent over the net surface in a mesh of 150–200 mm with bars of 6 mm. The surface of the shotcrete is smoothened by plastering with a strickle and by sponging.

Conclusion

The construction of the reinforcement is the most labour-intensive part of the process. One of the recommendations is to replace the traditional reinforcement by fibre reinforced concrete. As long as the reinforcement is only used to prevent shrinkage cracks, this is an option in larger projects.

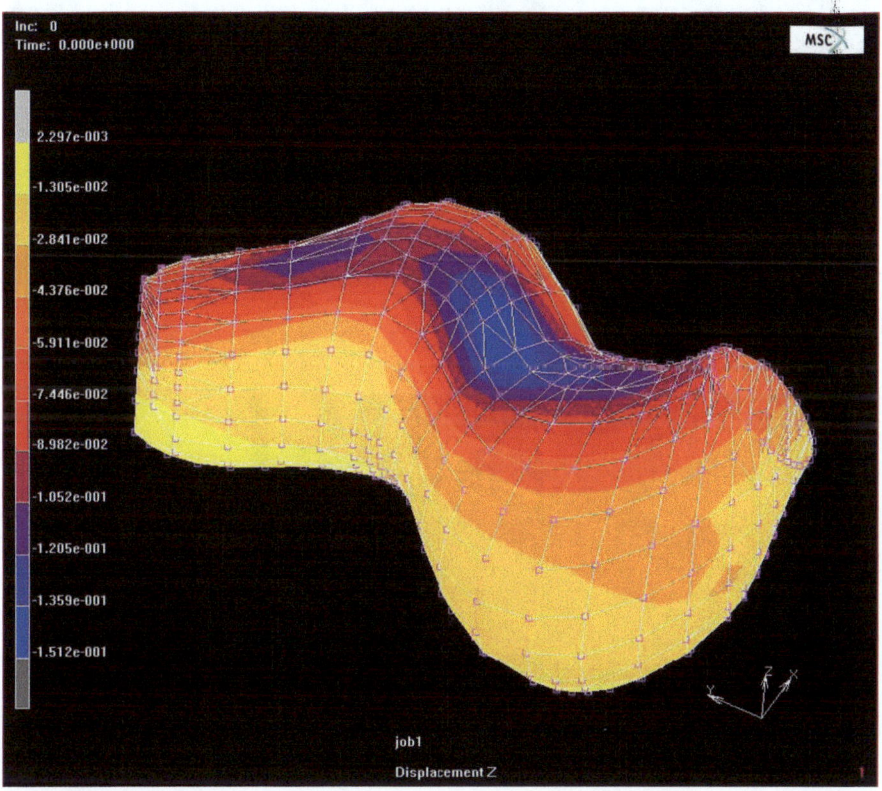

Fig. 11.129 Engineering of the structure

11.17 Studies for a Houseboat with Fluid Shell Structure and for a Bicycle Cellar

Flexible Mould

The form-active structure: net structure (Figs. 11.130, 11.131, and 11.132).

Fig. 11.130 Design of house boat with fluid shell roof

Fig. 11.131 Shell for Bicycle cellar in Eindhoven

Fig. 11.132 Treadmill at the entrance of the bicycle cellar

Form-active typology: single layer.
Way to stabilize the mould: prestress.

Composite Material
Technique to handle the rigidizing material: spraying.
Reinforcement of the material: bars.
Surface treatment: plastering.

Project
Name and location: Houseboat with fluid shell structure. Bicycle cellar Eindhoven.
Design and Engineering: Arno Pronk (houseboat), Massimiliano Fuksas (bicycle cellar).
Contractor: Bam Beton Technieken.
Year: 2010.

Introduction

In this section I will describe two different projects, one involving the designing of a houseboat in Amsterdam and one involving a design for a fluent architecture roof for a bicycle cellar in Eindhoven (Figs. 11.133 and 11.134).

Amsterdam has the ambition to be self-sufficient in its energy production and consumption. For a client we developed a houseboat with a combination of a windmill and photovoltaic cells facing south. Amsterdam is at 52 degrees north latitude.

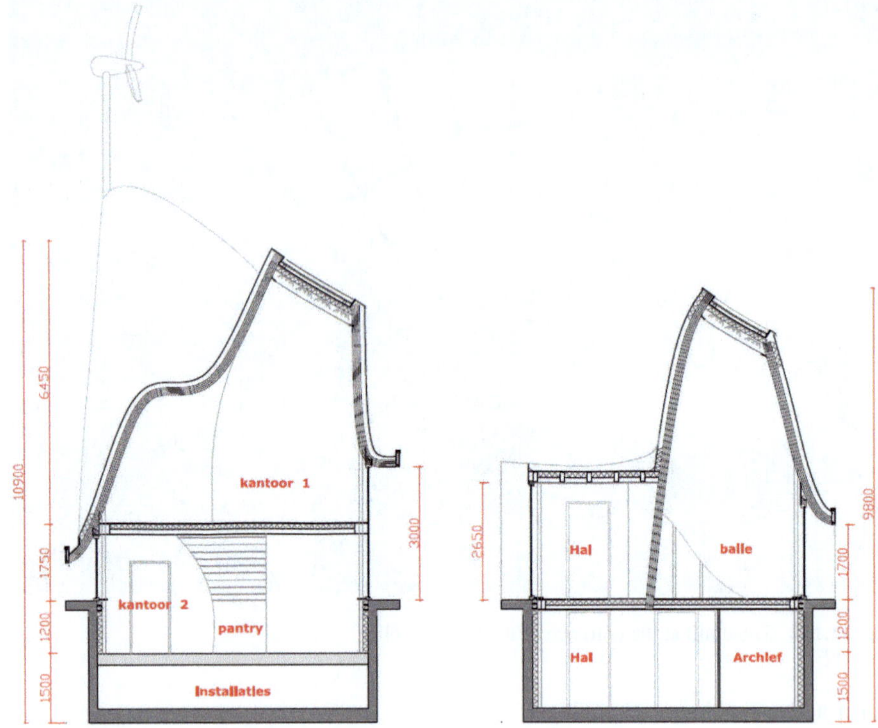

Fig. 11.133 Sections of the house boat

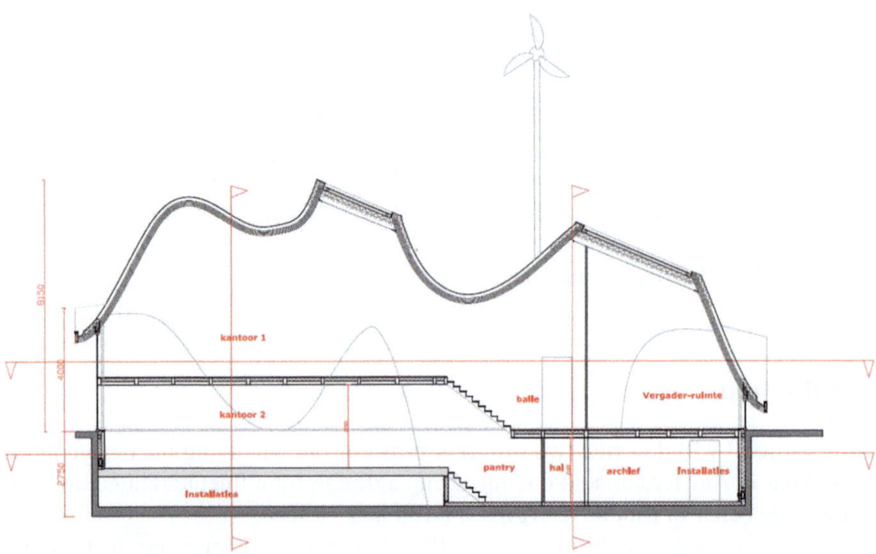

Fig. 11.134 Sections of the house boat

Therefore, the orientation of the solar cells is an important factor in its energy efficiency. The best way to avoid the need for a sun-tracker is a position of about 36 degrees. In winter, the efficiency of the solar cell is very low and will be compensated for by the wind that has a relatively strong force in the Netherlands and blows mostly from the south west or north west. The position and outlines of the houseboat are fixed. The houseboat is positioned north and south. The sawtooth roof has three plane surfaces facing the sun. The plane surfaces and foundation are connected by a fluid shell surface (Fig. 11.135).

The entrance of the Eindhoven bicycle cellar is designed by Massimiliano Fuksas. The project is located at the centre of Eindhoven. The cellar is below a shopping square. The two entrances are formed by two shell structures (Fig. 11.131). The structure is made with a wooden mould but the rigidizing technique of the surface is the same and made by the same contractor. Therefore, we describe the two projects in the same chapter.

To prove feasibility of the two projects we realized a prototype 1:1 of a representative part of the building. The main focus in the houseboat was more on the construction and insulation, whereas the main focus of the bicycle cellar was more on the surface and reliability. First the prototype of the houseboat is discussed and later the bicycle cellar.

Material

A part of the rectangular concrete caisson for the floating foundation of the houseboat was made. The biggest goal for this prototype was to design and test an insulated structure with a slender edge and continuous ceiling from the inside to the cantilevering outside in Fig. 11.138. The shell structure is made by shotcrete more or less referring to the ferrocement method. Ferrocement is the combination of cement with a woven reinforcement grid of steel. The method has been known since the beginning of the application of reinforced concrete in 1948. On the foundation a steel structure formed the outlines and supporting columns of the shell structure. A 100 by 100 mm^2 bar forms the structural edge of the shell (Figs. 11.136 and 11.137). Within the edge a chain linked net was stretched. At the outside of the square bar strips were welded to hold and reinforce the cantilevering edge of the shell structure. This cantilevering is structurally separated and insulated with foam to avoid thermal bridges and condensation at the inside of the shell during the winter. The outside of the shell is formed by spraying concrete onto a wooden formwork at the end of the curved shell. The wooden formwork is temporarily connected with the reinforcement of the cantilevering. At the bottom side of the steel bar wooden strips are screwed on to make a slit for the glass to be placed in the façade. This slit has been made extra deep to provide more space for differences in the curvature of the shell and the edge of the glass plate. In this way the edge of glass can be straight as usual. At the lower side of the chain linked network PVC tubes are placed to control the thickness of the concrete surface over the net structure. The PVC is just a spacer and removed during the construction and finishing of the shell. The stretched net is sprayed and plastered from two sides in layers of about 20 mm thickness. The sides

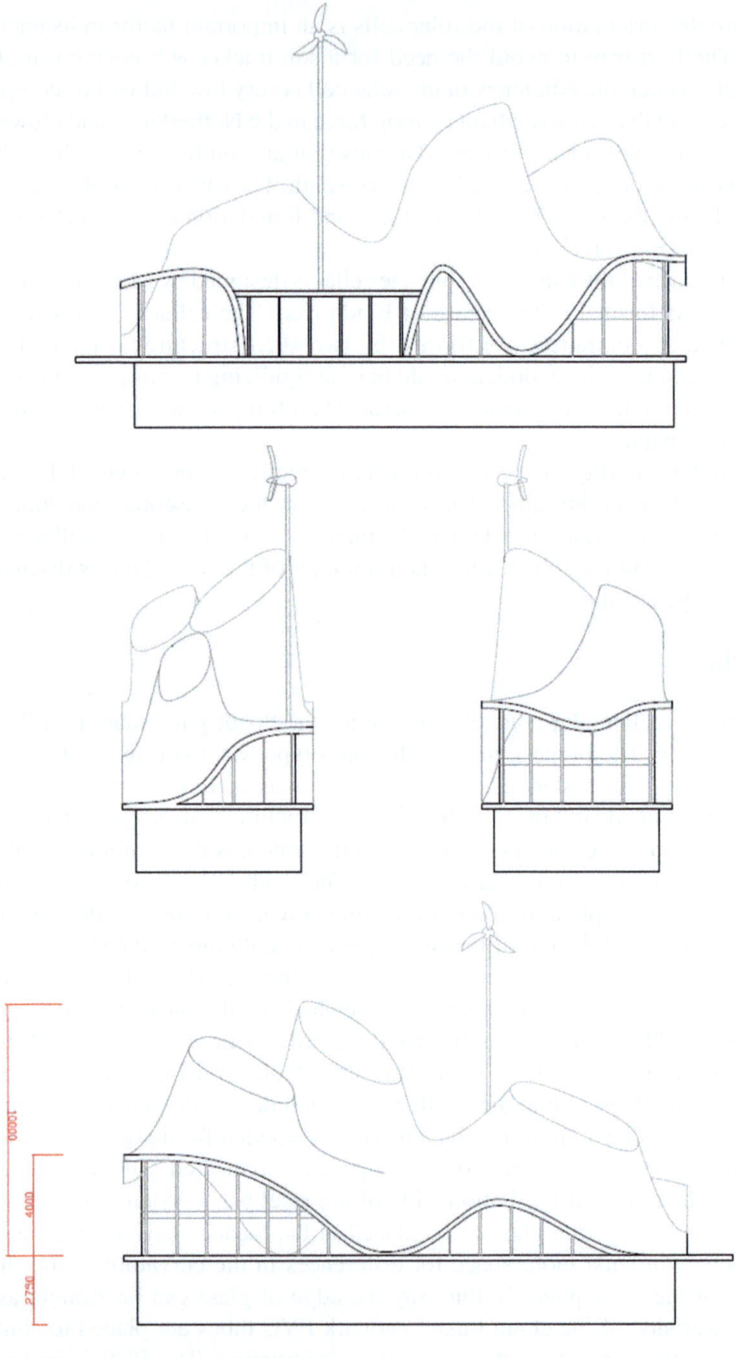

Fig. 11.135 Façade drawings of the houseboat

Fig. 11.136 Steel structure with net and wooden formwork at the edge of the shell

of the wooden strips (for the glass) and formwork at the edge are used to "strickle" the surface in a controlled way. After the realization of the concrete work it was plastered and painted with latex (Figs. 11.139, 11.140, and 11.141).

Eindhoven Shell

Material

For the shell in Eindhoven, the first 6 meters of the entrance were made as a prototype scale 1:1. In this case, the mould was made out of wood. To avoid a counter mould the same ferrocement based technique was chosen as was used in the prototype for the houseboat, Philips Pavilion and Barbapapa object. In this case there was less curvature of the surface and the curvature was almost single but still regular. The other difference was that this building had to be built and remain for a long time. The city of Eindhoven demanded a high flexible load on the shell to be able to resist jumping people having a celebration. The architect wanted the shell to be as slender as possible. The Movarus company who did the calculations came up with a thickness of 15 cm and a double layered and very dense grid of reinforcement (Fig. 11.142).

Fig. 11.137 Spraying shotcrete from above

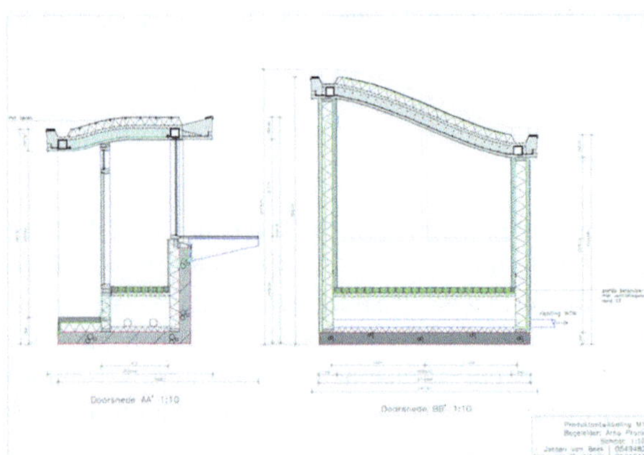

Fig. 11.138 Section of prototype

Fig. 11.139 Spraying of shotcrete to chain

Fig. 11.140 PVC distance holders. linked net structure

To avoid honeycombing behind the reinforcement, shotcrete had to be sprayed from different angles around the reinforcement bars. Shotcrete had to be sprayed in layers of about 20 mm. A double layered reinforcement grid cannot be sprayed in at once from one direction. Therefore, also the reinforcement had to be constructed in layers. At the outer layer of the reinforcement with a distance of 1,700 mm PVC tubes forming arches were bent. The PVC tubes were necessary as distance holders to strickle the shotcrete in the right form in alternating sections (Figs. 11.143 and 11.144). After the PVC tubes were removed, the sections in between were strickled. The outer surface of the shell was coated with a two-component polyurea spray resin. This coating has an excellent adhesion to concrete and steel, has a strength of (40 MPa) tensile and over 500% elongation. The inside of the shell was coated with latex paint to be permeable for aqueous vapour (Pronk 2008b: 27) (Fig. 11.145).

Fig. 11.141 The final prototype

Fig. 11.142 Reinforcement of the shell structure

Fig. 11.143 Spraying of shotcrete in sections

Fig. 11.144 Alternating sections

Fig. 11.145 Shell structure before coating

11.18 The Reconstruction of the 1958 Philips Pavilion

Flexible Mould

The form-active structure: textile composite/cable-net (Figs. 11.146 and 11.147).
Form-active typology: single layer.
Way to stabilize the mould: stressed surface.

Composite Material

Technique to handle the rigidizing material: spraying.
Reinforcement of the material: single fibres.
Surface treatment: plastering.

Project

Name and location: Reconstruction Philips Pavilion 1958, Eindhoven, the Netherlands.
Design and Engineering: 1958 design: Le Corbusier and Xenakis.

Fig. 11.146 The Philips Pavilion 1958 in Brussels

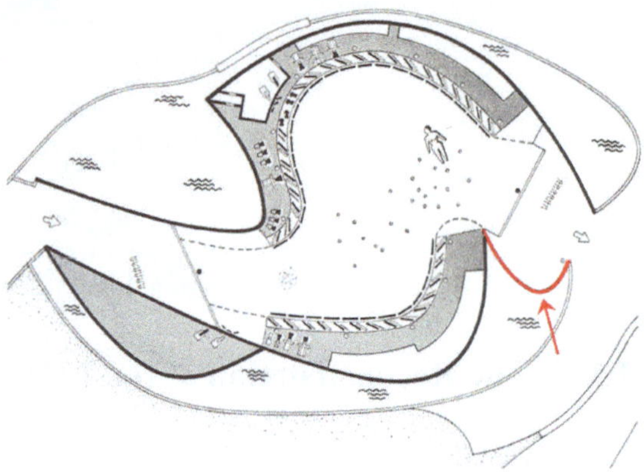

Fig. 11.147 Floor plan of Philips Pavilion

Reconstruction research: Arno Pronk.
Contractor: Stichting Reconstructie Philips Paviljoen 1958.
Year: 2012.

Introduction

In 1958 the Philips Pavilion was built according to a design of Xenakis (2008), an employee of Le Corbusier, for the world exhibition in Brussels. This pavilion is considered as one of the first examples of modern fluid architecture (Treib and Feliciano 1996). The idea behind the Philips Pavilion was to show the influence of modern technique on art. The pavilion was a symbiosis between architecture, music and visual art (Kalff et al. 1958). The techniques investigated were: structural design of a shell structure in a very large space frame, the possibilities of electrical amplifiers and synthesizers, and the projection of light and pictures. In Eindhoven, the Netherlands, there are plans to reconstruct the Philips Pavilion as a permanent building. In 1958 the pavilion was built as a temporary structure with limited demands. Nowadays the demands are much higher and the way the pavilion was built in 1958 is tremendously expensive nowadays. Therefore, a survey for cheaper alternatives was carried out. In this chapter the most promising technique is provided including the experiments we did and the realization of a part of the Philips Pavilion scaled 1:3. This chapter concludes with an overview of the problems that should be considered by engineering an anticlastic membrane as a mould for a concrete shell structure. The research could be carried out thanks to a subsidy of the municipality of Eindhoven and the participation of Stichting Alice, Stichting Reconstructie Philips Paviljoen 1958, Bam Betontechnieken, Tentech and Buitink Technology (Pronk et al. 2006, 2007).

Form

The pavilion was designed as a compilation of hyperbolic paraboloid shells in a main frame with a volume of 4,000 m^3. The floor plan resembled a stomach with a surface area of 1,000 m^2 (40×25 m^2). It has a maximum height of 22 metres. The construction method of the pavilion at the time was 50 mm thick prefabricated concrete panels of roughly 1.5 m^2 that were hung on 8 mm thick steel cables. The fabrication of the panels started with moulding sand hills to the desired shape. Next the sand hills were divided into a rhombic grid of casing planks. After placing reinforcement meshes the rhombi were cast in concrete to form the individual panels after hardening. The panels were numbered and transported to the construction site where they were assembled much like a big jig-saw puzzle. The resulting shells were stressed by steel cables on the outside of the shells and finally coated with aluminium paint.

The tiles forming the shell remained partially visible under the layer of cement and the silver coating. the distribution of tiles (please have a look at Fig. 11.148) was not consistent with the grid of the cables over the surface.

In Fig. 11.149, the distribution of the tiles is mapped onto a 3D model. In this way the exact boundary and curvature of the tiles and length of the cables can be measured.

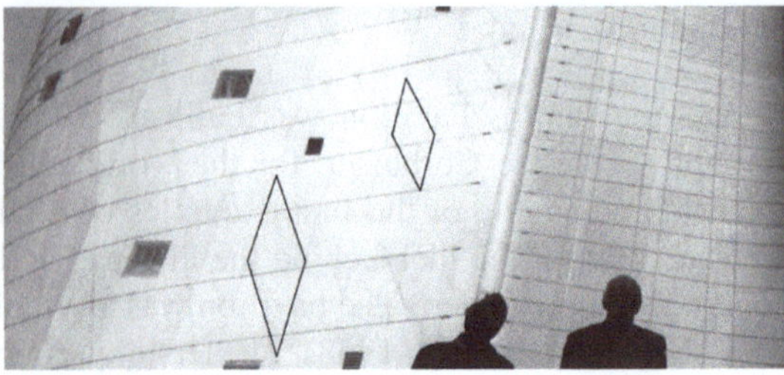

Fig. 11.148 Horizontal cables in combination with the rhombic tiles

Fig. 11.149 3D Rhino model of the Philips Pavilion

Technique

Considering the present labour costs and occupational health and safety laws the original construction method is not acceptable for the present time. With present technology it should be possible to construct the pavilion faster and more cost efficiently. In response, the Eindhoven University of Technology (TU/e) researched the possibilities for a better construction method for doubly curved surfaces in concrete. The techniques studied use a form-active mould on which concrete is sprayed or cast. Generally known are the methods where inflated moulds are used on which concrete is sprayed (Chilton 2000; Sobek 1987: 1–34). In our case, the desired anticlastic forms cannot be made with inflatable moulds. Therefore, we used anticlastic prestressed membranes referring to the experimental building methods by Gustav Lilienthal (1899) and James Hardress de Warenne Waller (1934). We assume that Le Corbusier and Xenakis were unfamiliar with the building methods of James Hardress de Warenne Waller. Mechanically prestressed membranes were not widely applied in 1958. Therefore, we assume that they saw their methods as the only possible way at that time, not considering any other method. We concluded that the shape of the original pavilion is made out of hyperboloids and can be manufactured with prestressed membranes or with a cable net. The membrane mould can be rigidized with shotcrete. The shotcrete is applied in layers of about 20 mm. A disadvantage of concrete is the high dead weight that results in the exposure of the membrane form to high forces. We realised an experiment by making a concrete shell next to the exit of the Pavilion. The original concrete shell structure can be reconstructed by shooting concrete onto these membranes. By adding a plasticizer and an accelerator to the mortar, the characteristics of the mortar are changed to eliminate the build-up of hydrostatic pressures in the wet mortar. Therefore, a reduction of the forces in the membrane structure can be achieved in comparison to the liquid mortars. To prove the feasibility of this technique a shell from the original pavilion was recreated. A tubular steel structure was placed on a concrete foundation and a membrane was tensioned in-between. This membrane was provided with insulation on the outside. Next, a concrete shell was shotcast on the inside of the membrane and the outside insulation was covered with a thin layer of concrete stucco (also shotcrete). The membrane is not removed and acts as a water-resistant layer. The shotcrete stucco on the outside has been finished with different techniques to demonstrate the possibilities. This shell has in the original pavilion a height of 18 metres and a width of 6.5 metres, but for budgetary reasons the shell was reduced to a scale of 1:2.5 for the main test and 1:7 for the demonstration model. The two versions of the shell have a height of 7 metres and 2.5 metres, respectively.

The transference of the internal forces within the concrete shell has been calculated with the help of a finite elements method program on the computer, in which, furthermore, the effects of the wind have been analysed. From these calculations it was found that the concrete shell does not require steel reinforcement and that the movement of the concrete under influence of its own weight and wind load is a mere 33 micrometres (wind load 0.75 kN/m^2, concrete thickness 50 mm). The maximum occurring internal force under the same conditions was 0.042 N/mm^2. An increased concrete thickness of 100 mm results in a displacement of 9 micrometres and an internal force of 0.031 N/mm^2. The sagging of the membrane during the shooting of the concrete has been calculated at 45 mm with a concrete thickness of 100 mm wet concrete. But in practice, the shotcrete will be applied in layers, where the first layer will reduce the sagging of the membrane when the next layers are added (Figs. 11.150 and 11.151).

Material

Shotcrete is concrete that is blown though a nozzle at relatively high pressures of 30–50 N/mm^2. Shotcrete has some admixtures to alter the characteristics of the mortar:

- accelerators are used to enhance the rigidity of the wet mortar for the support of new layers of mortar and for minimal flow of the shotcrete under gravity;
- plasticizers are used to reduce the amount of water needed to be able to shoot the mortar; too much water in the mortar results in a reduced final strength;

Fig. 11.150 Concrete displacement

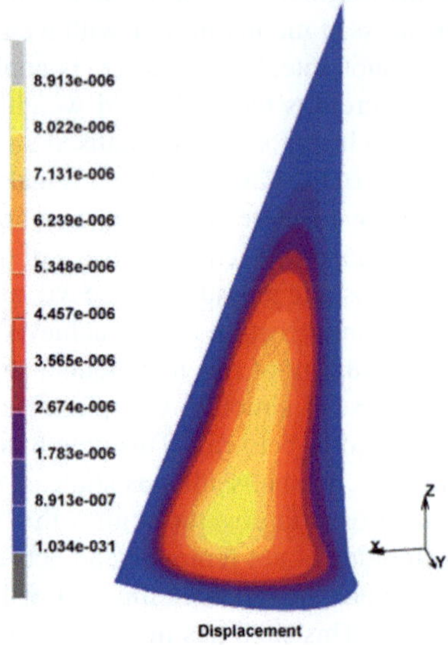

Fig. 11.151 Concrete equivalent Von Mises Stress

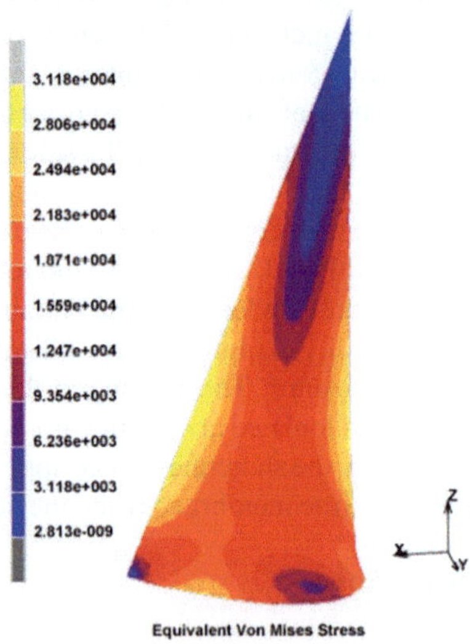

- fibres (steel/polymer) are added to increase the tensile strength of the concrete; with this addition the development of thermal and shrinkage cracks should be reduced (Figs. 11.152 and 11.153).

Before we realised the above-mentioned shell, we did two other experiments. The first experiment was the application of shotcrete on a vertical prestressed membrane of 1.5 × 1.5 m^2. Because the experimental concept (shotcrete on prestressed

Fig. 11.152 Applying shotcrete to a prestressed PVC membrane

Fig. 11.153 Applying shotcrete to a doubly curved prestressed PVC membrane

membranes) was not tested in this form before, the method of 'trial and error' was used to determine the behaviour of shotcrete on membranes. Eventually we were able to apply a thickness of 120 mm shotcrete to a vertical prestressed PVC membrane at the underside of a membrane.

The second experiment regarded the application of shotcrete on a doubly curved prestressed membrane (Fig. 11.154). Point of attention for this experiment was the behaviour of the membrane during the application of the shotcrete. The conclusion to this experiment was that the deformation of the membrane is controllable. It has to be stated that careful engineering and execution are essential. Other aspects of the experiment were the attachment of the reinforcement steel and the adhesion of

Fig. 11.154 Shotcrete on a doubly curved prestressed PVC membrane

the mortar to the steel during execution. The results were good as can been seen in Fig. 11.153, where the reinforcement steel and shotcrete are hanging at the underside of a horizontal stressed membrane.

For the creation of the shape in the scale of 1:2.5 a membrane was stretched between two steel columns and a steel arch was cast into the concrete foundation. This membrane is a woven polymer textile with a PVC top layer that is capable of absorbing great tension strengths. The membrane (Ferrari Precontraint 1502 Fluotop T2) has a maximum tensile strength of 200 kN/m. This is more than the expected strength required due to the weight of the concrete shell, nonetheless it was chosen to manufacture a double layer of the membrane for the first 1.5 metres from the ground. This double layering reduces sagging of the membrane by 50%. The company Tentech made the required calculations and the templates for manufacturing of the membrane.

The connection between the membrane and the steel columns is realized by an aluminium Keder Profile. This connection technique has been tested by the TU/e to analyse the maximum tension load possible. This maximum tension load has been measured at 95 kN/m, where a Ferrari 1202 membrane with a PVC tendon in the hem was used. In four out of five tests, the membrane itself failed before the connection with the aluminium profile failed. One of the tests resulted in the PVC tendon being pulled out of the aluminium profile. For the connection used in the reconstruction of the shell of the pavilion, a stronger membrane was used and a steel cable replaced the PVC tendon (Fig. 11.155).

The membrane has to be extremely well tensioned to reduce vibrations during the shooting of the concrete and sagging or other movement during the hardening of the concrete. To make this tensioning possible, a splice plate is mounted in each of the three corners of the membrane, so the membrane can be tightened with the help of threading and nuts. For applying the 70 mm concrete on the membrane the choice was made to apply it in three phases. The first phase was the shooting of the first two metres with concrete with a thickness of 30–40 mm. After shooting the first layer a glass fibre netting was pressed into the wet concrete to reduce the occurrence of shrink cracks. In the second phase a second layer of shotcrete was applied on the first two metres and in the same act a first layer on the rest of the membrane. Finally, a second layer was applied on the rest of the membrane and a third layer on the bottom

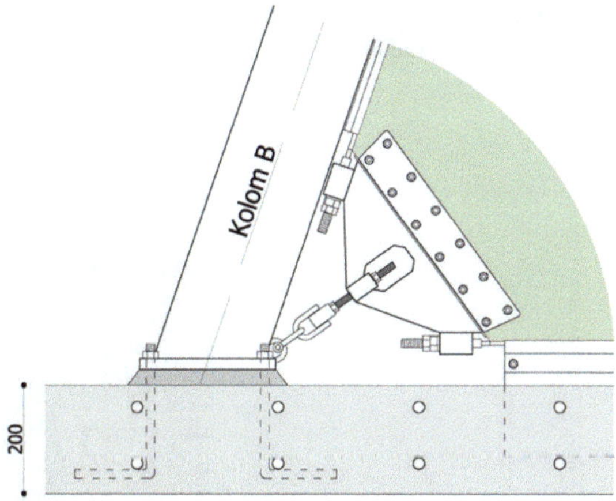

Fig. 11.155 Corner detail for tensioning membrane to steel column

two metres. Because of the increased tensions in the first two metres we decided to have the concrete in the first two metres be thicker than in the rest of the membrane (Figs. 11.156 and 11.157).

The concrete was shot with the wet method. This means that the mortar is mixed with water and pushed through a hose by a worm screw. Inside the nozzle, air under high pressure is added to shoot the concrete out the nozzle. With the dry method the mortar is blown through the hose by compressed air and the water is added in the nozzle. The wet method has as the advantage that the force of the shot is less than in the dry method. Therefore, a smaller percentage of gravel is being rebounded by the membrane and the creation of dust is reduced. The disadvantage of the wet method is that every interruption in the work means that the hose has to be emptied and cleaned. Because of the minimal surface area that had to be shot with concrete, during every phase a lot of concrete was wasted with cleaning the hose. This loss in relation to the area that is worked on is reduced with the increase of the volume and surface area (Fig. 11.158).

For the finishing of the inside surface, the concrete was smoothed with a trowel one hour after spraying and treated with a wet sponge (Fig. 11.159). This produces a relatively smooth, yet sandy texture to the concrete. Insulation is screwed onto the outer surface with a rhombic grid of 100 × 100 mm at one side to be able to bend the insulation plates in the desired curves. To realise a permanent contemporary building with sound and warmth insulation we came to a detail as drawn in Figs. 11.161 and 11.162. This structure was tested in this experiment. The benefit of this method of construction is that the structural shell is at the warm side. This avoids thermal tension and cracks. A concrete shell can be made extremely thin and provides an optimum application of the structurally used material. But in contemporary buildings the dimension of the shell is not determined by the structural demands but by the demands of sound insulation. This is the reason for the relatively thick dimensions (Fig. 11.160).

Fig. 11.156 Spraying concrete on a net structure

Fig. 11.157 Spraying concrete on the membrane

Fig. 11.158 Frame with membrane mould

At the other side of the insulation plate, a steel wire mesh serves as a bonding for a finishing layer of the shotcrete plaster. The outside layer is finished in 5 different ways: (1) roughening, (2) plastering and sponging (3) plastering (4) plastering and polishing, and (5) the raw spray structure (Fig. 11.159).

We also tried a combination of a chain linked wire mesh mould with shotcrete. The benefit of this method is the gentle way the mesh will adapt to a doubly curved form within the boundaries of the frame without making patterns as would be needed with membranes. The net can also be used to connect the reinforcement and will ensure a good connection with the shotcrete.

The construction of the reinforcement in combination with a curved shell and membrane mould is the next problem to be solved. Steel Fibre Reinforced Shotcrete (SFRS) might provide a solution to this problem. It has been applied for more than

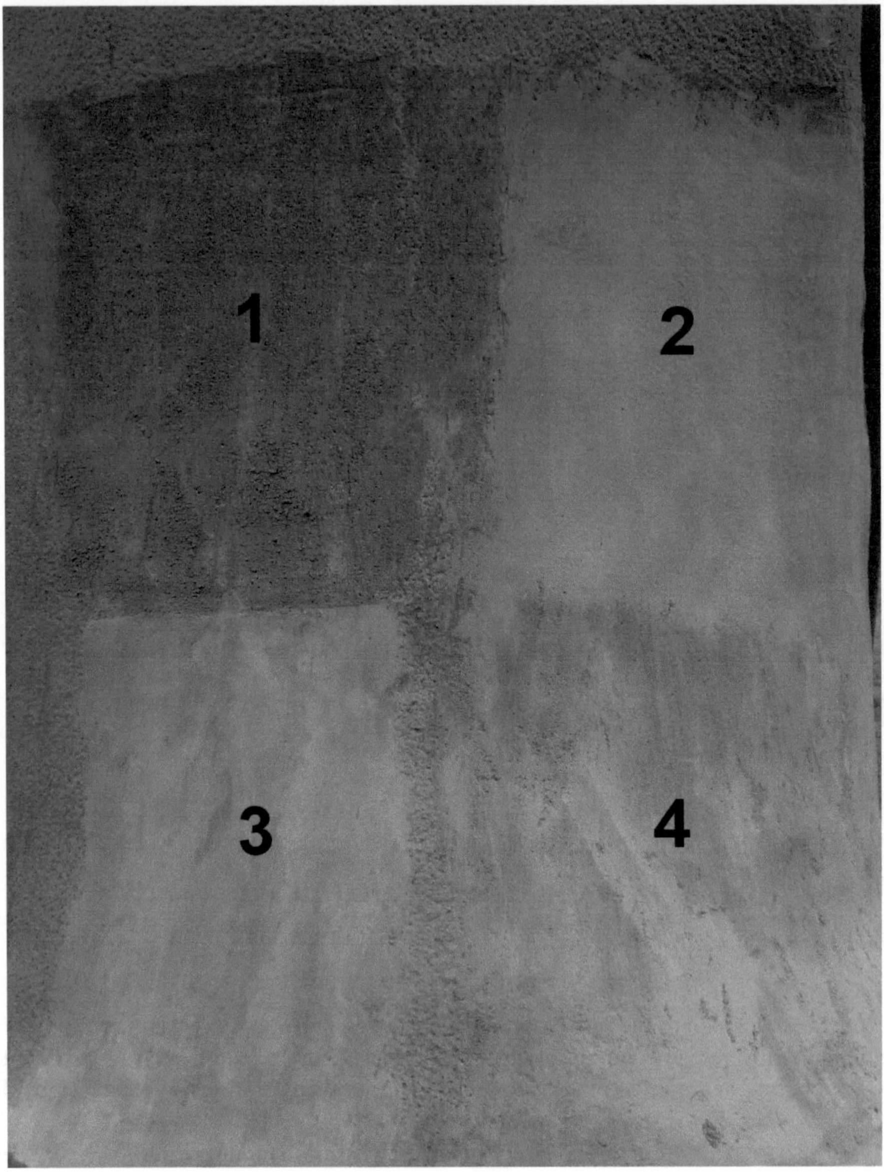

Fig. 11.159 Different surface treatments

Fig. 11.160 Final result

30 years to improve the ductility of concrete. This is always done in combination with conventional reinforcement. To improve the structural behaviour of Steel Fibre Reinforced Concrete, (Rossi et al. 2005: 27–33) came up with the concept of multi-modal fibre-reinforced concrete. The idea is that a hybrid mixture of short and long fibres will give a synergy between both with a better overall result. Markovic (Markovic 2006) proved this assumption in his dissertation (2006) about hybrid-fibre concrete with a concentration of 1–7% volume of fibres combined with high performance concrete which is cast in a conventional way. In the application of shotcrete, it is not possible to spray such a high volume of fibres and high-performance concrete, nevertheless, we

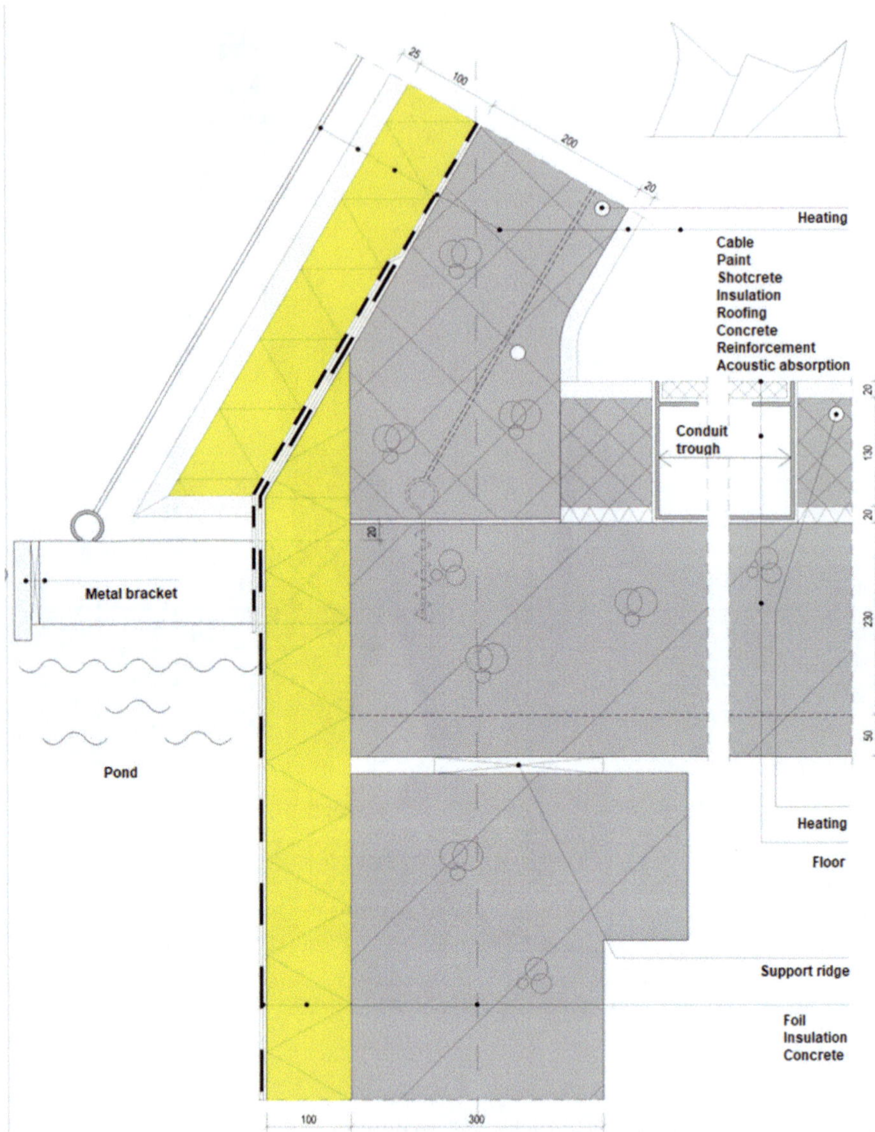

Fig. 11.161 Vertical section of shell at the connection of two shells

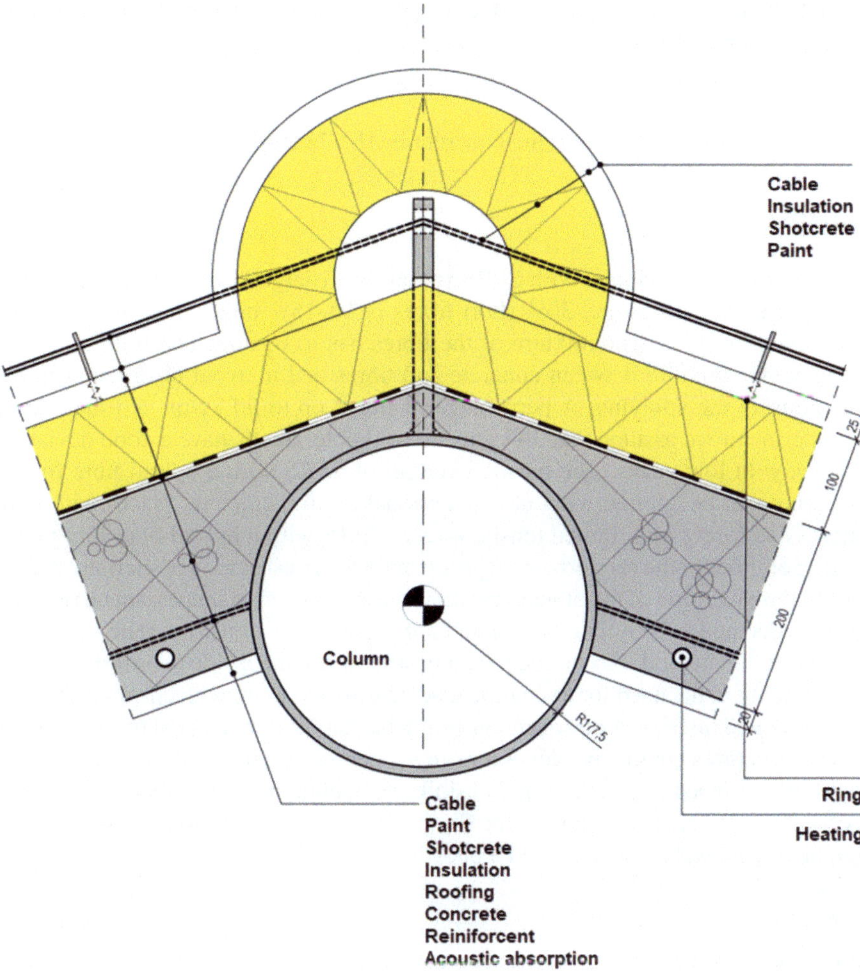

Fig. 11.162 Horizontal section of the foundation

(Hazenberg 2008) did some experiments. In cooperation with B.A.S. (B|A|S, www.basbv.com/index.php/en/) Putzmeister (Putzmeister Holding Gmbh, www.putzmeister.com/enu/index.htm) and Bekaert (Bekaert, www.bekaert.com) 20 tiles of SFRS were sprayed and tested. The fibres of Bekaert Dramix OL 13/0.20 (13 mm length and 0.2 mm diameter) and RC 65/35 BN (35 mm length and 0.55 mm diameter) were applied in several combinations:

- Batch A: 50 kg fibres RC 65/35 BN (reference) (50 kg/m^3 in total) tensile strength 3.82 N/mm^2.
- Batch B: 40 kg fibres RC 65/35 BN, 10 kg OL 13/.20 (50 kg/m^3 in total) tensile strength 3.98 N/mm^2.

- Batch C: 35 kg fibres RC 65/35 BN, 15 kg OL 13/.20 (50 kg/m^3 in total) tensile strength 4.39 N/mm^2.
- Batch D: 30 kg fibres RC 65/35 BN, 20 kg OL 13/.20 (50 kg/m^3 in total) tensile strength 3.87 N/mm^2.
- The compressive strength of the concrete is 61.5 N/mm^2.

Conclusion

The result of the experiment with SFRS is that the synergetic effect of a hybrid mix C with 70% long fibres and 30% short fibres is 10–15% better than the reference with only long fibres. The mixture of the fibres has to be done carefully to have a homogeneous mixture between concrete and fibres and to avoid obstruction in the nozzle during the spraying. A percentage of fibres up to 50 kg/m^3 is usual. After this experiment we assume that 70 kg/m^3 is workable. Shells have a good deviation of forces with low peaks. The tensile strength of SFRS with a hybrid fibre mix is limited but can be applied without conventional reinforcement at places with high compressive forces and limited tensile forces. SFRS with a hybrid fibre mix might give the solution for the reinforcement problems at the construction of shell structures.

With the execution of the other experiments it is shown that shapes can be realized that were deemed to be impracticable and too expensive to construct. The advantage of this building method is a confident lasting structure at the inside (warm side) of the building without the need for a separate load bearing structure with an equal shape at the exterior and interior. A disadvantage might be that the dimensional measurements of these structures could have deviations in the centre of a shell. At the borders, this deviation is reduced to a minimum. Feasibility calculations proved that it is possible to reconstruct the shell structure of the Philips Pavilion for 2 million Euros while the original way nowadays would need a budget of 10 million.

11.19 Seacrete

Flexible Mould
The form-active structure: cable net (Figs. 11.163 and 11.164).
Form-active typology: single/double layer.
Way to stabilize the mould: stressed surface/bending stiffness.

Composite Material
Technique to handle the rigidizing material: submersion.
Reinforcement of the material: ropes/bars.
Surface treatment: none.

Fig. 11.163 Reef construction, Pemuteran Indonesia (*Source* E. Morricone)

Project
Name and location: Karang Lestari Project, Pemuteran Indonesia, TU/e.
Design and Engineering: W. H. Hilbertz & T.J. Goreau, T.J. v. Engelen & A.D.C. Pronk.
Organisation: Global Coral Reef Alliance, TU/e.
Year: 2000, 2003.

Introduction

In Pemuteran Bay in Indonesia residents have created coral nurseries based on the seacrete technology: electrolytic mineral accretion. Corals grown on mineral accretion grow way faster than regular coral. The seacrete technology used in Pemuteran Bay is based on the ideas of Professor Wolf H. Hilbertz and Dr. Thomas J. Goreau (Hilbertz 1970: 98–103; 1979: 94–113).

In the 1970s article in Progressive Architecture Hilbertz explains that the world has been evolving for six or seven hundred million years and that mankind can learn from these continuously improved evolutionary processes and use them in innovative technologies. The seacrete technology, also known as seament and under the brand BioRock, arose from experiments in the 1970s when Hilbertz (1981; Hilbertz and Goreau 1996) was studying how seashells and reefs grow, by passing electrical currents through salt water. In an attempt to recreate this natural growth for

Fig. 11.164 Rings of seacrete formed on an iron tube

the use of protection of metals, he found that around the cathode in the seawater, calcium carbonate forms, leaving a coating which could grow stronger but lighter than concrete. Seacrete created in six weeks, on a 0.5-inch wire mesh, has a breaking strength of 4267 PSI, which is about 20% stronger than regular concrete. If it is deposited for longer periods, it produces an even stronger seacrete, with breaking strengths up to 8000 PSI obtained after one year.

At the Eindhoven University of Technology, T.J. van Engelen and A.D.C. Pronk have succeeded in experiments on quicker creation of seacrete by creating a controlled environment. The possibility of reaching higher speeds and the ability to control the environment make the use of seacrete as a building material more feasible.

Material

The result is a combination of materials. The basic shape is made out of rods of a conductive metal, for instance steel. The composition of the deposited material is a combination of calcium carbonate and magnesium hydroxide. The harder and therefore preferred component is calcium carbonate. Magnesium hydroxide is softer and is therefore the non-preferred component. The faster seacrete accretes, the higher the concentration of the softer magnesium hydroxide component is in the resulting material.

Form

One of the advantages of seacrete is the freedom of form. Because seacrete is formed by accretion, its building process can be described as a form of growth. A seacrete structure starts only with the basis and the seacrete is slowly deposited around it. This technique of accretion has the advantage that no external formwork is needed. New particles of calcium carbonate and magnesium hydroxide are only formed on top of the conductive metal or on already formed seacrete, so there is no extra need to keep the materials together. There is no requirement for sand, aggregate or cement. The only external material supply is the seawater itself. This freedom means that the only limit for the form of the construction is the limit on what can be created with steel rods, cables or nets. Examples of this freedom of creation can be seen in the Pemuteran Bay project (Fig. 11.165).

Fig. 11.165 Underwater base for seacrete (*Source* M. Pinta)

Technique

A pre-shaped form of electrically conductive material is disposed in a volume of electrolyte, such as seawater, to serve as a cathode and one or more anodes are placed in proximity to the form. By establishing a direct electrical current between the electrodes, calcium carbonates, magnesium hydroxides, and hydrogen are precipitated at the cathode, while at the anode, oxygen and chlorine are produced. For the creation of seacrete only the precipitation at the cathode is of direct interest. As long as the electrical current flows, calcium carbonate and traces of other minerals will continue to accrete and thicken on the metal indefinitely. An advance over regular concrete would be the possibility to repair damaged parts of seacrete, by placing it back in seawater and re-attaching an electrical power supply (Eashwar et al. 2013: 185–193; Iqbal et al. 2010).

In order to understand the process of the creation of seacrete, the principle of electrolysis must be understood. When an electrical current is passed through an electrolyte fluid, electrons flow from negative cathode to the positive anode. As the cathode is negatively charged it attracts positive ions in the fluid and the positively charged anode attracts negative ions.

Chemical reactions

The complete process that leads to the creation of seacrete is as follows:

- As the voltage across the electrode rises, there will be a point where the cathode will become electronegative enough to attract the hydrogen ions in the seawater, donate electrons to them and take them out of the solution by converting them into hydrogen gas which then bubbles up to the surface of the ocean.

$$2e^- + 2H^+ \Leftrightarrow H_2(\text{gas})$$

- As the hydrogen ions become depleted near the electrode, there is a chemical reaction involving the carbonic acid in seawater that will try to re-establish the old equilibrium concentration (Le Chatelier's Principle). Carbonic acid can disassociate to form bicarbonate ions and hydrogen ions. The decrease of hydrogen ions triggers the following reaction to move to the right side.

$$H_2CO_3 \Leftrightarrow H^+ + HCO_3^- \Leftrightarrow 2H^+ + CO_3^{2-}$$

- This reaction will also increase the concentration of carbonate ions in the solution. When the solubility product of calcium ions and carbonate ions exceeds the maximum amount that can be kept in the solution, calcium carbonate will be deposited as a solid on the cathode. This solid calcium carbonate, also called argonite, is the preferred material in seacrete: hard, strong and nearly insoluble.

$$Ca^{2+} + CO_3^{2-} \Leftrightarrow CaCO_3\ (\text{solid})$$

- The formation of $CaCO_3$ is the first accretion related reaction that occurs as the cathode voltage and current density rises. As the cathode voltage and current density is increased further another reaction starts to dominate.

This second reaction is what forms the majority of the seacrete at high current densities. As the hydrogen ions near the cathode are turned into hydrogen gas, the region near the cathode becomes depleted of H^+ ions and this increases the pH of the region near the cathode making the liquid into a base. This forces the following reaction to move to the right:

$$H_2O \Leftrightarrow H^+ + OH^-$$

- The result is not only that the hydrogen ion level goes up again, but also makes the local hydroxide ion concentration go up. When the solubility product of magnesium ions and hydroxide ions exceeds the maximum amount that can be kept in the solution, magnesium hydroxide will be deposited as a solid on the cathode. This solid magnesium hydroxide, also called brucite, is a softer material than calcium carbonate and is the non-preferred component in seacrete.

$$Mg^{2+} + 2OH^- \Leftrightarrow Mg(OH)_2 \text{ (solid)}$$

- Recent Indian studies show one of the elements that can increase the strength of seacrete is the proximity of sunlight and some symbiotic organisms. When the amount of carbon dioxide rises, the calcification reaction slows down. When a photosynthesis reaction is triggered, the amount of carbon dioxide drops. When this amount drops, the calcification reaction runs faster again.

$$Ca^{2+} + 2HCO_3^- \Leftrightarrow CaCO_3 + CO_2 + H_2O \text{ (calcification)}$$

$$CO_2 + H_2O \rightarrow CH_2O + O_2 \text{ (photosynthesis)}$$

Power

The amount of electrical current required to cause seacrete to accrue on the wire mesh is low. Only thirty amperes at six volts is needed to form a current density of 1.3 milliamperes per square inch and create a one-inch thickness of seacrete in approximately 6 weeks. For more rapid deposition of calcium carbonate onto the wire mesh, current densities as high as 50,000 milliamperes per square foot can be used, but the greater the current density, the weaker the seacrete.

Professor Hilbertz perfected the accretion of seacrete down to exact formulas. 189 milliamperes of current density through a 0.5-inch wire mesh will accrete 0.1 inch of seacrete in 170 h. This is an accretion rate of .0005 inches per hour. It takes three weeks, at that current density, to fill the 0.5-inch gaps in the wire mesh.

The amount that is extracted from the seawater and deposited is claimed to be as follows: one kilowatt hour of electricity, running through a wire mesh, will extract 1.9 kilograms of calcium carbonate out of the seawater and deposit it onto the wire mesh as a solid (Fig. 11.166).

Application

Although seacrete has not been used as such, it could possibly be used as a building product. Hilbertz in Toward Cybertecture already mentioned a vision of creating complete buildings under water. In order to make the creation of buildings or building components by accretion feasible, it is necessary to create seacrete faster. In order to do this, it could be a solution to create a controlled environment instead of using the sea.

T.J. van Engelen and A.D.C. Pronk performed some initial tests in this field at the Eindhoven University of Technology (van Engelen and Pronk 2003). They used

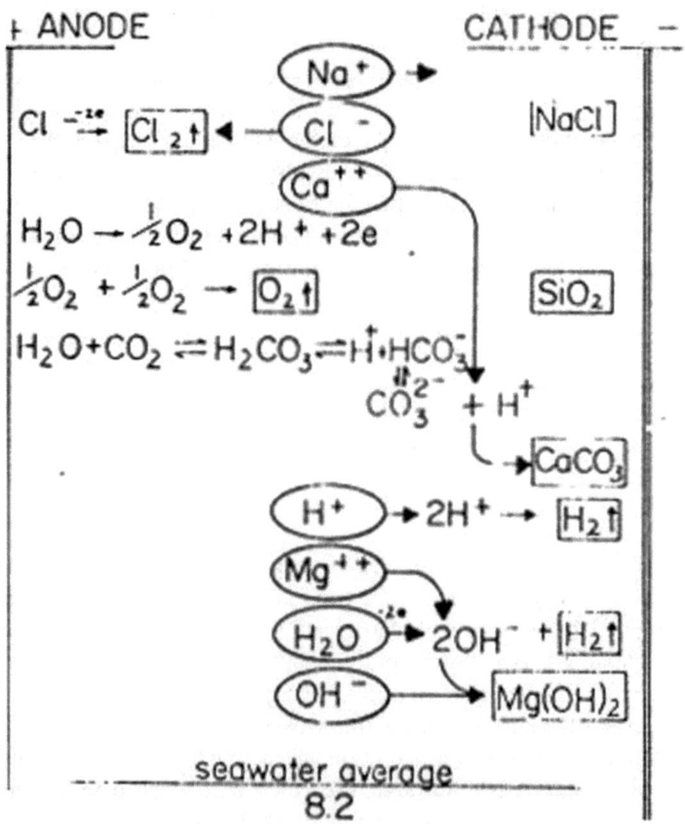

Fig. 11.166 Accretion process (*Source* W. Hilbertz)

a mixture of calcium chloride hexahydrate and sodium chloride in water and added extra carbon dioxide.

As mentioned, a higher electrical current through seawater leads to a higher formation of the softer and undesirable magnesium hydroxide. The absence of magnesium in the mixture of Van Engelen and Pronk makes it impossible to form the softer magnesium hydroxide. For this reason, the use of a higher electrical current, which leads to faster accretion, becomes possible. The sodium chloride in the solution serves as the oxidator in the electrolysis. The addition of carbon dioxide results in a faster formation of the necessary carbonate ions. This design leads to the possibility of faster accretion.

Several tests were done. Raising the electrical current up to 750 W with a maximum voltage of 60 V it was possible to create 4 mm of accreted material in 44 h. This is about 0.0035 inch per hour which is seven times as fast as the design by Hilbertz. In this test, the quality and exact composition of the accreted material was not tested, however, and the electrical current density that was used fluctuated.

The fluctuation of the process was caused by a drop in the pH level of the liquid to acid (5.13) and this will frustrate the chemical reaction. Within this limited research, this effect was not resolved. In the experiments with the higher electrical currents, the chemical process also produced H_2, CO_2, CL_2 beside the calcium carbonate. H2 gas can be used for the production of energy. CO_2 gas is a natural gas, CL_2 has to be collected and can be used as a commodity in all kinds of applications. Besides the non-toxic CO_2, all other side products can be used in useful applications. So, the results are promising, but because of the fluctuating process, more tests are desired.

Conclusion

Because of the relatively slow process of accretion, the logical manner for creating building components by accretion would be in prefabrication. One possible way is to create moulds in which a layer of metal is added to the surface of the mould, for instance by thin film deposition of a metal on a fabric. The metal surface makes it possible to create a wide variety of bases for seacrete to grow on. In this way it is also possible to rigidify a doubly curved flexible surface with seacrete. Although the chemical process and the electrical current may need looking after, the physical set up of an accretion process is quite simple. There is the metal base as a cathode and an anode and a mixture. Because the seacrete is accreted on the metal base without intervention and only needs waiting, it makes it possible to completely control and fine-tune the form of the metal base before applying the electrical current. This makes it possible, for instance, to create a "Gaudian" catenary system, making use of gravity, which can be placed in a basin. After the accretion process, the resulting building component can be placed upside down and due to the catenary form, there will be a perfect diversion of gravitational forces (Figs. 11.167 and 11.168).

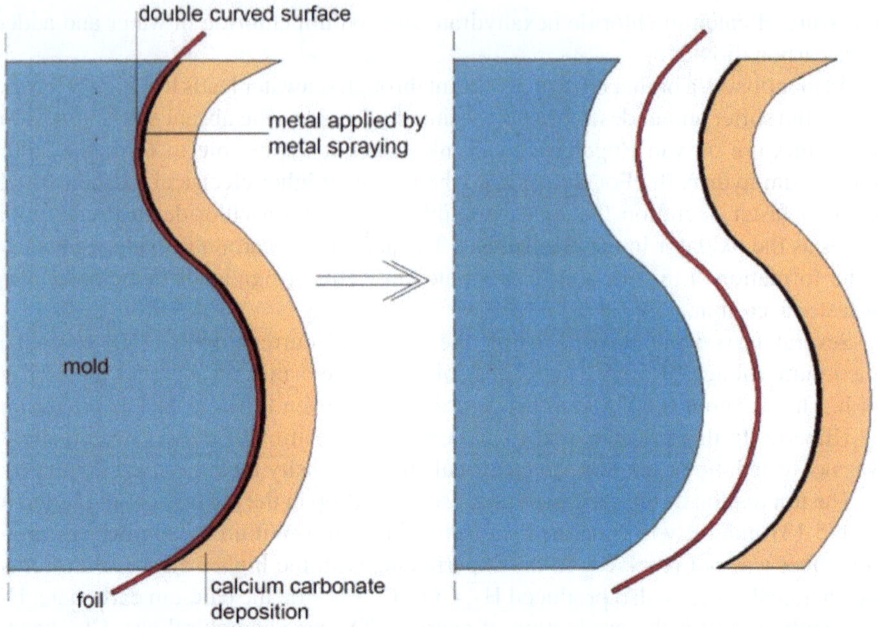

Fig. 11.167 Seacrete process on free-form mould

Fig. 11.168 Gaudian catenary system

11.20 Summary

In this chapter on concrete composites, several moulding systems with a fabric mould for the realization of the following projects have been researched:

i. a concrete bone structure based on a membrane mould on hydrostatic pressure;
ii. concrete structures in combination with distance fabric and additional reinforcement;
iii. a 16 × 16 m shell structure for a parking garage;
iv. a Barbapapa object based on a patent by Pronk on the combination of inflatables and wire frames with concrete;
v. the construction of a shell for a bicycle cellar in Eindhoven by Fuksas;
vi. a design and construction of a prototype fluid shell for a houseboat;
vii. the design and construction of a prototype for a recreation of the Philips Pavilion by Xenakis and Le Corbusier; and
viii. some experiments with Seacrete on a steel mesh.

Bibliography

BlAlSl. BlAlS Research & Technology in the concrete and asphalt. http://www.basbv.com/index.php/en/. Accessed 1 Aug 2013
Bakker ED (1951) Inflatable mattress. US Patent 2753573
Bekaert. http://www.bekaert.com/. Accessed 1 Aug 2013
Bini D (1967) A new pneumatic technique for the construction of thin shells. In: Proceedings of the first international association for shell structures (LASS), international colloquium on pneumatic structures, University of Stuttgart, Germany, May 1967
Bini D (1969) Method for erecting structures. US Patent 3,462,521
Bini D (1974) Concrete domes. Official J Masters Builders' Assoc NSW 3(7)
Brewin P, Crawford W (2005) WO2005124063A2. A prefabricated shelter
Brewin P, Crawford W (2010) Prefabricated shelter. US Patent 7,721,749
Brown G, Haggard R, Brook Norton B (2001) Inflatable structures for deployable wings. In: 16th AIAA aerodynamic decelerator systems technology conference and seminar
Bullens T, Folmer T (2007) Vrije vormen in beton, alternatieve bekistingsmaterialen voor dubbel gekromde vormen. Faculteit Bouwkunde, mastertrack Building Technology, chair Product Development
Cauberg N, Parmentier B, Janssen D, Mollaert M (2008) Fabric formwork for flexible Architectural concrete. Taylor & Francis Group, London
Cauberg N, Parmentier B, Vanneste M, Mollaert M (2009) Fabric formwork for flexible, architectural concrete. In: Proceedings of the international symposium on fibre-reinforced polymer reinforcement for concrete structures, FRPRCS-9, Sydney
Chandler A (2008) Fabric-formed walls-Kenzo Unno. In: Fabric formwork conference for architectural structures, C.A.S.T., University of Manitoba
Chenoweth AC (1916) Protective reinforced concrete construction. US Patent 1,187,502
Chilton JC (2000) Heinz Isler. The engineer's contribution to contemporary Architecture, Thomas Telford Ltd, London

Cocke Jr. BW (1958) Wind-tunnel Investigation of the Aerodynamic and Structural Deflection Characteristics of the Goodyear Inflatoplane. In: National advisory committee for aeronauts, memorandum. Langley Aeronautical Laboratory, Langley Field VA
Concrete Canvas. Concrete Canvas: Erosion control & containment. http://www.concretecanvas.co.uk/. Accessed 28 Feb 2017
Condie C (1911) Revetment-mattress. US Patent 984,121
Cross WB (1965) Method of increasing stiffness of an inflated structure. US Patent 3,205,106
Domebuilder.com. http://www.domebuilder.com/. Accessed 28 February 2017
Dominicus MMT, Hoppe LD ten, Beetz J, Pronk ADC, Mohammadi M (2012) Facade elements based on optimised open cell structures: An Experimental modelling process for the designer. Design and Decision Support Systems in Architecture and Urban Planning Conference (DDSS2012), 27–29 August 2012, Eindhoven, The Netherlands, pp 1–8
Eashwar M, Sathish Kumar P, Ravishankar R, Subramanian G (2013) Sunlight enhances calcareous deposition on cathodic stainless steel in natural seawater. Biofouling 29(2):185–193
van Engelen TJ, Pronk ADC (2003) Seament: calciumcarbonaat als bouwproduct. TU Eindhoven, Eindhoven
Florian Schmid. Stitching Concrete. http://www.florian-schmid.com/stitching_concrete_collection.html. Accessed 28 Feb 2017
Garcia M (2010) Amanda Levete Architects (AL_A). Architectural Design 80(1):106–113
Gilbert J, Patton M, Mullen C, Black S (1970) Vacuumatics. Queen's University, Belfast
Hardress de Warenne Waller J (1934) Method of building with cementitious material applied to vegetable fabrics. US Patent 1,955,716
Hazenberg B (2008) Hybrid-fibre shotcrete. TU Eindhoven, Eindhoven
Herzog T, Bacon S (1977) Pneumatic structures: a handbook for the architect and engineer. Crosby Lockwood Staples, London
Hilbertz WH (1970) Toward Cybertecture. In: Progressive architecture, a reinhold publication: 98–103
Hilbertz WH (1979) Electrodeposition of minerals in sea water: experiments and applications. IEEE J Oceanic Eng 4(3):94–113
Hilbertz WH (1981) Mineral accretion of large surface structures, building components and elements. US Patent 4,246,075
Hilbertz WH, Goreau TJ (1996) Method of enhancing the growth of aquatic organisms, and structures created thereby. US Patent 5,543,034
Huijben FAA (2009) Vacuumatics 3D formwork systems: customised concrete architecture. The Masterbuilder 11(6):78
Huijben FAA (2014) Vacuumatics 3D formwork systems: customised concrete architecture. NPN Drukkers, Breda
Iqbal M, Subhan B, Madduppa HH, Zamani NP, Isnul J, Bachtiar R, Adi JW (2010) Study on biorock® technique using three different anode materials (magnesium, aluminium, and titanium). Jurnal Ilmu dan Teknologi Kelautan Tropis 2(1)
Kalff LC, Tak W, Bruin SL de (1958) The "electronic poem" performed in the Philips pavilion at the 1958 Brussels World Fair". Philips Tech Rev 20(2–3)
Kassel Institut für Konstruktiven Ingenieurbau, Universität Kassel (2007) Ultra high performance concrete: (UHPC); 10 years of research and development at the University of Kassel (No 7). Kassel University Press GmbH, Kassel
Knaack U, Klein T, Bilow M (2008) Imagine 02: Deflateables. 010 Publishers, Rotterdam
Lilienthal LWG (1899) Fireproof Ceiling. US Patent 619,769, 1899
Manelius AM (2012) Fabric formwork for concrete—investigations into formwork tectonics and stereogentity. Royal Danish Academy of Fine Arts, Copenhagen
Markovic I (2006) High-performance hybrid-fibre concrete. TU Delft, Delft
Milliken. Concrete cloth. http://www.foundationperformance.org/pastpresentations/MillikenPres Slides-Nov12.pdf. Accessed 28 Feb 2017
Monolithic. http://www.monolithic.com. Accessed 10 June 2013

Neff W (1942) Building construction. US Patent 2270229A
Neff W (1945) Construction of concrete buildings. GB Patent 603655A
Neff W (1953) Improvements in concrete building construction. GB Patent 1672953A
Oesterle S, Vansteenkiste A, Mirjan A (2012) Zero waste free-form formwork. In: Proceedings of the second international conference on flexible formwork, ICFF. CICM and University of Bath, Dept. of Architecture and Civil Engineering, vol 30, p 52
Pedreschi R (2011) The use of fabrics as formwork for concrete structures and elements. In: International conference on textile composites and inflatable structures, Structural Membranes, Barcelona
Phane WJ (1954) Inflatable fabric structural element. US Patent 2,698,020
Pronk ADC (2008a) A spatial structure and a method to build such a structure. EP Patent 2007007587
Pronk ADC (2008b) Fuksas in Eindhoven, underground bicycle storage. Stedenbouw en Architectuur, 2008(3): 27
Pronk ADC, Bullens T, Folmer T (2007a) A feasible way to make freeform shell structures. Proceedings of IASS, Venice
Pronk ADC, Haas T de (2007b) Heat-transmitting membrane. International Conference on Textile Composites and Inflatable Structures
Pronk ADC, Houtman R (2004) Making blobs with textile mould. TU Delft, Delft
Pronk ADC, Houtman R, Afink M (2006) The Philips Pavilion. In: Extended abstracts of international symposium on new Olympics new shell and spatial structures
Pronk ADC, Houtman R, Afink M (2007) The reconstruction of the Philips-pavilion. Sources of architectural form, theory and practice, Kuwait City, vol. 1
Pronk ADC, Maffei R, Martin H (2009) Water as a sound insulation in a membrane structure. In: International conference on textile composites and inflatable structures
Putzmeister Holding GmbH. Index. http://www.putzmeister.com/enu/index.htm. Accessed 1 Aug 2013
Rossi P, Arca A, Parant E, Fakhri P (2005) Bending and compressive behaviour of a new cement composite. Research: 27–33
Rozvany GIN (2001) Aims, scope, methods, history and unified terminology of computer-aided topology optimization in structural mechanics. Struct Multidisc Optim 21:90–108
Sobek W (1987) Auf pneumatisch gestutzten Schalungen hergestellte Betonschalen. Stuttgart, pp 1–34
South DB (1979) Building structure and method of making same. US Patent 4155967A
Store J (1922) Method of constructing subaqueous concrete structures. US Patent 1,421,857
Technet Gmbh. http://technet-gmbh.de. Accessed 28 Feb 2017
Treib M, Feliciano R (1996) Space calculated in seconds: The Philips Pavilion, Le Corbusier, Edgard Varese, vol 41. Princeton University Press, Princeton
University of Manitoba. Kenzo Unno. http://www.arch.umanitoba.ca/cast/2.kenzounno/. Accessed 5 June 2013
University of Manitoba. Mark West. http://umanitoba.ca/faculties/architecture/facstaff/faclist/west.html. Accessed 28 Feb 2017
Unno K (1998) 'Curvy' concrete. Nikkei Archit Mag 608:124–127
Unno K (1999) 'Curvy' concrete. Toto Mag 4:1–9
Unno K (2007) Fabric-formed walls. Quilt-point Method 4(11)
Verhaegh RWA (2010) Free forms in concrete: the fabrication of free-form concrete segments using fabric formwork. Technische Universiteit Eindhoven, Department of the Built Environment, Eindhoven
Walraven JC (2004) Designing with ultra-high strength concrete: basics, potential and perspectives. In: Proceedings of the international symposium on ultra-high performance concrete, Kassel, Germany, Sept 13, vol 15, pp 853–864
Walter J (2012) Paper 16: Crushedwall, Heartlands. In: Proceedings of ICFF 2012
Weiss AC (1956) Fiber-reinforced structural panel and method of making same. US Patent 2,762,739

West M (1995) Method of forming a concrete column capital in a standard flat plate concrete slab. US Patent 5,390,464
West M (2001) Fabric-formed concrete structures. University of Manitoba, Winnipeg
West M (2004) Fabric-formed Concrete Columns. Concr Int 26(6):42–45
West M (2007) Method of casting a concrete truss. US Patent 11/770, 174
West M (2008) Kenzo unno, fabric formed walls. University of Manitoba, Winnipeg. Available via: http://fabwiki.fabric-formedconcrete.com/lib/exe/fetch.php?media=unno:kenzo_unno_article.pdf. Accessed 5 January 2017
West M (2016) The fabric formwork book: methods for building new architectural and structural forms in concrete. Routledge, New York
West M. Thin-Shell Concrete From Fabric Moulds. http://fabwiki.fabric-formedconcrete.com/lib/exe/fetch.php?media=chile:thin_shell_panels.pdf. Accessed 28 Feb 2017
West M, Arraya R (2009) Fabric formwork for concrete structures and architecture. Centre for Architectural Structures and Technology / University of Manitoba, Faculty of Architecture
West M, Arraya R (2010) Fabric-formwork for reinforced concrete structures and architecture. Centre for Architectural Structures and Technology (CAST), University of Manitoba
Xenakis I (2008) Music and architecture: architectural projects, texts, and realizations. Pendragon Press, Sheffield, MA
Yang CY, Li WZ (2011) The design principle and weaving technique of distance fabric. Adv Mater Res, 332: 647–650. Trans Tech Publications, Baech
Yokota T, Horiya S, Uno S, Katayama K, Tanabe D (1989) Concrete form shuttering having double woven fabric covering. US Patent 4,856,754
Zimmerman J (1988) Air formed concrete dome structures. Spec Publ 107:41–58
Zimmermann G (2007) Membran-Beton-Gitterschalen-Tragwerke: Entwicklung und Vorbemessung. Books on Demand GmbH
Zimmermann G, Teichmann T (2004) Membrane concrete grid shells—UHCP grid shells. Ultra high performance concrete, p. 839. In: Proceedings of the international symposium on ultra high performance concrete, Kassel, pp. 13–15

Chapter 12
Future Developments

In Chap. 2, the necessity to combine synclastic and anticlastic curvature for free-form surfaces was demonstrated. By combining synclastic and anticlastic curvature, an endless horizon of possibilities and freedom in design can be achieved, i.e. the truly "free" form. However, this does not mean that there are no limitations for realizing any form that is designed. In "free" form architecture, technique and material have become the limiting factors. Because of these limitations, the word "free" is misleading. Therefore, it is replaced here by "fluid".

A methodological overview of the ways a membrane can be manipulated is presented together with the major geometrical effects of those manipulations. This overview aims to cover all the possibilities.

In fluid architecture the forms are not dominated by structural optimization but by other reasons. Release of the structural domination was achieved for the first time in the late French gothic style known as "flamboyant". The historical development of fluid architecture can be classified in two ways. In the first classification, the relation between structure and form is the determining factor. In the second, the different style periods within fluid architecture are classified. In this book, the following style periods of fluid architecture have been determined: precursors, early, modern, deconstructive and individual fluid architecture. Important buildings in this overview are the San Carlo alle Quattro Fontane, the Statue of Liberty, TWA terminal, the church in Ronchamp, the Philips Pavilion and the Guggenheim Museum in Bilbao. The complete separation of building envelope and structure in fluid architecture was realized for the first time in the Statue of Liberty (early fluid architecture) and comes back in the fluid architecture at the end of the twentieth century in the Guggenheim museum in Bilbao (deconstructive fluid architecture). The architectural shape of the 1958 Philips pavilion, the 1956 TWA terminal and the 1954 church in Ronchamp are not dominated by the actual structure. The perceived structural elements are architecturally applied and are far from efficient structural applications of doubly curved shell structures. These buildings can be seen as the first examples of "modern" fluid architecture.

A. Pronk, *Flexible Forming for Fluid Architecture*,
https://doi.org/10.1007/978-3-030-71551-9_12

What can be concluded from the historical development of styles and the position of fluidity in Architecture is that each important style period ends in a 'superfluous' practice of irregularly curved surfaces. The four main styles are: Gothic, Renaissance, Romantic and Modern.

- The gothic style evolved into the flamboyant style (1350–1500) (precursors of fluid architecture),
- The renaissance style evolved into the rococo (1700–1800) (early fluid architecture),
- The romantic style evolved into the Art Nouveau (1890–1910) and (early fluid architecture),
- The modern style evolved into fluid de-constructivism (since 1988) (modern and deconstructive fluid architecture).

From this list one might conclude that at this moment we are at the end of an important style period. It is uncertain how long this period of deconstructive fluid architecture will last. Fluid architecture in different style periods has always been supported by good economic prosperity and optimism. It is possible that for reasons like the increase in international terrorism, the exhaustion of natural resources and overpopulation, we will face a long-lasting global recession. This will have consequences for our society, art and architecture. Considering the death of one of the most important representatives of fluid architecture, Zaha Hadid, as well, I think we are at a turning point and very soon we will face a new style period.

The relation between architecture and organic shapes substantially influences fluid architecture. The reference to nature in the form of flames (flamboyant) and Shells (rococo) were the motivation to make asymmetrical forms in architecture. With regard to membranes the relation between the skin of a body and architecture is discussed. Also, the human skin can be compared to the surface of fluid architecture and placed in a long historical tradition where human measurements are used in architecture. The skin of a body behaves like a membrane. It is doubly curved, irregular, smooth and elastic, just like the appearance of fluid architecture.

There are no functional reasons for fluid architecture. Fluid architecture refers to organic shapes and is suggestive. The reasoning for this kind of architecture can be found in its reference to organic attractive shapes and appeal to erotic shapes. Objects or bodies in nature are rarely mimicked explicitly in architecture. Only in other (visual) art and in primitive cultures is there an explicit reference to naked bodies and sexuality. In art as well as in advertising, naked bodies (Balgaradean and Babut 2015: 63–77; Reichert and Lambiase 2003) draw almost all the attention away from the surrounding environment. In fluid architecture the building is also a stand-alone, not—or poorly—connected to the surrounding area, taking all the attention and creating a world in itself. The implicit reference to erotic shapes in fluid architecture explains why it is still a point of discussion. Like the use of erotic features in advertising, this reference will cause more attention and publicity. This will be seen as positive as long as the reference is not too explicit. This can be a reason for clients of highbrow buildings, to order a stadium, museum or other public building in this style.

The analogy between organic surfaces and the building envelope of fluid architecture explains why it is valid to research flexible moulding based on the manipulation of membrane surfaces. Structures realized with flexible moulding techniques based on the manipulation of membranes are one of the keys to a better integration of the building envelope and structure within fluid architecture.

Three examples of fluid architecture were described and analysed. The three architects reviewed in this book, Oosterhuis (The web of North Holland), Gehry (Guggenheim) and Cook/Fournier (Kunsthaus) strive in their designs for the integration of the façade and the structure. However, they used techniques that have been developed for orthogonal buildings in which the traditional separation between the main load bearing structure and other building elements prevails. In two cases, the outer surface is not the water-resistant layer. Oosterhuis and Cook/Fournier used the façade as a doubly curved second-skin façade. In all cases, the use of 3D computer software made it possible to analyse, optimize and realize the form designed by the architect, which has led to high-profile buildings. The increased complexity of the buildings also led to problems concerning the measurements of elements and the connection between the support structure and façade panels.

The deviations of complex curved elements in a building façade need extra attention. In the three case studies above, these problems were handled differently. Eric Vastert researched the perceptibility of deviations between building elements in his dissertation. He concluded that the perceived deviation is based on three aspects:

1. the pattern of the elements:
   - size of the panel;
   - width of the join;
   - number of joins;
   - regularity; and
   - complexity and proportion;
2. the surface of the elements; and
3. the observer.

The perceived deviation is very much related to the architectural style. Clean, simple, regular, rectangular, smooth buildings with big elements and a small width between the elements are the most sensitive to deviations in the size of the elements and voids. The perceptibility of deviations is mostly influenced by the proportion of the elements and the joins. In the Web of North Holland, the width of the space between the elements is relatively big. This is also the most important aspect when aiming to avoid perception of the deviations in the façade. The joins form a pattern of continuous lines that give the object a kind of kinetic gesture. In the Kunsthaus in Graz, the architects had a much more difficult job due to the small width of the joins and the transparent smooth material. The joins where not supposed to be perceived as a pattern. In the Guggenheim building it was much easier to avoid the perceptibility of deviation in the size of elements due to the choices in material, size and pattern of the façade.

The morphology of the three buildings above was analysed in relation to the Gestalt Theory of Max Wertheimer, as well as to the 15 basic properties necessary to create spaces that contain a certain aspect of "life" as distinguished by Christopher Alexander.

The conclusion is that in all cases, fluid architecture has the following features:

1. it is irregular;
2. it is perceived as smooth;
3. it is perceived as doubly curved;
4. it is built according to the principles of Vitruvius: Utilitas, Firmitas and Venustas;
5. it has a multi-faced body with a combination of perceived synclastic and anticlastic surfaces;
6. it avoids strong centres;
7. it avoids clear boundaries;
8. it encloses an undefined space;
9. it has an undefined shape;
10. it has main boundaries without symmetry. If symmetry is found it is a local symmetry;
11. it merges different volumes; if the size of the building is big enough, the different spaces within the building are connected by a void;
12. it has a complex geometry;
13. it presents a unique object; there is no relation or a very limited relation with other buildings on that site;
14. it has smaller building elements, gradually changing in size and shape, which do not call for attention themselves, but contribute to the building surface as a whole;
15. it has a perceived fluidity of the surface that refers to organic forms.

The most important conclusion is that in fluid architecture one has to *perceive* fluidity, i.e. that there is no absolute demand for a physical fluidity as long as the audience will perceive the building as fluid.

The possibilities for flexible and form-active structures as a mould to make other typologies were illustrated in diagrams. This demonstrates that by using form-active mouldings, mass-, vector- and surface-active structures can be made. Form-active structures and surface-active structures have a strong similarity, with one exception, namely the difference in construction: in situ vs. prefabricated (off site). The ability of form active structures to act as the prefabricated mould for surface-active structures is the best known application and it is practised in the construction of domes and shells.

In order to demonstrate that flexible moulding has the capacity to be used for the construction of fluid architecture, case studies of three artists who made a pavilion with a given flexible moulding technique were presented. The technique did not prevail over their own input. They were all able to make their designs without giving up their own artistic style.

To answer the question above, an overview is given of flexible moulding through history in different materials and techniques. The materials that are to be moulded need to be able to make a phase change from liquid to rigid. Therefore, the materials water/ice, polymer composite, concrete and glass were chosen for further inspection. All known techniques in relation to flexible moulding are summarized and categorized according to six different aspects in such a way that they can be compared:

1. form-active structure
2. form-active typology
3. way to stabilize the mould
4. technique to handle the rigidizing material
5. reinforcement
6. surface treatment.

This overview provides an opportunity to look for new combinations. This is not a definitive list. As new techniques or materials are developed, the list can be extended and new combinations can be added. Flexible moulding is compared to other (moulding) techniques feasible in case of:

1. Curved surfaces
2. Unique products, or very small series
3. Large tolerances.

The following conclusions can be drawn:

1. The morphology of surface-active and form-active structures is close or similar to the morphology of fluid architecture.
2. Form-active structures are an appropriate typology for long span structures, lightweight structures and temporary buildings.
3. The ability of flexible and form-active structures to be the prefabricated mould for all other structures is demonstrated by many projects and companies, although most of the projects are focused on the making of surface-active structures such as dome or dome-like structures.
4. The experiments by Linkwitz and Otto show the basic principles and endless possibilities concerning the manipulation of pre-stressed membrane structures, catenaries and inflatable structures.
5. Surface-active structures are the most material-efficient structures compared to vector-active and mass-active structures.
6. The ability to manipulate flexible and form-active structures and adapt other structural typologies leads to an endless number of different forms.
7. Surface-active structures are the most efficient typology to make fluid architecture.
8. If a pure surface-active structure is not possible, a hybrid combination of section-active and vector-active structures has to be considered.

This leads to the final conclusion that flexible and form-active structures in many cases are appropriate for the construction of fluid architecture.

Case studies to be developed for the combination of ice with:	
Form-active structure:	1. Knitted fabric.
Form-active typology	2. Double layer.
Way to stabilize the mould:	3. Hydraulic pressure, 4. Underpressure.
Rigidizing technique:	5. Hand lay-up, 6. Vacuum injection, 7. Pumping, 8. Inflating.
Reinforcement:	9. Ropes, bars.
Surface treatment:	10. Plastering, 11. Polishing, 12. Coasting, 13. Drape.

Fig. 12.1 New developments for ice

Concrete, composites, glass and ice have been combined with flexible moulding in an overview. This list of techniques, materials and case studies is based on the matrix in Chap. 7. This matrix gives 25,000 theoretical possibilities for each material. 49 case studies have been described. About 200 of the 25,000 theoretical possibilities might be interesting/possible. This list is an open system: more materials, techniques and case studies can be added. For researchers and designers (structural, industrial and architectural designers) this list of case studies, combined with all the theoretical options, is intended as a helpful instrument. For designers it might help to find the ultimate combination to construct their concept of fluid architecture with a flexible moulding technique.

Below, the possible combinations of these materials that have not been researched so far will be discussed. The purpose of this evaluation is to develop a new research agenda.

Concerning technique, the combination of 3D printing on a flexible mould was not researched in this PHD. The TU-Delft and TU-Eindhoven have recently done some experiments on this topic (Fig. 12.1).

Not researched:

1: Knitting is making a fabric that consists of consecutive rows of interlocking loops. A knitted fabric is known for its elasticity and the freedom to make any shape and texture that is needed. Changing the curvature is done by changing the number of loops on a line, changing the texture is done by influencing the way the loop is connected to the other loops. Knitting can be done with "yarns" of all kinds of material. The combination of ice with knitted fabric seems interesting to me and might result in all kinds of new possibilities, too many to be aware off. The reinforcement of ice with knitted yarns might also be an option. The bonding of ice with natural fibres will be good and it will increase the strength and ductility of the ice along with the artistic input of the knitted element. Machine-knitted elements might also open the way to seamless fabric moulds for complex geometries (Fig. 12.2).

2: Double-layered fabric, such as distance fabric, in combination with ice might be a good combination. This has already been researched for concrete. We have recently begun to research this combination together with students at the Eindhoven University of Technology.

Fig. 12.2 Complex knitting

3, 7, 8: Inflating a fabric mould with water, later to be frozen, seems to be an interesting option that might give interesting results. Especially rubber moulds will provide a good application. Also, the inflation of a hull of distance fabric with water is thought to be interesting.

4: Underpressure for the stabilization of a fabric was done by Frank Huyben for concrete. It is assumed that with ice this is also possible.

5: The hand layup method for rigidizing the surface of a fabric mould is possible but spraying of water is much more effective.

6: The vacuum infusion of water in fabric might result in an ice composite with a very dense amount of fibres. Since the maximum amount of fibres in cast and sprayed ice composite is limited, this might result in an even stronger material.

9: The reinforcement of ice with natural fibres is possible, the reinforcement of ice with ropes seems to be interesting if there is bonding with the ice. With students we did some experiments. To be sure of the positive effects more research needs to be done. The freezing of wood and twigs is also possible and might give some positive effect. Research has proven that the smaller the particles of the wood, the better the structural results. Therefore, this was not researched further.

10, 11: The plastering/polishing of ice or snow by melting of the surface is a common technique in ice/snow sculpturing. For larger applications in ice buildings this is possible but assumed to be too labour intensive and therefore too expensive. In case of unreinforced ice shells there might be problems with thermal shock as result of the sudden rise in temperature.

Case studies to be developed for the combination of concrete with:	
Form-active structure:	1. The material of the shell in fluid condition.
Form-active typology	No
Way to stabilize the mould:	No
Rigidizing technique:	2. Vacuum injection, 3. Inflate.
Reinforcement:	None
Surface treatment:	4. Tension of the material, 5. Polishing

Fig. 12.3 New developments for concrete

12: The colouring of ice is possible and done for sculpturing applications. In many applications, the colouring is done in combination with the lighting on the white or transparent ice.

13: Draping the surface of an ice structure with a wet fabric or a net structure is possible but seems to give a limited structural contribution, it is only done to raise the ductility and the surface quality of the structure (Fig. 12.3).

1: Because of the low surface tension of the material it is not possible to make a continuous unsupported surface in the air. Unless a way is found to increase this surface tension of concrete, it is assumed that it is not possible to rigidize a concrete surface only supported by the air pressure or tension in the boundary cables like it is possible with glass or a soap film.

2: Vacuum injection of cement is possible, Frank Huijben did some successful experiments in 2014 and 2015. Therefore, it is proven to be possible. More research has to be done to make it an economically feasible way for making fluidly formed elements.

3: The inflation of concrete is only possible if the limited surface tension of the concrete can be increased.

4: The surface tension of the material is too low to form a smooth surface. Deforming the surface of a horizontal cast element is an option like is done by Bini, Kolleger and Schipper & Eigenraam.

5: Polishing of concrete is mostly done for indoor applications like the terrazzo system. Terrazzo consists of marble, quartz, granite, glass or other chips bound with cement. There is also Terrazzo with an epoxy binder. Terrazzo is cured, ground and polished. Technically it is possible to have the same treatment on the outside of the surface of a building but it will be very expensive (Fig. 12.4).

1: Knitting with composite/polymer yarns is quite possible and done. The combination of composites with a knitted fabric seems to be interesting and might result in all kinds of new possibilities, too many to be aware of. The reinforcement of polymers with knitted yarns will increase the strength and ductility of the composite. Machine knitted elements might also open the way to seamless fabric moulds with a smooth polymer surface for complex geometries.

Case studies to be developed for the combination of composites with:	
Form-active structure:	1. Knitted fabric, 2. The material in fluid condition, 3. Hinged plate structure.
Form-active typology	No
Way to stabilize the mould:	4. Underpressure.
Rigidizing technique:	5. Casting, 6. Prefab elements, 7. Inflate.
Reinforcement:	8. No, 9. Bars.
Surface treatment:	10. Tension of the material, plastering, polishing, drape.

Fig. 12.4 New developments for composites

2, 7: The thermoplastic polymers in composites have good properties for injection moulding by inflation. It is assumed that injection against a flexible surface is possible. The investments for this tool will be high. Because of the decentralized organization structure of the building industry this might be a problem and a reason why it has not been developed up till now.

3, 6: Lightweight foldable plate structures for formwork onsite can be interesting. Composite will be the best material to realize that kind of formwork because of the high strength and limited mass. The application of origami for building purposes has become a popular research item that might find its application in this way. Up till now it seems to have been an area waiting to be researched.

4: Vacuum injection and under-/overpressure for the stabilization of a fabric was first developed for composites. It was transferred to concrete by Huijben (2014) and by Verhaegh (2010). It is to be expected that the results achieved there are also possible with polymer composites.

5: Open air casting is unhealthy because of the emission of, for instance, styrene. There are better alternatives like vacuum injection and injection moulding.

8: A polymer is an unreinforced composite. Because of the size of elements, the high price of polymers and the low price of reinforcement fibres, non-reinforced composite structures only have an application were transparency is needed like in daylight domes. The curvature of that kind of material for fluid architecture has been realized in for instance the Kunsthaus in Graz. The limited lifetime properties of polymers are problematic for building application and might cause high maintenance cost.

9: The reinforcement of polymers with bars is possible but for fluid architecture those have to be bent in the curvature of the element. Woven reinforcement will not have this problem and so it is commonly used for the reinforcement of composites.

10: The tensile strength of the material is sufficient to create smooth surfaces without any other treatment. This way of surface treatment could be interesting to research.

11, 12, 13: Plastering, polishing or draping with composites is done to repair and smoothen surfaces. A well-known application is in the car industry. So, it can be used, but there is not much to be researched, since this is already a well-known application (Fig. 12.5).

Case studies to be developed for the combination of *glass* with:	
Form-active structure:	1. Knitted fabric, 2. Textile composite, 3. Foil.
Form-active typology	4. Connected double layer.
Way to stabilize the mould:	5. Hydraulic pressure, 6. Underpressure.
Rigidizing technique:	7. Hand lay-up, 8. Spraying, 9. Submersion, 10. Casting 11. Pumping.
Reinforcement:	12. Single fibres, 13. Ropes, 14. Bars.
Surface treatment:	15. Plastering, 16. Polishing, 17. Coating, 18. Drape.

Fig. 12.5 New developments for Glass

1, 4, 12, 13: Knitting with steel yarns is quite possible and done. The combination of steel with a knitted fabric seems to be interesting. The combination of weaving for a glass mould was done, knitting might result in all kinds of new possibilities. The "reinforcement" of steel yarns, ropes and bars with glass will influence the behaviour of the material. Machine-knitted elements and complex 3D woven surfaces might also open the way to a seamless fabric hull of steel in which a glass bulb can be inflated.

2, 3: Because of the substantial difference in melting points and elasticity of glass and polymers, a combination of these will not work. The lamination of polymers and glass is possible and done.

5: The blow moulding of glass is an old technique, working with a hydraulic medium might be possible.

6: The blow moulding of glass is an old technique, vacuum sucking into a mould might also be possible.

7: The fusion of glass in different layers by hand layup has been done in artistic applications of glass, it might be interesting to research this method for architecture and (civil)engineering.

8: The spraying of air into hot glass is done for the production of glass wool. It might be interesting to do some experiments with the spraying of glass on flexible moulds.

9: The submersion of glass is used to collar it. No other possibilities found.

10: The casing of glass for curved applications might be possible.

11: Because of the high melting temperature it is not possible to pump glass. It will be possible to press glass in a mould by an over-/underpressure.

15, 16, 17, 18: It is possible to paint glass to colour it or to protect it. Other applications to have layers on top of the glass surface seem to be useless.

For fluid architecture I am less optimistic. Because of the economic downturn and environmental problems we have to deal with, (I believe) that fluid architecture only for architectural reasons is over the hill. Nevertheless, curved structures will be needed for all kinds of reasons. The building industry is one of the largest consumers of resources. Curved structures are in most cases more efficient and can save a lot of material. I expect that fluid forms will be dominated by functional reasons again in the future. Fluidity will remain in structural optimization as long as the laws of physics do not change.

The fluid style will pass but fluidity will remain!

Bibliography

Balgaradean CM, Babut, R (2015) Young people's perceptions towards sexuality in advertising. Manag Mark J:63–77. University of Craiova, Faculty of Economics and Business Administration

Huijben FAA (2014) Vacuumatics 3D formwork systems

Reichert T, Lambiase J (2003) Sex in advertising: perspectives on the erotic appeal. Lawrence Erlbaum Associates Inc, Publishers, Mahwah NJ

Verhaegh RWA (2010) Free forms in concrete: The fabrication of free-form concrete segments using fabric formwork

Chapter 13
Conclusions

The goal of this book is:

A. to improve the realization of the architectural intention of fluid architecture.
B. to improve integrated solutions between structure and envelope.
C. to improve the quality of fluid buildings in general.
D. to have a more efficient construction method with less environmental impact.

Below is a bullet list with the achieved results in this research:

- Overview of the ways membranes can be manipulated.
- A historical overview of form and form-active structures.
- The conclusion that in Fluid architecture curved surfaces in opposite directions will almost always be combined.
- An improved definition for fluid architecture.
- The finding that every important style period will end in a practice of superfluous irregular curved surfaces.
- Fluid architecture will find its source in erotic curvatures although designers avoid to make this too explicit in their architecture.
- The perceived curvature is more important than the actual "curvature".
- The proof why flexible moulding is appropriate for fluid architecture.
- The proof that it is possible to make all kinds of structural typologies with flexible moulding.
- The proof, by case studies, that designers can work with flexible moulding while maintaining their own stylistic identity.
- An open schedule with materials and techniques in combination with flexible form-active structures and case studies.
- A research agenda with a list of new combinations of materials and techniques with flexible moulding.

With the list of achievements above, I presume that the envisaged goals have been reached. Only goal D: the efficiency of construction with flexible moulding, has not been proved in this research but was already proved for columns, beams and floors

A. Pronk, *Flexible Forming for Fluid Architecture*,
https://doi.org/10.1007/978-3-030-71551-9_13

by the research of John Orr and Mark West. The general developments in flexible on-demand production techniques will also influence the production of flexible material and therefore also the possibilities for flexible moulding. The research interest in flexible moulding has increased enormously over the last 10 years. I foresee this trend will be continued in the next decade.

MIX
Papier aus verantwortungsvollen Quellen
Paper from responsible sources
FSC® C105338

If you have any concerns about our products,
you can contact us on
ProductSafety@springernature.com

In case Publisher is established outside the EU,
the EU authorized representative is:
Springer Nature Customer Service Center GmbH
Europaplatz 3, 69115 Heidelberg, Germany

Printed by Libri Plureos GmbH
in Hamburg, Germany